AF479324

COSMIC EXPLOSIONS

COSMIC EXPLOSIONS

Tenth Astrophysics Conference

College Park, Maryland   11–13 October 1999

EDITORS

Stephen S. Holt

William W. Zhang

NASA/Goddard Space Flight Center
Greenbelt, Maryland

Melville, New York
AIP CONFERENCE PROCEEDINGS ■ 522

Editors:

Stephen S. Holt
NASA/Goddard Space Flight Center
Code 600
Greenbelt, MD 20771
USA

E-mail: steve.holt@gsfc.nasa.gov

William W. Zhang
NASA/Goddard Space Flight Center
Code 662
Greenbelt, MD 20771
USA

E-mail: zhang@xancus10.gsfc.nasa.gov

L.C. Catalog Card No. 00-103470
ISBN 1-56396-943-2
ISSN 0094-243X
Printed in the United States of America

CONTENTS

GAMMA RAY BURSTS

JETS

NOVAE AND X-RAY BURSTS

STELLAR EXPLOSIONS

RAPPORTEUR SUMMARY

APPENDICES

PREFACE

This is the tenth installment of the current series of annual October Astrophysics Conferences in Maryland. These conferences are organized by astrophysicists at the Goddard Space Flight Center and the University of Maryland. The topic for each conference is selected with the help of an International Advisory Committee, the current membership of which is:

Marek Abramowicz, Gotenborg	*Sir Martin Rees*, Cambridge
Roger Blandford, Pasadena	*Vera Rubin*, Washington
Claude Canizares, MIT	*Joseph Silk*, Berkeley
Arnon Dar, Haifa	*David Spergel*, Princeton
Alan Dressler, Pasadena	*Rashid Sunyaev*, Moscow
Guenther Hasinger, Potsdam	*Alex Szalay*, Budapest
Steve Holt, Greenbelt	*Yasuo Tanaka*, Tokyo
Dick McCray, Boulder	*Scott Tremaine*, Princeton
Jim Peebles, Princeton	*Simon White*, Garching

In the spirit of this series, where we attempt to identify "hot" topics with broad appeal to both observers and theoreticians, the sudden release of energy on scale sizes ranging from stellar flares to galaxies clearly qualifies as an appropriate conference theme. The programme of the conference was developed by the Scientific Organizing Committee comprised of:

Neil Gehrels	*Virginia Trimble*
Guenther Hasinger	*Nick White*
Steve Holt	*Stan Woosley*
Bob Kirshner	*Will Zhang*
Joe Silk	

The conference began with a welcome from Steve Halperin, the new Dean of the College of Computer, Mathematical and Physical Sciences at the University of Maryland, College Park. The welcoming address was followed by introductory invited reviews by Virginia Trimble on the history of phenomena that might be characterized as cosmic explosions, and by Roger Blandford on the outstanding issues that confront our current understanding of such explosive phenomena. The programme then proceeded through the next two days with a series of non-paralleled sessions. Each session was devoted to a specific topic, with two or three invited talks and an extensive discussion period that included the opportunity for each conference attendee to advertise his/her poster paper in one-minute with one viewgraph.

The banquet at the conclusion of the second day of these conferences generally features a distinguished speaker. This year our speaker was Steve Maran, who shared some of his more humorous experiences as the press officer of the American Astronomical Society. The conference concludes with a rapporteur summarizing some of the more important issues discussed during the previous two days, and Craig Wheeler provided us with an excellent review this year.

As usual, John Trasco and Susan Lehr made sure that all the logistics were handled flawlessly. Thanks also to Will Zhang, the co-editor of these Proceedings, for doing such an efficient job of handling the details associated with producing this volume.

Steve Holt
January 2000

Introductory Overview

The First Explosions

Virginia Trimble

Department of Astronomy, University of Maryland, College Park, MD 20742
and Physics Department, University of California, Irvine, CA 92697

Abstract. Events that can be (at least broadly) described as explosions have joined the astrophysical inventory in many different ways. Some were predicted and then discovered. Some were predicted, often as a tentative explanation for some specific sort of event, and have not yet been seen. Others were seen and understood quite promptly, or very slowly. And a good many had been in the inventory for years to millenia before their explosive nature was recognized.

INTRODUCTION: HOW TO ORGANIZE THE EVENTS

Since we will be looking, at least briefly, at nearly 40 sorts of events, how they are ordered is of some importance. An obvious choice might be in the order in which they occurred in the universe, from the Big Bang at 16,833,203,494 BC (at 9 o'clock in the morning) to the collapse of a white dwarf to a black hole by tunneling or barrier penetration, for which the time scale is roughly $10^{10^{76}}$ in the future. And with this sort of number, it does not matter whether the units are years or seconds.

Another possibility is by order of discovery. This, however, invites numerous ambiguities. Were supernovae really discovered by a Zinjanthropan named Og, by Tycho Brahe (who understood that his event of 1572 was outside the orbit of the moon), or by Baade and Zwicky, who separated them off from common novae in 1933-34?

One might consider the order in which the events were understood. This unfortunately leads to a large pile-up in 2001, or perhaps later. Grouping them by the physical force responsible seems too coarse-grained. Gravitation, nuclear reactions, and electromagnetic processes are really the only games in town. At any rate, I am not aware of any phenomenon that can be attributed to explosive release of weak interaction energy. Ordering by time or size or mass scale all also present problems.

Eventually I decided that the right heirarchy was in terms of the total energy released in the event, from large to small. Curiously, Roger Blandford had independently settled on the same scheme, though from small to large, at least for his oral presentation.

CP522, *Cosmic Explosions: Tenth Astrophysical Conference,*
edited by Stephen S. Holt and William W. Zhang
© 2000 American Institute of Physics 1-56396-943-2/00/$17.00

Each of the following sections begins with the following information. Logarithm (base 10) of total energy in ergs (log E =), followed by the years in which the events were Predicted (P), Seen (S), and Understood (U), with occasional brief comments. The main text of each section is then a discussion of the history of the phenomenon. Where no "P" is indicated, there was no obvious prediction that I have been able to find. Missing "U"s indicate that no obvious further understanding seems to be called for. U = ? or "multiple" indicates issues that are still under discussion, some of them later in these proceedings.

THE BIG BANG

Log E = 77 (or possibly minus infinity, if the universe is Hawking's "free lunch"). P 1921-31, S 1965 (and earlier), U 2001 (possibly as scalar field).

Einstein himself, notoriously, wrote down only a static solution to his own equations as applied to the universe. Non-static, in some sense expanding, solutions came from de Sitter (1). Friedmann (2, 3), and Lemaitre (4) who also demonstrated that the Einstein static solution was unstable. De Sitter space is empty (though a test particle would show something like a quadratic velocity-distance relation). The others have non-zero mass-energy density and related geometry. The best evidence for an expanding universe is, of course, the Hubble or velocity-distance linear relationship, put forward in 1929 (5). The alternative explanation called tired light (due to Zwicky along with many other items) was decisively ruled out only in the last few years by the observation of time dilation in the light curves of distant supernovae (whose observers use them for other purposes later in this volume).

George Gamow is generally credited with being the first to take seriously the idea of an early universe different from the present, beginning in 1935 (7,8), though in fact his "ylem" was scooped by Georges Lemaitre's "cosmic egg" in 1931 (6). Testable consequences of a hot, dense early universe were invented under Gamow's influence and described by his colleagues Alpher and Herman (9.10). As early as 1954, at least one of the proponents of the Steady State alternative (Hermann Bondi, 11) had recognized that a universal abundance of helium as large as 25% would present serious difficulties to steady state while it could be regarded as a relic of a big bang. Difficulties in measuring helium abundances in appropriate sites with sufficient precision delayed its recognition as the definitive relic.

Instead, early counts of radio sources from Ryle's group in Cambridge began to reveal that galaxy populations were different in the past (12). Quasars were more readily understood as saying the same thing, because they had measured redshifts (for instance the V/V_{max} test of Schmidt, 15, although this came only after the discovery of the microwave background). Eventually die-hard steady staters were pushed into a corner called "non-cosmological redshifts" in order to maintain that the universe has not changed in the face of changing radio source populations.

The definite demonstration of the hot, dense past was, for most people, the measurement of the residual radiation (13, 14) predicted by Alpher and Herman. As for seeing clearly that normal galaxies were different in the past, this has really taken the power of 8-10 meter class telescopes and HST and cannot be described as history, even by the very young.

I have discussed this material in greater detail elsewhere (16).

ACTIVE GALACTIC NUCLEI (QUASARS, RADIO GALAXIES, AND THE REST)

Log E = 61. S 1909-63, U 1964 (and many thousands of papers thereafter). Of the wild richness of sorts of active galaxies (Urry elsewhere in this volume), the first into the inventory are called Seyfert galaxies, because they were discovered by Fath in 1908 (17), and again by Seyfert in 1943 (18). My prejudice is, however that the real "discovery" goes with the recognition that a central mass of about $10^8 M_\odot$ is required to confine the high-velocity central gas and that the lifetime must not be much longer than 10^8 yr by Woltjer in 1959 (19).

Radio galaxies joined the parade with an accurate radio position for Cygnus A (Smith 1951, 20) leading to the optical identification by Baade and Minkowski in 1954 (21). The identifiers themselves suggested colliding galaxies as the energy source. Among the other imaginative suggestions were the collisions of anti-matter with matter galaxies (22, 23), which helped speed up the development of gamma ray astronomy to all our benefits, multiple supernovae (24), and supermassive objects (25). What would undoubtedly have been further proliferation in any case was greatly accelerated by the discovery of quasars (26) in 1963.

Since supermassive objects had proven unstable, the first entry was inevitably gravitational collapse (27). Nuclear star bursts, stellar collisions, monster magnetic fields, and many other suggestions appeared soon after (28). But, so did the essence of the right idea, a massive central black hole with continuous energy extraction. Both Salpeter (29) and Zeldovich (30, 31) suggested accretion, and a magnetic mechanism, called Blandford-Znajek (32) followed. For further discussion see an earlier conference in this series (33).

NUCLEAR STAR BURSTS

Log E = 60. P 1963, S 1973, U 1975. When quasars proved (to most of us at least) to be running on black hole accretion energy, Gold's 1963 (34) suggestion of nuclear star bursts (also Woltjer 1964, 35) acquired the status of a prediction. The objects now most closely associated with the idea are the infra-red bright, star-forming (or IRAS) galaxies. Infrared emission from the core of NGC 253, a prototype, was recorded in 1973 (36). Riecke and Low (37) recognized soon after that star formation there must be episodic, since there wasn't enough gas in the whole galaxy to keep up what its center was doing for very long. M82 is a somewhat similar beast, and Solinger, Morrison, and Markert (38) deduced from looking at it and its surroundings that the enhanced star formation was very probably due to interaction with another galaxy. Riecke and Lebofsky spoke specifically of "bursts" of star formation in 1978 (39), though the key phrase "starburst" did not make the index of Astronomy and Astrophysics Abstracts until 1985, about the time Roberto Terlevich and his colleagues were reviving the idea that many or most active galaxies might be powered primarily by explosive star formation. I suppose one could regard this entire territory as having been staked out by Fritz Zwicky with his 1958 suggestion that new dwarf galaxies are made from tidal debris dragged out by interacting galaxies (40).

MILKY WAY HITS M31 AND OTHER GALAXY-GALAXY COLLISIONS

Log E = 60, P 1959, S 1954 (false alarm). A galaxy collision was Baade and Minkowski's (21) candidate for explaining Cygnus A. A dust lane does indeed make it look rather like two spirals meeting face to face. That the Milky Way must inevitably encounter M31 after a few Hubble times is implicit in Kahn and Woltjer's 1959 (41) discussion of the total mass of the Local Group. In recent years, galaxy mergers, cannibalism and, so forth have become part of the main-stream inventory. The "current event" for the Milky Way is the gradual disruption of the dwarf spheroidal in Sagittarius, which I still think should be called IGI for its discoverers, Ibata, Gilmore, and Irwin (42). None of these processes now looks particularly explosive because, although there may be a total kinetic energy of log E = 60 (ergs) to dissipate, there is a billion years in which to do it. All in all, these collisions are less exciting than you might have expected.

THE 3 KPC ARM

Log E = 54. S 1957, U ?. This cock-eyed and, it would seem, rapidly-moving part of the spiral structure of the Milky Way carries a name based on a now-outmoded scale of galactic distances, in which it was 3 kpc from the galactic center. Its discovery in 21 cm emission maps was reported from Oort's group in 1957 (43). Oort himself thought it had probably been sent on its way by an explosive event connected with a black hole at the galactic center (for which, to him at the time, it formed part of the evidence, 44). Later, Yuan suggested that the velocity pattern was simply part of the rotating spiral with a central bar that perhaps describes our galaxy (45). This is probably now the majority view; see the earlier conference in this series (46).

HIGH VELOCITY CLOUDS

Log E = 53, S 1963, U multiple. The high velocity clouds of neutral hydrogen in (or near) the Milky Way were first identified in 21 cm maps from Oort's group (47). The energy quoted represents the arrival (or departure) of a million solar masses at a relative speed of 100 km/sec which are typical values. Large negative velocities at high galactic latitudes had already been reported for optical absorption lines in halo stars by Munch and Zirin (48), but only the 21 cm work could provide masses for the gas clouds and thus some indication of the energies concerned. Three major interpretations are in the literature. Oort (49) regarded the clouds as virgin intergalactic gas being added to the Milky Way. W.A. Fowler must have agreed, because I can remember him saying, "Here these stars are burning their hearts out trying to make heavies, and those b–'s keep diluting it." A second possibility is that the speeds we see are just projections of spiral arm motion well outside the galactic plane (50). The third interpretation ties the clouds to a "fountain model" of halo gas, where they are the return part of the flow driven upward by supernova bubbles (51).

Tests that ought to distinguish the possibilities include the distances, compositions, and ionization states of the clouds. None is as clean as you might suppose. If you do see the absorption due to a particular cloud in the light of a star, you are guaranteed that the cloud is between you and the star and contains some of the element that makes that absorption feature. But if you don't see the absorption, the cloud might be beyond the star, lacking the element, or with the element in some other ionization state, or two of the three. I think the jury is still out on this one. Assorted giant HII regions, non-nuclear bursts of star formation, and superbubbles in the interstellar medium of other galaxies have all been blamed by somebody on the arrival of high velocity clouds (though reasonably careful searches do not require that all spirals have them). My guess is that some of them are and some of the aren't (and also that the clouds are not necessarily all the same kind of beast).

QUASAR FLARES AND SUPERLUMINAL MOTIONS

Log E = 50-52. P 1966, S 1963, U 1964-76. Very shortly after the 1963 optical identification of 3C 273, Smith and Hoffleit (52) and Sharov (53) realized that it had been widely and erratically variable, sometimes on very short time scales. Smith and Hoffleit used the historic collection of plates at Harvard; I have not checked the Sharov paper to find out what his sources were. Shklovsky was immediately on the spot, suggesting that we were going to need relativistic gas motions, perhaps beamed, to account for the data (54). The need became acute when Dent and his colleagues reported in 1965 that the radio flux from 3C 273 was also varying rapidly (55).

At this point, non-Soviet astronomers pointed out that relativistic motions would both prevent an inverse Compton crisis in the sources and probably lead to rapid changes in the source structure on the scales probed by very long baseline interferometry (56, 57). Such changes were seen for both a radio galaxy and a quasar in 1971 (58, 59) and they are now regarded as commonplace. There was, however, a last flurry of non-cosmological redshifts as a way to bring the sources closer and make the transverse motions correspond to smaller space velocities. Of greater interest, perhaps, was a revival of the multiple-supernova mechanism for active galaxies, in which motion was mimicked by one radio-bright SN fading and another appearing a little ways away. In retrospect, we who remember that epoch will be surprised to realize that the entire "Christmas tree" fuss was over by about 1976, before our European colleagues had had the opportunity to figure out that our Xmas lights flash on and off by design and not just when the power fails. This same period saw the first model for "extended radio sources" in which most of the energy transport was in Poynting flux rather than mass motions (Rees 1971, 60) and the more elaborate twin-exhaust model (Rees and Blandford 1974, 61). For more about these events and the history of AGNs in general, see proceedings of an earlier conference in this series (62).

GAMMA RAY BURSTS

Log E $= 52 \pm 2$. P 1968, S 1972-74, U 1997. Two kinds of gamma ray bursters were predicted, and two kinds have been seen. Unfortunately, they are not the same two kinds. In 1968, Stirling Colgate was modeling how the shock driving a supernova should emerge from the star and concluded it should break out at sufficiently high energy to radiate X- and gamma-rays (ultraviolet and soft X-rays is what the observations eventually revealed). And, in 1974, Stephen Hawking proposed that there might be black holes with masses as small as 10^{15} g, which would be evaporating about now, ending in a chirp of high energy photons (and other particles).

Discovery was announced in 1973 (64, 65), based on several years' worth of photon collecting by sets of satellites launched by the US (Vela series) and USSR (Kosmos series). The satellites had been put in orbit to look for neutrons and gamma rays from atmospheric bomb tests violating the treaties then in effect. Whether they ever detected any such tests depends on whether you think countries that have not signed a treaty should be bound by it. The events were, in any case an illustration of a theorem due to Philip Morrison, that it is hard to waste 10^8 dollars. Recent programs suggest that the number should be inflated to 10^9.

At the 1974 Texas Symposium, everybody had a GRB model, although the printed version of Mal Ruderman's talk (66) does not include the overhead showing the list of "theorists who do not have a gamma ray, burst model." (The name on the list was Jerry Ostriker.) Very few of the models were for extragalactic sources, which remained a distinct minority taste at least up until the launch of CGRO, and perhaps until the first optical identification with a redshift. These events are thoroughly covered elsewhere in this volume, as is the second class of GRBS, the soft repeaters. I note only that the last paper Fritz Zwicky submitted (67) was a GRB model. He had died by the time his suggestion of "nuclear gobblins" had been published.

SUPERNOVAE

Log E = 53, 51, 49. S 1572-1933, U 1933. The collections of events called supernovae are slightly difficult to fit into the present format. For instance, the total energy from a core-collapse event is about 10^{53} ergs, but it comes out largely in neutrinos (and perhaps gravitational radiation), so that one might perhaps claim that supernovae were not truly discovered until 1987. The kinetic energy of the ejecta amounts to roughly 10^{51} ergs and is approximately the same for all types. I still think this is just a little mysterious. Typical Type II's (core collapse events with appreciable hydrogen) eject more mass than classical Ia's (CO explosive burning events) but at lower speed. Finally, the total light output has log E near 49 for both (on somebody's distance scale), but even this was completely unknown until 1924, although the very wide (and then unidentified) spectral features made it clear much earlier that large velocities must be involved.

Events on the observational front that I think belong in any precis include (a) the 1572 demonstration by Tycho that "his" nova stella was beyond the orbit of the moon (because its geocentric parallax was undetectable), (b) the gradual realization that such "novae" belong to two discrete luminosity classes by Lundmark (67a) and Curtis (68) around 1920, (c) the firm statements by Baade and Zwicky (69) 1933 that supernovae are different, come from collapses to neutron star configurations, and accelerate cosmic rays, and (d) the separation of types I and II by Minkowski in 1941 (70). Broadness of the lines and their incomprehensibility were noted in a number of papers in 1936-37. I mention here only the work of Payne Gaposchkin (71), because I borrowed a good deal of information on novae and supernovae from her book (72) [my copy carries her signature and is a gift from Jim Wright], Popper (73) because he was so up-front about saying that he couldn't make heads or tails out of the spectral features of 1937C (it was a type I), and Humason (74) because he seems to have been the first to declare "hydrogen" for 1936A (it was a type II). Minkowski probably made the first mass estimate (85).

On the theoretical front, the two ideas we now regard as most germane came from Baade and Zwicky (69 - core collapse) and from Colgate (75) and Hoyle and Fowler (76) who, in 1960, all said "nuclear explosion." The wrong ideas (births of double stars, rotational instabilities, stellar collisions, and many others) are at least as interesting, see (72, chapter 11) and (77). The trouble with a single explosive event as the essence of supernovae is that thief, adiabatic expansion, who puts all the energy into coherent expansion, leaving none to be radiated over the decline phase of months to years. As long ago as 1940, Zwicky (78) had fastened on a nuclear process with appropriate half-life as the way to go. Lyle Borst, more recently professor emeritus at SUNY-Buffalo, revived the idea in 1950 (79) suggesting Be as a plausible decayer. Burbidge, Burbidge, Fowler, and Hoyle in their classic 1957 compendium of element production in stars (80) invoked Cf^{254}. Fowler still had that in mind in 1966-67, when he taught nuclear astrophysics at Caltech because he began the school year with hydrogen and ended in May when he got to Californium. The right answer, nuclides of mass 56, decaying from nickel to cobalt to iron appears in an otherwise obscure 1962 thesis (81). Credit for persuading the rest of us goes rightly to Colgate (82).

Finally, we can connect up with much of the discussion at the present meeting by noting two 1939 papers advocating supernovae as standard candles for cosmological applications by Olin Wilson (83) and, again, Zwicky (84).

LUMINOUS BLUE VARIABLES

Log E = 49. S 1943, U ?. The remarkable range of phenomena associated with the 1843 outburst of Eta Carinae and its subsequent behavior are fully described elsewhere in this volume. The general category of luminous blue variables or Hubble-Sandage variables (86) includes the brightest stars in our own and other galaxies, some of which have undergone or will undergo similar traumatic events. P Cygni, which brightened to 3rd magnitude in 1600 and some of the other "rejected" early novae in Payne Gaposchkin's book (72) may have been related. Q Cygni, on the other hand, despite its odd name, was a perfectly respectable nova and is still around. To a considerable extent, our reliable knowledge of the 1843 outburst derives from the good luck of Herschel (meaning John) having been "at the Cape" at the time (87). Many mechanisms have been proposed, none confidently.

STAR-STAR COLLISIONS

Log E = 49, P 1929, 1939, ff$_v$, S 1997, U? This particular sort of violence has been invoked several times to explain known events. Jeffreys used one to make the solar system in 1929 (88), in a modification of the Chamberlain-Moulton tidal encounter scenario. Whipple proposed them for supernovae in 1939 (89), and several early quasar models mentioned above included stellar collisions as an energy source. It remains unclear whether we have ever actually seen such a collision (something transiently nasty in the core of a globular cluster might be a good bet, with thanks to Ted Stecher for reminding me of this) and unclear also whether the sorts of merger processes expected in common envelope binaries and more vigorously in some models for gamma ray bursts should be described as stellar collisions.

HELIUM FLASH IN LOW-MASS STARS

Log E = 48. P 1958. That helium will ignite explosively in the cores of low-mass stars that have exhausted their central supply of hydrogen fuel is clear from the description in Schwarzschild's 1958 text (90). Implicitly, we see the result in the greatly modified structure of horizontal brach (core helium burning) stars in globular clusters. But direct detection will require a neutrino telescope of considerable sensitivity and angular resolution.

LAST HELIUM SHELL FLASH IN AN AGB STAR

Log E = 48. S 1968, U 1984. Stars whose location on the HR diagram have changed in historic times are relatively rare. One of the most spectacular is FG Sge, which started galloping around at the beginning of the century, although the observations were not summarized until 1968, by Herbig and Boyarchuk (91). A series of papers by Icko Iben and his co-workers, of which the 1984 one (92) is the first clear statement, gradually established that what we are seeing is one last flash of degenerate helium burning in a star that was already well on its way to becoming a nucleus of a planetary nebula. Indeed Herbig and Boyarchuk noted that FG Sge was already surrounded by a faint PN and seemed to be trying to form another one. The star catalogued as V 605 Aql, whose spectrum was recorded in 1920 by Lundmark (93), is another likely candidate, and the most recent is variously called V4334 Sgr and Sakurai's object. All have in common the gradual changes in atmospheric composition resulting from carbon and s-process production and probable transformation to R CrB stars. One of the early supposed novae has also been suggested as a member of this class.

EJECTION OF PLANETARY NEBULAE

Log E = 45. S 1918, U 1967? In 1918 Campbell and Moore (94) reported line profiles as a function of position across several PNe that made it perfectly clear that the basic structure was an expanding shell. Russell, Dugan, and Stewart in 1926 (95) nevertheless describe the gas as rotating around the central star, for which they then not unnaturally derive a very large mass. Modern pictures of how PNe are ejected include a complex mix of fast and slow winds, radiation pressure, and effects of close binary systems, but the first key idea, that the total energy of a very extended AGB envelope is positive, seems to have occurred almost simultaneously to three people (96, 97, 98). In the Soviet Union, understanding of PNe was impeded by the influence of the ideas of Ambarsumian, who firmly believed that they were the first phase in the evolution of a star, expanding from dense, "prestellar" material and were followed in time by structures like the Orion Nebula, with normal stars coming later. He held similar expansive ideas about active galactic nuclei, clusters of galaxies, and much else. Without this background information, one is greatly puzzled to find Shklovsky, in a semi-popular book as late as 1979 devoting many paragraphs to the evidence for planetaries coming late, not early, in the lives of stars (99). He had also said it much earlier (100).

NOVAE

Log E = 44-45. S 1670, U 1956+. The question of who discovered novae presents some of the same problems as the discovery of supernovae. Og of Olduvai Gorge with a date of 1 million BC or thereabouts (and portrayed by Victor Mature) is one possibility. Tycho is another, but of course we now know that his nova stella was really a supernova. The first catalogued event still generally counted among the true novae was the "nova sub capite Cygni" of 1670 (now called CK Vul, because the constellations have moved, a point discussed in some detail by Griffin this year, 100a). Useful guides to the early novae include Payne Gaposchkin (71), McLaughlin in the Kuiper Compendium (101), and the paper by Shara and his colleagues (102) that describes the recovery of CK Vul. Its total amplitude was about 17 mag. The second known nova (WY Sge, seen in July 1783 by D'Agelet and soon after by Hevelius) was almost as agressive, rising some 15 magnitudes. It was recovered in 1951. These recoveries of the very faint remnants were possible because the historic observers had recorded very accurate positions relative to nearby more stable stars.

The next three novae were 1848 (V841 Oph which rose only 11 mag. and was never lost), 1860 (T Sco), and 1866 (T CrB, retrospectively recognized as the first of the recurrent novae and also the first to have its spectrum described). One has the impression that astronomers from 1600 to 1850 must have been busy doing other things than seeking out non-recurrent variable stars, although a good many novae stellae with inadequate data for recovery or characterization date from this period, as do P Cygni and Eta Carinae. T Aur in 1891 was the first to have its spectrum photographed. Nova Persei was the first to be seen expanding (indeed to be seen expanding superluminally, but it was understood long before the discovery of AGNs that the phenomenon was like a flashlight shining across a distant wall, and due to the illumination of material already in place (103, and see 104 for further details). Ambartsumian and Kosirev were able to estimate the ejected mass for several novae in 1933 (105).

On the theoretical side, we find a number of interesting false starts. RDS (95) had in mind episodic release of "subatomic energy" due to a comet impact. Although it was already known that the post-nova star was normally a lot like the pre-nova star, Milne (106) none the less proposed the collapse of a normal star to a white dwarf as the basic energy source. Notice that this sort of collapse idea was, therefore, already in the air when Baade and Zwicky started thinking about supernovae (69). Biermann (107) realized that you couldn't afford to shoot the whole star on one event and attempted to figure out a way in which the collapse occurred in many small steps while the atmosphere cycled between radiative and convective equilibrium.

Explosive release of nuclear energy was put forward several times by Schatzman (108-110), and at least twice by Mestel (111-112) in the years just after the war, although reading the papers from a 1990s perspective, one finds it hard to decide whether they were really making novae or supernovae. It is no use asking them; in response to a query last year, Mestel claimed to have forgotten that he had thought seriously about the issue at all.

Final understanding waited upon the realization that all novae and related objects are close binaries (113, 114). This permitted the final synthesis of explosive fusion of hydrogen on the surface of a white dwarf, which has accreter gas from a close, non-degenerate companion (115).

BLACK HOLE X-RAY TRANSIENTS

Log E = 44-45. S 1975, U 1986+. The first of these (at least the first to be seen rising and falling) was 0620-00 recorded by Ariel 5 in 1975 (116), even as its builders were hosting the first all European astronomy meeting in Leicester. Not until 1986 was there a good enough orbit for the system (117) that most of us were reasonably confident the compact mass was too large for a neutron star. Just how enough systems evolve to this condition, what the instabilities are, and so forth counts as "work in progress."

PULSAR GLITCHES

Log E = 43. S 1969, U multiple. The first of these, in the Vela and Crab pulsars, turned up in 1969, just about as soon as they could have (118-120) and were quickly modeled by Ruderman (121) as star quakes. This is still the easiest model to envision, and the required crustal strength was not implausible, or at least not implausible relative to the models involving eccentric planets, comet collisions, magnetospheric flares, and so forth. But fashion soon swung in the direction of vortex unpinning in the superfluid interior (122), though in recent years there has been some carping about whether the changes in rotation rate will really be as large as the ones seen. Proposed replacements also invoke various phenomena of condensed matter physics (which can no longer claim the status as the one piece of important physics that astronomers didn't need to know about), and I will not attempt to explain them here, in accordance with a theorem due to Ehrenfest, that it is difficult to explain something even when you understand it and almost impossible if you don't. All the other proposed mechanisms are therefore available to explain new phenomena as and when they surface.

DWARF NOVAE

Log E = 41. S 1896, U 1956, 1970-80. SS Cygni was the first variable recognized as doing something rather like what classical novae do, only much oftener and with much less vigor, about a century ago (123). Basic understanding, in the form of recognizing that the systems are binaries, in which material accretes episodically onto a white dwarf came about 60 years later (124). Whether the instability responsible resides in the donor star (125) or the accretion disk (126) has been under recurrent discussion for 30 years, and it is probable that not all the systems are the same. It is certainly true that some disks heat inside out and others outside in. In case you wondered, U Gem was the second dwarf nova (126a), and it was Joy who showed that SS Cyg is a binary (127).

RECURRENT NOVAE

Log E = 44. S 1919, U 1972+. T Pyx went off for the second time in 1919. The person who recognized that "that star hasn't been seen for almost 30 years" was somewhat unusually allowed to report the results in her own name, Henrietta Swan Leavitt (128) rather than, as all too often happened in those days, as "Pickering 1912." The physical event is essentially like that of a classical nova, except that a giant donor and massive white dwarf allow an explosive amount of hydrogen to accumulate in 100 years or less, rather than in 10^4 - 10^5 years. They are one of several current candidates for progenitors of Type Ia supernovae (elsewhere in this volume).

T TAURI AND FU ORI OUTBURSTS

Log E = 39-41. S 1937, 1945 U. gradual but essentially accomplished. The class of T Tauri variables was defined by Alfred Joy in 1945 (129) and the demonstration that they are young pre-main-sequence stars came from Herbig (130). The botany of rotation, activity, disks, bipolar outflow and so forth associated with T Tauri variability is a ferment of modern research.

FU Orionis rose (from a faint but somewhat variable state,131) by six magnitudes in 90 days and faded only very gradually after 1937 (132). Payne Gaposchkin regarded it as a VERY slow nova, but Wachmann (132) described it as "an extrinsic nebular variable." The implication that it is probably young rather than old is probably correct. It and related stars are discussed elsewhere in this volume. The underlying energy source is apparently accretion, but a very non-steady sort.

X-RAY BURSTS IN NEUTRON STAR BINARIES

Log E 39-41, S 1976, U 1976-78. What we now call Type I bursts were recognized in data from ANS in 1976 (133) and modeled as explosive burning of helium (after

the hydrogen had burned steadily) in 1978 (134). A very similar calculation had
been done two years earlier by Woosley and Taam (135), but they thought they
were modeling gamma ray bursts, and should perhaps be given credit for predicting
the X-ray ones. You will find later in this volume that the fate of the hydrogen is
actually somewhat more complex, and indeed speakers on this subject intimated
that my standards of "understood" had been set too low. It is not the first time I
have been accused of having low standards.

The other sort, type 2 bursts. were, for many years confined to the "rapid
burster" in which they had been recognized in 1976 by Walter Lewin and colleagues
(136). The discoverers proposed unsteady accretion as the mechanism and this
remains the best buy model.

SS 433 AND THE GALACTIC SUPERLUMINALS (MICRO-QUASARS)

Log E = 39-41. S 1979, 1994; U 1979, 1994. The remarkable pattern of shifting
wavelengths in the optical spectrum of SS 433 was recorded by Bruce Margon in
1979. It was quickly modeled as the emission from a pair of oppositely directed,
precessing jets, whose velocity of 0.26 c was set by the locking of a hydrogen line on
to the continuum edge (138, 139). Several other "subluminal" sources (discussed
elsewhere in this volume) have jet speeds of 10-20% that of light.

The first two galactic superluminals were recognized within about six months of
each other in 1994 (140, 141) as a result of radio monitoring of X-ray transient
sources of various kinds. I said at the time (142) that such things must be fairly
common, once you know how and where to look, but the inventory has, at most,
doubled since then (Greiner, elsewhere in this volume). A good many details remain
to be sorted out.

POSITRON EVENT AT THE GALACTIC CENTER

Log E = 37. S 1978, U (probably false alarm). At one time, it looked as if the
flux of the 511 keV line coming for us from the general direction of the center of
the Milky Way had flared (143). It now seems more likely (M. Leventhal, personal
communication 1999) that the source or sources is/are reasonably steady, but that
different instruments picked up different amounts of the total flux at different times.

THE UNSEEN

An assortment of potentially violent astrophysical processes have been suggested
in the past as explanations for observed phenomena, including pulsar glitches and
gamma ray bursters. These are currently available as predictions, to be associated
with some future set of observations, since they are not needed in their original

contexts. References are above, with the events they were invented for. Examples include neutron star quakes, comets hitting neutron stars (anti-comets are even more spectacular), and evaporation of primordial black holes. Impacts of large asteroids on the earth (log E = 35) may also belong in this category, though it is widely believed that observers of the year 65,000,000 BC recorded one very shortly before they became extinct.

SOLAR AND STELLAR FLARES

Log E = 32 (sun) up to 38 (other stars). S 1859, 1949, U 1960-70s. A white light flare was spotted for about five minutes on 1 September 1859 by English amateur astronomers Carrington (144) and Hodgson (145), though it seems likely that earlier sky-watchers must have seen flares occasionally. The variability of the H-alpha emission from the sun was seen in the 1890s by Hale and Delandres (who spoke of eruptions chromospheriques), but not until 1926 did Hale, with his spectroheliograph, catch one in real time. Carrington had already noted the simultaneous occurrence of a terrestrial magnetic storm.

The second flare star was L726-8, generally credited to Joy and Humason in 1949 (146), who in turn credit Luyten (147). The "standard model" of solar and stellar flares has developed over a number of years (e.g. 148), but the first significant contribution was probably that of Gold and Hoyle, who discussed magnetic loop interactions in 1960 (149). For more on solar and stellar flares, past and present, see 150, 151, and elsewhere in this volume.

MINOR EXPLOSIONS

We are rapidly approaching the realm of the whimper and terrestrial events, including:

COMET HITS EARTH. Log E = 30. S = 1908? One could perhaps claim that the book of Revelations and other apochalyptic literature predict such events. The Tunguska impact may have been an example. There will surely be others.

COMET HITS SUN. Log E = 33. P 1749, S 1979. A glancing blow from a comet to drag out pre-planetary material was the suggestion of the French naturalist Buffon for the origin of the solar system. I think no one really expected to see such events. But the satellite Solwind recorded several in 1979, with confirmations and later observations from Solar Maximum Mission.

THUNDERSTORMS. Log E = 22, S prehistoric, U 1990s - 2001. Actually the hardest part here was getting a good number for the total energy. Arthur H. Few of Rice University indicates that the largest number involved is that of the latent heat of the condensing water. It gets fed, in varying proportions, into kinetic energy of updraft, electrostatic energy, lightening, and thunder (a mere 10^{13-14} ergs). The very largest tropical storms are about 100 times more energetic than typical ones.

FISSION AND FUSION BOMBS. Log E = 21-24. P 1938, S 1945. The literature on this topic is too voluminous (and too erratically classified) for me to attempt any summary.

SOLAR RADIO BURSTS. Log E = 20. Radio noise from the sun was recognized during World War II (as interference in radar work) though not published until 1946 (152). The second radio star was UV Ceti (153), which is still the prototype of spotted, binary flare stars.

METEORS AND FIREBALLS. Log E = 12-13. S ancient, U 2001. The biggest ones, of course, become meteorites, and there is presumably in fact a continuum of energies up to impacts that may cause global extinction events.

ULTRA-HIGH-ENERGY COSMIC RAYS. Log E = 8 (comparable as another speaker noted, with energies of terrestrial projectiles like golf balls). P 1968, S 1980s, U 2001. The "prediction" is an implication of the theory of cosmic ray acceleration put forward by the Barnothys, in which gamma rays travel all the way around the universe, coming back with higher energy each time. The idea never really caught on. In fact, most people would now say that the main problem is getting HeV and more cosmic rays here (though the various photon backgrounds of the universe) from even the nearest likely sources (154).

DECAYS OF SINGLE WIMPS AND INOs. Log E = -11 to +1. P 1970s, S 1991? A significant minority of the 100+ candidates proposed over the years for dark matter are unstable on varying time scales, and indeed detection of the decay products is one way we might identify the DM. Sciama (155) has suggested that the effects of a decaying 23 eV neutrino have already been seen in the form of otherwise inexplicable ionization of galactic and intergalactic gas. Most of the other decay products are expected to be of considerably higher energy (possibly even including the ultra-high energy cosmic rays).

REFERENCES

[1] de Sitter, W. 1917, MNRAS 76, 49 & 77, 155.

[2] Friedmann, A. 1922, Z. f. Phys. 10, 377.

[3] Friedmann, A. 1924, Z. f. Phys. 21, 326.

[4] Lemaitre, G. 1927, Ann. Soc. Sci. Bruxelles 47A, 49.

[5] Hubble, E.P. 1929, Proc. NAS 15, 168.

[6] Lemaitre, G. 1931, MNRAS 91, 483 & 470.

[7] Gamow, G. 1935, Ohio J. Sci. 35, 406.

[8] Gamow, G. 1949, Rev. Mod. Phys. 21, 367.

[9] Alpher, R.A. and Herman, R.C.. 1950. Rev. Mod. Phys. 22, 153.

[10] Alpher, R.A. and Herman, R.C. 1953, Ann. Rev. Nucl. Sci, 2, 1.

[11] Bondi, H. 1954. Public discussion with McKellar at Victoria, BC, Canada (April).

[12] Scheuer, P.A.G. 1957, Proc. Cam. Phil. Soc. 53, 764.

[13] Penzias, A. and Wilson, R.W. 1965, ApJ, 142, 420.

[14] Dicke, R.H. et al. 1965, ApJ 142, 414.

[15] Schmidt, M. 1968, ApJ 151, 393.

[16] Trimble, V..1996, PASP 108, 1073.

[17] Fath, E.A. 1908, Lick Obs. Bull. 5, 71.

[18] Seyfert, C.K. 1943, ApJ 97, 28.

[19] Woltjer, L. 1959, ApJ 130, 138.

[20] Smith, F.G. 1951, Nature 168, 555.

[21] Baade, W. and Minkowski, R.,1959, ApJ 119, 206 & 215.

[22] Burbidge, G.R. 1956, ApJ 124, 416.

[23] Burbidge, G.R. and Hoyle, F. 1956, Nuovo Cimento 4, 558.

[24] Shklovsky, J.S. 1960, Sov. AJ 4, 885.

[25] Hoyle, F. and Fowler, W.A. 1963, MNRAS 125, 169.

[26] Schmidt, M. 1963, Nature 197, 1040.

[27] Hoyle, F. and Fowler, W.A. 1963, Nature 197, 533.

[28] Robinson, I., Schild, A., and Schucking, E., Eds.1965, Quasi-Stellar Sources and Gravitational Collapse, U. Chicago Press.

[29] Salpeter, E.E. 1964, ApJ,140, 796.

[30] Zeldovich, Ya.B. 1964, DAN 155, 67.

[31] Zeldovich, Ya.B. and Novikov, I.D. 1964, DAN 158, 811.

[32] Blandford, R.D. and Znajek, R.1977, MNRAS 197, 433.

[33] Trimble, V. 1992, in S.S. Holt et.al. Eds. Testing the AGN Paradigm, AIP Conf. Proc. 254, 647.

[34] Gold, T., 1965 in (28) p. 97.

[35] Woltjer, L. 1964, Nature 201, 245.

[36] Becklin, E. et al. 1973, ApJ 181, L27.

[37] Riecke, G.H. and Low, F. 1975, ApJ 197, 17.

[38] Solinger, A., Morrison, P., and Markert, T. 1977, ApJ 211, 707.

[39] Riecke, G. and Lebofsky, M.J. 1978, ApJ 211, 707.

[40] Zwicky, F., 1959, Handb. Phys. 53, 373.

[41] Kahn, F. and Woltjer, J.1959, ApJ 130, 105.

[42] Ibata, R.A., Gilmore, G., and Irwin, M.J. 1994, Nature 370, 184.

[43] van Woerden, H., Rougoor, G.W., and Oort, J.H. 1957, CR Acad. Sci. Paris 244, 1691.

[44] Oort, J.H. 1977, ARA&A 15, 295.

[45] Yuan, C. 1984, ApJ 281, 600.

[46] Holt, S.S. and Verter, F. (Eds.) 1993, Back to the Galaxy, AIP Conf. Proc. 278.

[47] Muller, C.A., Oort, J.H., and Raimond, E. 1963, CR Acad. Sci. Paris 257, 1661.

[48] Münch, G. and Zirin, H. 1961, ApJ 133, 11.

[49] Oort, J.H. 1970, A&A 7, 381.

[50] Davies, R.D. 1972, MNRAS 160, 381.

[51] Shapiro, P.R. and Field, G.B. 1976, ApJ 205, 762.

[52] Smith, A. and Hoffleit, D.,1963 Nature 198, 650.

[53] Sharov, A.S. 1963, A. Zh. 40, 956.

[54] Shklovsky, J.S. 1964, A. Zh. 41, 176 & 42, 345.

[55] Dent, W.A. 1965, Science 148, 145.

[56] Rees. M.J. 1966, Nature 211, 468 & MNRAS 135, 345.

[57] Woltjer, J. 1966, ApJ 146, 597.

[58] Whitney, A.R. et al. 1971, Science 173, 225.

[59] Cohen, M.H. et al. 1971, ApJ 170, 207.

[60] Rees, M.J. 1971 Nature 229, 312.

[61] Blandford, R.D. and Rees, M.J. 1974, MNRAS 169, 395.

[62] Colgate, S.A. 1968, Can. J. Phys. 46, S476.

[63] Hawking, S.W. 1974, Nature 248, 30.

[64] Klebesadel, R.W. et al. 1973, ApJ 182, L85.

[65] Mazets, E.P. et al. 1974, JETP Lett.19, 126.

[66] Ruderman, M.A. 1975, Ann. NY Acad. Sci. 262, 164.

[67] Zwicky, F. 1974, A&SS 28, III.

[67a] Lundmark, K.1920, Sven Vetekapsakad. Hand. 60, No. 8.

[68] Curtis, H.D. 1921, Bull. Natl. Res. Council 2, part 3, p. 171.

[69] Baade, W. and Zwicky, F. 1934, Proc. NAS 20, 254 & 259.

[70] Minkowski, R. 1941, PASP 53, 224.

[71] Payne Gaposchkin, C.H. l936, PNAS 22, 228 &.ApJ 82, 173.

[72] Payne Gaposchkin, C.H. 1957, The Galactic Novae (North Holland and Dover 1964).

[73] Popper, D.M. 1937 PASP 49, 283.

[74] Humason, M.L. 1936, PASP 48, 110.

[75] Colgate, S.A. 1960.

[76] Hoyle. F., and Fowler, W.A.1960, ApJ 132, 565.

[77] Trimble, V. 1982, RMP 43, 1185.

[78] Zwicky, F. 1940, RMP 12, 66.

[79] Borst, L.B. 1950, PR 78, 807.

[80] Burbidge, E.M., Burbidge, G.R., Fowler, W.A., and Hoyle, F. 1957, RMP 29, 547.

[81] Pankey, T. 1962, PhD dissertation, Howard University.

[82] Colgate S.A. 1968, ApJ 153, 335.

[83] Wilson, O.C. 1939, ApJ 90, 634.

[84] Zwicky, F. 1939, PR 55, 726.

[85] Minkowski, R. 1942, ApJ 96, 199.

[86] Hubble, E., and Sandage, A. 1953, ApJ 118, 353.

[87] Herschel, J.F. 1847, Results of Astronomical Observations made at the Cape of Good Hope, p. 32-37.

[88] Jeffreys, H. 1929, MNRAS 89, 636 & 731.

[89] Whipple, F. 1939, Proc. NAS 25, 118.

[90] Schwarzschild, M. 1958, Structure and Evolution of the Stars, Princeton Univ. Press, p. 228.

[91] Herbig, G. and Boyarchuk, A. 1968, ApJ 153, 397.

[92] Iben, I., 1984, ApJ 277, 333.

[93] Lundmark, K. 1921, PASP, 33, 314.

[94] Campbell, W.W., and Moore, J.H. 1918, Pub. Lick Obs. 13.

[95] Russell, H.N., Dugan, J.A., and Stewart, J.Q. 1926, Astronomy (Boston, Ginn & Co.).

[96] Lucy, L.B. 1967, AJ 72, 813.

[97] Paczynski, B. and Ziolkowski, J. 1968, IAUS 34, 396.

[98] Roxburgh, I.W. 1967, Nature 215, 838.

[99] Shklovsky, I.S. 1979, Stars, Their Birth, Life, and Death.

[100] Shklovsky, I.S. 1956, A. Zh. 33, 315.

[100a] Griffin, R.F. 1999, Obs. 119, 272.

[101] McLaughlin, D.B. 1960 in J.L. Greenstein (ed.) Stellar Atmospheres, U. Chicago Press, p. 585.

[102] Shara, M.M. 1989, PASP 101, 5 and references therein.

[103] Couderc, L. 1939, Ann. d'Ap. 2, 271.

[104] Felten, J.E 1988, in M. Kafatos & A. Michalitsianos (Eds.) Supernova 1987A in the Large Magellanic Cloud, Cambridge Univ. Press, p. 932.

[105] Ambartsumian, V.A. and Kosirev, N. 1933, Zs. f. Ap. 18, 320.

[106] Milne, E.A. 1931, MNRAS 91, 54.

[107] Biermann, L. 1939, Zs. f. Ap. 18, 344.

[108] Schatzman, E. 1946, Ann. d'Ap. 9, 199.

[109] Schatzman, E., 1950, Ann. d'Ap. 12, 16.

[110] Schatzman, E. 1951, Ann. d'Ap. 14, 294.

[111] Mestel, L. 1952, Obs. 72, 184.

[112] Mestel, L. 1952, MNRAS 112, 583 & 598.

[113] Struve, O. 1955, Sky & Tel. 14, 275.

[114] Kraft, R.P. 1962, ApJ 135, 408.

[115] Kraft, R.P., 1963, Adv. A&A 2, 43.

[116] Elvis, M. et al. 1975, Nature 257, 656.

[117] McClintock, J.E. and Remillard, R.A., 1986, ApJ 308, 110.

[118] Richards, D. et al. 1969, IAUC 2181.

[119] Reichley P.E. and Downs, G.S, 1969, Nature 222, 229.

[120] Radhakrishnan.,V. and Manchester, R.N. 1969, Nature 222, 228.

[121] Ruderman, M.A. 1969, Nature 223, 597.

[122] Packard, R.E., 1972, PRL 28, 1080.

[123] Wells, L.D. 1896, HCO Circ. No. 12.

[124] Crawford, J.A. and Kraft, R.P. 1956, ApJ 123, 44.

[125] Paczynski, B. 1968, Acta Astron. 15, 89.

[126] Osaki, Y. 1974, PASJ 26, 429.

[126a] Van der Bilt, J. 1908, The Variable Star U Geminorum, Rech. Ast. III.

[127] Joy, A.H.,1943 PASP 55, 283.

[128] Leavitt, H.S., 1920, Harvard Annals 84,121.

[129] Joy, A.H., 1945 ApJ 112, 168.

[130] Herbig, G.H. 1952, JRASV 46, 223.

[131] Hoffleit, D. 1939, HOB 911.

[132] Wachmann, A.A. 1954, Zs. f. Ap. 35, 231.

[133] Grindlay, J.E. et al. 1976, ApJ 205, L127.

[134] Joss, P.C. 1978, ApJ 225, L123.

[135] Woosley, S.E. and Taam, R.E. 1976, Nature 263, 101.

[136] Lewin, W.H.G. et al. 1978, ApJ 225, L11.

[137] Margon, B. 1979, ApJ 230, L41 and 233, L63.

[138] Milgrom, M. 1979, A&A 76, L3; 78, L9, and 78, 617.

139 Fabian, A.C. and Rees, M.J. 1979, MNRAS 187, 1p.

140 Mirabel, I.F. and Rodriguez, L.F. 1994, Nature 371, 46.

141 Reynolds, J. and Jauncey, D. 1994, IAUC 6063.

142 Trimble, V. and Leonard, P.J.T. 1995, PASP 107, 1.

143 Leventhal, M. et al. 1978, ApJ 225, L1.

144 Carrington, R.D. 1860, MNRAS 20, 13.

145 Hodgson, R. 1860, MNRAS 20, 15.

146 Joy, A.H. and Humason, M.L. 1949, PASP, 61, 133.

147 Luyten, W.J. 1949, Havard Announcement Card 990.

148 Golub, L. and Pasachoff, J.M. 1997, The Solar Corona, Cambridge University Press, Sect. 9.2.

149 Gold, T. and Hoyle, F. 1960, MNRAS 120, 89.

150 Haisch, B., Strong, K.T., and Rodono, M. 1991, ARA&A 29, 275.

151 Hudson, H.S. 1987, Sol. Phys. 113, 1.

152 Appleton, E.V. and Hey, J.S. 1946, Phil Mag. Ser. 7, 37, 73.

153 Lovell, B. et al. 1963, Nature 198, 228.

154 Stecker, F.W. 1998, PRL 80, 1816.

155 Sciama, D.W. 1991, Comm. on Ap. 15, 71.

Current Issues

R. D. Blandford

130-33 Caltech
Pasadena
CA 91125

Abstract. Cosmic explosions are observed in many astrophysical environments. They range in scale from hydromagnetic instabilities in the terrestrial magnetotail and solar "nanoflares" to cosmological gamma ray bursts, supernovae and the protracted intervals of nuclear activity that produce the giant quasars. There are many parallels in the analysis of the explosion sites that are highlighted at this workshop, specifically stellar coronae, accretion disks, supernovae and compact objects. In this introductory talk, some general issues are discussed and some more specific questions relating to the individual sites are raised.

INTRODUCTION

A star lives its life, from the time that it is born out of a loose assembly of molecular gas till it makes its quietus as a white dwarf, neutron star or black hole, fighting gravity. Although it eventually looses the fight (unless it manages to make a Type Ia supernova), it does not concede gracefully. Time and again it finds itself transitioning from a metastable equilibium to a state of lower energy on a dynamical timescale. Similar principles govern the evolution of galaxies where there can be runaway formation of massive stars or episodic accretion onto the central, massive black hole. These are "Cosmic Explosions" - the topic of this workshop. When I was first asked to introduce "Current Issues", I thought the title curiously apposite because it is ultimately currents - electrical, weak (charged and neutral) and (with some license) strong - that are responsible for these impulsive releases of energy. Out of the many astrophysical sites that could have been included on the program, the organizers have chosen to concentrate on a few of the most interesting ones, that I shall consider in turn - the solar corona, accretion disks surrounding young stellar objects, novae, supernovae, "hypernovae" and jets. As I am neither competent nor patient enough to describe these in any detail, I have chosen to list some recent advances in our observational and theoretical understanding in each case and to pose a few questions some of which may already have answers which I hope subsequent speakers will provide. In view of the large range of topics reviewed I cannot hope to give a representative or even useful bibliography, and so I shall

CP522, *Cosmic Explosions: Tenth Astrophysical Conference,*
edited by Stephen S. Holt and William W. Zhang

give none and defer to subsequent speakers.

SOLAR AND STELLAR FLARES

The combined observations of the YOHKOH, SOHO and TRACE satellites are transforming our view of the solar corona and, consequently, of the surface activity of other stars. In particular they have given us an appreciation of the dynamics of magnetic field lines as they are gently shuffled by underlying convective motions. The whole region above the photosphere is permeated by a *magnetic carpet* which is re-woven every couple of days. The solar prominences and coronal arches, prominent in X-ray images are just the regions where the plasma happens to be hottest and, contrary to what might have been thought, the magnetic field is *weakest*. This magnetic activity is intrinsically dissipative and this keeps the corona at million degree temperatures and launches the solar wind.

The quiet solar wind appears to be a simple and quasi-steady flow at least at high latitude (as measured by the Ulysses spacecraft) with poloidal and toroidal magnetic field components declining as $\propto r^{-2}, r^{-1}$, respectively. By contrast, the equatorial outflow appears to be dominated by unsteady coronal mass ejections. The equatorial current sheet is naturally unstable and develops its characteristic "ballerina skirt" sector structure. (Perhaps something similar has been observed by Chandra in the Crab Nebula.) Solar physics has much to teach us about accretion disks, where the underlying motions are much faster and, necessarily, supersonic. It should be no surprize that they are often accompanied by hot coronae, that dissipate a large fraction of the gravitational energy release, and powerful outflows.

My list of questions includes:

- What are the true laws of astrophysical MHD? Traditional, global MHD has been based upon analytic solutions of the equations of conservation of mass, momentum and flux under conditions of high symmetry ignoring dissipation. However, real MHD is heavily influenced by the microphysical behavior of current sheets, tiny reconnecting regions, shock fronts etc in much the same way that hydrodynamic flows are beholden to boundary layers. Perhaps there are simple, phenomenological rules which can reconcile these two approaches.

TABLE 1. Observed Characteristics of Some Cosmic Explosions

Explosion	Energy (erg)	Timescale (s)	Power (erg s^{-1})
Solar Flares	10^{32}	10^{4}	10^{28}
FUORs	10^{45}	10^{9}	10^{36}
Novae	10^{44}	10^{6}	10^{38}
Supernovae	10^{50}	10^{6}	10^{44}
Hypernovae (GRBs)	10^{53}	10^{2}	10^{51}
Jets	10^{61}	10^{14}	10^{47}

- What is a solar flare? We know of many examples of magnetostatic configurations that can be slowly altered until they become unstable and release a large amount of magnetic energy. However we do not understand which of these are most likely to occur in practice and what is the partition of the release of energy between local heating and the bulk kinetic energy that drives outgoing shock waves. (Similar questions exist in earthquake studies.)

- What is the structure of shock fronts? Simulation and *in situ* measurement has greatly improved our understanding of collisionless shocks. It appears that thermal electrons are commonly transmitted with sub-equipartition energies, as is also found to be the case with supernova shock waves. The detailed plasma physics still eludes us, though. This issue is related to the question of the injection of suprathermal ions into the first order Fermi acceleration process that appears to be responsible for producing most Galactic cosmic rays.

- What determines the energy and length scales that dominate coronal heating? The form of this dissipation appears to be primarily reconnection and to be dominated by frequent "nanoflares", (although this conclusion is still controversial). This realization has, in turn, stimulated analysis of new modes of magnetic reconnection. Still, it is the occasional giant flare that commands our observational attention and provides the most detailed diagnostics.

- How is the solar wind launched? The observed coronal temperature is insufficient to give the gas its 700-800 km s^{-1} outflow speed as measured by Ulysses. This leaves hydromagnetic wave acceleration as the prime suspect. Understanding the acceleration and stability of the solar wind is highly relevant to the study of jets.

YOUNG STELLAR OBJECTS

Accretion disks and bipolar outflows appear to be a standard feature of star formation. The optical jets can propagate through the interstellar medium over distances more than ~ 10 pc, quickly polluting it with magnetic field and metals (an observation of some cosmological importance). However, this outflow is not steady. In particular, thermal instability of the accretion disk produces "FU Ori" outbursts where perhaps ~ 0.01 M$_\odot$ of gas are expelled with comparatively high speed over a decade or so every ten thousand years. Smaller scale explosions create the "Herbig-Haro" objects which are presumably traveling forward-reverse shocks. These are sometimes observed in matched pairs, one in each jet confirming that they originate at the disk. The morphological similarity to, for example, the knots in the M87 jet is clear.

Some important questions include:

- How much of the mass in the original protostellar disk accretes onto the central protostar and how much is lost in the form of a wind or jet? A related question is how much of the angular momentum is removed in this manner as opposed to being transported radially in the disk to large radii where it can, supposedly be extracted by large tidal torques.

- How much mass and energy is associated with the FUOR outbursts and how much is transported in the long intervals between outbursts? (It is not clear how far one can push the dynamical analogy but similar questions have been raised in trying to understand the Galactic microquasar GRS 1915+105.)

- How are the optical jets collimated and confined? Magnetic collimation is commonly invoked, but even here, several alternatives have been discussed. The field may be primarily vertical near the disk and shape the outflow through magnetic pressure. Alternatively, the field may have a significant radial component so that the jet can be launched centrifugally so that the hoop stress associated with toroidal field may be largely responsible for the collimation. A third possiblity, that has been discussed, is that the magnetic field be mostly toroidal near the disk and wound up like a coiled spring so that it can push the gas away vertically. Observations of protostellar outflows have as good a chance as those of any jets of measuring the magnetic structure. Ultimately the flow must be confined laterally at large cylindrical radius. It is not clear whether this is achieved by the ram pressure of infalling gas or through the application of a quasi-static thermal pressure.

- What is the dynamical structure of Herbig-Haro objects and what can they tell us about the explosions that cause them? Forward-reverse shocks arise naturally if the velocity with which the jet is launched varies so that the faster moving gas overtakes the slower outflow and forms a shock. The pressure behind this shock front may be sufficient to form a reverse shock giving a characteristic dynamical structure. The high pressure inter-shock gas will expand transversely, weakening the shock strengths. Again, observations should help us to understand what is really happening.

NOVAE

Classical novae, by contrast, are thermonuclear explosions which arise when hydrogen-rich gas from a companion accumulates on the surface of a C-O or O-Ne-Mg white dwarf and then detonates, initially uncontrollably, under degenerate conditions. The energy release per nucleon is enough to heat the gas above the Fermi temperature, causing it to expand, and then above the escape energy. As several of the nuclear reactions involve weak interactions that take place on timescales that are long compared with the dynamical timescales, the ejected gas is believed to contain many prominent radioactive species that can act as monitors of stellar activity.

X-ray novae involve similar processes occuring on the surface of a neutron star. Here, the reactions occur much faster but the gravitational potential well is so deep that the gas cannot escape using its own thermal pressure. (It may be expelled by radiation pressure, however.) Naturally, this burning will not occur uniformly over the surface of the star and rotational modulation of the X-ray emission was predicted and is observed. (Curiously, the rotational frequency is observed to vary slightly, which may be due to elevation of the X-ray photosphere by radiation pressure with approximate conservation of angular momentum.

My personal question list for novae is:

- What can be learned by observing radioactive nuclei and positron annihilation from classical novae? Novae are prime targets for missions like INTEGRAL and HESSI that promise to open up the new field of MeV spectroscopy. We need to go beyond mere detection of radioactive nuclei and use measurements of line strengths and widths to learn about the underlying explosion.

- What is the status of the beat-frequency model of QPOs? This posited that the neutron star was rotating with a period similar to that of the inner regions of the accretion disk and that the observed, varying frequencies were a beat rather than a fundamental. The first part of this hypothesis has been vindicated, but I wonder about the evidence for the second part?

- Are *any* QPO modes due to neutron star osciallations? The problem here is that some of the modes that had been attributed to neutron star oscillations are also found in black hole systems. (Neutron star modes can only provide the clock because the energies associated with them are necessarily quite small.)

- Can we measure the neutron star mass-radius relation? One of the best ways for high energy astrophysics to repay its immense debt to nuclear physics is to measure the equation of state of cold nuclear matter (in contrast to the hot nuclear matter that will be explored by heavy ion colliders). This may be possible through measuring the gravitational redshift of atomic and nuclear lines from the surface of hot neutron stars, however it is not clear what it will take to do this in practice.

SUPERNOVAE

Supernovae are once again at center stage. In cosmology, Type Ia explosions have been modeled empirically as one parameter standard candles, and if this is the case, they suggest that the universe is entering a (second?) epoch of inflationary expansion. This is a remarkable discovery, if true. In addition, there is circumstantial evidence that at least some types of γ-ray bursts are associated with supernovae, both through the suggested identification of GRBs with star forming regi ons and the possible discovery of supernova light curves in a few instances. For both lines of research to advance, it is imperative to develop a far better understanding of the physics and the astrophysics of supernova explosions.

The blast waves that result from these explosions are not always well-described by Sedov point explosions in uniform media. Even if the energy release is fairly isotropic (and there are several reasons for suspecting that it is not) the external medium is likely to be anisotropic. The beautiful images of η Car and SN 1987a, the former being an accident waiting to happen and the latter being one that we are still witnessing, explain why so many mature supernova remnants are quite non-circular despite having essentially isobaric interiors. These supernova remnants are excellent laboratories for studying particle acceleration and magnetic field amplification at shock fronts that provide a bridge between heliospheric studies and more energetic phenomena associated with AGN and GRBs. Non-relativistic shocks behave quite differently from relativistic shocks and so it is fortunate that we have plerions like the Crab Nebula and classic remnants like Tycho so close to home to study.

The questions:

- How important are Type Ia supernova evolutionary corrections? The big concern, as always with cosmographic studies of the expansion of the universe, is whether or not we are confusing kinematics with physical evolution. This is particularly troubling here because there is no commonly agreed identity for the progenitors of these explosions and, I believe, no consensus yet on the reason for the "Phillips" correction although some promising suggestions have been made. There are internal consistency checks and some of these have already been satisfied but more will be needed before we can sign off on the result.

- How do we classify supernovae observationally? I doubt that I am alone in not understanding the spectroscopic and physical distinctions between the various types of supernova Type Ibc, Type IIn etc. I hope we can have a primer on the subject here.

- What are Type Ia supernovae anyway? Single degenerate and double degenerate models have their advocates. Likewise for detonation of Chandrasekhar mass CO white dwarf versus an off center explosion in a lighter star with a helium envelope.

- When do Type II supernovae form black holes as opposed to neutron stars and what are the associate d rates? This question is timely because Chandra has just discovered a point source inside Cas A, and as of now, the odds are about evens for it being a black hole or a neutron star.

HYPERNOVAE

Gamma ray bursts continue to amaze. There has been direct verification that the long duration bursts are located at cosmological distances through the measurement of redshifts. (The same is probably true for the short duration bursts, although HETE2 is probably going to be necessary to verify this.) This leads to an impressively broad range of isotropic burst energies, from $\sim 10^{-6}$ $M_\odot c^2$ in the case of GRB 980425 to ~ 2 $M_\odot c^2$ for GRB 990123 - hardly standard candles (though this has not prevented some from trying to use them for cosmography). GRBs are now widely interpreted as optically thick fireballs created with large entropies per baryon, like the universe itself. The actual γ-ray emission, lasting for up to a few minutes, is commonly thought to be produced by internal shocks in the expanding ejecta and this accounts for the great heterogeneity in observed γ-ray burst time profiles. The ninth magnitude optical burst, seen by ROTSE from GRB 990123 (with an isotropic energy $\sim 10^{-3}$ of the total) may be caused by a reverse shock. Studying the afterglows is proving to be interesting in its own right, for what it has to say about the behavior of relativistic shocks, as a probe of the environment in which the burst occurs, as a measure of the explosion energy and as an indicator of beaming. Broken power law spectra are observed and these have been variously interpreted as being due to cut-offs in the electron distribution function, radiative cooling and self-absorption.

A recent development is the circumstantial evidence for the association of GRBs 970228, 980326, 980425 with supernovae, albeit of different types. If the association is also with young stars, then GRBs will be invaluable probes of the early universe and galaxy formation. Another somewhat more secure story is that soft γ-ray repeaters are "magnetars". That is to say, their outbursts are magnetically powered and originate on the surfaces of young neutron stars with surface magnetic fields $B \gtrsim 10^{14}$ G.

It is hard to limit the number of questions in this subject.

- Are GRBs beamed? In my view, although the jet hypothesis is eminently reasonable and fits in with some source models, especially collapsars, we are really only interpreting occasional steepening in the light curves in this manner, rather than seeing the clear evidence that was provided by VLBI in the case of AGN. The argument that the bursts must be beamed, otherwise they would have energies in excess of a stellar rest mass, reminds me of a similar argument in favor of them being local!

- Are there γ-ray quiet afterglows? These are surely a prediction of beaming. At present the observational constraints are surprisingly poor.

- Is magnetic field amplified at external shocks? We know from observations of young supernova remnants that relativistic protons and electrons are accelerated at non-relativistic shock fronts and that thermal electrons are transmitted with temperatures below the equipartition value. However, in a source like Cas A, it appears that the magnetic field only becomes strong in the interaction zone between the shocked interstellar medium and the explosion debris. (It is noteworthy that even as impressive a radio source as Cas A is four orders of magnitude under-luminous relative to a homogenous, maximally emitting synchrotron source with the same total pressure.) What is assumed makes a big difference. When Chris McKee and I computed the nonthermal emission that would be observed from decelarating, relativistic blast waves, we assumed that the magnetic field is just compressed along with the gas in passing through the relativistic shock front. In this case, ϵ_{mag}, the ratio of the magnetic to total energy density is only $2v_A^2/c^2 \sim 10^{-9}$ in the interstellar medium, where v_A is the Alfvén speed ahead of the shock front. More recent calculations, that are applied specifically to GRB afterglows, generally assume that $\epsilon_{mag} \sim 10^{-2}$ which is necessary to fit the fluxes (although the scaling laws are unchanged). For this reason and because the best studied GRB afterglow, GRB 970508, shows no sign of a mildly relativistic transition in the particle acceleration efficiency, I still suspect that the afterglow emission originates well downstream from the outer shock.

- Are GRBs really associated with supernovae? The late time light curve seen in GRB 980326 could have a different explanation. In particular, as Ann Esin and I have been considering, it fits rather well with scatttering of the initial optical burst by dust just outside the sublimation radius. This occurs typically at a distance $\sim 10^{18}$ cm. As refractory dust has high albedo and is forward scattering, the characteristic delay is plausibly a few months, as observed.

- How easy is it to have an ultrarelativistic jet emerge from inside a collapsing star? Entrainment of gas might easily occur and prevent the flow from attaining bulk Lorentz factors ~ 300.

- Are the "cyclotron" lines real and the ~ 300 keV "breaks" generic and, if so, can they be formed in ultrarelativistic outflows? Existing explanations seem a little contrived.

- What is the underlying physical mechanism for creating the fireball? Most models now seem to involve black holes, magnetic field and wishful thinking. The problem is hard. T he main challenge is to amplify the magnetic field fast enough to make an electromagnetic bomb. Fortunately there are new ingredients in the strongly curved spacetime around a black hole or a pair of orbiting neutron stars. The orbits can precess differentially at near relativistic speed and this can lead to an extremely rapid field growth - faster than conventional dynamos and, indeed, faster than exponential. This, in turn, induces $\gtrsim 10^{22}$ V EMFs which cannot be shorted out and accelerate pairs directly.

JETS

The black hole model of AGN has been vindicated observationally, and, beyond all reasonable doubt, most normal galaxies, like our own contain central black holes with masses in the million to billion solar mass range. There are also at least nine well measured compact object masses in excess of ~ 2.5 M$_\odot$ that are surely also black holes. There are promising, but so far less compelling indications that some of these holes spin rapidly.

Jet are a common, though not universal, accompaniment of accretion which suggests that they are involved in carrying off some of the energy and angular momentum released by the infalling gas. However, we do not understand how they are formed or even if there is a universal mechanism at work. We do know that magnetic field must grow to dynamically significant levels in accretion disks and most jet formation models now involve magnetic field except perhaps when the mass accretion rate greatly exceeds the Eddington rate. Our understanding of the emission from jets has also advanced. The discovery of rapidly variable GeV and TeV γ-rays from blazars shows that they can be extremely luminous. Radio astronomers have been mapping the smoke not the fire. The standard radio synchrotron model is also under assault. There is increasing evidence that compact components have brightness temperatures well in excess of the inverse Compton limit, even allowing for plausible Lorentz factors. Large degrees of circular polarization are also being reported. All of this suggests that some alternative, possibly coherent emission mechanism is at work at least in the compact cores. (Note, that it is not sufficient to explain how high brightness radio emission is *emitted*. It is also necessary to explain how it is *transmitted* out of the nucleus when there are many potential non-linear scattering mechanisms which will degrade the brightness temperature.)

In an impressive display of the power of VLBI, the radio astronomers have been able to show that the M87 jet is collimated within $\sim 100m$. If relativistic jets are powered by the black hole itself or the gas flow around the black hole then their energy has to be carried *initially* in some form other than electron-positron pairs, which are subject to catastrophic radiative losses. Electromagnetic Poynting flux is the prime suspect. In other words, relativistic jets are starting to resemble pulsars. (Contrariwise, the Crab pulsar now appears to form a pair of "jets").

My final list of questions is:

- Are AGN jets hyper-relativistic? Radio jets exhibit bulk Lorentz factors $\Gamma \sim$ 10; γ-ray burst models have taken us over the psychological hurdle to $\Gamma \sim 300$ which might just account for the reported radio variability of jets under the synchrotron model were it not to imply *steady* γ-ray burst level powers in AGN jets. Hence the appeal to coherent processes. Before we solve this problem, though, we must identify the jet working substance and if, and where, Poynting flux is transformed to plasma. (This last is still an interesting question in the case of the Crab pulsar wind.)

- Are jets better approximated as episodic or steady? Traditionally, we have modeled jets as stationary flows upon which have been imprinted perturbative disturbances which form shock fronts - the emitting elements. However, GRS 1915+105 suggest a quite different model - jets as a sequence of small explosions, perhaps associated with intermittent flow in the accretion disk, that expand into and keep open an evacuated channel. (YSO jets offer support to both views.)

- How are jets collimated? Ordered and disordered, poloidal and toroidal field have all been proposed for launching and collimating AGN jets from disks, just as with the YSO jets. 3D MHD global simulations are becoming increasingly ambitious and ever more relevant.

- Are relativistic jets powered by the spin energy of the hole or the binding energy of the accreting gas? The former seems more likely to form ultrarelativistic outflows; the latter may, on average, release more power. General relativistic numerical simulations are starting to guide our intuition.

- Why are there no gamma-ray megabursts? GRBs are thought to be associated with the birth of stellar black holes and produce powers of up to $\sim 10^{-7} c^5/G$ for $\sim 10^6 m$. If massive black holes are formed with masses $\sim 10^6$ $M_\odot$ at a rate of several per year, we might expect to see megabursts with similar powers but lasting for months. We don't. Perhaps, instead, massive black holes grow from much smaller holes, which, themselves, might be relics of the first generation of stars which may have masses $\sim 10^3 - 10^4$ $M_\odot$ at $z \sim 30$.

- How much do AGN contribute to the luminosity density of the universe? This is a closely related question. The measurement of the far infared background, the discovery of hard X-ray emission from some Seyfert galaxies and the spectrum of the X-ray background all point to AGN power being a significant fraction of the stellar luminosity density. If so, then there are probably implications for galaxy formation and development. For example, elliptical galaxies may result when a black hole grows rapidly and early in the life of the galaxy so that it is capable of blowing away late infalling gas before it can form a disk.

CONNECTIONS

As I hope this brief introduction has brought out, there are strong interconnnections between our studies of these different types of cosmic explosion. Accretion disk coronae can look quite like their solar counterpart. Novae have some dynamical similarities to miniature supernovae whose remnants, in turn, behave quite like aging γ-ray burst afterglows. Similar electromagnetic processes are at work around pulsars and black holes. γ-ray bursts themselves have some similarities, at least radiatively, with the early universe. And so on.

ACKNOWLEDGEMENTS

I acknowledge support under NASA grant 5-2837 and NSF grant AST 99-00866 and Re'em Sari for comments.

Supernovae Ia

Type Ia Supernovae: Progenitors and Evolution with Redshift

Ken'ichi Nomoto[1], Hideyuki Umeda[1], Chiaki Kobayashi[1], Izumi Hachisu[2], Mariko Kato[3], & Takuji Tsujimoto[4]

[1] *Department of Astronomy and Research Center for the Early Universe, University of Tokyo, Japan*
[2] *Department of Earth Science and Astronomy, College of Arts and Sciences, University of Tokyo, Japan*
[3] *Department of Astronomy, Keio University, Hiyoshi, Yokohama, Japan*
[4] *National Astronomical Observatory, Mitaka, Japan*

Abstract.
Relatively uniform light curves and spectral evolution of Type Ia supernovae (SNe Ia) have led to the use of SNe Ia as a "standard candle" to determine cosmological parameters. Whether a statistically significant value of the cosmological constant can be obtained depends on whether the peak luminosities of SNe Ia are sufficiently free from the effects of cosmic and galactic evolutions.

Here we first review the single degenerate scenario for the Chandrasekhar mass white dwarf (WD) models of SNe Ia. We identify the progenitor's evolution and population with two channels: (1) the WD+RG (red-giant) and (2) the WD+MS (near main-sequence He-rich star) channels. In these channels, the strong wind from accreting WDs plays a key role, which yields important age and metallicity effects on the evolution.

We then address the questions whether the nature of SNe Ia depends systematically on environmental properties such as metallicity and age of the progenitor system and whether significant evolutionary effects exist. We suggest that the variation of the carbon mass fraction $X(C)$ in the C+O WD (or the variation of the initial WD mass) causes the diversity of the brightness of SNe Ia. This model can explain the observed dependences of SNe Ia brightness on the galaxy types and the distance from the galactic center.

Finally, applying the metallicity effect on the evolution of SN Ia progenitors, we make a prediction of the cosmic supernova rate history as a composite of the supernova rates in different types of galaxies.

I INTRODUCTION

Type Ia supernovae (SNe Ia) are good distance indicators, and provide a promising tool for determining cosmological parameters (e.g., [1]). SNe Ia have been discovered up to $z \sim 1.32$ [2]. Both the Supernova Cosmology Project [3,4] and the High-z Supernova Search Team [5,6] have suggested a statistically significant value for the cosmological constant.

CP522, *Cosmic Explosions: Tenth Astrophysical Conference,*
edited by Stephen S. Holt and William W. Zhang
© 2000 American Institute of Physics 1-56396-943-2/00/$17.00

However, SNe Ia are not perfect standard candles, but show some intrinsic variations in brightness. When determining the absolute peak luminosity of high-redshift SNe Ia, therefore, these analyses have taken advantage of the empirical relation existing between the peak brightness and the light curve shape (LCS). Since this relation has been obtained from nearby SNe Ia only [7–9], it is important to examine whether it depends systematically on environmental properties such as metallicity and age of the progenitor system.

High-redshift supernovae present us very useful information, not only to determine cosmological parameters but also to put constraints on the star formation history in the universe. They have given the SN Ia rate at $z \sim 0.5$ [10] but will provide the SN Ia rate history over $0 < z < 1$. With the Next Generation Space Telescope, both SNe Ia and SNe II will be observed through $z \sim 4$. It is useful to provide a prediction of cosmic supernova rates to constrain the age and metallicity effects of the SN Ia progenitors.

SNe Ia have been widely believed to be a thermonuclear explosion of a mass-accreting white dwarf (WD) (e.g., [11] for a review). However, the immediate progenitor binary systems have not been clearly identified yet [12]. In order to address the above questions regarding the nature of high-redshift SNe Ia, we need to identify the progenitors systems and examine the "evolutionary" effects (or environmental effects) on those systems.

In §2, we summarize the progenitors' evolution where the strong wind from accreting WDs plays a key role [13–15]. In §3, we address the issue of whether a difference in the environmental properties is at the basis of the observed range of peak brightness [17]. In §4, we make a prediction of the cosmic supernova rate history as a composite of the different types of galaxies [18].

II EVOLUTION OF PROGENITOR SYSTEMS

There exist two models proposed as progenitors of SNe Ia: 1) the Chandrasekhar mass model, in which a mass-accreting carbon-oxygen (C+O) WD grows in mass up to the critical mass $M_{\mathrm{Ia}} \simeq 1.37 - 1.38 M_\odot$ near the Chandrasekhar mass and explodes as an SN Ia (e.g., [19,20]), and 2) the sub-Chandrasekhar mass model, in which an accreted layer of helium atop a C+O WD ignites off-center for a WD mass well below the Chandrasekhar mass (e.g., [21]). The early time spectra of the majority of SNe Ia are in excellent agreement with the synthetic spectra of the Chandrasekhar mass models, while the spectra of the sub-Chandrasekhar mass models are too blue to be comparable with the observations [22,23].

For the evolution of accreting WDs toward the Chandrasekhar mass, two scenarios have been proposed: 1) a double degenerate (DD) scenario, i.e., merging of double C+O WDs with a combined mass surpassing the Chandrasekhar mass limit [24,25], and 2) a single degenerate (SD) scenario, i.e., accretion of hydrogen-rich matter via mass transfer from a binary companion (e.g., [26,20]). The issue of DD vs. SD is still debated (e.g., [12]), although theoretical modeling has indicated that the merging of WDs leads to the accretion-induced collapse rather than SN Ia explosion [27–29].

In the SD Chandrasekhar mass model for SNe Ia, a WD explodes as a SN Ia only when its rate of the mass accretion ($\dot{M}$) is in a certain narrow range (e.g., [26,30]). In particular, if $\dot{M}$ exceeds the critical rate $\dot{M}_{\rm b}$, the accreted matter extends to form a common envelope [31]. This difficulty has been overcome by the WD wind model (see below). For the actual binary systems which grow the WD mass ($M_{\rm WD}$) to $M_{\rm Ia}$, the following two systems are appropriate. One is a system consisting of a mass-accreting WD and a lobe-filling, more massive, slightly evolved main-sequence or sub-giant star (hereafter "WD+MS system"). The other system consists of a WD and a lobe-filling, less massive, red-giant (hereafter "WD+RG system").

A   White dwarf winds

Optically thick WD winds are driven when the accretion rate $\dot{M}$ exceeds the critical rate $\dot{M}_{\rm b}$. Here $\dot{M}_{\rm b}$ is the rate at which steady burning can process the accreted hydrogen into helium as $\dot{M}_{\rm b} \approx 0.75 \times 10^{-6} \left(\frac{M_{\rm WD}}{M_\odot} - 0.40 \right) M_\odot \ {\rm yr}^{-1}$.

With such a rapid accretion, the WD envelope expands to $R_{\rm ph} \sim 0.1 R_\odot$ and the photospheric temperature decreases below $\log T_{\rm ph} \sim 5.5$. Around this temperature, the shoulder of the strong peak of OPAL Fe opacity [32] drives the radiation-driven wind [13,15]. The ratio of $v_{\rm ph}/v_{\rm esc}$ between the photospheric velocity and the escape velocity at the photosphere depends on the mass transfer rate and $M_{\rm WD}$. (see Fig.6 in [15]). We call the wind *strong* when $v_{\rm ph} > v_{\rm esc}$. When the wind is strong, $v_{\rm ph} \sim 1000$ km s^{-1} being much faster than the orbital velocity.

If the wind is sufficiently strong, the WD can avoid the formation of a common envelope and steady hydrogen burning increases its mass continuously at a rate $\dot{M}_{\rm b}$ by blowing the extra mass away in a wind. When the mass transfer rate decreases below this critical value, optically thick winds stop. If the mass transfer rate further decreases below $\sim 0.5\,\dot{M}_{\rm b}$, hydrogen shell burning becomes unstable to trigger very weak shell flashes but still burns a large fraction of accreted hydrogen.

The steady hydrogen shell burning converts hydrogen into helium atop the C+O core and increases the mass of the helium layer gradually. When its mass reaches a certain value, weak helium shell flashes occur. Then a part of the envelope mass is blown off but a large fraction of He can be burned to C+O [33] to increase the WD mass. In this way, strong winds from the accreting WD play a key role to increase the WD mass to $M_{\rm Ia}$.

B   WD+RG system

This is a symbiotic binary system consisting of a WD and a low mass red-giant (RG). A full evolutionary path of the WD+RG system from the zero age main-sequence stage to the SN Ia explosion is described in [15,16]. The occurrence frequency of SNe Ia through this channel is much larger than the earlier scenario, because of the following two evolutionary processes, which have not considered before.

(1) Because of the AGB wind, the WD + RG close binary can form from a wide binary even with such a large initial separation as $a_i \lesssim 40,000 R_\odot$. Our earlier estimate [13] is constrained by $a_i \lesssim 1,500 R_\odot$.

(2) When the RG fills its inner critical Roche lobe, the WD undergoes rapid mass accretion and blows a strong optically thick wind. Our earlier analysis has shown that the mass transfer is stabilized by this wind only when the mass ratio of RG/WD is smaller than 1.15. Our new finding is that the WD wind can strip mass from the RG envelope, which could be efficient enough to stabilize the mass transfer even if the RG/WD mass ratio exceeds 1.15. If this mass-stripping effect is strong enough, though its efficiency η_{eff} is subject to uncertainties, the symbiotic channel can produce SNe Ia for a much (ten times or more) wider range of the binary parameters than our earlier estimation.

With the above two new effects (1) and (2), the WD+RG (symbiotic) channel can account for the inferred rate of SNe Ia in our Galaxy. The immediate progenitor binaries in this symbiotic channel to SNe Ia may be observed as symbiotic stars, luminous supersoft X-ray sources, or recurrent novae like T CrB or RS Oph, depending on the wind status.

C   WD+MS system

In this scenario, a C+O WD is originated, not from an AGB star with a C+O core, but from a red-giant star with a helium core of $\sim 0.8 - 2.0 M_\odot$. The helium star, which is formed after the first common envelope evolution, evolves to form a C+O WD of $\sim 0.8 - 1.1 M_\odot$ with transferring a part of the helium envelope onto the secondary main-sequence star. A full evolutionary path of the WD+MS system from the zero age main-sequence stage to the SN Ia explosion is described in [14].

This evolutionary path provides a much wider channel to SNe Ia than previous scenarios. A part of the progenitor systems are identified as the luminous supersoft X-ray sources [34] during steady H-burning (but without wind to avoid extinction), or the recurrent novae like U Sco if H-burning is weakly unstable. Actually these objects are characterized by the accretion of helium-rich matter.

D   Realization frequency

For an immediate progenitor system WD+RG of SNe Ia, we consider a close binary initially consisting of a C+O WD with $M_{\mathrm{WD},0} = 0.6 - 1.2 M_\odot$ and a low-mass red-giant star with $M_{\mathrm{RG},0} = 0.7 - 3.0 M_\odot$ having a helium core of $M_{\mathrm{He},0} = 0.2 - 0.46 M_\odot$. The initial state of these immediate progenitors is specified by three parameters, i.e., $M_{\mathrm{WD},0}$, $M_{\mathrm{RG},0} = M_{\mathrm{d},0}$, and the initial orbital period P_0 ($M_{\mathrm{He},0}$ is determined if P_0 is given).

We follow binary evolutions of these systems and obtain the parameter range(s) which can produce an SN Ia. In Figure 1, the region enclosed by the thin solid line produces SNe Ia for several cases of the initial WD mass, $M_{\mathrm{WD},0} = 0.75 - 1.1 M_\odot$. For smaller $M_{\mathrm{WD},0}$, the wind is weaker, so that the SN Ia region is smaller. The

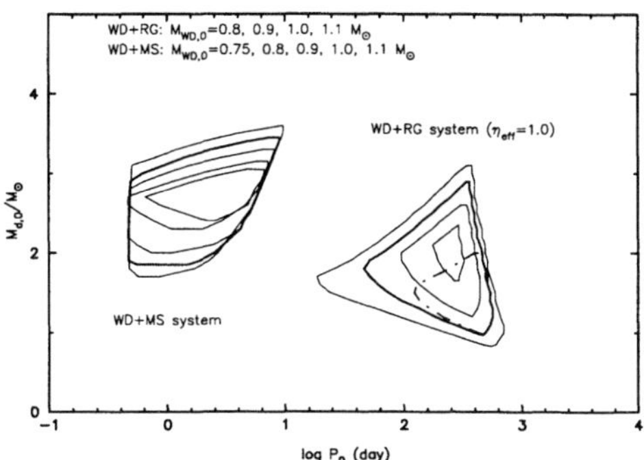

FIGURE 1. The region to produce SNe Ia in the $\log P_0 - M_{d,0}$ plane for five initial WD masses of $0.75 M_\odot$, $0.8 M_\odot$, $0.9 M_\odot$, $1.0 M_\odot$ (heavy solid line), and $1.1 M_\odot$. The region of $M_{WD,0} = 0.7 M_\odot$ almost vanishes for both the WD+MS and WD+RG systems, and the region of $M_{WD,0} = 0.75 M_\odot$ vanishes for the WD+RG system. Here, we assume the stripping efficiency of $\eta_{eff} = 1$. For comparison, we show only the region of $M_{WD,0} = 1.0 M_\odot$ for a much lower efficiency of $\eta_{eff} = 0.3$ by a dash-dotted line.

regions of $M_{WD,0} = 0.6 M_\odot$ and $0.7 M_\odot$ vanish for both the WD+MS and WD+RG systems.

In the outside of this region, the outcome of the evolution at the end of the calculations is not an SN Ia but one of the followings: (i) Formation of a common envelope for too large M_d or $P_0 \sim$ day, where the mass transfer is unstable at the beginning of mass transfer. (ii) Novae or strong hydrogen shell flash for too small $M_{d,0}$, where the mass transfer rate becomes below 10^{-7} $M_\odot$ yr^{-1}. (iii) Helium core flash of the red giant component for too long P_0, where a central helium core flash ignites, i.e., the helium core mass of the red-giant reaches $0.46 M_\odot$. (iv) Accretion-induced collapse for $M_{WD,0} > 1.2 M_\odot$, where the central density of the WD reaches $\sim 10^{10}$ g cm^{-3} before heating wave from the hydrogen burning layer reaches the center. As a result, the WD undergoes collapse due to electron capture without exploding as an SN Ia [30].

It is clear that the new region of the WD+RG system is not limited by the condition of $q < 1.15$, thus being ten times or more wider than the region of [13]'s model (depending on the the stripping efficiency of η_{eff}).

The WD+MS progenitor system can also be specified by three initial parameters: the initial C+O WD mass $M_{WD,0}$, the mass donor's initial mass $M_{d,0}$, and the orbital period P_0. For $M_{WD,0} = 1.0 M_\odot$, the region producing an SN Ia is bounded by $M_{d,0} = 1.8 - 3.2 M_\odot$ and $P_0 = 0.5 - 5$ d as shown by the solid line in Figure 1. The upper and lower bounds are respectively determined by the common envelope formation (i) and nova-like explosions (ii) as above. The left and right bounds are determined by the minimum and maximum radii during the main sequence of the donor star [14].

We estimate the rate of SNe Ia originating from these channels in our Galaxy

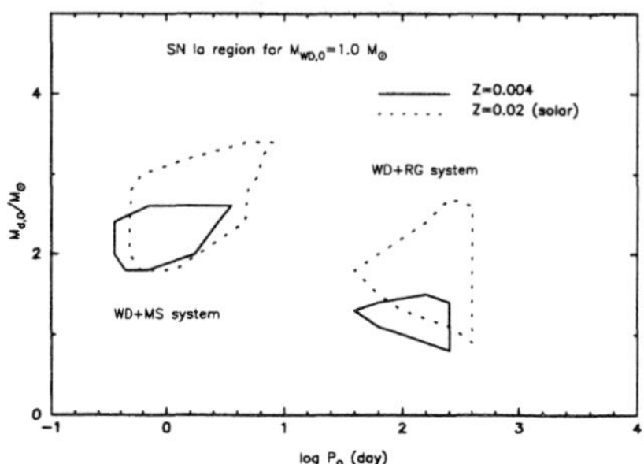

FIGURE 2. The regions of SNe Ia is plotted in the initial orbital period vs. the initial companion mass diagram for the initial WD mass of $M_{\mathrm{WD},0} = 1.0 M_\odot$. The dashed and solid lines represent the cases of solar abundance ($Z = 0.02$) and much lower metallicity of $Z = 0.004$, respectively. The left and the right regions correspond to the WD+MS and the WD+RG systems, respectively.

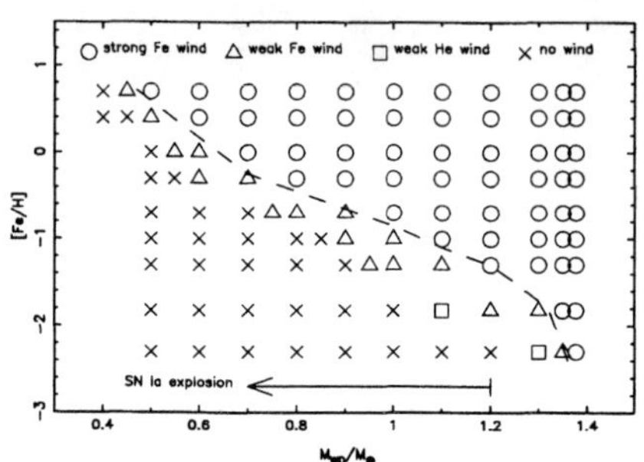

FIGURE 3. WD mass vs. metallicity diagram showing the metallicity dependence of optically thick winds. We regard the wind as "strong" if $v_{\mathrm{w}} > v_{\mathrm{esc}}$ but "weak" if $v_{\mathrm{w}} < v_{\mathrm{esc}}$. The term of "He" or "Fe" wind denotes that the wind is accelerated by the peak of iron lines near $\log T(\mathrm{K}) \sim 5.2$ or of helium lines near $\log T(\mathrm{K}) \sim 4.6$. The dashed line indicates the demarcation between the "strong" wind and the "weak" wind.

by using equation (1) of [24]. The realization frequencies of SNe Ia through the WD+RG and WD+MS channels are estimated as ~ 0.0017 yr^{-1} (WD+RG) and ~ 0.001 yr^{-1} (WD+MS), respectively. The total SN Ia rate of the WD+MS/WD+RG systems becomes ~ 0.003 yr^{-1}, which is close enough to the inferred rate of our Galaxy.

E   Low metallicity inhibition of type Ia supernovae

The optically thick winds are driven by a strong peak of OPAL opacity at $\log T(\mathrm{K}) \sim 5.2$ (e.g., [32]). Since the opacity peak is due to iron lines, the wind velocity v_{w} depends on the iron abundance [Fe/H] ([35,36]), i.e., v_{w} is higher for larger [Fe/H]. The metallicity effect on SNe Ia is clearly demonstrated by the size of the regions to produce SNe Ia in the diagram of the initial orbital period versus initial mass of the companion star (see Fig. 2). The SN Ia regions are much smaller for lower metallicity because the wind becomes weaker.

The wind velocity depends also on the luminosity L of the WD. The more massive WD has a higher L, thus blowing higher velocity winds ([15]). In order for the wind velocity to exceed the escape velocity of the WD near the photosphere, the WD mass should be larger than a certain critical mass for a given [Fe/H]. This implies that the initial mass of the WD $M_{\mathrm{WD},0}$ should already exceed that critical mass in order for the WD mass to grow to the Ch mass. This critical mass is larger for smaller [Fe/H], reaching $1.1 M_\odot$ for [Fe/H] $= -1.1$ (Fig. 3). Here we should note that the relative number of WDs with $M_{\mathrm{WD},0} \gtrsim 1.1 M_\odot$ is quite small in close binary systems ([47]). And for $M_{\rm WD,0} \gtrsim 1.2 M_\odot$, the accretion leads to collapse rather than SNe Ia ([30]). Therefore, no SN Ia occurs at $[{\rm Fe/H}] \leq -1.1$ in our model.

It is possible to test the metallicity effects on SNe Ia with the chemical evolution of galaxies.

In the one-zone uniform model for the chemical evolution of the solar neighborhood, the heavy elements in the metal-poor stars originate from the mixture of the SN II ejecta of various progenitor masses. The abundances averaged over the progenitor masses of SNe II predicts $[{\rm O/Fe}] \sim 0.45$ (e.g., [37,38]). Later SNe Ia start ejecting mostly Fe, so that $[{\rm O/Fe}]$ decreases to ~ 0 around $[{\rm Fe/H}] \sim 0$. The low-metallicity inhibition of SNe Ia predicts that the decrease in $[{\rm O/Fe}]$ starts at $[{\rm Fe/H}] \sim -1$. Such an evolution of $[{\rm O/Fe}]$ well explains the observations ([35]).

However, we should note that some anomalous stars have $[{\rm O/Fe}] \sim 0$ at $[{\rm Fe/H}] \lesssim -1$. The presence of such stars, however, is not in conflict with our SNe Ia models, but can be understood as follows: The formation of such anomalous stars (and the diversity of $[{\rm O/Fe}]$ in general) indicates that the interstellar materials were not uniformly mixed but contaminated by only a few SNe II (or even single SN II) ejecta. This is because the timescale of mixing was longer than the time difference between the supernova event and the next generation star formation. The iron and oxygen abundances produced by a single SN II vary depending on the mass, energy, mass cut, and metallicity of the progenitor. Relatively smaller mass SNe II $(13 - 15 M_\odot)$ and higher explosion energies tend to produce $[{\rm O/Fe}] \sim 0$ ([38,39]). Those metal poor stars with $[{\rm O/Fe}] \sim 0$ may be born from the interstellar medium polluted by such SNe II.

The metallicity effect on SNe Ia can also be checked with the metallicity of the host galaxies of nearby SNe Ia. There has been no evidence that SNe Ia have occurred in galaxies with a metallicity of $[{\rm Fe/H}] \lesssim -1$, although host galaxies are detected only for one third of SNe Ia and the estimated metallicities of host galaxies are uncertain. Three SNe Ia are observed in low-metallicity dwarf galaxies; SN1895B and SN1972E in NGC 5253, and SN1937C in IC 4182. Metallicities of these galaxies are estimated to be $[{\rm O/H}] = -0.25$ and -0.35, respectively [40]. If $[{\rm O/Fe}] \sim 0$ as in the Magellanic Clouds, $[{\rm Fe/H}] \sim -0.25$ and -0.35 which are not so small. Even if these galaxies have extremely SN II like abundance as $[{\rm O/Fe}] \sim 0.45$, $[{\rm Fe/H}] \sim -0.7$ and -0.8 (being higher than -1), respectively. Since these host galaxies are blue ($B - V = 0.44$ for NGC 5253 and $B - V = 0.37$ for IC 4182 according to RC3 catalog), the MS+WD systems are dominant progenitors for the present SNe Ia. The rate of SNe Ia originated from the MS+WD systems is not so sensitive to the metallicity as far as $[{\rm Fe/H}] > -1$ ([36]). Even if $[{\rm Fe/H}] \sim -0.7$ in such blue galaxies, therefore, the SN Ia rate is predicted to be similar to those in more metal-rich galaxies.

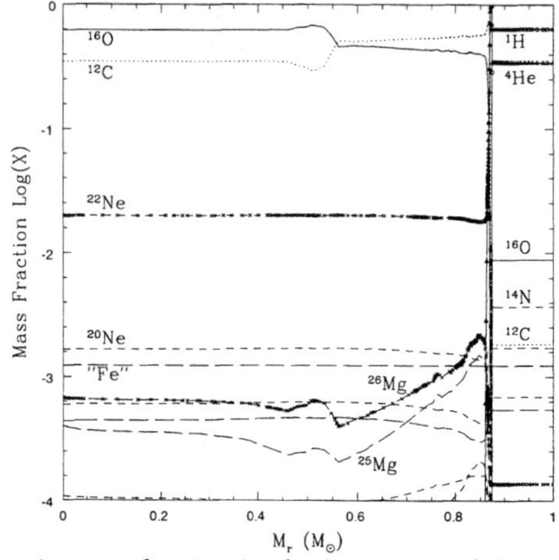

FIGURE 4. Abundances in mass fraction in the inner core of the 6 $M_\odot$ star for $Y = 0.2775$ and $Z = 0.02$ at the end of the second dredge-up.

III THE ORIGIN OF DIVERSITY OF SNE IA AND ENVIRONMENTAL EFFECTS

There are some observational indications that SNe Ia are affected by their environment. The most luminous SNe Ia seem to occur only in spiral galaxies, while both spiral and elliptical galaxies are hosts for dimmer SNe Ia. Thus the mean peak brightness is dimmer in ellipticals than in spiral galaxies [41]. The SNe Ia rate per unit luminosity at the present epoch is almost twice as high in spirals as in ellipticals [42]. Moreover, [43,44] found that the variation of the peak brightness for SNe located in the outer regions in galaxies is smaller.

[45,46] examined how the initial composition of the WD (metallicity and the C/O ratio) affects the observed properties of SNe Ia. [47] obtained the C/O ratio as a function of the main-sequence mass and metallicity of the WD progenitors. [17] suggested that the variation of the C/O ratio is the main cause of the variation of SNe Ia brightness, with larger C/O ratio yielding brighter SNe Ia. We will show that the C/O ratio depends indeed on environmental properties, such as the metallicity and age of the companion of the WD, and that our model can explain most of the observational trends discussed above. We then make some predictions about the brightness of SN Ia at higher redshift.

A C/O ratio in WD progenitors

In this section we discuss how the C/O ratio in the WD depends on the metallicity and age of the binary system. The C/O ratio in C+O WDs depends primarily on the main-sequence mass of the WD progenitor and on metallicity.

We calculated the evolution of intermediate-mass $(3 - 9M_\odot)$ stars for metallicity Z=0.001 – 0.03. In the ranges of stellar masses and Z considered in this paper, the most important metallicity effect is that the radiative opacity is smaller for lower Z. Therefore, a star with lower Z is brighter, thus having a shorter lifetime than a star with the same mass but higher Z. In this sense, the effect of reducing metallicity for these stars is almost equivalent to increasing a stellar mass.

For stars with larger masses and/or smaller Z, the luminosity is higher at the same evolutionary phase. With a higher nuclear energy generation rate, these stars have larger convective cores during H and He burning, thus forming larger He and C-O cores.

As seen in Figure 4, the central part of these stars is oxygen-rich. The C/O ratio is nearly constant in the innermost region, which was a convective core during He burning. Outside this homogeneous region, where the C-O layer grows due to He shell burning, the C/O ratio increases up to C/O $\gtrsim 1$; thus the oxygen-rich core is surrounded by a shell with C/O $\gtrsim 1$. In fact this is a generic feature in all models we calculated. The C/O ratio in the shell is C/O $\simeq 1$ for the star as massive as $\sim 7M_\odot$, and C/O > 1 for less massive stars.

When a progenitor reaches the critical mass for the SNe Ia explosion, the central core is convective up to around 1.1 $M_\odot$. Hence the relevant C/O ratio is between the central value before convective mixing and the total C/O of the whole WD. Using the results from the C6 model [19], we assume that the convective region is 1.14 $M_\odot$ and for simplicity, C/O $= 1$ outside the C-O core at the end of second dredge-up. Then we obtain the C/O ratio of the inner part of the SNe Ia progenitors (Fig. 5).

From this figure we find three interesting trends. First, while the central C/O is a complicated function of stellar mass [47], as shown here the C/O ratio in the core before SNe Ia explosion is a decreasing monotonic function of mass. The central C/O ratio at the end of second dredge-up decreases with mass for $M_{\rm ms} \gtrsim 5M_\odot$, while the ratio increases with mass for $M_{\rm ms} \gtrsim 4M_\odot$; however, the convective core mass during He burning is smaller for a less massive star, and the C/O ratio during shell He burning is larger for smaller C+O core. Hence, when the C/O ratio is averaged over 1.1 $M_\odot$ the C/O ratio decreases with mass. Second, as shown in [47], although the C/O ratio is a complicated function of metallicity and mass, the metallicity dependence is remarkably converged when the ratio is seen as a function of the C+O core mass $(M_{\rm CO})$ instead of the initial main sequence mass.

According to the evolutionary calculations for 3–9 $M_\odot$ stars by [47], the C/O ratio and its distribution are determined in the following evolutionary stages of the close binary.

(1) At the end of central He burning in the 3–9 $M_\odot$ primary star, C/O< 1 in the convective core. The mass of the core is larger for more massive stars.

(2) After central He exhaustion, the outer C+O layer grows via He shell burning, where C/O$\gtrsim 1$ [47].

(3a) If the primary star becomes a red giant (case C evolution; e.g., [48]), it then undergoes the second dredge-up, forming a thin He layer, and enters the AGB

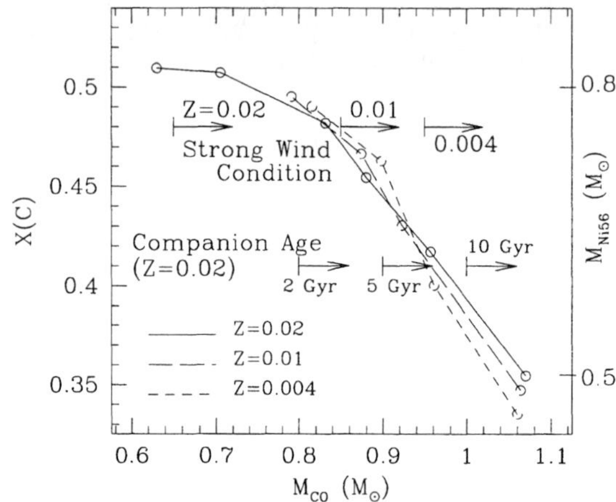

FIGURE 5. The total ^{12}C mass fraction included in the convective core of mass, $M = 1.14 M_\odot$, just before the SN Ia explosion as a function of the C+O core mass before the onset of mass accretion, M_{CO}. The lower bounds of M_{CO} obtained from the age effects and the conditions for strong wind to blow are also shown by arrows.

phase. The C+O core mass, M_{CO}, at this phase is larger for more massive stars. For a larger M_{CO} the total carbon mass fraction is smaller.

(3b) When it enters the AGB phase, the star greatly expands and is assumed here to undergo Roche lobe overflow (or a super-wind phase) and to form a C+O WD. Thus the initial mass of the WD, $M_{WD,0}$, in the close binary at the beginning of mass accretion is approximately equal to M_{CO}.

(4a) If the primary star becomes a He star (case BB evolution), the second dredge-up in (3a) corresponds to the expansion of the He envelope.

(4b) The ensuing Roche lobe overflow again leads to a WD of mass $M_{WD,0} = M_{CO}$.

(5) After the onset of mass accretion, the WD mass grows through steady H burning and weak He shell flashes, as described in the WD wind model. The composition of the growing C+O layer is assumed to be C/O=1.

(6) The WD grows in mass and ignites carbon when its mass reaches $M_{Ia} = 1.367 M_\odot$, as in the model C6 of [19]. Because of strong electron-degeneracy, carbon burning is unstable and grows into a deflagration for a central temperature of 8×10^8 K and a central density of 1.47×10^9 g cm^{-3}. At this stage, the convective core extends to $M_r = 1.14 M_\odot$ and the material is mixed almost uniformly, as in the C6 model.

In Figure 5, we show the carbon mass fraction $X(C)$ in the convective core of this pre-explosive WD, as a function of metallicity (Z) and initial mass of the WD before the onset of mass accretion, M_{CO}. Figure 5 reveals that: 1) $X(C)$ is smaller for larger $M_{CO} \simeq M_{WD,0}$. 2) The dependence of $X(C)$ on metallicity is small when

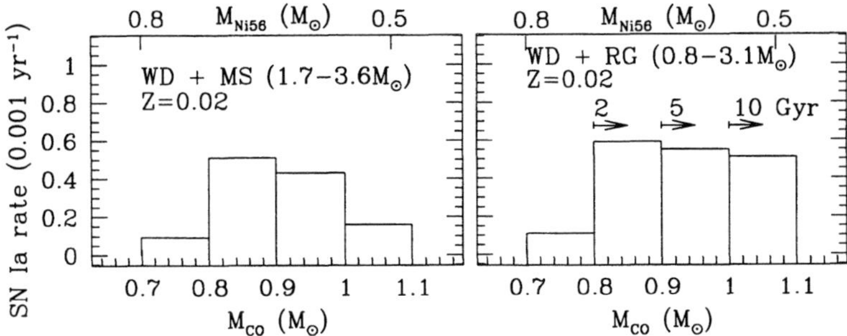

FIGURE 6. SN Ia frequency for a galaxy of mass $2 \times 10^{11} M_\odot$ as a function of M_{CO} for Z=0.02. For the WD+RG system, constraints from the companion's age are shown by the arrows. SNe Ia from the WD+MS system occur in spirals but not in ellipticals because of the age effect. M_{CO} and M_{Ni56} is assumed to be related as shown here.

plotted against M_{CO}, even though the relation between M_{CO} and the initial stellar mass depends sensitively on Z [47].

B Brightness of SNe Ia and the C/O ratio

In the Chandrasekhar mass models for SNe Ia, the brightness of SNe Ia is determined mainly by the mass of ^{56}Ni synthesized (M_{Ni56}). Observational data suggest that M_{Ni56} for most SNe Ia lies in the range $M_{Ni56} \sim 0.4 - 0.8 M_\odot$ (e.g., [49]). This range of M_{Ni56} can result from differences in the C/O ratio in the progenitor WD as follows.

In the deflagration model, a larger C/O ratio leads to the production of more nuclear energy and buoyancy force, thus leading to a faster propagation. The faster propagation of the convective deflagration wave results in a larger M_{Ni56}. For example, a variation of the propagation speed by 15% in the W6 – W8 models results in M_{Ni56} values ranging between 0.5 and $0.7 M_\odot$ [19], which could explain the observations.

In the delayed detonation model, M_{Ni56} is predominantly determined by the deflagration-to-detonation-transition (DDT) density ρ_{DDT}, at which the initially subsonic deflagration turns into a supersonic detonation [50]. As discussed in [17], ρ_{DDT} could be very sensitive to $X(C)$, and a larger $X(C)$ is likely to result in a larger ρ_{DDT} and M_{Ni56}.

Here we postulate that M_{Ni56} and consequently brightness of a SN Ia increase as the progenitors' C/O ratio increases (and thus $M_{WD,0}$ decreases). As illustrated in Figure 5, the range of $M_{Ni56} \sim 0.5 - 0.8 M_\odot$ is the result of an $X(C)$ range $0.35 - 0.5$, which is the range of $X(C)$ values of our progenitor models. The $X(C)$ – M_{Ni56} – $M_{WD,0}$ relation we adopt is still only a working hypothesis, which needs to be proved from studies of the turbulent flame during explosion (e.g., [51]).

C   Metallicity and age effects

1   Metallicity effects on the minimum $M_{WD,0}$

As mentioned in §2.5, M_w is the metallicity-dependent minimum $M_{WD,0}$ for a WD to become an SN Ia (*strong wind condition* in Fig. 5). The upper bound $M_{WD,0} \simeq 1.07 M_\odot$ is imposed by the condition that carbon should not ignite and is almost independent of metallicity. As shown in Figure 5, the range of $M_{CO} \simeq M_{WD,0}$ can be converted into a range of $X(C)$. From this we find the following metallicity dependence for $X(C)$:

(1) The upper bound of $X(C)$, which is determined by the lower limit on M_{CO} imposed by the metallicity-dependent conditions for a strong wind, e.g., $X(C)$ $\lesssim 0.51$, 0.46 and 0.41, for $Z=0.02$, 0.01, and 0.004, respectively.

(2) On the other hand, the lower bound, $X(C) \simeq 0.35 - 0.33$, does not depend much on Z, since it is imposed by the maximum M_{CO}.

(3) Assuming the relation between M_{Ni56} and $X(C)$ given in Figure 5, our model predicts the absence of brighter SNe Ia in lower metallicity environment.

2   Age effects on the minimum $M_{WD,0}$

In our model, the age of the progenitor system also constrains the range of $X(C)$ in SNe Ia. In the SD scenario, the lifetime of the binary system is essentially the main-sequence lifetime of the companion star, which depends on its initial mass M_2. [14,15] have obtained a constraint on M_2 by calculating the evolution of accreting WDs for a set of initial masses of the WD ($M_{WD,0} \simeq M_{CO}$) and of the companion (M_2), and the initial binary period (P_0). In order for the WD mass to reach M_{Ia}, the donor star should transfer enough material at the appropriate accretion rates. The donors of successful cases are divided into two categories: one is composed of slightly evolved main-sequence stars with $M_2 \sim 1.7 - 3.6 M_\odot$ (for $Z=0.02$), and the other of red-giant stars with $M_2 \sim 0.8 - 3.1 M_\odot$ (for $Z=0.02$) (Fig. 1).

If the progenitor system is older than 2 Gyr, it should be a system with a donor star of $M_2 < 1.7 M_\odot$ in the red-giant branch. Systems with $M_2 > 1.7 M_\odot$ become SNe Ia in a time shorter than 2 Gyr. Likewise, for a given age of the progenitor system, M_2 must be smaller than a limiting mass. This constraint on M_2 can be translated into the presence of a minimum M_{CO} for a given age, as follows: For a smaller M_2, i.e. for the older system, the total mass which can be transferred from the donor to the WD is smaller. In order for M_{WD} to reach M_{Ia}, therefore, the initial mass of the WD, $M_{WD,0} \simeq M_{CO}$, should be larger. This implies that the older system should have larger minimum M_{CO} as indicated in Figure 5. Using the $X(C)$-M_{CO} and M_{Ni56}-$X(C)$ relations (Fig. 5), we conclude that WDs in older progenitor systems have a smaller $X(C)$, and thus produce dimmer SNe Ia.

D   Comparison with observations

The first observational indication which can be compared with our model is the possible dependence of the SN brightness on the morphology of the host galaxies.

[41] found that the most luminous SNe Ia occur in spiral galaxies, while both spiral and elliptical galaxies are hosts to dimmer SNe Ia. Hence, the mean peak brightness is lower in elliptical than in spiral galaxies.

In our model, this property is simply understood as the effect of the different age of the companion. In spiral galaxies, star formation occurs continuously up to the present time. Hence, both WD+MS and WD+RG systems can produce SNe Ia. In elliptical galaxies, on the other hand, star formation has long ended, typically more than 10 Gyr ago. Hence, WD+MS systems can no longer produce SNe Ia. In Figure 6, we show the frequency of the expected SN I for a galaxy of mass $2 \times 10^{11} M_\odot$ for WD+MS and WD+RG systems separately as a function of $M_{\rm CO}$. Here we use the results of [15,14], and the $M_{\rm CO} - X({\rm C})$ and $M_{\rm Ni56} - X({\rm C})$ relations given in Figure 5. Since a WD with smaller $M_{\rm CO}$ is assumed to produce a brighter SN Ia (larger $M_{\rm Ni56}$), our model predicts that dimmer SNe Ia occur both in spirals and in ellipticals, while brighter ones occur only in spirals. The mean brightness is smaller for ellipticals and the total SN Ia rate per unit luminosity is larger in spirals than in ellipticals. These properties are consistent with observations.

The second observational suggestion is the radial distribution of SNe Ia in galaxies. [43,6] found that the variation of the peak brightness for SNe Ia located in the outer regions in galaxies is smaller. This behavior can be understood as the effect of metallicity. As shown in Figure 5, even when the progenitor age is the same, the minimum $M_{\rm CO}$ is larger for a smaller metallicity because of the metallicity dependence of the WD winds. Therefore, our model predicts that the maximum brightness of SNe Ia decreases as metallicity decreases. Since the outer regions of galaxies are thought to have lower metallicities than the inner regions [52,53], our model is consistent with observations. [43] also claimed that SNe Ia may be deficient in the bulges of spiral galaxies. This can be explained by the age effect, because the bulge consists of old population stars.

E   Evolution of SNe Ia at high redshift

We have suggested that $X({\rm C})$ is the quantity very likely to cause the diversity in $M_{\rm Ni56}$ and thus in the brightness of SNe Ia. We have then shown that our model predicts that the brightness of SNe Ia depends on the environment, in a way which is qualitatively consistent with the observations. Further studies of the propagation of the turbulent flame and the DDT are necessary in order to actually prove that $X({\rm C})$ is the key parameter.

Our model predicts that when the progenitors belong to an old population, or to a low metal environment, the number of very bright SNe Ia is small, so that the variation in brightness is also smaller, which is shown in Figure 7. In spiral galaxies, the metallicity is significantly smaller at redshifts $z \gtrsim 1$, and thus both the mean brightness of SNe Ia and its range tend to be smaller (Fig. 7). At $z \gtrsim 2$ SNe Ia would not occur in spirals at all because the metallicity is too low. In elliptical galaxies, on the other hand, the metallicity at redshifts $z \sim 1 - 3$ is not very different from the present value. However, the age of the galaxies at $z \simeq 1$ is

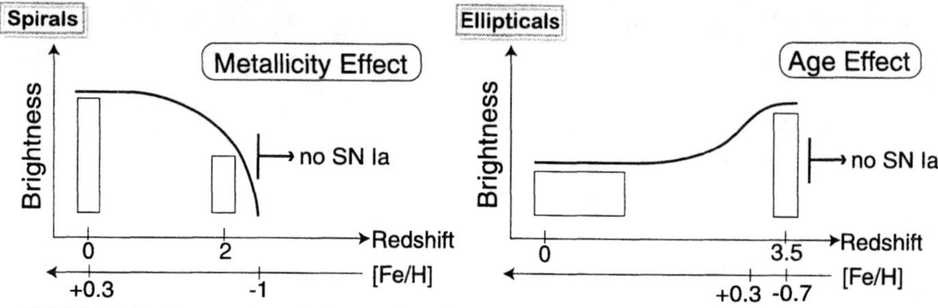

FIGURE 7. Illustration of the predicted variation in SN Ia brightness with redshift.

only about 5 Gyr, so that the mean brightness of SNe Ia and its range tend to be larger at $z \gtrsim 1$ than in the present ellipticals because of the age effect.

We note that the variation of $X(C)$ is larger in metal-rich nearby spirals than in high redshift galaxies. Therefore, if $X(C)$ is the main parameter responsible for the diversity of SNe Ia, and if the LCS method is confirmed by the nearby SNe Ia data, the LCS method can also be used to determine the absolute magnitude of high redshift SNe Ia.

F   Possible evolutionary effects

In the above subsections, we consider the metallicity effects only on the C/O ratio; this is just to shift the main-sequence mass - $M_{\rm WD,0}$ relation, thus resulting in no important evolutionary effect. However, some other metallicity effects could give rise to evolution of SNe Ia between high and low redshifts (i.e., between low and high metallicities).

Here we point out just one possible metallicity effect on the carbon ignition density in the accreting WD. The ignition density is determined by the competition between the compressional heating due to accretion and the neutrino cooling. The neutrino emission is enhanced by the *local* Urca shell process of, e.g., ^{21}Ne–^{21}F pair [54]. (Note that this is different from the *convective* Urca neutrino process). For higher metallicity, the abundance of ^{21}Ne is larger so that the cooling is larger. This could delay the carbon ignition until a higher central density is reached [55].

Since the WD with a higher central density has a larger binding energy, the kinetic energy of SNe Ia tends to be smaller if the same amount of ^{56}Ni is produced. This might cause a systematically slower light curve evolution at higher metallicity environment. The carbon ignition process including these metallicity effects as well as the convective Urca neutrino process need to be studied (see also [56] for nucleosynthesis constraints on the ignition density).

IV   COSMIC SUPERNOVA RATES

Attempts have been made to predict the cosmic supernova rates as a function of redshift by using the observed cosmic star formation rate (SFR) [57–59]. The

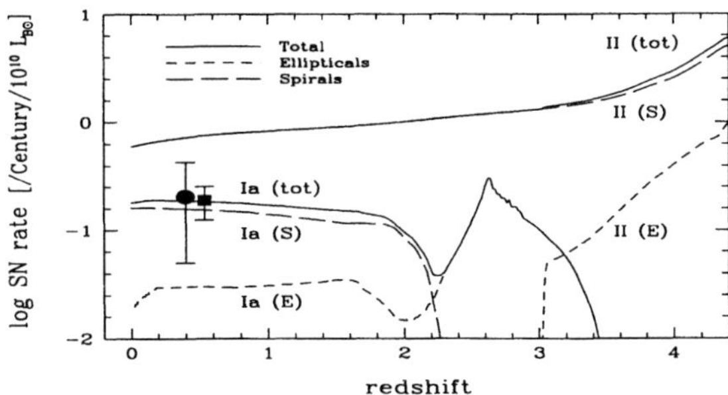

FIGURE 8. The cosmic supernova rates (solid line) as the composite of ellipticals (short-dashed line) and spirals (long-dashed line). The upper three lines show SN II rates, the lower three lines show SN Ia rates. Observational data sources: circle, [63]; square, [10].

observed cosmic SFR shows a peak at $z \sim 1.4$ and a sharp decrease to the present [60]. However, UV luminosities which is converted to the SFRs may be affected by the dust extinction [61]. Recent updates of the cosmic SFR suggest that a peak lies around $z \sim 3$.

[35] predicts that the cosmic SN Ia rate drops at $z \sim 1-2$, due to the metallicity-dependence of the SN Ia rate. Their finding that the occurrence of SNe Ia depends on the metallicity of the progenitor systems implies that the SN Ia rate strongly depends on the history of the star formation and metal-enrichment. The universe is composed of different morphological types of galaxies and therefore the cosmic SFR is a sum of the SFRs for different types of galaxies. As each morphological type has a unique star formation history, we should decompose the cosmic SFR into the SFR belonging to each type of galaxy and calculate the SN Ia rate for each type of galaxy.

Here we first construct the detailed evolution models for different type of galaxies which are compatible with the stringent observational constraints, and apply them to reproduce the cosmic SFR for two different environments, e.g., the cluster and the field. Secondly with the self-consistent galaxy models, we calculate the SN rate history for each type of galaxy and predict the cosmic supernova rates as a function of redshift.

A   In Clusters

Galaxies that are responsible for the cosmic SFR have different timescales for the heavy-element enrichment, and the occurrence of supernovae depends on the metallicity therein. Therefore we calculate the cosmic supernova rate by summing up the supernova rates in spirals (S0a-Sa, Sab-Sb, Sbc-Sc, and Scd-Sd) and ellipticals with the ratio of the relative mass contribution. The relative mass contribution is obtained from the observed relative luminosity proportion and the calculated mass to light ratio in B-band [18]. The photometric evolution is calculated with the

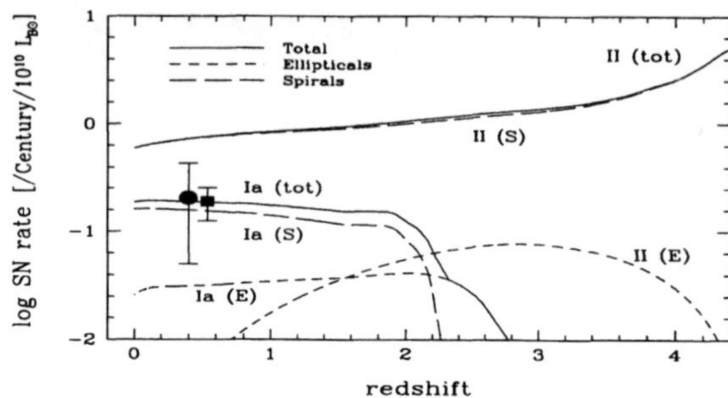

FIGURE 9. The same as Figure 8, but for the formation epochs of ellipticals span at $1 \lesssim z \lesssim 4$, which is corresponding to field ellipticals.

spectral synthesis population database taken from [62]. We adopt $H_0 = 50$ km s^{-1} Mpc^{-1}, $\Omega_0 = 0.2$, $\lambda_0 = 0$, and the galactic age of $t_{\text{age}} = 15$ Gyr.

First, we make a prediction of the cosmic supernova rates by using the galaxy models which are constructed to meet the observational constraints of cluster galaxies. We assume that elliptical galaxies are formed by a single star burst and stop the star formation at $t \sim 1$ Gyr due to a supernova-driven galactic wind, while spiral galaxies are formed by a relatively continuous star formation. The infall rates and the SFRs are given by [18]. These models are constructed to meet the latest observational constraints such as the present gas fractions and colors for spirals, and the mean stellar metallicity and the color evolution from the present to $z \sim 1$ for ellipticals (see [18] for the figures).

The synthesized cosmic SFR has an excess at $z \gtrsim 3$ due to the early star burst in ellipticals and a shallower slope from the present to the peak at $z \sim 1.4$, compared with Madau's plot [60]. Figure 8 shows the cosmic supernova rates in cluster galaxies. The SN Ia rate in spirals drops at $z \sim 1.9$ because of the low-metallicity inhibition of SNe Ia. We can test the metallicity effect by finding this drop of the SN Ia in spirals, if high-redshift SNe Ia at $z \gtrsim 1.5$ and their host galaxies are observed with the Next Generation Space Telescope. In ellipticals, the chemical enrichment takes place so early that the metallicity is large enough to produce SNe Ia at $z \gtrsim 2$. The two peaks of SN Ia rates at $z \sim 2.6$ and $z \sim 1.6$ come from the MS+WD and the RG+WD systems, respectively. The SN Ia rate in ellipticals decreases at $z \sim 2.6$, which is determined from the shortest lifetime of SNe Ia of ~ 0.5 Gyr. Thus, the total SN Ia rate decrease at the same redshift as ellipticals, i.e., $z \sim 2.6$. (Note, the decrease of the SN Ia rate at $z \sim 1.6$ disappears if we adopt $z_{\text{f}} \sim 3$, because the peak from the MS+WD systems moves to lower redshifts.)

B   In Field

We also predict the cosmic supernova rates assuming that the formation of ellipticals in field took place for over the wide range of redshifts, which is imprinted in the observed spectra of ellipticals in the Hubble Deep Field [64]. The adopted

SFRs are the same as the case of cluster galaxies, but for the formation epochs z_f of ellipticals distribute as $\propto \exp(-((z-2)/2)^2)$ in the range of $0 \leq z \leq 5$,

The synthesized cosmic SFR has a broad peak around $z \sim 3$, which is in good agreement with the recent sub-mm observation [65]. Figure 9 shows the cosmic supernova rates in field galaxies. As in Figure 8, the SN Ia rate in spirals drops at $z \sim 1.9$. The averaged SN Ia rate in ellipticals decreases at $z \sim 2.2$ as a result of ~ 0.5 Gyr delay of the decrease in the SFR at $z \gtrsim 3$. Then, the total SN Ia rate decreases gradually from $z \sim 2$ to $z \sim 3$.

The rate of SNe II in ellipticals evolves following the SFR without time delay. Then, it is possible to observe SNe II in ellipticals around $z \sim 1$. The difference in the SN II and Ia rates between cluster and field ellipticals reflects the difference in the galaxy formation histories in the different environments.

C Summary

(1) In the cluster environment, the predicted cosmic supernova rate suggests that in ellipticals SNe Ia can be observed even at high redshifts because the chemical enrichment takes place so early that the metallicity is large enough to produce SNe Ia at $z \gtrsim 2.5$. In spirals the SN Ia rate drops at $z \sim 2$ because of the low-metallicity inhibition of SNe Ia.

(2) In the field environment, ellipticals are assumed to form at such a wide range of redshifts as $1 \lesssim z \lesssim 4$. The SN Ia rate is expected to be significantly low beyond $z \gtrsim 2$ because the SN Ia rate drops at $z \sim 2$ in spirals and gradually decreases from $z \sim 2$ in ellipticals.

REFERENCES

1. Branch, D. 1998, ARA&A, 1998, 36, 17
2. Gilliland, R. L., Nugent, P. E., & Phillips, M. M. 1999, ApJ, in press (astro-ph/9903229)
3. Perlmutter, S. et al. 1997, ApJ, 483, 565
4. Perlmutter, S. et al. 1999, ApJ, 517, 565
5. Garnavich, P. et al. 1998, ApJ, 493, 53
6. Riess, A. G. et al. 1998, AJ, 116, 1009
7. Phillips, M. M. 1993, ApJ, 413, L75
8. Hamuy, M., et al. 1995, AJ, 109, 1
9. Riess, A. G., Press, W. H., & Kirshner, R. P. 1995, ApJ, 438, L17
10. Pain, R. 1999, Talk at the Type Ia supernova workshop in Aspen
11. Nomoto, K., Iwamoto, K., & Kishimoto, N. 1997a, Science, 276, 1378
12. Branch, D., Livio, M., Yungelson, L. R., Boffi, F. R., & Baron, E. 1995, PASP, 107, 717
13. Hachisu, I., Kato, M., & Nomoto, K., 1996, ApJ, 470, L97
14. Hachisu, I., Kato, M., Nomoto, K., & Umeda, H. 1999a, ApJ, 519, 314
15. Hachisu, I., Kato, M., & Nomoto, K. 1999b, ApJ, 522, 487
16. Tutukov, A.V., & Yungelson, L. 1977, Astrophysics, 12, 342
17. Umeda, H., Nomoto, K., Kobayashi, C., Hachisu, I, & Kato, M. 1999b, ApJ, 522, L43
18. Kobayashi, C., Tsujimoto, T., & Nomoto, K. 2000, ApJ, submitted (astro-ph/9908005)
19. Nomoto, K., Thielemann, F. -K., & Yokoi, K., 1984, ApJ, 286, 644
20. Nomoto, K., Yamaoka, H., Shigeyama, T., Kumagai, S., & Tsujimoto, T. 1994, in Supernovae, Les Houches Session LIV, ed. S. A. Bludman et al. (North-Holland), 199
21. Arnett, W. D. 1996, Nucleosynthesis and Supernovae (Princeton: Princeton Univ. Press)

22. Höflich, P., & Khokhlov, A., 1996, ApJ, 457, 500

23. Nugent, P., Baron, E., Branch, D., Fisher, A., & Hauschildt, P. H. 1997, ApJ, 485, 812

24. Iben, I. Jr., & Tutukov, A. V. 1984, ApJS, 54, 335

25. Webbink, R. F. 1984, ApJ, 277, 355

26. Nomoto, K. 1982, ApJ, 253, 798

27. Saio, H., & Nomoto, K. 1985, A&A, 150, L21

28. Saio, H., & Nomoto, K. 1998, ApJ, 500, 388

29. Segretain, L., Chabrier, G., & Mochkovitch, R. 1997, ApJ, 481, 355

30. Nomoto, K., & Kondo, Y. 1991, ApJ, 367, 119

31. Nomoto, K., Nariai, K., & Sugimoto, D. 1979, PASJ, 31, 287

32. Iglesias, C. A., & Rogers, F. 1993, ApJ, 412, 752

33. Kato, M., & Hachisu, I. 1999, ApJ, 513, L41

34. van den Heuvel, E.P.J., Bhattacharya,D., Nomoto,K., & Rappaport,S. 1992, A&A, 262, 97

35. Kobayashi, C., Tsujimoto, T., Nomoto, K., Hachisu, I, & Kato, M. 1998, ApJ, 503, L155

36. Hachisu, I., & Kato, M., 2000, in preparation

37. Tsujimoto, T. et al. 1995, MNRAS, 277, 945

38. Nomoto, K. et al. 1997c, Nuclear Physics, A621, 467c

39. Umeda, H., Nomoto, K., & Nakamura, T. 2000, in The First Stars, ed. A. Weiss et al. (Berlin: Springer), in press (astro-ph/9912248)

40. Kochanek, C. S. 1997, ApJ, 491, 13

41. Hamuy, M., Phillips, M. M., Schommer, R. A., & Suntzeff, N. B., 1996, AJ, 112, 2391.

42. Cappellaro, E. et al. 1997, A&A, 322, 431

43. Wang, L., Höflich, P., & Wheeler, J. C. 1997, ApJ, 483, L29

44. Riess, A. G. et al. 1999, AJ, 117, 707

45. Höflich, P., Wheeler, J. C., & Thielemann, F. -K., 1998, ApJ, 495, 617

46. Höflich, P., Nomoto, K., Umeda, H., & Wheeler, J. C., 2000, ApJ, in press

47. Umeda, H., Nomoto, K., Ymamaoka, H., & Wanajo, S. 1999a, ApJ, 513, 861

48. van den Heuvel, E. P. J., 1994, in Interacting Binaries, eds. S.N. Shore et al. (Berlin: Springer-Verlag), 263

49. Mazzali, P.A., Cappellaro,E., Danziger,I.J., Turatto,M., & Benetti,S., 1998, ApJ, 499, L49·

50. Khokhlov, A. 1991, A&A, 245, 114

51. Niemeyer J.C., & Hillebrandt W. 1995, ApJ, 452, 769

52. Zaritsky, D., Kennicutt, R. C., & Huchra, J. P. 1994, ApJ, 420, 87

53. Kobayashi, C., & Arimoto, N. 1999, ApJ, 527, 573

54. Paczyński, B. 1973, Acta Astr. 23, 1

55. Nomoto, K. et al. 1997d, in Thermonuclear Supernovae, Eds. P.Ruiz-Lapuente et al. (Dordrecht: Kluwer), 349

56. Iwamoto, K. et al. 1999, ApJS, 125, 439

57. Ruiz-Lapuente, P., & Canal, R. 1998, ApJ, 497, L57

58. Sadat, R., Blanchard, A., Guiderdoni, B., & Silk, J. 1998, A&A, 331, L69

59. Yungelson, L., & Livio, M. 1998, ApJ, 497, 168

60. Madau, P. et al. 1996, MNRAS, 283, 1388

61. Pettini, M. et al. 1998, ApJ, 508, 539

62. Kodama, T., & Arimoto, N., 1997, A&A, 320, 41

63. Pain, R. et al. 1996, ApJ, 473, 356

64. Franceschini, A. et al. 1998, ApJ, 506, 600

65. Hughes, D. et al. 1998, Nature, 394, 241

Type Ia Supernova Explosion Models: Homogeneity versus Diversity

Wolfgang Hillebrandt*, Jens C. Niemeyer†, and Martin Reinecke*

*Max-Planck-Institut für Astrophysik, Garching, Germany
†Enrico-Fermi-Institute, University of Chicago, Chicago, IL 60637

Abstract. Type Ia supernovae (SN Ia) are generally believed to be the result of the thermonuclear disruption of Chandrasekhar-mass carbon-oxygen white dwarfs, mainly because such thermonuclear explosions can account for the right amount of ^{56}Ni, which is needed to explain the light curves and the late-time spectra, and the abundances of intermediate-mass nuclei which dominate the spectra near maximum light. Because of their enormous brightness and apparent homogeneity SN Ia have become an important tool to measure cosmological parameters.

In this article the present understanding of the physics of thermonuclear explosions is reviewed. In particular, we focus our attention on subsonic ("deflagration") fronts, i.e. we investigate fronts propagating by heat diffusion and convection rather than by compression. Models based upon this mode of nuclear burning have been applied very successfully to the SN Ia problem, and are able to reproduce many of their observed features remarkably well. However, the models also indicate that SN Ia may differ considerably from each other, which is of importance if they are to be used as standard candles.

INTRODUCTION

Type Ia supernovae, i.e. stellar explosions which do not have hydrogen in their spectra, but intermediate-mass elements, such as silicon, calcium, cobalt, and iron, have recently received considerable attention because it appears that they can be used as "standard candles" to measure cosmic distances out to billions of light years away from us. Moreover, observations of type Ia supernovae seem to indicate that we are living in a universe that started to accelerate its expansion when it was about half its present age. These conclusions rest primarily on phenomenological models which, however, lack proper theoretical understanding, mainly because the explosion process, initiated by thermonuclear fusion of carbon and oxygen into heavier elements, are difficult to simulate even on supercomputers, for reasons we shall discuss in this article.

The most popular progenitor model for the average type Ia supernovae is a massive white dwarf, consisting of carbon and oxygen, which approaches the Chan-

CP522, *Cosmic Explosions: Tenth Astrophysical Conference,*
edited by Stephen S. Holt and William W. Zhang
© 2000 American Institute of Physics 1-56396-943-2/00/$17.00

drasekhar mass, $M_{Ch} \simeq 1.39\ M_\odot$, by a yet unknown mechanism, presumably accretion from a companion star, and is disrupted by a thermonuclear explosion (see, e.g., [56] for a review). Arguments in favor of this hypothesis include the ability of the models to fit the observed spectra and light curves rather well.

However, not only is the evolution of massive white dwarfs to explosion very uncertain, leaving room for some diversity in the allowed set of initial conditions (such as the temperature profile at ignition), but also the physics of thermonuclear burning in degenerate matter is complex and not well understood. The generally accepted scenario is that explosive carbon burning is ignited either at the center of the star or off-center in a couple of ignition spots, depending on the details of the previous evolution. After ignition, the flame is thought to propagate through the star as a sub-sonic deflagration wave which may or may not change into a detonation at low densities (around $10^7 g/cm^3$). Numerical models with parameterized velocity of the burning front have been very successful, the prototype being the W7 model of [37]. However, these models do not solve the problem because attempts to determine the effective flame velocity from direct numerical simulations failed and gave velocities far too low for successful explosions [25,18,1]. This has led to some speculations about ways to change the deflagration into a supersonic detonation [14,15]. All these aspects of supernova models will be discussed in the following sections.

NUCLEAR BURNING IN DEGENERATE C+O MATTER

Owing to the strong temperature dependence of the nuclear reaction rates nuclear burning during the explosion is confined to microscopically thin layers that propagate either conductively as subsonic deflagrations ("flames") or by shock compression as supersonic detonations [6,21]. Both modes are hydrodynamically unstable to spatial perturbations as can be shown by linear perturbation analysis. In the nonlinear regime, the burning fronts are either stabilized by forming a cellular structure or become fully turbulent – either way, the total burning rate increases as a result of flame surface growth [24,53,60]. Neither flames nor detonations can be resolved in explosion simulations on stellar scales and therefore have to be represented by numerical models.

When the fuel exceeds a critical temperature T_c where burning proceeds nearly instantaneously compared with the fluid motions a thin reaction zone forms at the interface between burned and unburned material. It propagates into the surrounding fuel by one of two mechanisms allowed by the Rankine-Hugoniot jump conditions: a deflagration ("flame") or a detonation.

If the overpressure created by the heat of the burning products is sufficiently high, a hydrodynamical shock wave forms that ignites the fuel by compressional heating. Such a self-sustaining combustion front that propagates by shock-heating is called a detonation. If, on the other hand, the initial overpressure is too weak, the temperature gradient at the fuel-ashes interface steepens until an equilibrium between heat diffusion and energy generation is reached. The resulting combustion front consists of a diffusion zone that heats up the fuel to T_c, followed by a thin reaction layer where the fuel is consumed and energy is generated. It is called a deflagration or simply a flame and moves subsonically with respect to the unburned material [21]. Flames, unlike detonations, may therefore be strongly affected by turbulent velocity fluctuations of the fuel. Only if the unburned material is at rest, a unique laminar flame speed S_l can be found which depends on the detailed interaction of burning and diffusion within the flame region, e.g. [60]. Following [21], it can be estimated by assuming that in order for burning and diffusion to be in equilibrium, the respective time scales, $\tau_b \sim \epsilon/\dot{w}$ and $\tau_d \sim \delta^2/\kappa$, where δ is the flame thickness and κ is the thermal diffusivity, must be similar: $\tau_b \sim \tau_d$. Defining $S_l = \delta/\tau_b$, one finds $S_l \sim (\kappa\dot{w}/\epsilon)^{1/2}$, where $\dot{w}$ should be evaluated at $T \approx T_c$ [50]. This is only a crude estimate due to the strong T-dependence of $\dot{w}$. Numerical solutions of the full equations of hydrodynamics including nuclear energy generation and heat diffusion are needed to obtain more accurate values for S_l as a function of ρ and fuel composition. Laminar thermonuclear carbon and oxygen flames at high to intermediate densities were investigated by [4,12,54], and, using a variety of different techniques and nuclear networks, by [50]. For the purpose of SN Ia explosion modeling, one needs to know the laminar flame speed $S_l \approx 10^7 \ldots 10^4$ cm s^{-1} for $\rho \approx 10^9 \ldots 10^7$ g cm^{-3}, the flame thickness $\delta = 10^{-4} \ldots 1$ cm (defined here as the width of the thermal pre-heating layer ahead of the much thinner reaction front), and the density contrast between burned and unburned material $\mu = \Delta\rho/\rho = 0.2 \ldots 0.5$ (all values quoted here assume a composition of $X_C = X_O = 0.5$ [50]). The thermal expansion parameter μ reflects the partial lifting of electron degeneracy in the burning products, and is much lower than the typical value found in chemical, ideal gas systems [53].

HYDRODYNAMIC INSTABILITIES AND TURBULENCE

The best studied and probably most important hydrodynamical effect for modeling SN Ia explosions is the Rayleigh-Taylor (RT) instability resulting from the buoyancy of hot, burned fluid with respect to the dense, unburned material [29,30,25,17,18,33], and after more than five decades of experimental and numerical work, the basic phenomenology of nonlinear RT mixing is fairly well understood [8,22,47,43,57]: Subject to the RT instability, small surface perturbations grow until they form bubbles (or "mushrooms") that begin to float upward while spikes of dense fluid fall down. In the nonlinear regime, bubbles of various sizes interact and create a foamy RT mixing layer whose vertical extent h_{RT} grows with time t according to a self-similar growth law, $h_{RT} = \alpha g(\mu/2)t^2$, where α is a dimensionless constant ($\alpha \approx 0.05$) and g is the background gravitational acceleration.

Secondary instabilities related to the velocity shear along the bubble surfaces [34] quickly lead to the production of turbulent velocity fluctuations that cascade from the size of the largest bubbles ($\approx 10^7$ cm) down to the microscopic Kolmogorov scale, $l_k \approx 10^{-4}$ cm where they are dissipated [33,18]. Since no computer is capable of resolving this range of scales, one has to resort to statistical or scaling approximations of those length scales that are not properly resolved. The most prominent scaling relation in turbulence research is Kolmogorov's law for the cascade of velocity fluctuations, stating that in the case of isotropy and statistical stationarity, the mean velocity v of turbulent eddies with size l scales as $v \sim l^{1/3}$ [20]. Given the velocity of large eddies, e.g. from computer simulations, one can use this relation to extrapolate the eddy velocity distribution down to smaller scales under the assumption of isotropic, fully developed turbulence [33]. Knowledge of the eddy velocity as a function of length scale is important to classify the burning regime of the turbulent combustion front [36,35,19]. The ratio of the laminar flame speed and the turbulent velocity on the scale of the flame thickness, $K = S_l/v(\delta)$, plays an important role: if $K \gg 1$, the laminar flame structure is nearly unaffected by turbulent fluctuations. Turbulence does, however, wrinkle and deform the flame on scales l where $S_l \ll v(l)$, i.e. above the *Gibson scale* l_g defined by $S_l = v(l_g)$ [40]. These wrinkles increase the flame surface area and therefore the total energy generation rate of the turbulent front [7]. In other words, the turbulent flame speed, S_t, defined as the mean overall propagation velocity of the turbulent flame front, becomes larger than the laminar speed S_l. If the turbulence is sufficiently strong, $v(L) \gg S_l$, the turbulent flame speed becomes independent of the laminar speed, and therefore of the microphysics of burning and diffusion, and scales only with the velocity of the largest turbulent eddy [7,5]:

$$S_t \sim v(L) \ . \tag{1}$$

Because of the unperturbed laminar flame properties on very small scales, and the wrinkling of the flame on large scales, the burning regime where $K \gg 1$ is called the corrugated flamelet regime [42,5].

As the density of the white dwarf material declines and the laminar flamelets become slower and thicker, it is plausible that at some point turbulence significantly alters the thermal flame structure [19,36]. This marks the end of the flamelet regime and the beginning of the distributed burning, or distributed reaction zone, regime, e.g. [42]. So far, modeling the distributed burning regime in exploding white dwarfs has not been attempted explicitly since neither nuclear burning and diffusion nor turbulent mixing can be properly described by simplified prescriptions (see, however, [27]). Phenomenologically, the laminar flame structure is believed to be disrupted by turbulence and to form a distribution of reaction zones with various lengths and thicknesses. In order to find the critical density for the transition between both regimes, we need to formulate a specific criterion for flamelet breakdown. A criterion for the transition between both regimes is discussed in [36,35] and [19]:

$$l_{\text{cutoff}} \leq \delta \ . \tag{2}$$

Inserting the results of [50] for S_l and δ as functions of density, and using a typical turbulence velocity $v(10^6 \text{cm}) \sim 10^7$ cm s^{-1}, the transition from flamelet to distributed burning can be shown to occur at a density of $\rho_{\text{dis}} \approx 10^7$ g cm^{-3} [35].

MODELING TURBULENT THERMONUCLEAR COMBUSTION

Next we will outline a way by which several of the ideas discussed in the previous sections can be incorporated into a numerical scheme to model thermonuclear combustion in SN Ia.

Numerical simulations of any kind of turbulent combustion have always been a challenge, mainly because of the large range of length scales involved. In type Ia supernovae, in particular, the length scales of relevant physical processes range from 10^{-3}cm for the Kolmogorov-scale to several 10^7cm for typical convective motions. In the currently favored scenario the explosion starts as a deflagration near the center of the star. Rayleigh-Taylor unstable blobs of hot burnt material are thought to rise and to lead to shear-induced turbulence at their interface with the unburnt gas. This turbulence increases the effective surface area of the flamelets and, thereby, the rate of fuel consumption; the hope is that finally a fast deflagration might result, in agreement with phenomenological models of type Ia explosions [37].

Despite considerable progress in the field of modeling turbulent combustion for astrophysical flows (see, e.g., [33]), the correct numerical representation of the thermonuclear deflagration front is still a weakness of the simulations. Methods used up to now are based on for the reactive-diffusive flame model [16], which artificially stretches the burning region over several grid zones to ensure an isotropic flame propagation speed. However, the soft transition from fuel to ashes stabilizes the front against hydrodynamical instabilities on small length scales, which in turn results in an underestimation of the flame surface area and – consequently – of the total energy generation rate. Moreover, because nuclear fusion rates depend on temperature nearly exponentially, one cannot use the zone-averaged values of the temperature obtained this way to calculate the reaction kinetics.

Therefore we have decided to use a front tracking method to cure some of these weaknesses. The method is based on the so-called *level set technique* which was originally introduced by Osher and Sethian [38]. They used the zero level set of a n-dimensional scalar function to represent $(n-1)$-dimensional front geometries. Equations for the time evolution of such a level set which is passively advected by a flow field are given in [49]. The method has been extended to allow the tracking of fronts propagating normal to themselves, e.g. deflagrations and detonations [48,45,44]. In contrast to the artificial broadening of the flame in the reaction-diffusion-approach, this algorithm is able to treat the front as an exact hydrodynamical discontinuity. We will demonstrate that such a method can be applied to the supernova problem and, in addition, we will show that even if one attempts to model the physics of thermonuclear burning on unresolved length scales well

57

by physically motivated "Large Eddy Simulations" (LES), one still has to perform calculations with very high spatial resolution in order to be at least near to a converged solution. The numerical results presented here were obtained by performing 2D simulations, but there is no problem extending them to 3D. Such simulations are under way.

APPLICATION TO THE SUPERNOVA PROBLEM

We have carried out numerical simulations in cylindrical rather than in spherical coordinates, mainly because it is much simpler to implement the level set on a Cartesian (r,z) grid. Moreover, the CFL condition is somewhat relaxed in comparison to spherical coordinates. The grid we used in most of our simulations maps the white dwarf onto 256×256 mesh points, equally spaced for the innermost 226×226 zones by $\Delta = 1.5 \cdot 10^6$cm, but increasing by 10% from zone to zone in the outer parts. The white dwarf, constructed in hydrostatic equilibrium for a realistic equation of state, has a central density of $2.9 \cdot 10^9$g/cm^3, a radius of $1.5 \cdot 10^8$cm, and a mass of $2.8 \cdot 10^{33}$g, identical to the one used by [33]. The initial mass fractions of C and O are chosen to be equal, and the total binding energy turns out to be $5.4 \cdot 10^{50}$erg. At low densities ($\rho \leq 10^7$g/cm^3), the burning velocity of the front is set equal to zero because the flame enters the distributed regime and our physical model is no longer valid. However, since in reality some matter may still burn the energy release obtained in the simulations is probably somewhat too low.

Here we present only the results for two models, one in which nuclear burning was ignited off-center in a blob and a second one in which initially five blobs were burned as an initial condition, and refer to [44] for simulations with other initial conditions.

First, in Fig.1 a snapshot of the "five-blob" model B5 is shown at t = 0.4s. The left panel gives the position of the burning front, the temperatures (in grayshading), and the expansion velocities. The right panel shows the distribution of turbulent velocity fluctuations. We find that, in accord with one's intuition, most of the turbulence is generated in a very thin layer near the front. Since in the limit of high turbulence intensity the nuclear flames propagate with the turbulent velocity it is obvious that this propagation velocity exceeds the laminar flame speed. However, for most of the initial conditions we have investigated this increase was not sufficient to unbind the star. In fact, model B5 of [44] was the only one of the set that did explode.

Moreover, we show the results obtained with 3 times higher resolution for comparison. Fig.2 gives a snapshot taken at 1.05s after ignition for the low-resolution run (left figure) and the high-resolution run, respectively. Although the increase in spatial resolution is only a factor of 3, the right panel shows clearly more structure. This is an important effect because in the flamelet regime the rate of fuel consumption, in first order, increases proportional to surface area of the burning front. The net effect is that the low-resolution model stays bound at the end of

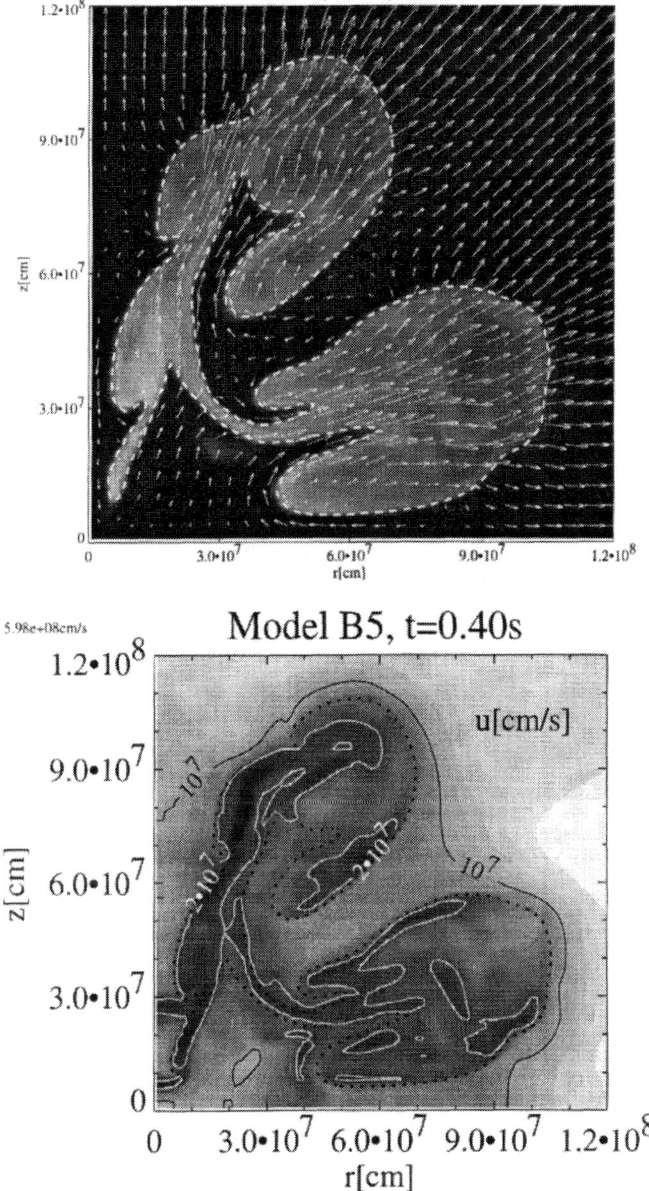

FIGURE 1. Snapshots of the temperature and the front geometry at 0.4s for model B5 of Reinecke et al. (1999a) (bottom figure) and turbulent velocity fluctuations on the grid scale (top panel). The position of of the front is indicated by the dotted curve.

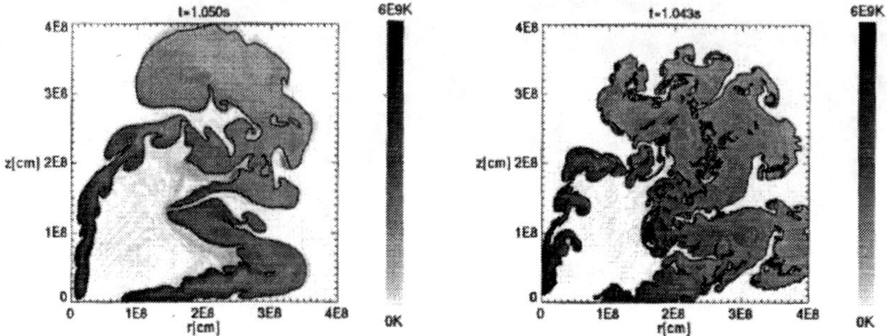

FIGURE 2. Snapshots of the temperature and the front geometry at 1.05s, taken for the low-resolution model (left figure) and high-resolution run, respectively.

the computations, whereas the better resolved model explodes with an explosion energy of about 2×10^{50}erg. Fig.2 also demonstrates that the level-set prescription allows to resolve the structure of the burning front down almost to the grid scale, thus avoiding artificial smearing of the front which is an inherent problem of front-capturing schemes. We want to stress that this gain of accuracy is not obtained at the expense of smaller CFL time steps because in our hybrid scheme the hydrodynamics is still done with cell-averaged quantities.

We also want to stress again, however, that the numerical simulations described here, although leaving behind unbound white dwarfs in several cases, would not fit the observed spectra of SN Ia, mainly because the expansion velocity of the partially burnt gas is too low. Whether this problem is cured once the model calculations are extended to three spatial dimensions has to be seen, but one can expect significant changes because of the different ratios of volume to surface area of burning blobs in 3D.

DELAYED DETONATIONS?

Turbulent deflagrations can sometimes be observed to undergo spontaneous transitions to detonations (deflagration-detonation transitions, DDTs) in terrestrial combustion experiments [53]. Thus inspired, it was suggested that DDTs may occur in the late phase of the explosion, providing an elegant explanation for the initial slow burning required to pre-expand the star, followed by a fast combustion mode that produces large amounts of high-velocity intermediate mass elements [14,55]. Many 1D simulations have meanwhile demonstrated the capability of the delayed detonation scenario to provide excellent fits to SN Ia spectra and light curves [56,11], as well as reasonable nucleosynthesis products with regard to solar abundances [15,13]. In the best fit models, the initial flame phase has a rather slow velocity of roughly one percent of the sound speed and transitions to detonation at a density of $\rho_{\mathrm{DDT}} \approx 10^7$ g cm^{-3} [11,13]. The transition density was also found to be a convenient parameter to explain the observed sequence of explosion strengths [11].

Various mechanisms for DDT were discussed in the early literature on delayed detonations (see [36] and references therein). Recent investigations have focussed on the induction time gradient mechanism [59,23], analyzed in the context of SNe Ia by [2] and [3]. It was realized by [19] and [36] that a necessary criterion for this mechanism is the local disruption of the flame sheet by turbulent eddies, or, in other words, the transition of the burning regime from "flamelet" to "distributed" burning. Simple estimates [35] show that this transition should occur at roughly 10^7 g cm^{-3}, providing a plausible explanation for the delay of the detonation.

The critical length (or mass) scale over which the temperature gradient must be held fixed in order to allow the spontaneous combustion wave to turn into a detonation was computed by [19] and [36]; it is a few orders of magnitude thicker than the final detonation front and depends very sensitively on composition and density. Recently, this conclusion was confirmed by [27], who computed the interaction between turbulence and the nuclear flame in the distributed regime by means of a particular turbulence model, and by [26].

The virtues of the delayed detonation scenario can be summarized as follows. It is undisputed that suitably tuned delayed detonations satisfy all the constraints given by SN Ia spectra, light curves, and nucleosynthesis. If ρ_{DDT} is indeed determined by the transition of burning regimes – which in turn might be composition dependent [52] – the scenario is also fairly robust and ρ_{DDT} may represent the explosion strength parameter. Note that in this case, the variability induced by multi-point ignition needs to be explained away. If, on the other hand, thermonuclear flames are confirmed to be almost unquenchable, the favorite mechanism for DDTs becomes questionable [31]. Moreover, should the mechanism DDT rely on rare, strong turbulent fluctuations one must ask about those events that fail to ignite a detonation following the slow deflagration phase which, on its own, cannot give rise to a viable SN Ia explosion. They might end up as pulsational delayed detonations or as unobservably dim, as yet unclassified explosions. Multidimensional simulations of the turbulent flame phase may soon answer whether the turbulent flame speed is closer to 1 % or 30 % of the speed of sound and hence decide whether DDTs are a necessary ingredient of SN Ia explosion models.

SUMMARY AND CONCLUSIONS

In this review we have outlined our present understanding of Type Ia supernovae explosions. From the tremendous amount of work carried out over the last couple of years it has become obvious that the physics of SNe Ia is very complex, ranging from the possibility of very different progenitors to the complexity of the physics leading to the explosion and the complicated processes which couple the interior physics to observable quantities. None of these problems is fully understood yet, but what one is tempted to state is that, from a theorist's point of view, it appears to be a miracle that all the complexity seems to average out in a mysterious way to make the class so homogeneous. In contrast, as it stands, a save prediction from theory seems to be that SNe Ia should get more divers with increasing observed sample sizes. If, however, homogeneity would continue to hold this would certainly add support to the Chandrasekhar-mass scenario discussed in this article. On the other hand, even an increasing diversity would not rule out Chandrasekhar-mass progenitors for most of them. In contrast, there are ways to explain how the diversity is absorbed in a one parameter family of transformations, such as the Phillips-relation [41,10] or modifications of it [46,39].

As far as the explosion/combustion physics and the numerical simulations are concerned significant recent progress has made the models more realistic (and reliable). Thanks to ever increasing computer resources 3-dimensional simulations have become feasible which treat the full star with good spatial resolution and realistic input physics. Already the results of 2-dimensional simulations indicate that pure deflagrations waves in Chandrasekhar-mass C+O white dwarfs can lead to explosions, and one can expect that going to three dimensions, because of the increasing surface area of the nuclear flames, should add to the explosion energy. If confirmed, this would eliminate pulsational detonations from the list of potential models. On the side of the combustion physics, the burning in the distributed regime at low densities needs to be explored further, but it is not clear anymore whether a transition from a deflagration to a detonation in that regime is needed for successful models. In fact, according to recent studies such a transition appears to be rather unlikely.

Acknowledgments: This work was supported in part by the Deutsche Forschungsgemeinschaft under Grant Hi 534/3-1, the DAAD, and by DOE under contract No. B341495 at the University of Chicago. The computations were performed at the Rechenzentrum Garching on a Cray J90.

REFERENCES

1. Arnett, W. D., and Livne, E., *ApJ*, 427:31, 1994.
2. Blinnikov, S. I. and Khokhlov, A. M., *Soviet Astronomy Letters*, 12:131, 1986.
3. Blinnikov, S. I. and Khokhlov, A. M., *Soviet Astronomy*, 13:364, 1987
4. Buchler, J. R.,, Colgate, S. A., and Mazurek, T. J., *Journal de Physique*, 41:2, 1980.
5. Clavin, P., *Annual Rev. Fluid Mech.*, 26:321, 1994.
6. Courant, R., and Friedrichs, K. O., *Supersonic Flow and Shock Waves*. Springer, New York, 1948.
7. Damköhler, G., *Z. Elektrochem.*, 46:601, 1940.
8. Fermi, E., In E. Segre, editor, *Collected Works of Enrico Fermi*, pages 816–821. 1951.
9. Fryxell, B. A., Müller, E., and Arnett, W. D., *MPA Preprint*, 449, 1989.
10. Hamuy M., Phillips M. M., Suntzeff N. B., Schommer R. A., et al., *AJ* 109:1, 1996.
11. Hoeflich, P., and Khokhlov, A., *ApJ*, 457:500, 1996.

12. Ivanova, L. N., Imshennik, V. S., and Chechetkin, V. M., *Pis ma Astronomicheskii Zhurnal*, 8:17, 1982.

13. Iwamoto, K., Brachwitz, F., Nomoto, K., Kishimoto, N., Umeda, H., Hix, W. R., and Thielemann, F.-K., *ApJ Suppl.*, in press, 1999.

14. Khokhlov, A. M., *A&A*, 245:114, 1991.

15. Khokhlov, A. M., *A&A*, 245:L25, May 1991.

16. Khokhlov, A. M., *ApJL*, 419:L77, 1993.

17. Khokhlov, A. M., *ApJL*, 424:L115, 1994.

18. Khokhlov, A. M., *ApJ*, 449:695, 1995.

19. Khokhlov, A. M., Oran, E. S., and Wheeler, J. C., *ApJ*, 478:678, 1997.

20. Kolmogorov, A. N., *Dokl. Akad. Nauk SSSR*, 30:299, 1941.

21. Landau, L. D., and Lifshitz, E. M., *Fluid Mechanics*. Butterworth-Heinemann, xxx, 1995.

22. Layzer, D., *ApJ*, 122:1, 1955.

23. Lee, J. H. S., Knystautas, R., and Yoshikawa, N. *Astronaut. Acta*, 5:971, 1978.

24. Lewis, B., and von Elbe, G., *Combustion, Flames, and Explosions of Gases*. Academic Press, New York, 2nd edition, 1961.

25. Livne, E., *ApJL*, 406:L17, 1993.

26. Lisewski, A. M., Hillebrandt, W., and Woosley, S. E., *ApJ*, in print, 2000.

27. Lisewski, A. M., Hillebrandt, W., Woosley, S. E., Niemeyer, J. C., and Kerstin, A. R., *ApJ*, in print, 2000.

28. Mulder, W., Osher, S., and Sethian, J. A., *JCP*, 100:209, 1992.

29. Müller, E., and Arnett, W. D., *ApJL*, 261:L109, 1982.

30. Müller, E., and Arnett, W. D., *ApJ*, 307:619, 1986.

31. *ApJL*, 523:L57, 1999.

32. Niemeyer, J. C., and Hillebrandt, W., *ApJ*, 452:779, 1995.

33. Niemeyer, J. C., and Hillebrandt, W., *ApJ*, 452:769, 1995.

34. Niemeyer, J. C., and Hillebrandt, W., Microscopic and macroscopic modeling of thermonuclear burning fronts. In P. Ruiz-Lapuente, R. Canal, and J. Isern, editors, *Thermonuclear Supernovae*, pages 441–456, Dordrecht, 1997. Kluwer.

35. Niemeyer, J. C., and Kerstein, A. R., Burning regimes of nuclear flames in sn ia explosions. *New Astronomy*, 2:239, 1997.

36. Niemeyer, J. C., and Woosley, S. E., *ApJ*, 475:740, 1997.

37. Nomoto, K., Thielemann, F. K., and Yokoi, K., *ApJ*, 286:644, 1984.

38. Osher, S., and Sethian, J. A., *JCP*, 79:12, 1988.

39. Perlmutter S., Gabi S., Goldhaber G., Goobar A, Groom D. E., et al., *ApJL* 483:565, 1997.

40. Peters, N., In *Symp. (Int.) Combust., 21st*, pages 1232, Pittsburgh, 1988. Combustion Institute.

41. Phillips M. M., *ApJL*, 413:L105, 1993.

42. Pope, S. B., *Annual Rev. Fluid Mech.*, 19:237, 1990.

43. Read, K. I., *Physica D*, 12:45, 1984.

44. Reinecke, M., Hillebrandt, W., and Niemeyer, J. C., *A&A*, 347:739, 1999.

45. Reinecke, M., Hillebrandt, W., Niemeyer, J. C., Klein, R., and Gröbl, A., *A&A*, 347:724, 1999.

46. Riess A. G., Press W. H., Kirshner R. P., *ApJ* 473:88, 1996.

47. Sharp, D. H., *Physica D*, 12:3, 1984.

48. Smiljanovski, V., Moser, V., and Klein, R., *Combust. Theory Modelling*, 1:183, 1997.

49. Sussman, M., Smereka, P., and Osher, S., *JCP*, 114:146, 1994.

50. Timmes, F. X., and Woosley, S. E., *ApJ*, 396:649, 1992.

51. Timmes, F. X., *ApJ Suppl.*, 124:241, 1999.

52. Umeda, H., Nomoto, K., Kobayashi, C., Hachisu, I., and Kato, M., *ApJL*, 522:L43, 1999.

53. Williams, F. A., *Combustion Theory.* Benjamin/Cummings, Menlo Park, 2nd edition, 1985.

54. Woosley, S. E., and Weaver, T. A., The physics of supernovae. In Dimitri Mihalas and Karl-Heinz A. Winkler, editors, *Radiation Hydrodynamics in Stars and Compact Objects, Lecture Notes in Physics*, volume 255, pages 91, Berlin, 1986. Springer.

55. Woosley, S. E., and Weaver, T. A., In J.Audouze, S. Bludman, R. Mochovitch, and J. Zinn-Justin, editors, *Supernovae, Les Houches*, Session LIV, pages 63, Paris, 1994. Elsevier.

56. Woosley, S. E., Type Ia supernovae: Carbon deflagration and detonation. In A. G. Petschek, editor, *Supernovae*, pages 182, Berlin, 1990. Springer.

57. Youngs, D. L., *Physica D*, 12:32, 1984.

58. Zeldovich, Y. B., *J. Appl. Mech. and Tech. Phys.*, 7:68, 1966.

59. Zeldovich, Y. B., Librovich, V. B., Makhviladze, G. M., and Sivashinsky, G. L., *Astronaut. Acta*, 15:313, 1970.

60. Zeldovich, Y. B., Barenblatt, G. I., Librovich, V. B., and Makhviladze, G. M., *The Mathematical Theory of Combustion and Explosions.* Plenum, New York, 1985.

The Observations of Type Ia Supernovae

Nicholas B. Suntzeff*

*Cerro Tololo Inter-American Observatory, NOAO[1]
Casilla 603, La Serena, Chile

Abstract.

The past ten years have seen a tremendous increase in the number of Type Ia supernovae discovered and in the quality of the basic data presented. The cosmological results based on distances to Type Ia events have been spectacular, leading to statistically accurate values of the Hubble constant and Ω_M and Ω_Λ.

However, in spite of the recent advances, a number of mysteries continue to remain in our understanding of these events. In this short review, I will concentrate on unresolved problems and curious correlations in the data on Type Ia SNe, whose resolution may lead to a deeper understanding of the physical mechanism of the Type Ia supernova explosions.

INTRODUCTION

The study of supernovae is certainly in vogue today. Type II plateau and Type Ia supernovae are used to measure distances to galaxies into the quiet Hubble flow out past 10,000 km s^{-1} to determine the Hubble constant (see for instance [1–3] for Type Ia distance scales, and [4] for Type II Baade-Wesselink distances). Type Ib/c in a number of cases, and possibly IIn supernovae are associated with gamma-ray bursts and it is now argued that a small percentage or even a majority of GRB's are associated with core-collapse supernovae with small or non-existent hydrogen envelopes.

Distances to Type Ia supernovae are claimed to be as accurate as 7%, based on the dispersion in the quiet Hubble flow ([5,6]). With such accurate distances, a deviation from the linear Hubble flow can be measured at higher redshifts which is related to the (de)acceleration of the local Universe, For a precision of Δm in distance modulus at redshift z, the deceleration can be measured to an accuracy of $\Delta q_0 \approx 0.9 \Delta m / z$. Statistical accuracies of $0^m_.1$ in an ensemble of supernovae at $z \approx 0.5$ can lead to a error in q_0 of less than 0.2 units, which is adequate to determine the sign of the acceleration.

---

[1] NOAO is operated under cooperative agreement with the National Science Foundation.

CP522, *Cosmic Explosions: Tenth Astrophysical Conference,*
edited by Stephen S. Holt and William W. Zhang

Spectacular, and probably even believable, claims have been made about the detection of an acceleration away from the linear local Hubble flow, which is consistent with a flat universe ($\Omega_M + \Omega_\Lambda = 1$) and a positive cosmological constant Ω_Λ ([7–9]). It is my own peculiar viewpoint that these results have not experienced the level of withering attacks, such as those directed towards [10], only because the results agree with the expectations of theory which almost unanimously require a flat universe. Some important criticisms have been made to the interpretation of the results ([11–14]) but the small number of papers in response to the results from the Supernova Cosmology Project (Saul Perlmutter, PI) and the High-Z Supernova Search Team (Brian Schmidt, PI) lead me to believe that these papers have not been carefully read by those who use the results in support of their own Universe. As a co-founder of the High-Z Supernova Search Team, I do believe our results (see the excellent summary by Bob Kirshner in this volume), but there are still some puzzling effects in the data, as well as the rather loose arguments on dust and evolution that we need to continue to perfect. Craig Wheeler, in his summary notes to this conference, has eloquently expressed some of the weak points of the use of Type Ia supernovae at high redshifts.

Few people question the use of Type Ia supernovae as standard candles, corrected to a standard luminosity by some modified form of the Phillips relationship ([15]) corrected for reddening ([6,16]). The empirical evidence from the very small scatter in the Hubble diagram convinces us that we can use Type Ia supernovae as standard candles. But we must remember that for these standards candles, we are unsure as to what the progenitors are, we are still not clear on where the explosion happens in the white dwarf nor how the flame propagates (but see article by Hillebrandt in this conference for the progress in understanding the physics of the flame), we do not understand the effects of age or abundance on the explosion, and we do not know the underlying physical reason for the Phillips relationship.

In many ways, the interpretation of what we observe in Type Ia supernovae is more complicated than in Type II events. In Type II supernovae, there is a nice, optically thick hydrogen envelope which hides lots of physics around the time of maximum light, and the resulting observable "uvoir" or ultraviolet+optical+infrared radiation is the scattered and thermalized gamma rays during the transfer through the photosphere. For Type Ia supernovae, the uvoir optical depths in mass are much larger, and in some wavelength regions, the ejecta are transparent at maximum light. The mass encountered by an escaping gamma ray is smaller in the Type Ia supernovae compared to Type II , and the gamma rays leak out in ever larger quantities near maximum light ([17] and this volume.) This also causes the local ionization and excitation equilibria to be affected by more distant parts of the ejecta leading to much more complicated NLTE effects than in Type II supernovae near maximum light.

One important simplifying condition in a Type Ia supernova is that the ejecta must expand from the size of a white dwarf to the size of a star before it is bright enough for us to observe it (via $L \propto T^4 R^2$ with T about 10000K). This expansion takes about 20 days to optical maximum (see the article by Aldering in this volume), and with such a long expansion time, many of the detailed effects on the explosion from different progenitor models get washed away over the large expansion. This is partly why the lack of an agreed-upon progenitor model is less embarrassing than it otherwise would be.

Rather than review the properties of Type Ia supernovae for this conference, I would rather discuss a few areas of topical interest in the observations of Type Ia's. These are areas in which I think there will be progress in the next five years, and the resolution of some of these topics will help us in understanding the physics of the explosions and ultimately the utility of Type Ia supernovae as standard candles.

Those seeking a review of supernovae, and of Type Ia supernovae in particular, should consider the following resources. An recent excellent and exhaustive book discussing supernovae from a theoretical perspective has been written by Arnett ([18]). Another excellent theoretical perspective is given by Wheeler et al.([19]). The physics of the explosion is a rapidly evolving subject and the most up-to-date review is [20]. Recent books covering the general field of supernovae from the perspective of invited authors are: [21–23]. The spectral properties of supernovae are given by Filippenko ([24]). An excellent general review of the optical properties of Type Ia supernovae has been written by Leibundgut ([25]). Unfortunately, there is no recent monograph on the photometry of Type Ia supernovae to my knowledge.

I TOPICS OF INTEREST IN TYPE IA SUPERNOVAE

A Age, Metallicity, and Reddening

The basic observational result of the Type Ia supernova searches at $z \approx 0.5$ is that the corrected peak brightness is fainter by about $0.^{m}25$ magnitude than predicted by a cosmological model with $\Omega_M = 0.2$ and $\Omega_\Lambda = 0$. Adding matter to the universe only makes the predicted magnitude brighter. What can cause this? The three suggestions are: evolution, cosmology, or reddening.

In terms of evolution, the look-back time to this redshift is ~ 5Gyrs. Specific predictions about the effects of evolution have been discussed by [11]. They consider the age and metallicity difference and how it may affect the peak brightness of Type Ia supernovae. However, the look-back time of 5Gyrs is a rather modest look-back in terms of the evolution of galaxies. The evolution in metallicity in a typical galaxy is very small over this time - after all, this is less than the age of the sun. For instance, the metallicity evolution for our Galaxy over this look-back time was only $\delta[\mathrm{Fe/H}] \sim -0.2$. This is even *smaller* than the natural real scatter in the metallicity of $\sigma[\mathrm{Fe/H}] \sim 0.25$ among stars at a given age in the Galactic disk ([26]). Metallicity evolution in the $z \sim 0.5$ sample is probably not an issue.

The issue of age is less clear. As discussed by [11], the C/O ratio of the soon-to-explode white dwarf is a function of age, and since the fuel for explosion comes the carbon and oxygen, the peak luminosity is a function of age. They find that a decreasing C/O ratio implies a brighter peak luminosity for a Type Ia. In an

attempt to parameterize this, [27] consider the evolution of a white dwarf population
a function of stellar population age. Using the sub-Chandrasekhar mass explosion
models of [28], they are able to recover the approximate range in luminosity seen in
the Calán/Tololo supernova sample ([5]). The supernova rates are higher in star-
bursting galaxies which are producing white dwarfs at higher masses which require
less accretion. In fact, their simple modeling shows that for modest increases in the
space density of star-burst galaxies at $z \sim 0.5$ which is well below any observational
limits on the enhancement of star formation at this redshift, the star-burst galaxies
can produce one half of the supernovae in that volume.

Certainly the anecdotal evidence supports this. As pointed out by [27], the
modest starburst galaxy M100 has had four supernovae (one was a Type II) whereas
the huge elliptical M87 has had none. More striking is the case of the amorphous
(and active star forming) low-mass galaxy NGC 5253 has had two SNe in the last
century, but the whole Coma cluster which has 10,000 gas-poor L_* galaxies, has
had only a few supernova the last 25 years.[2] More rigorous tests ([29]) show that
spiral galaxies have higher Type Ia rates per unit luminosity than elliptical galaxies
by roughly a factor of two. It has even been claimed that Type Ia supernovae are
associated with spiral *arms* ([30].

Given the fact that Type Ia supernovae are more likely to occur in spiral galaxies,
yet they also appear in normal elliptical galaxies, implies that both young and old
populations can produce Type Ia supernovae. Average ages for elliptical galaxies
are probably at least as old as 5Gyrs which means that the *local* sample of host
galaxies of supernovae probably cover the same range in average population age as
the look-back time to $z \sim 0.5$. *Thus the local supernovae used in the calibrations of
the Phillips effect probably cover the parameter range in age and metallicity implied
by a look-back time of 5 Gyrs.* In detail, it is important to check this by comparing
the slope and zero-point of the relationship for the nearby and $z \sim 0.5$ samples,
and by checking whether the ellipticals and the spirals give the same relationship
locally.

The final question is that of reddening. The cosmological result depends on a
measurement of $0^m\!\!.25$, which corresponds to about $E(B - V) = 0.08$. Apparently
the mean colors of the nearby and distant samples are inconsistent with the distant
sample being reddened by 0.08. A novel form of dust has been invented by [13,14]
which, by the way of formation, is more grey than typical dust in the disk of
a galaxy. Clearly multicolor observations of supernovae at $z \sim 0.5$ (which will
require near-infrared photometry) are needed to sort this out. But a more mundane
question arises - do we understand the intrinsic colors of nearby supernovae?

In a recent paper ([16]), we have reexamined the reddening of Type Ia supernovae
based on supernovae we suspect of having little or no reddening due to material
in the host galaxy. Choosing "unreddened" supernovae to define a baseline sample

---

[2] It is not clear, however, how carefully we are looking for supernovae in Coma, and certainly
this cluster should be targeted by northern hemisphere observers involved in CCD searches. The
comparison of supernovae in Coma and a nearby cluster such as Fornax would be an important
test for distance-scale calibrations.

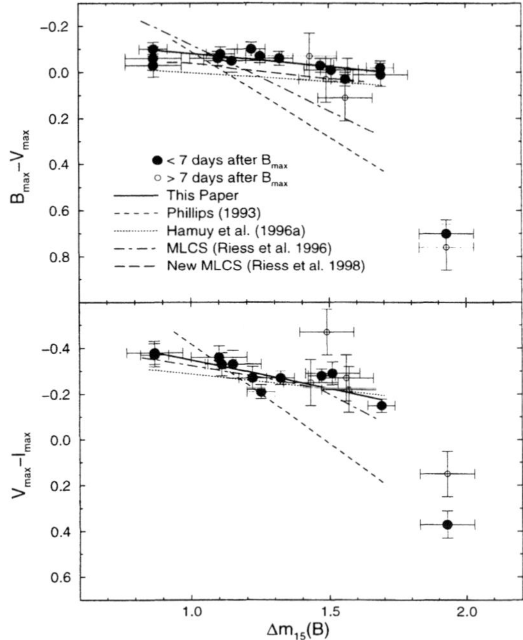

FIGURE 1. The $(B_{max} - V_{max})$ and $(V_{max} - I_{max})$ "color" is plotted versus the decline rate parameter $\Delta m_{15}(B)$ for 20 events with $E(B-V) < 0.06$. These SNe are likely to have little or no host galaxy reddening. Note that these colors are not the BVI colors at B maximum light, but are calculated from the maximum light magnitudes for each broad-band magnitude. For instance, the V_{max} occurs about 1.5 days after B_{max}. Figure from [16].

is not easy. At low dispersion, the absence of interstellar lines at the 0.1A level still does not rule out significant reddening. Only with high S/N high-dispersion spectra, which are rare for SNe, can one really directly verify if there is reddening. One can also chose SNe in elliptical or S0 galaxies away from obvious dust lanes as probably unreddened. One must worry, however, that the properties of the "unreddened" sample are not biased by chosing the sample based on galaxy type. In Figure 1 I show our best guess for the intrinsic color range for Type Ia supernovae as a function of the luminosity parameter Δm_{15} used by our Cálan/Tololo group as published in [16].

Figure 1 shows how radically our idea of the unreddened colors of Type Ia supernovae have changed in the last 7 years. The early results suggested a very strong color dependence as a function of $\Delta m_{15}(B)$. Curiously, this color dependence mimicked the reddening correction such that the reddening and the assumed correction to standard luminosity were degenerate, a fact pointed out by Sidney van den Bergh. However, it now appears that Type Ia supernovae, except for the very underluminous events like SN1991bg, are very similar in $B_{max} - V_{max}$ and $V_{max} - I_{max}$ color at maximum light despite the range of over 1 magnitude in intrinsic brightness. Theory must explain why the range in Type Ia supernovae are so "grey" at maximum light.

In the early years of the use of Type Ia supernovae to measure the Hubble constant, reddening was not even considered. It was assumed that the distribution of reddening in the local and Hubble flow sample was the same. This can't be true in general, since the calibration sample comes from host spiral galaxies with Cepheids (where there can be reddening) while the field sample contains both ellipticals and spirals. It is obviously important to correct for reddening and the more modern techniques, such as used by [6,7,16], solve for reddening as part of the least-squares technique. But if the true unreddened colors of Type Ia supernovae remain somewhat uncertain as I believe, the Baysean assumption of no negative reddening is not valid.

Much observational work on the intrinsic colors of Type Ia supernovae needs to be done. The inclusion of infrared photometry will aid greatly in the definition of intrinsic supernova colors and reddening. At this time, there are only a very few papers on infrared photometry of Type Ia supernovae ([31–35]).

B   Are We Doing Photometry Yet?

Most people who are interested in using supernovae for cosmology will go to the literature and read photometric numbers out of a table. If they actually look at the paper, they will usually see some dismal discussion about local standards, "natural" versus "standard" system, color terms, shutter corrections, non-linearity in the detector, and the like. One is usually reassured that the local standards repeat to some level like $0\overset{m}{.}02$ and therefore the photometry is great.

Having done supernova photometry for over 15 years, I can assure you that my local standards agree to $0\overset{m}{.}02$ and that the photometry is great. But there are some subtle problems that are not usually revealed. In Figure 2 I show the data for the Type Ia supernova SN 1998bu as published in [1] based on CTIO 0.9m CCD data. Included in this diagram are photometric points from the CTIO YALO 1m telescope. The YALO data were reduced using color terms appropriate for that telescope/detector and the same local standards as used for the reductions of the data from the 0.9m telescope. Said another way, *the two data sets come from the facility instruments on two telescopes side-by-side on CTIO and reduced using the same local standards by me.* The YALO data did not start until 35 days after maximum light.

Plotted at the typical scale of a supernova light curve, the data seem to be similar, but actually they are quite different. The differences between the data sets are: $\delta(U, B, V, R, I) = (-0.41, -0.7, 0.06, 0.00, 0.05)$ in the sense of (0.9m - YALO). The

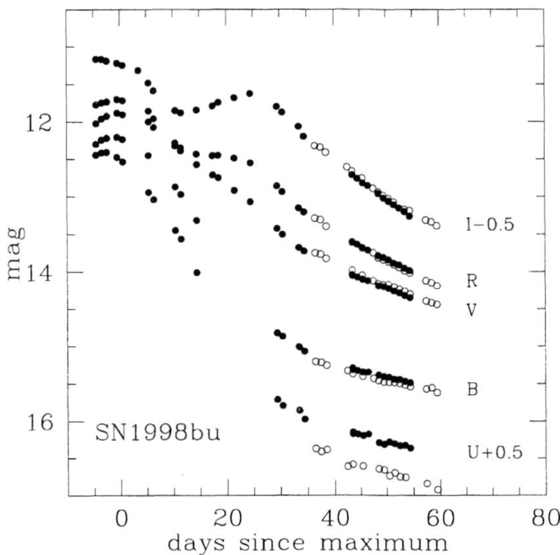

FIGURE 2. The $UBVRI$ photometry for SN 1998bu. The solid circles refer to data taken with the CTIO 0.9m telescope as published by Suntzeff, et al. (1998). The open circles are data taken with the CTIO YALO 1m telescope. All the data have been reduced using the same local standards.

color difference is $\delta(B - V) = -0.12$. The large difference in the U photometry is not surprising. The 0.9m data were taken with a Tek (now SITe) CCD which has no response shortward of 3500A whereas the YALO telescope CCD is a Loral device which has excellent ultraviolet response. In general, most CCD U photometry for supernovae is suspect at the $0{.}^{m}1$ level, unless it was taken with a CCD with a flat quantum efficiency in UB, such as coated GEC, EEV, or TI chips.

The large difference in the $B - V$ colors can be easily reproduced by performing synthetic photometry using the actual filter/detector transmission curves convolved with spectrophotometry of standard stars and the supernova. These errors come from the fact that the supernova flux distribution is very different than the defining stellar fluxes of the standard stars. Near maximum light where a supernova spectrum is nearly stellar (except in $U - B$), the differences between data sets are very small, with $\delta(B - V) \leq 0.04$. Similar differences in photometry have been reported for SN 1987A ([36,37])

However, the point to be made here is neither the 0.9m nor the YALO transmissivities are the correct transmissivities as defined by Bessell ([38]) for the Kron-Cousins $UBVRI$ photometric system. The transformed supernova photometry will have systematic errors even though we have performed the usual color and extinction corrections. In the future, photometrists will need to correct their photometry not only to the standard Kron-Cousins $UBVRI$ as done now, but also correct the photometry to a standard filter transmissivity curve using spectrophotometry, in a manner identical to K-corrections. The Cálan/Tololo group has made a large atlas of spectrophotometry of supernova spectra and Landolt standard stars which will be made available to the community to perform these corrections.

C   Other Topics

1   What is Causing the Secondary Maximum?

The secondary maximum at about 30 days past blue maximum is a prominent feature in Type Ia light curves. It is seen in $IJHK$ and the precise date of secondary maximum is not a constant, but is roughly a function of $\Delta m_{15}(B)$. It gets much smaller for the low luminosity supernovae and is missing in SN 1991bg. A small inflection can be seen in the R light curve, and even in the V curve. The secondary peak has also been seen in the I band light curve of a supernova at $z \approx 0.5$ ([41]). The "uvoir" bolometric light curves also include this inflection point ([39,40]) which seems to indicate that flux has been redistributed from the blue to the red. Due to its presence across many photometric bandpasses, it is obvious this feature is not a spectral feature but is likely due to the time evolution of the line opacities ([42,43]). Since the peak is so prominent and a function of luminosity, it is important that the precise physical reason for its variability be established.

2   Where is the Hydrogen?

One thing we can all agree upon at this conference is that Type Ia supernovae are characterized by a lack of hydrogen in the observed spectrum. This is curious because most of the models require a companion star to dump material onto the white dwarf to bring it to a critical mass for explosion. It is difficult to imagine such mass transfer without large amounts of hydrogen being present, unless the explosion is due to the merger of a double degenerate white dwarf pair. The DD scenario is less attractive at this time due to the observed lack of DDs that are close enough to merge in a Hubble time. As discussed in [44–46], the interaction of the supernova ejecta and the secondary is expected to produce 0.1-$0.5 M_\odot$ at velocities below 1000 km s^{-1}, depending on the nature (dwarf/giant) of the progenitor, with a small high-velocity tail. This hydrogen (and smaller amounts of helium) will be present a few months after maximum mixed in with the low velocity iron layer and may be detectable with high S/N high-dispersion spectra. A number of attempts have been made to find this hydrogen and helium, but there has been no convincing evidence found yet, partly because of the confusion in the spectra, and partly due to the lack of high quality data on Type Ia supernovae. This is a project that needs to be done right now with an echelle spectrograph on an 8-10m class telescope.

It is important not only to observe Type Ia's a few months after maximum to search for hydrogen, but also at the time of maximum. It is possible that transient narrow H_α may be present from a relic wind from the secondary ([45]), either in emission or absorption.

3 Velocity Structure at Maximum Light

Nugent et al ([47]) have shown that the spectral diversity of Type Ia supernovae at maximum light can be explained by a small range in T_{eff} from 7500-11000K. The close relationship between the spectral features and the luminosity (as measured by $\Delta m_{15}(B)$ or its equivalent) has been used to create spectral methods for correcting peak luminosity into a standard value. However, while the spectral features may be closely related to $\Delta m_{15}(B)$ the velocity at maximum light is not clearly related to $\Delta m_{15}(B)$. Wells et al ([48]) have shown that the velocity at maximum light for the SiII line at 6355Å is basically uncorrelated with the intrinsic luminosity. The CaII lines show a weak correlation with $\Delta m_{15}(B)$. It has been claimed that the velocity is correlated with galaxy type, with the early-type galaxies showing lower peak velocities ([49]). A large number of supernovae (M. Phillips, private communication) does not show this trend.

It is rather worrisome that we are presently using a single physical parameter such as $\Delta m_{15}(B)$ to correct to a standard luminosity, while another fundamental parameter, velocity, which related to the kinetic energy of expansion and therefore the overall energy budget of the explosion, seems to be uncorrelated with standard luminosity. What, then, are the line velocities telling us about the explosion? This is an important question where the observers need some guidance from theory.

I would like to thank Mario Hamuy, Mark Phillips, Phillip Pinto, Brian Schimdt, and Robert Schommer for our many discussions about supernovae. If there are errors in this paper, it can only be my fault.

REFERENCES

1. Suntzeff, N., et al., *Astron. J.*, **117**, 1175, (1999).
2. Jha, S., et al., *Astroph. J. Sup.*, **125** , 73, (1999).
3. Gibson, B., *astro-ph9906487*, (1999).
4. Schmidt, B., *Astroph. J.*, **432** , 42, (1994).
5. Hamuy M., et al., *Astron. J.*, **112** , 2398, (1996).
6. Riess A., et al., *Astroph. J.*, **473** , 588, (1996).
7. Riess A., et al., *Astron. J.*, **116** , 1009, (1998).
8. Garnavich P., et al., *Astroph. J.*, **493** , L53, (1998).
9. Perlmutter S., et al., *Astroph. J.*, **517** , 565, (1999).
10. Lauer T., and Postman M., *Astroph. J.*, **400** , 47, (1992).
11. Hoeflich P., et al., *Astroph. J.*, **521** , 179, (1999).

12. Drell P., et al., *astro-ph/9905027*, (1999).

13. Aguirre A., *Astroph. J.*, **512** , 19, (1999).

14. Aguirre A., *astro-ph/990439*, accepted ApJ, (1999).

15. Phillips M., *Astroph. J.*, **413** , L105, (1993).

16. Phillips M., et al., *Astron. J.*, **118**, 1766, (1999).

17. Leibundgut B., and Pinto P., *Astroph. J.*, **401** , 49, (1992).

18. Arnett D., *Supernovae and Nucleosynthesis : an Investigation of the History of Matter, From the Big Bang to the Present*, Princeton: Princeton University Press, (1996).

19. Wheeler J., et al., *Phys. Rep.*, **256**, 211, (1995).

20. Hillebrandt W., and Niemeyer J., *ARA&A*, **38**, in press, (2000).

21. McCray R., and Wang Z., *Supernovae and Supernova Remnants*, Cambridge: Cambridge University Press, (1996).

22. Ruiz-Lapuente P., et al., *Thermonuclear Supernovae*, Dordrecht: Klkuwer, (1997).

23. Niemeyer J., and Truran J., *Type Ia Supernovae: Theory and Cosmology*, Cambridge: Cambridge Univ. Press, (1999).

24. Filippenko, A., *ARA&A*, **35**, 309, (1997).

25. Leibundgut, B., *Astron. and Astroph. Rev.*, in press, (2000).

26. Edvardsson, B., et al., *Astron. Astroph.*, **275** , 101, (1993).

27. von Hippel, T., et al. *Astron. J.*, **114** , 1154, (1997).

28. Woosley S., and Weaver T. *Astroph. J.*, **423** , 371, (1994).

29. Cappelaro, E., et al., *Astron. Astroph.*, **322** , 383, (1997).

30. Bartunov, O., et al., *Thermonuclear Supernovae*, Dordrecht: Kluwer, 87, (1997).

31. Elias J., et al., *Astroph. J.*, **251** , L16, (1981).

32. Elias J., et al., *Astroph. J.*, **296** , 379, (1985).

33. Frogel, J., et al., *Astroph. J.*, **315** , L129, (1987).

34. Meikle, P., *astro-ph/9912123*, (1999).

35. Krisciunas K., et al., *astro-ph/9912219*, (1999).

36. Suntzeff N., et al., *Astron. J.*, **96** , 1864, (1988).

37. Menzies J., *M.N.R.A.S.*, **237** , 21p, (1989).

38. Bessell, M., *P.A.S.P.*, **102** , 1181, (1990).

39. Suntzeff N., *Supernovae and Supernova Remnants*, Cambridge: Cambridge University Press, 41, (1996).

40. Contardo G., et al., *Astron. Astroph.*, submitted , (2000).

41. Riess, A., et al., submitted to Ap.J., (2000).

42. Spyromilio, J., et al., *M.N.R.A.S.*, **266** , 17p, (1994).

43. Wheeler, J., *Astroph. J.*, **496** , 908, (1998).

44. Chugai N., *Sov. Ast.*, **30** , 562, (1986).

45. Wheeler, J., *IAU Symp. 151, Evolutionary Processes in Interacting Binary Stars*, Dordrecht: Kluwer , 225, (1992).

46. Marietta E., et al., *astro-ph/9908116*, (2000).

47. Nugent P., et al., *Astroph. J.*, **455** , L147, (1995).

48. Wells L., et al *Astron. J.*, **108** , 2233, (1994).

49. Branch D., and van den Bergh S., *Astron. J.*, **105** , 2231, (1993).

Type Ia Supernovae &
Cosmic Acceleration

Greg Aldering

Lawrence Berkeley National Lab, One Cyclotron Rd., Berkeley, CA, 94720

Abstract. The current application of Type Ia supernovae to the determination of the expansion history of the universe is discussed. Work by two separate research teams — the Supernova Cosmology Project, and the High-Z Team — has shown that the most straightforward interpretation of observations of Type Ia supernova at $\langle z \rangle \sim 0.5$ is that the expansion of the universe is accelerating. The form of this acceleration is consistent with a positive value of Einstein's "cosmological constant." Various forms of systematic error, both intrinsic and extrinsic to Type Ia supernovae have been proposed as alternative explanations. I review the constraints on systematic errors provided by current observations, and then discuss improvements which will be possible with a new sample of nearby supernovae and planned future observations of $z > 1$ supernovae. Assuming that systematic errors are negligible, or can be converted into statistical errors, large homogeneous samples of very distant Type Ia supernovae hold the potential to determine several key cosmological parameters to an accuracy rivaling or exceeding that of planned CMB observations. A satellite concept designed to bring this promise to fruition is outlined.

INTRODUCTION

Type Ia supernovae have proven to be an extremely valuable class of cosmic explosion. The copious luminosity of these events has long held great potential as cosmological distance indicators. It is now known how to standardize the peak brightnesses of these events using information from the timescale of the explosion [1–3]. This knowledge has been exploited by the Supernova Cosmology Project, as well as the High-Z Team (see Kirshner, these proceedings) to measure the expansion history of the universe. Using independent methods of correcting Type Ia supernova peak brightnesses for the width/decline rate of the lightcurve, and using mostly separate sets of supernovae, both groups have come to the conclusion that the expansion of the universe is accelerating.

CP522, *Cosmic Explosions: Tenth Astrophysical Conference,*
edited by Stephen S. Holt and William W. Zhang
© 2000 American Institute of Physics 1-56396-943-2/00/$17.00

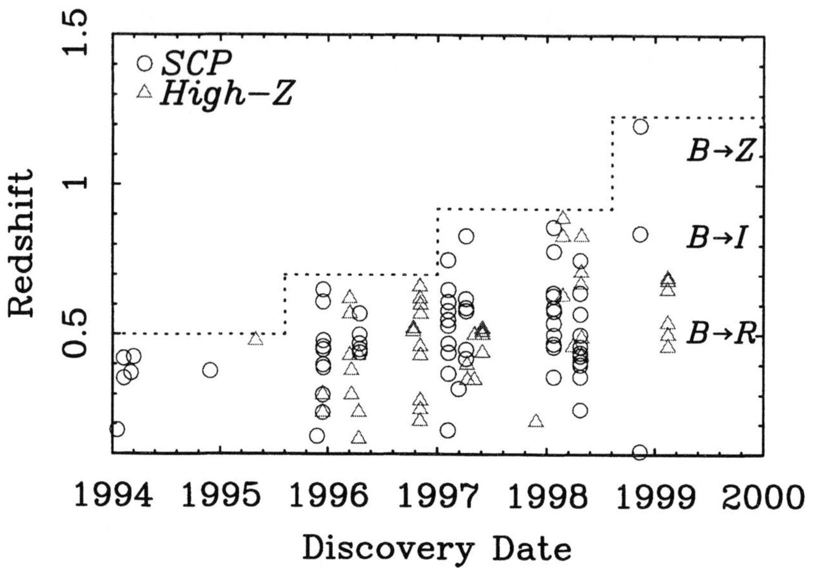

FIGURE 1. Discoveries of spectroscopically confirmed Type Ia supernovae at cosmological distances up through the time of this meeting, compiled from the IAU Circulars. The contributions of each of the research groups, the Supernova Cosmology Project (SCP) and the High-Z Team, are shown separately. The observer-frame filter covering the rest-frame B-band is shown on the right for several redshift ranges. As of this writing the most distant of these is SN1998eq, discovered at Keck II by the SCP [4].

CURRENT RESULTS

Unlike other cosmological distance indicators, SNe are not static, cataloged objects. The effort to discover them requires resources comparable to the cosmologically relavent follow-up observations, and entails significant risk. Ingredients for successful supernova searching include a 3-week gap between reference and discovery images, with full moon between, providing discoveries near maximum light with follow-up in dark time; large CCD mosaics ($\sim$ 1/4 sq deg per field); use of sites and season with clearest weather (Chilean summer); and Keck spectroscopy + CTIO/WIYN/HST follow-up photometry (with VLT/Gemini/Subaru/MMT/Magellan valuable new additions). Special challenges include fast ($<$ 12 hrs) reduction & subtraction of search images, scheduling coordination of different telescopes (esp. *HST*), and attention to bad-weather back-up scenarios. Progress in discovering more spectroscopically confirmed Type Ia supernovae — and at ever increasing distances — is portrayed in Fig. 1.

The confidence regions in the Ω_M,Ω_Λ plane have tightened from the wide band of [5], which was consistent with the then-expected Einstein-de Sitter model, to a proper ellipse with the addition of a Type Ia supernovae (SN1997ap) at $z = 0.83$

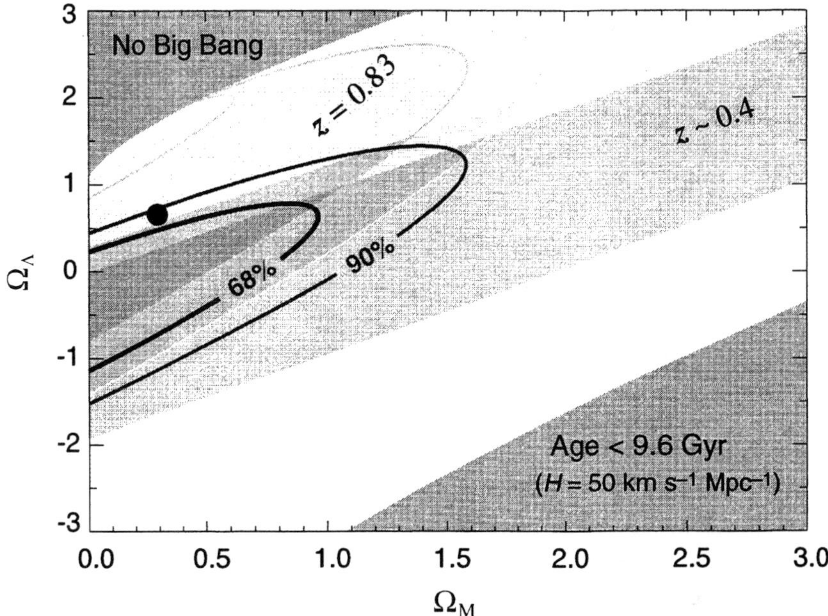

FIGURE 2. The first Type Ia supernova constraints on cosmological parameters. The blue band is from the seven supernovae of [5], the blue region is the constraint from the $z = 0.83$ supernova, SN1997ap, and their combined constraints for 68% and 90% probability are drawn as the red ellipses (from [6]). The black circle indicates the current best-fit values of Ω_M and Ω_Λ for a flat universe from [7]. The presence of a probable, but spectroscopically unconfirmed, Type Ic supernova in the dataset of [5] at that time lead to a weak preference for an Einstein-de Sitter universe. However, those early results are still consistent with the best current results.

and observed with HST [6] — both shown in Fig. 2— and then to the smaller error ellipse rejecting an Einstein-de Sitter model and favoring a (flat) universe with a positive cosmological constant [7,8], shown in Fig. 3. Two independent research groups reached this same result, and the indications of low Ω_M from galaxy cluster dynamics and abundances, combined with indications of a flat universe from CMB measurements, are also consistent with this result. The discovery of such "dark energy" — now playing a dominant role in the expansion of the universe — presents an exciting new puzzle for fundamental physics to explain.

CONSTRAINING SYSTEMATIC UNCERTAINTIES

A more prosaic interpretation of the Type Ia supernova observations is that some systematic error leads to an apparent dimming of supernova at higher redshift,

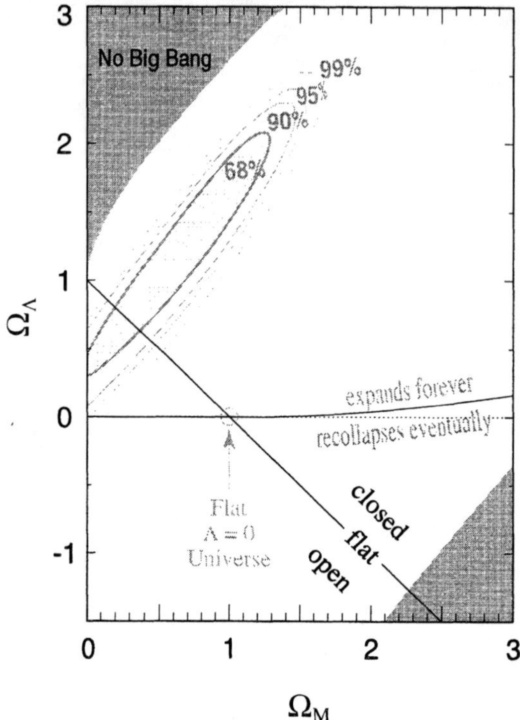

FIGURE 3. The most recent Type Ia supernova constraints on cosmological parameters, from [7]. These results, based on 42 high-redshift and 18 low-redshift suerpnovae rule-out the Einstein-de Sitter model. They also require a cosmological constant if the curvature of the universe is zero. Even after marginalizing the probability over Ω_M, there is still strong evidence for a cosmological constant from these data.

thereby mimicking a cosmological constant. The history of astronomy is filled with examples where dust extinction or evolution of a "standard candle" has lead to misinterpretation of observations, so such concerns cannot be treated lightly. Several classical systematics, such as Malmquist bias, host galaxy extinction, and gravitational lensing are examined in [7] and found to be unlikely as alternative explanations. Another aspect of the high-redshift dataset which many find puzzling is the small amount of host galaxy extinction inferred from the supernovae colors in concert with standard dust extinction laws. [9] have modeled the extinction expected for Type Ia supernovae observed through galaxy halos and randomly oriented disks. I have then taken their simulations and applied the additional bias against heavily extincted supernovae which arises in a flux-limited search. These two cases are shown in Fig. 4; the disk and halo results are shown with their

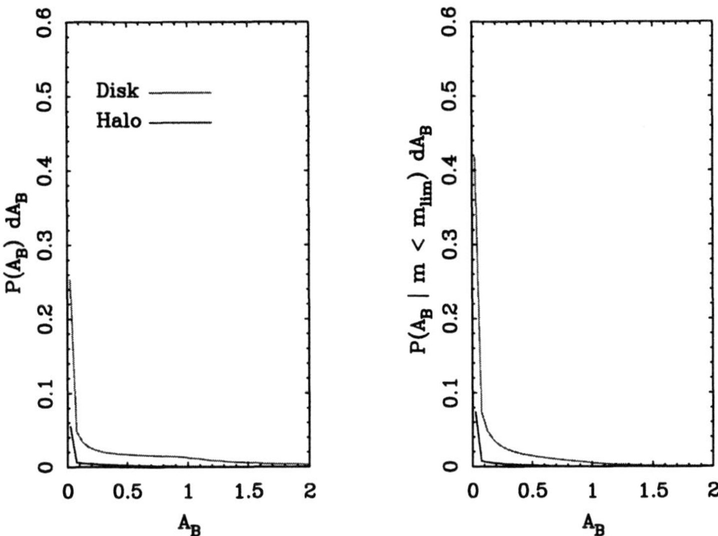

FIGURE 4. The expected extinction probability distribution of Type Ia supernovae, with (right) and without (left) inclusion of bias due to flux-limited selection. The extinction distributions are shown separately for halo and disk populations, according to their relative frequencies in the galaxy population.

correct relative normalization. The surprising result is that almost half of the Type Ia supernovae discovered in the high-redshift searches should suffer less than 0.025 magnitudes of extinction in the rest-frame B-band. This is consistent with the actual observations of [7] and [8]. Extinction of gravitational lenses, as in [10], have very different selection biases and geometries which favor the higher observed extinctions. Similarly, some of the differences in the radial distribution of various low- and high-redshift supernova samples seen by [11] also include a contribution due to the selection against AGN used by the SCP — namely, a small brightness change at the center of a galaxy is generally not followed with spectroscopy at Keck due to the high probability it will be due to variability of an AGN.

Recently, [12,13] have claimed there is a difference in the time from explosion to peak between low-redshift and high-redshift supernovae, and have speculated that such a difference in rise-times might translate into a difference in peak brightnesses. As discussed at length at the conference, a more complete analysis of the data finds that these time scales are in fact reasonably consistent and that features of the current datasets can easily bias attempts to measure a difference [14]. Thus, rise-time differences are no longer a likely alternative to a cosmological constant, although for completeness this question will have to be revisited once more suitable datasets become available.

Dust extinction has always been the astronomer's nemesis, and [15–17] (and these

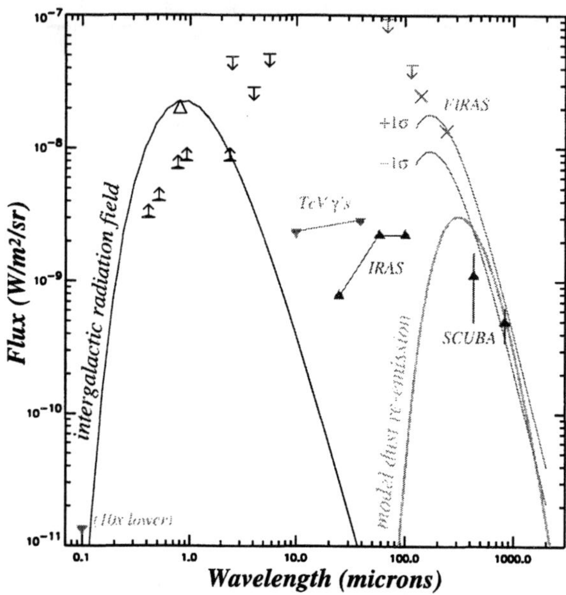

FIGURE 5. Gray dust emission and constraints. The intergalactic radiation field (optical-NIR) from galaxies will heat an gray dust, which will then re-emit in FIR. The dust re-emission shown is that expected if there is enough gray dust to explain the Type Ia supernova observations without a cosmological constant, and is adapted from [17]. I have added lower limits from IRAS and SCUBA galaxy counts from [19], and upper limits in the EUV from [20] (drawn 10× too high so as to fit on graph), and MIR upper limits from TeV γ-ray absorption from [21]. The SCUBA galaxy counts nearly account for the background observed by FIRAS [18], leaving limited room for intergalactic gray dust.

proceedings) has proposed that intergalactic dust with an extinction law which is nearly flat (gray) at optical wavelengths could explain the Type Ia supernovae observations without a cosmological constant. Such dust doesn't strongly redden supernova light, and there are few other observational constraints on whether such dust could exist. Perhaps the most promising direct means of limiting the amount of such dust is through observations of the cosmic far-infrared background (CIB). As Fig. 5 illustrates, the intergalactic radiation field from galaxies will heat the dust to a temperature slightly warmer than the CMB. FIRAS has measured such an excess, but galaxy counts at 850μm seem able to account for most of this background at wavelengths relavent for re-radiation by intergalactic dust [18,19]. Future deep sub-mm observations resolving the CIB should settle this questions. In the meantime, we can state that a preliminary analysis of SN1999eq at $z = 1.20$ indicates that it has a peak brightness expected from extrapolation of the best-fit cosmological

TABLE 1. Nearby Supernova Campaign: Survey Parameters

Search	Type	Aper (m)	FOV (deg^{-2})	Scale ("/pixel)	Exp (sec)	Filter
EROS	Staring	1.0	1.00	0.60	300	B & R
MOSAIC	Staring	0.9	1.00	0.43	240	R
NEAT	Staring	1.0	2.54	1.40	60	open
Spacewatch	Scanning	0.9	$0.57{\times}t$	1.05	430	OG515
QUEST	Scanning	1.0	$2.30{\times}t$	1.00	550	V

parameters of [7]. If gray dust were present in quantities sufficient to explain the observations of [7,8], SN1999eq should have been dimmer.

THE SPRING 1999 NEARBY CAMPAIGN

Another potential systematic error would occur if Type Ia supernova explosions are different at low- and high-redshift. Such an error — sometimes referred to as evolution — could arise it the current peak luminosity - lightcurve width relation does not already correct for variations in Type Ia supernova initial conditions and if the mean value of some initial parameter changes systematically with redshift. Such changes in initial conditions must be well synchronized with redshift or else the luminosity dispersion for high- and low-redshift supernovae would be different, contrary to what is observed [7].

Type Ia supernovae in nearby galaxies can help settle the question of whether the luminosity-width relation works over a wide range of initial conditions. Supernova spectroscopic features, as well as stellar population indicators from the host galaxies, can serve as initial condition indicators. Since the nearby universe contains stellar populations that are both young and old, metal-poor and metal-rich, it should be possible to find local counterparts to high-redshift supernovae. With this in mind, the SCP coordinated an extensive campaign to find a large number of nearby Type Ia supernovae and obtain extensive follow-up observations of them during the spring of 1999. In this endeavor, the SCP was joined by members of the EROS and QUEST collaborations, the Nearby Galaxies Supernova Search Team [22], and received early alerts of supernova candidates from LOSS [23], the MSSSO Abell Cluster Search [24], the Wise Observatory supernova search [25], and the Tenegra Observatory search [26]. The NEAT [27] and Spacewatch [28]asteroid search teams also contributed data which was used for searching.

Table 1 gives the parameters of the searches, broken down by facility, and then totaled at the bottom. In all cases CCD detectors were used, and in most cases the same SCP search software used for high-redshift searching was also used for the Nearby Campaign. This largely eliminates several systematic bias which plagued large photographic supernova searches [30,29]. Over 1300 sq. deg. of sky was

TABLE 2. Nearby Supernova Campaign: Survey Results

Search	Time (hrs)	Coverage (deg^2)	Ia	II	Ic	Untyped (faint)
EROS	125	~450	11	3	0	10
MOSAIC	48	~175	6	2	0	1
NEAT	15	~425	3	1	1	0
Spacewatch	140	~150	3	0	0	2
QUEST	4	~140	0	0	0	1
Other Searches[a]			6	2	0	0
Totals		~1340	29	8	1	14

[a] LOSS, MSSSO Abell Cluster Search, Wise Obs., etc, discussed elsewhere in these proceedings

searched in this campaign. Table 2 presents a summary of the supernovae discovered by the Nearby Campaign, and Fig. 6 shows the redshift distribution of the discoveries. Of the Type Ia supernova, 20 were discovered early enough and at low enough redshift to warrant intensive follow-up observations. The redshift distribution of these "select" supernovae is also given in Fig. 6. Follow-up observations consisted of $UBVRI$ photometry every 3–7 days, and spectrophotometry once per week. In all, some 100 nights of telescope time were used to search, and another 100 nights were used for follow-up. The amount of spectroscopic coverage equals that of all Type Ia supernova spectra published to date, and the temporal coverage is far better than the average (e.g., a discovery spectrum near peak). Analysis of this dataset is underway, and we expect it to yield much firmer constraints on how well the luminosity-width relation accounts for different initial conditions amongst Type Ia supernovae.

ON-GOING AND FUTURE ADVANCES

While concerns regarding systematic uncertainties are being addressed, the SCP is also pushing forward with new observations of high-redshift supernovae to better constrain Ω_M, Ω_Λ, and the curvature, Ω_k. Observations of over a dozen more Type Ia supernovae having exquisite observations from HST PC2 and NICMOS are being analyzed, including SN1999eq at $z = 1.20$ which was one of the last science targets observed by NICMOS before its cryogen ran out.

Even with Keck and HST, progress in increasing the high-redshift Type Ia supernovae samples is slow going (see Fig. 1). A sample of 2000 Type Ia supernovae out to redshift $z \sim 1.7$ would enable measurements of Ω_M to ±0.02, Ω_Λ to ±0.05, and Ω_k to 0.05 from supernova alone. Adding the constraint of a flat universe ($\Omega_k = 0$) would enable such a dataset to measure Ω_M and Ω_Λ to ±0.01. Moreover, since Type Ia supernovae can map the expansion history of the universe explicitly,

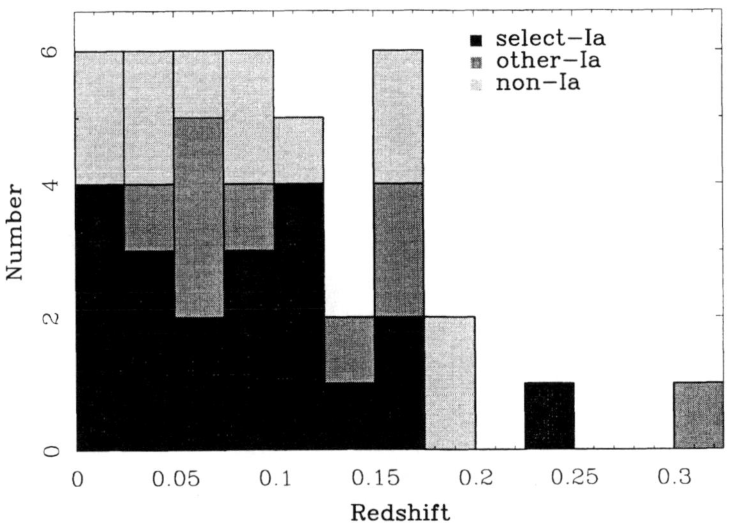

FIGURE 6. Redshift distribution of the Spring 1999 Nearby Campaign.

such a dataset would allow a determination of whether the "dark energy" is really a cosmological constant of whether an different (perhaps time varying) equation of state applies.

Given this potential for making Type Ia supernovae a pillar of observational cosmology — and an important complement and check for measurements of the CMB — and our assessment that obtaining such a large dataset with excellent control over systematics from the ground is difficult at best, the SCP has developed a concept for a dedicated wide-field imager and optical/NIR spectrograph in space. In one year such a satellite could discover and obtain photometric and spectroscopic follow-up for 2000 Type Ia supernovae out to $z \sim 1.7$. The supernovae would be discovered very early, eliminating Malmquist bias and allowing measurement of their rise-times. Optical through NIR spectroscopy at peak would allow checks on several important indicators of explosion luminosity, kinetic energy, metallicity, etc., while also allowing correction for host galaxy extinction for a variety of extinction laws. Although NGST has plans to observe Type Ia supernovae as well, constraining the nature of the "dark energy" requires observations at $z < 1$, where searching with NGST would be prohibitive and the budget for follow-up observations would be dominated by the slew time rather than time on target. With dedicate space-based facility, the number of well observed high-redshift a decade from now could be $100\times$ larger than it is today.

ACKNOWLEDGMENTS

Most of the work described herein are products of the Supernova Cosmology Project or the Spring 1999 Nearby Campaign. I would like to thank my collaborators in these projects for their hard work and support. Observations described herein derive from Keck, YALO, CTIO, KPNO, ESO, Lick, NOT, WIYN, WHT, INT, JKT, APO, and Mt. Laguna Observatory facilities, and the allocation committees and staff of these observatories are thanked for their continuing support.

REFERENCES

1. Phillips, M., *et al.* 1993, *ApJ*, 413, L105.
2. Riess, A., *et al.* 1996, *ApJ*, 473, 88.
3. Phillips, M., *et al.* 1999, *AJ*, 118, 1766.
4. Aldering, G., *et al.* 1998, IAUC 7046.
5. Perlmutter, S., *et al.* 1997, *ApJ*, 483, 565.
6. Perlmutter, S., *et al.* 1998, *Nature*.
7. Perlmutter, S., *et al.* 1999, *ApJ*, 517, 565.
8. Riess, A., *et al.* 1998, *AJ*, 116, 1009.
9. Hatano, K., Branch, D., and Deaton, J. 1998, *ApJ*, 501, 177.
10. Falco, E., *et al.* 1999, *ApJ*, 523, 617.
11. Howell, A., *et al.* 1999, astro-ph/9908127.
12. Riess, A., *et al.* 1999a (astro-ph/9907037).
13. Riess, A., *et al.* 1999b (astro-ph/9907038).
14. Aldering, G., Knop, R., Nugent, P., AJ, submitted.
15. Aguirre, A. 1999, *ApJ*, 512, L19.
16. Aguirre, A. 1999, *ApJ*, 525, 583.
17. Aguirre, A., and Haiman, Z. 1999, (astro-ph/9907039).
18. Fixsen, D. J., *et al.* 1996, *ApJ*, 473, 576.
19. Blain, A. W., *et al.* 1999, (astro-ph/9908024).
20. Donahue, M, Aldering, G., and Stocke, J. T. 1995, *ApJ*, 450, L45.
21. Vassiliev, V. V. 1999, (astro-ph/9908088).
22. Strolger, L., *et al.* (see IAUC 7125).
23. Li, W., *et al.* (these proceedings).
24. Reiss, D., *et al.* 1998, *AJ*, 115, 26.
25. Moaz, D., *et al.* (these proceedings).
26. Schwartz, M., private communications.
27. Pravdo, S., *et al.* 1999, *AJ*, 117, 1616.
28. McMillan, R., Larsen, J., *et al.* (see IAUC 7134).
29. Hamuy, M., *et al.* 1993, *AJ*, 106, 2392.
30. Hamuy, M., *et al.* 1999, *AJ*, 117, 1185.

Late Light Curves of SN Ia

Peter A. Milne[1], Lih-Sin The[2], and Mark D. Leising[2]

[1] *NRC/NRL Research Associate, Code 7650, Washington, D.C., 20375*
[2] *Clemson University, Clemson, SC 29631*

Abstract. A "positron phase" exists in type Ia supernovae, during which the energy deposition into the SN ejecta is dominated by the slowing of $^{56}Co \rightarrow {}^{56}Fe$ β^+ decay positrons. Comparisons of model-generated bolometric light curves with V and B band photometry during this phase suggests that the magnetic field is either too weak to confine positrons, or is radially combed. In each case, positrons escape the ejecta leading to a deficit of energy deposition. Ramifications of this escape upon other studies of SN Ia emission is discussed.

INTRODUCTION

Within a few weeks after the explosion of a type Ia supernova (SN), the energy deposition is dominated by the products of $^{56}Co \rightarrow {}^{56}Fe$ decays: photons & positrons. The photons are far more energetic than the positrons and dominate the energy deposition at early times, when both are efficiently trapped. Though more energetic, the photons are less likely to interact with the ejecta. At later times, this leads to photons preferentially escaping the SN ejecta and positrons dominating the energy deposition. It is during this "positron phase" that positrons can be used to probe the magnetic field and to a lesser extent the level of ionization of the SN ejecta. Certain magnetic field configurations permit significant numbers of positrons to escape the ejecta leaving a deficit of energy deposition relative to configurations which permit no escape. If this escape were confirmed by observations, the ramifications would extend beyond understanding magnetic fields and ionization levels. To gamma-ray astrophysics, it would motivate type Ia SNe to be a major contributor to galactic annihilation radiation (seen at 511 keV). To SN modelling, it would necessitate a re-analysis of nebular spectra studies which purport to differentiate SN model mass (Liu et al. 1997) [1] and the nickel production (Ruiz -Lapuente & Fillipenko 1996, Bowers et al. 1997) [2] [3]. In a recently published study, Milne, The & Leising (1999) [4] generated light curves for 21 SN models and fit these light curves to observations of 10 type Ia SNe. This work is extracted from that study, the emphasis in this paper is on what observations say about type Ia SNe.

CP522, *Cosmic Explosions: Tenth Astrophysical Conference,*
edited by Stephen S. Holt and William W. Zhang
© 2000 American Institute of Physics 1-56396-943-2/00/$17.00

ENERGY DEPOSITION

There are a number of explosion scenarios suggested to explain type Ia SNe. For our purposes, SN models can be categorized as Chandrasekhar mass and sub-Chandrasekhar mass. Peak apparent magnitudes combined with distance estimates to the host galaxies indicate that there may be normal-, super- and sub-luminous sub-classes of type Ia SNe. Within each explosion scenario, multiple models have been suggested to account for this inhomogeneity. This paper uses the same prescriptions for photon and positron transport as was employed in Milne, The & Leising (1999) [4], but displays one Chandrasekhar mass model and one sub-Chandrasekhar mass model fit to each SN. By doing this, the subtle differences between model light curves can be seen, as well as the extent to which the results are independent of model mass.

To probe the magnetic field and level of ionization, light curves were generated for extreme assumptions. For the magnetic field three scenarios were simulated. First, is was assumed that the magnetic field was strong enough to confine positrons to the field lines and the field lines have been combed to become essentially radial. This scenario was suggested by Colgate et al. (1980) [5], and was treated by Chan & Lingenfelter (1993) [6], and features mirroring/beaming which bends trajectories towards radial escape (enhancing positron escape). The second scenario also assumes a strong field, but in this case the alignment is turbulent. Positrons spiral along these field lines, but accomplish no net diffusion through the ejecta. This scenario is refered to as a "trapping scenario". Positrons can develop significant lifetimes as the ejecta rarefies, but no escape occurs. This scenario was suggested by Axelrod (1980) [7], and was also treated by Chan & Lingenfelter (1993) [6]. The third scenario assumes that the field is too weak to confine the positrons, and the positrons travel along straight line trajectories. Colgate et al. (1980) [5] argued that this "weak" scenario is synonymous with the first, radial scenario, a result approximately confirmed by simulations by Milne, The & Leising (1999) [4]. Thus, while only radial and trapping curves are shown, radial≈weak is implied. For the level of ionization, two extremes were chosen. The low ionization extreme assumes 1% ionization (or 1 of every 100 nuclei is singly ionizaed). The high ionization extreme assumes that every nucleus is triply ionized. As shown in Figure 1, the ranges of ionization lead to the radial and trapping curves having finite thicknesses. Though the separation between low and high ionization light curves becomes considerable, the separation between field alignments is greater.

LIGHT CURVE CASE STUDY

Shown in Figure 1 are model generated light curves fit to the V and B band observations of six type Ia SNe.[1] The left panels are Chandresekhar mass models,

---

[1] Discussions regarding the validity of fitting model-generated energy deposition rates to observed band photometry and/or "uvoir" bolometric light curves are given in Milne et al. (1999).

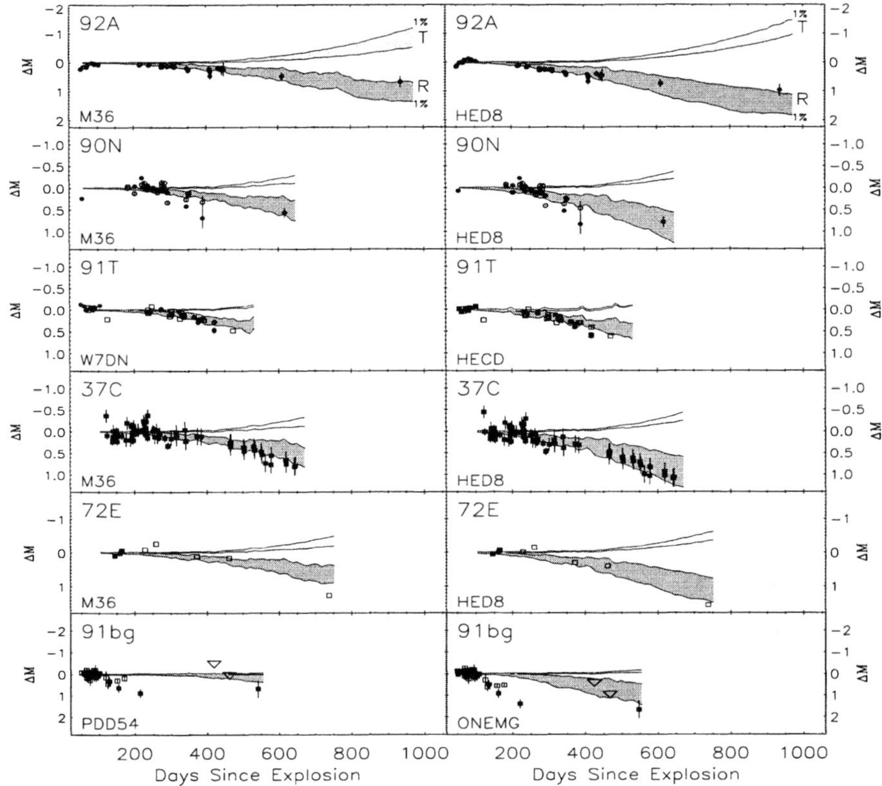

FIGURE 1. Model-generated bolometric light curves fit to the V and B band light curves of six type Ia SNe; SN 1992A (V: Suntzeff 1996), SN 1990N (B&V: Lira et al. 1998), SN 1991T (V: Lira et al. 1998, Cappellaro et al. 1997), SN 1937C (B: Schaefer 1994), SN 1972E (V: Ardeberg & de Groot 1973, B: Kirshner & Oke 1975), & SN 1991bg (V: Turatto et al. 1996, Leibundgut et al. 1993, Filippenko et al. 1992)(The inverted triangles shown for SN 1991bg are upper limits). The left panels show Chandrasekhar mass models, the right panels show sub-Chandrasekhar mass models. All curves are shown in units of magnitude relative to the instantaneous deposition assumption. For SN 1992A, the radial curves are labelled (R), trapping curves are labelled (T). For the upper five SNe (all normally- or super-luminous), the light curves are better explained by radial escape (dark shading) than by positron trapping (light shading). The sub-luminous SN 1991bg is not explained by either magnetic field scenario, for either model type. Model references are: M36 (Höflich 1995), PDD54 (Höflich, Khokhlov & Wheeler 1995), W7DT (Yamaoka et al. 1992), HED8 (Höflich & Khokhlov 1996), HECD (S. Kumagai 1997).

the right panels are sub-Chandrasekhar mass models. All curves are shown in units of magnitude relative to the light curve that would result if 100% of the positron decay energy were deposited instantaneously, in-situ (an assumption made in some spectral studies). The curves and data are normalized to fit the earliest portion of the plotted data. As seen for all cases, at 60 days positrons have negligible lifetimes and the in-situ assumption is valid. By 200^d, positron lifetimes become important. For the radial scenario (dark shading), positrons increasingly escape the ejecta resulting in a deficit of energy deposition. For the trapping scenario (light shading), positrons develop increased lifetimes. This energy does not leave the ejecta, however, it is simply stored and deposited after a delay. The end result is that radial curves become increasingly fainter relative to the in-situ assumption, while trapping curves become increasingly brighter. Positrons have larger ranges in low ionization media, so the 1% ionized approximations (labeled for SN 1992A) feature the largest deviations from the in-situ assumption.

This study does not treat photon diffusion timescales, thus curves are not fit earlier than 60^d. The B band is not shown until after $\sim 120^d$, because that band does not trace the bolometric luminosity at earlier epochs. The upper five SNe are normally- or super-luminous and are all better explained by positron escape than by positron trapping. In-situ and trapping scenarios are strongly rejected for sub-Chandrasekhar mass models. The level of ionization is poorly constrained. SN 1992A implying a transition from low-to-high ionization, SNe 1991T & 1972E suggesting the opposite transition, and SNe 1990N & 1937C suggesting intermediate ionizations without transitions.[2] Ruiz -Lapuente & Spruit (1997) [8] argued for a transition of the magnetic field from trapping to weak scenarios, which would mimic the effect of lowering the ionization. The current observations do not present compelling arguments for the dominance of any of these interpretations.

SN 1991bg, shown in the lower panel, is not acceptably fit by either model, for all field and ionization scenarios. There is an apparent deficit of luminosity after 100^d, but detectable emission after 500^d. This tendency is not unique, the other well-observed sub-luminous SNe Ia, 1992K [9]& 1997cn [10] possess similar deficits after 100^d. Whether this suggests a failure of the V (or B) band to accurately trace the bolometric luminosity, or rather that sub-luminous SNe Ia are an entirely different event than modelled is unclear. These findings are in disagreement with the findings of Ruiz -Lapuente & Spruit (1997) [8], who fit a low mass model to bolometric data derived from the Turatto data. These findings do agree with the findings of Cappellaro et al. (1997) [11] who found that only complete transparency to positrons could approximate the 100^d-230^d data. It is clear that there is much to be learned about this sub-class of type Ia SNe.

---

[2] Observations made of SN 1991T after 500^d were not shown because that emission is dominated by a light echo.

DISCUSSION

The light curves of normally- and super-luminous SNe Ia suggest that positrons escape the ejecta in quantity. This suggests that either the magnetic field is radially combed, or is too weak to confine the positrons. Most previous studies that treated the emission after 100^d from SN Ia have assumed either instantaneous stopping of positrons, or positron trapping. Although more observations and a more detailed treatment of the conversion of deposited energy to observable emission are required, the indications presented in this paper suggest the need to re-scrutinize those previous studies. Sub-luminous SN Ia can not be explained by any of the models tested. This does not reject these models as potential solutions to this SN Ia sub-class, but the failure to explain this sub-class while explaining the other sub-classes must be justified. Late observations of SNe Ia are a promising avenue through which aspects of the SN event can be understood, and they are a valid use of telescope time.

REFERENCES

1. Liu, W., and Jeffery, D.J., and Schultz, D.R., *ApJ* **483**, L107 (1997).
2. Ruiz -Lapuente, P., and Filippenko, A.V., *in IAU Colloq. 145, Supernovae and Supernova Remnants, ed. R. McCray & Z. Wang (Cambridge: Cambridge Univ. Press)*, 33 (1996).
3. Bowers, E.J.C. et al., *LANL* **9707119**, (1997).
4. Milne, P.A., and The, L.-S., and Leising, M.D., *ApJS* **124**, 503 (1999).
5. Colgate, S., and Petschek, A.G., and Kreise, J.T., *ApJ* **237**, L81 (1980).
6. Chan, K.-W., and Lingenfelter, R.E., *ApJ* **405**, 614 (1993).
7. Axelrod, T.S., *Ph.D.thesis, Univ. California at Santa Cruz*, (1980).
8. Ruiz -Lapuente, P., and Spruit, H., *ApJ* **500**, 360 (1997).
9. Hamuy, M. et al., *AJ* **108**, 2226 (1994).
10. Turatto, M. et al., *AJ* **116**, 2431 (1998).
11. Cappellaro, E. et al., *A&A* **328**, 203 (1997).
12. Sunzteff, N., *in IAU Colloq. 145, Supernovae and Supernova Remnants, ed. R. McCray & Z. Wang (Cambridge: Cambridge Univ. Press)*, 41 (1996).
13. Lira, P. et al., *AJ* **115**, 234 (1998).
14. Schaefer, B.E., *ApJ* **426**, 493 (1994).
15. Ardeberg, A.L., and de Groot, M.J., *A&A* **28** 295 (1973).
16. Kirshner, R.P., and Oke, J.B., *ApJ* **200**, 574 (1975).
17. Turatto, M. et al., *MNRAS* **283**, 1 (1996).
18. Leibundgut, B. et al., *AJ* **105**, 301 (1993).
19. Filippenko, A.V. et al., *AJ* **104**, 1543 (1992).
20. Höflich, P., *ApJ* **443**, 89 (1995).
21. Höflich, P., and Khokhlov, A., and Wheeler, J.C., *ApJ* **444**, 831 (1995).
22. Yamaoka, H. et al., *ApJ* **393**, L55 (1992).
23. Höflich, P., and Khokhlov, A., *ApJ* **457**, 500 (1996).
24. Kumagai, S., *private communication*, (1997).

A High Peculiarity Rate for Type Ia SNe

W. D. Li[*], A. V. Filippenko[*], A. G. Riess[*,1], R. R. Treffers[*],
J. Y. Hu[†], and Y. L. Qiu[†]

[*]*Department of Astronomy, University of California, Berkeley, CA 94720-3411 USA*
email: (wli, alex)@astro.berkeley.edu
[†]*Beijing Astronomical Observatory, Chinese Academy of Sciences, Beijing 100080 China*

Abstract. We have compiled a sample of 90 SNe Ia from 1997 to 1999 (up to SN 1999da) and studied the peculiarity rate of SN 1991T, SN 1991bg and SN 1986G-like objects. A Monte Carlo code was written to study the observational biases involved in the evaluation of the intrinsic peculiarity rate of SNe Ia. We find that SNe Ia have a high peculiarity rate ($> 30\%$) and a flat luminosity function.

INTRODUCTION

Type Ia supernovae (SNe Ia) are not perfectly homogeneous. There are peculiar ones: SN 1991T-like (overluminous), SN 1986G-like (subluminous), and SN 1991bg-like (very subluminous) objects. Figure 1 shows a comparison of the spectra of peculiar SNe Ia with that of a relatively normal SN Ia, SN 1994D.

The peculiarity rate, however, is not well established. An estimate (less than 10%) by Branch et al. (1993) [1] is limited by the small number of peculiar SNe Ia known at that time.

In the past 3 years, a number of peculiar SNe have been discovered in the course of several successful nearby SN surveys, and we try to update the peculiarity rate here.

THE SN IA SAMPLE

We have compiled a sample of 90 nearby ($Z < 0.1$) SNe Ia from 1997 to 1999 (up to SN 1999da), and subclassified them as normal or as one of the peculiar SNe Ia: SN 1991T, which had prominent Fe III absorption lines and weak Si II lines prior to and near maximum brightness [2, 3]; SN 1991bg, which had an enhanced Si II 5700Å absorption, and a broad absorption trough extending from about 4150 to 4400Å due to Ti II lines [4, 5]; SN 1986G, which also had Ti absorption but less

---

[1] Now at Space Telescope Science Institute, Baltimore, MD 21218

prominent than in SN 1991bg [6, 7]. Classification is done based on information in the International Astronomical Union Circulars (IAUC) and our SN spectrum and photometry database.

THE OBSERVED PECULIARITY RATE

We have divided our SN Ia sample into several subsamples and reported the observed peculiarity rates as follows.

In the total sample, all 90 SNe are considered. 17 (18.9%) SNe are peculiar, among which 11 (12.2%) SNe are SN 1991T-like and 6 (6.7%) SNe are SN 1991bg/1986G-like.

In the near-maximum sample, only the SNe Ia that were spectroscopically classified at no later than a week after maximum are considered. 61 SNe are in the sample, among which 17 (27.9%) are peculiar. 11 (18.0%) SNe are SN 1991T-like and 6 (9.9%) SNe are SN 1991bg/1986G-like.

In the Lick-Beijing (LB) sample, only the SNe that were discovered in the sample galaxies of the Lick Observatory Supernova Search (LOSS) and the Beijing Astronomical Observatory Supernova Survey (BAOSS) are considered. 35 SNe are in the sample, among which 13 (37.1%) are peculiar. 7 (20.0%) SNe are SN 1991T-like and 6 (17.1%) SNe are SN 1991bg/1986G-like.

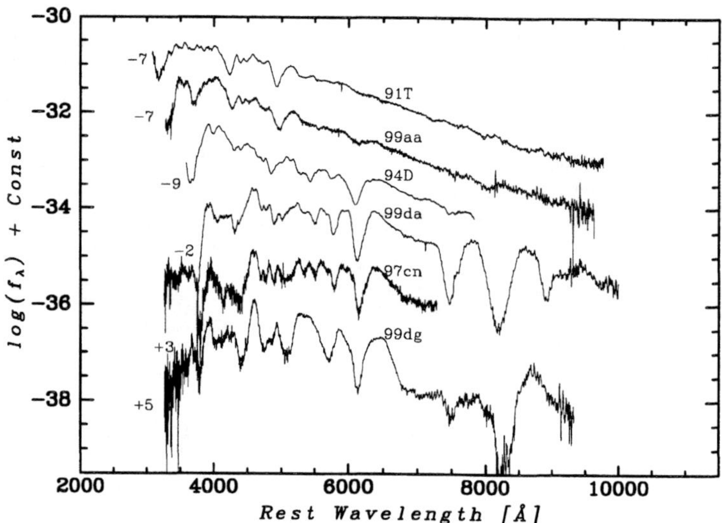

FIGURE 1. Spectra of two SN 1991T-like objects (99aa, 91T), one normal object (94D), and three SN 1991bg-like objects (99da, 97cn, 99dg). Days are relative to maximum.

THE OBSERVATIONAL BIASES

There are various observational biases that make the observed peculiarity rate deviate from its intrinsic value.

1. The maximum-only bias – caused by the fact that SN 1991T-like objects can best be spectroscopically identified prior to or near maximum brightness. It is thus unknown whether a SN discovered well after maximum is normal or SN 1991T-like. Classifying them as normal causes the maximum-only bias that underestimates the peculiarity rate.

2. The Malmquist bias – caused by the difference in luminosity among SNe Ia.

3. The light-curve shape bias – caused by the difference in light-curve shape among SNe Ia.

MONTE CARLO SIMULATIONS

We have done Monte Carlo simulations to study the effects of the observational biases. Simulations are done for magnitude-limited SN surveys, with the baseline as the only input parameter. Simulations are also done for distance-limited SN surveys, with the baseline and the limiting magnitude of the survey as parameters.

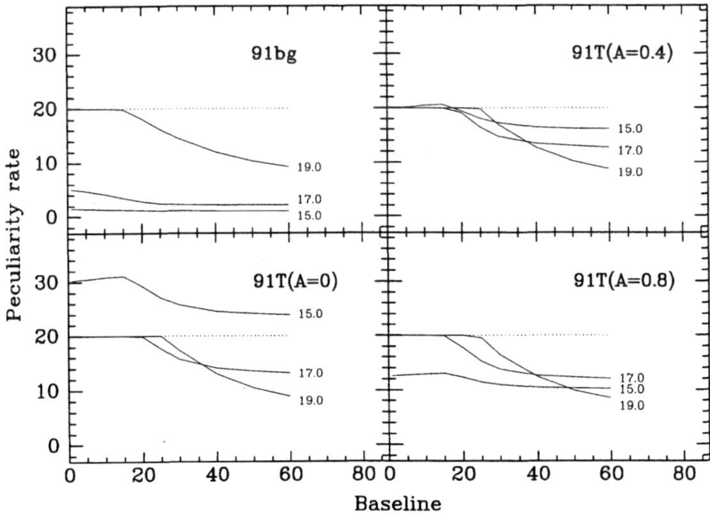

FIGURE 2. The peculiarity rate found in the Monte Carlo simulations for distance-limited SN surveys. Results are shown for different baselines and different limiting magnitudes.

All observational biases are well accounted for in the Monte Carlo simulations. We also studied the role of extinction of SN 1991T-like objects in determining the rate of those objects. There are speculations that SN 1991T-like objects are more likely to occur in dusty, star-forming regions and thus may suffer more extinction than the normal and SN 1991bg/1986G-like objects.

An example of the results of the Monte Carlo simulations is shown in Figure 2.

THE INTRINSIC PECULIARITY RATE

Our simulations indicate that a high peculiarity rate (more than 30%) and a flat luminosity function for SN Ia (e.g., the rate of SN 1991bg-like and SN 1991T-like objects are comparable) are consistent with the observed peculiarity rates in all three (total, near-maximum, and LB) samples.

These results have important implications for studies of high-redshift SNe Ia and of SN Ia progenitor systems.

1. The high-redshift results: The high peculiarity rate for nearby SNe Ia ($>30\%$) and the small apparent (but preliminary) peculiarity rate for the several dozen high-redshift SNe Ia studied thus far [8, 9] may indicate a systematic difference between them. To reconcile the absence of SN 1991T-like objects found at high redshifts with the peculiarity rate found at low redshifts, an extra extinction of more than ~ 1 mag for the SN 199T-like objects is needed, which is not supported by the observations. However, the existing spectral studies of high-redshift SNe Ia are not very detailed.

2. The progenitor systems of SNe Ia: The high peculiarity rate, together with other evidence, favors the existence of multiple progenitor systems for SNe Ia (e.g., single-degenerate systems and double-degenerate systems).

Our supernova research at UC Berkeley is supported by NSF grant AST-9417213 and NASA grant GO-7434.

REFERENCES

1. Branch, D., Fisher, A., and Nugent, P., *A.J.* **106**, 2383 (1993).
2. Filippenko, A. V., et al., *Ap.J.* **384**, L15 (1992a).
3. Phillips, M. M., et al., *A.J.* **103**, 1632 (1992).
4. Filippenko, A. V., et al., *A.J.* **104**, 1543 (1992b).
5. Leibundgut, B., et al., *A.J.* **105**, 301 (1993).
6. Phillips, M. M., et al., *P.A.S.P.* **99**, 592 (1987).
7. Cristiani, S., et al., *A&A* **259**, 63 (1992).
8. Riess, A. G., et al., *A.J.* **116**, 1009 (1998).
9. Perlmutter, S., et al., *Ap.J.* **517**, 565 (1999).

New Models for X-Ray Synchrotron Radiation from the Remnant of Supernova 1006 AD

Kristy K. Dyer, Stephen P. Reynolds, and Kazimierz J. Borkowski

North Carolina State University
Physics Department, Raleigh NC 27695-8202

Abstract. Galactic cosmic rays up to energies of around 10^{15} eV are assumed to originate in supernova remnants (SNRs). The shock wave of a young SNR like SN 1006 AD can accelerate electrons to energies greater than 1 TeV, where they can produce synchrotron radiation in the X-ray band. A new model (*SRESC*) designed to model synchrotron X-rays from Type Ia supernovae can constrain values for the magnetic-field strength and electron scattering properties, with implications for the acceleration of the unseen ions which dominate the cosmic-ray energetics. New observations by ASCA, ROSAT, and RXTE have provided enormously improved data, which now extend to higher X-ray energies. These data allow much firmer constraints. We will describe model fits to these new data on SN 1006 AD, emphasizing the physical constraints that can be placed on SNRs and on the cosmic-ray acceleration process.

X-RAY SYNCHROTRON EMISSION FROM SNRS

Young supernova remnants are in a period of transition to the Sedov-Taylor phase, sweeping up several times their original ejected mass as they expand into a circumstellar medium. The shocked gas at temperatures of 10^7 K emits a thermal spectrum (bremsstrahlung and lines) through interactions of electrons with ions. However, the synchrotron emission, which produces the radio spectrum, persists through the X-ray regime (and in some case dominates the thermal X-rays [1]).

In order to study the energetics of the accelerated electrons and protons in the shock, as well as to obtain accurate abundances from thermal models, it is imperative to understand this synchrotron spectrum. SNR observations show that the synchrotron emission drops *significantly* below a powerlaw at X-ray frequencies, implying a curved synchrotron spectrum. To accurately describe the synchrotron spectrum we have developed the following models:

- **SRCUT** A homogeneous population of electrons, with an exponentially cut off power-law energy distribution, radiating in a constant magnetic field, produces the sharpest physically plausible cutoff in the emitted spectrum. Given

CP522, *Cosmic Explosions: Tenth Astrophysical Conference,*
edited by Stephen S. Holt and William W. Zhang

a radio flux and spectral index, the models depend on a single parameter: the frequency at which the spectrum has dropped by a factor of 10 below the power-law extrapolation. If a remnant emits primarily thermal X-rays, *SRCUT* will give the maximum nonthermal-electron energies allowed, so as not to exceed observed X-ray fluxes. *SRCUT* fits to 14 galactic SNRs give electron upper limits of 100 TeV or less, well below the 1000 TeV "knee" in the cosmic-ray (ion) spectrum. [2]

- **SRESC** In some remnants the energy of the most energetic electrons will be limited because the shock cannot effectively scatter particles above a certain gyroradius. The *SRESC* model describes Sedov expansion into a uniform magnetic field, appropriate for Type Ia and late core-collapse SNR. Details are discussed in Reynolds (1998). [3]

These models are currently available and will be distributed in the next release of XSPEC. The XSPEC model formats should also be compatible with CXC, the Chandra software. More sophisticated models, as described in Reynolds (1998) [3], will be released in the future.

THE ESCAPE MODEL

The electron scattering required for acceleration to high enough energies to produce synchrotron radiation may become less efficient above some energy. Or, in the limiting case, electrons above some energy may freely escape the SNR. In the escape model we assume that magnetohydrodynamic scattering waves are much weaker above some wavelength λ_{max}. Since electrons with gyroradius r_g scatter resonantly with waves of wavelengths $\lambda = 2\pi r_g$, electrons will escape once their energy reaches E_{max} given by:

$$E_{max} = \lambda_{max} B_1 / 4 \qquad (1)$$

where B_1 is the upstream magnetic field strength.

Reynolds (1998) produced a detailed model for SNR emission with electron energies limited by escape, which includes correct accounting for variation of shock-acceleration efficiency and post-shock radiative and adiabatic losses, assuming Sedov dynamics.

The model has three parameters: 1) the radio flux measurement at 1 GHz, 2) α, the radio spectral index (flux density $\propto \nu^{-\alpha}$), 3) a characteristic rolloff frequency, the frequency at which the spectrum has dropped by approximately 10 below a straight powerlaw. The spectral index and 1 GHz flux for SNR are fixed by observations. For Galactic SNRs, they can be found at Green's website. [4]

SN 1006 AD

SN 1006 AD is an example of a young remnant where the synchrotron emission dominates in the X-ray regime [1]. Early observations, with low signal to noise, could be adequately fit by curved synchrotron models [5]. Observations from RXTE are a 10-fold improvement in sensitivity and provide a more sensitive test.

ASCA and RXTE observations of SN 1006 AD can be adequately fit with two components: the escape model and a thermal plane-parallel nonequilibrium-ionization shock model (*VSHOCK*) with variable abundances (Borkowski et al., in preparation). The model fit depends relatively weakly on the flux at 1 GHz. For SN 1006 AD the flux was fixed at the observed value of 19 Jy [4]. The column density was fixed at 5×10^{20} cm^{-2}, a value consistent with optical and ROSAT PSPC observations. The best fit, shown in Figure 1, has a α=0.58 and the rolloff frequency of 1.7×10^{17} Hz with a χ^2 of 665 for 376 degrees of freedom. The *VSHOCK* obtains abundances (relative to solar) of O 0.27, Ne 0.16, Mg 1.56, Si 6.43 and S 8.16, compared in Figure 2 to predictions for Type Ia supernova by Nomoto et al.(1984). [6] From the synchrotron formula, $\nu_m \propto E^2 B$, and equation 1, assuming a magnetic field of 3μG, we calculate a λ_{max} of 1.2×10^{17} cm.

X-ray emission from SN1006 AD is well fit by synchrotron models with electron energies limited by escape. Comparing the escape model to future models that

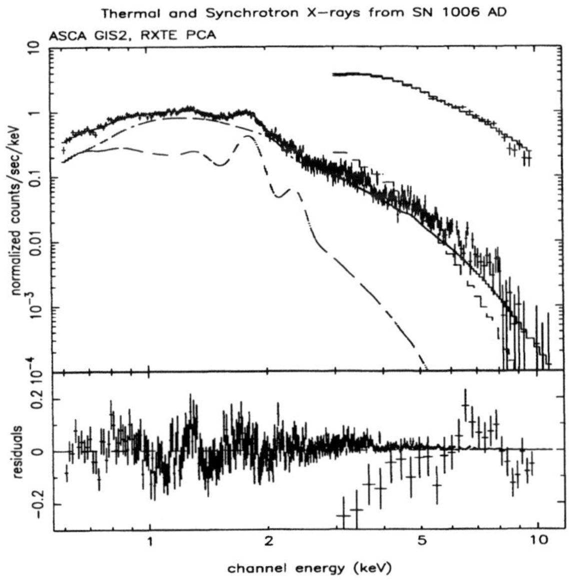

FIGURE 1. X-ray observations of SN 1006 by RXTE PCA (upper) and ASCA GIS2 (lower) fit with *SRESC* and *VSHOCK*. The separate contributions of the *VSHOCK* and *SRESC* models are shown (broken lines) below the sum (solid lines).

are limited by radiation losses or SNR age will give us a better understanding of synchrotron emission at X-ray energies. Our future work involves applying these models to SNRs and making them widely available to the community by distribution through XSPEC.

Thanks to G. Allen for sharing RXTE X-ray observations of SN1006 AD and J. Keohane for research notes and advice. This research is supported by NASA grant NAG5-7153 and NGT5-65 through the Graduate Student Researchers Program.

REFERENCES

1. Koyama, K., Petre, R., Gotthelf, E.V., Hwang, U., Matsura, M., Ozaki, M. & Holt, S.S. *Nature*, **378**, 255. (1995)
2. Reynolds, S.P., Keohane, J.W. *ApJ* **525**, 368 (1999)
3. Reynolds, S.P., *ApJ* **493**, 357 (1998)
4. Green D.A.,'A Catalogue of Galactic Supernova Remnants (1998 September version)', Mullard Radio Astronomy Observatory, Cambridge, United Kingdom (available on the World-Wide-Web at "http://www.mrao.cam.ac.uk/surveys/snrs/"). (1998)
5. Reynolds, S.P., *ApJL* **459**, L13 (1996)
6. Nomoto, K., Thielemann, F-K, Yokoi, K., *ApJ* **286**, 644 (1996)

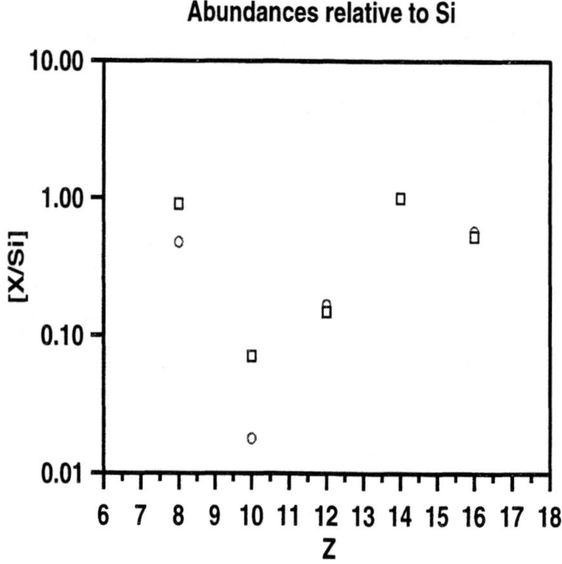

FIGURE 2. Abundances (by mass, normalized to silicon) from a *VSHOCK + SRESC* fit (circles) compared to theoretical predictions (squares) from Nomoto, Thielemann & Yokoi (1984) for a Type I supernova model (W7).

Constraining SN Ia Models Using X-ray Spectra of Clusters of Galaxies

Renato A. Dupke[1] and Raymond E. White III[2,3]

[1]*Dept. of Astronomy,University of Michigan, Ann Arbor, MI 48109-1090*
[2]*Code 662 NASA/GSFC Greenbelt, MD 20771*
[3]*Dept. of Physics & Astronomy,University of Alabama, Tuscaloosa, AL 35487*

Abstract. We present constraints on theoretical models of Type Ia SNe using spatially resolved ASCA X-ray spectroscopy of four galaxy clusters: Abell 496, Abell 2199, Abell 3571 & Perseus. All four clusters have central Fe abundance enhancements and an ensemble of abundance ratios are used to show that most of the Fe in the central regions of the clusters comes from SN Ia. At the center of each cluster, simultaneous analysis of spectra from all ASCA instruments shows that the Ni to Fe abundance ratio (normalized by the solar ratio) is $\approx$ 4. We use the Ni/Fe ratio as a discriminator between SN Ia models: the Ni/Fe ratio of ejecta from the Convective Deflagration model W7 is consistent with the observations, while those of delayed detonation models are not consistent at the 90% confidence level.

INTRODUCTION

Type Ia SNe are thought to be generated by thermonuclear explosions of carbon-oxygen white dwarfs undergoing accretion in stellar binary systems[1]. The nature of the progenitor binary systems, the masses of the white dwarfs and the explosion mechanism(s) (e.g. convective deflagration, delayed-detonation, etc.) for SN Ia are still open questions [2,3]. The explosion of a Ch mass C-O white dwarf is thought to be initiated by carbon ignition at the center, followed by a subsonic nuclear flame (deflagration wave) propagating outwards. In the classical W7deflagration model [4], the propagation speed of the flame front is relatively high ($\approx$ 15-25% of the c_s), but remains subsonic. In delayed detonation models(WDD), the flame speed is initially much lower ($\approx$ 1-3% c_s), but rises to become supersonic [5,6].

The existence of heavy elements in the intracluster medium (ICM) indicates that part of the ICM is not primordial, i.e. was processed in stars and injected into the ICM. The inferred ICM Fe abundances are typically in the range 0.3-0.4 Solar. X-ray spectroscopy of the ICM has indicated a dominance of SN II [7,8]. However, theoretical uncertainties in the elemental yields from SN Ia & II allow ASCA spectroscopy to also be interpreted as showing that as much as 50% of the Fe in clusters comes from SN Ia [9,10,11] and may rise to $\approx$ 70% in the vicinity

CP522, *Cosmic Explosions: Tenth Astrophysical Conference,*
edited by Stephen S. Holt and William W. Zhang
© 2000 American Institute of Physics 1-56396-943-2/00/$17.00

of its central cD galaxy [12,13,14], which may be due to a secondary SN Ia-driven wind (post-protogalactic wind), partially suppressed in the vicinity of the cD.

Despite the general agreement on the predicted elemental mass yields among different SN Ia models, W7 & WDD models predict significantly different mass yields for some elements (e.g. Ni). Accurately determining the elemental yields from SN Ia is crucial to determining the relative contribution of different SNe types to the metal enrichment of galaxies and ICM. Thus, determining which enrichment mechanism(s) were most dominant in contaminating the ICM (e.g. protogalactic, ram pressure stripping, SN Ia-driven winds). Furthermore, by comparing the observed abundance ratio values to those predicted by different SN Ia models, we can constraint theoretical SN Ia models .

We analyze ASCA X-ray spectra of the central regions of 4 galaxy clusters: A496, A2199, A3571 & Perseus. The central regions of these cluster are shown to be dominated by SN Ia ejecta ($\approx$ 65-80%). The abundance measurement uncertainties allow to discriminate between competing SN Ia models. We use an *ensemble* of different elemental abundance ratios to determine the SN Ia Fe mass fraction. This allows us to check for self-consistency when comparing different SN Ia models.

SN Ia Fe MASS FRACTION & DELAYED DETONATION MODELS

Radial distributions of global abundances are shown in Fig. 1a. All clusters show mild, but significant, central abundance enhancements. The best-determined abundances in all four clusters are those of Fe, Si & Ni. O, Ne & S abundances are not well constrained in Abell 3571. Ni abundances are about twice solar in all four clusters. To estimate the SN Ia/II mass fraction, we compare various observed abundance ratios to the theoretical predictions of specific models for SN Ia & II. For SN Ia we initially adopt the updated W7 model of Nomoto et al. [15], while for SN II we use the calculations of Nomoto et al. [15,16], who adopt a Salpeter initial mass function over a SN II progenitor mass range of 10-50 $M_\odot$.

Dupke & White [12] and Dupke & Arnaud [14], using different theoretical models for SN II yields, found that the theoretical S yield would have to be reduced by a factor of 3 and that the theoretical Ne yields would have to be increased by a factor of 2.7 to be consistent with results for most other elemental ratios. Therefore, to be conservative, we avoid including abundance ratios involving these two elements in the calculation of SN Ia Fe mass fraction. We use only abundance ratios involving the better constrained (observationally and theoretically) abundances of Fe, Si, Ni & O. The SN Ia Fe mass fraction derived from abundance ratios involving the above elements is listed in Table 1 for A496 and Perseus, where weighted averages are denoted WA. Errors are propagated 90% confidence errors for Perseus and 1-σ for A496. Dashed missing values indicate that the abundance ratio values is outside the boundaries between SN Ia & II. The results for A3571 and A2199 are similar.

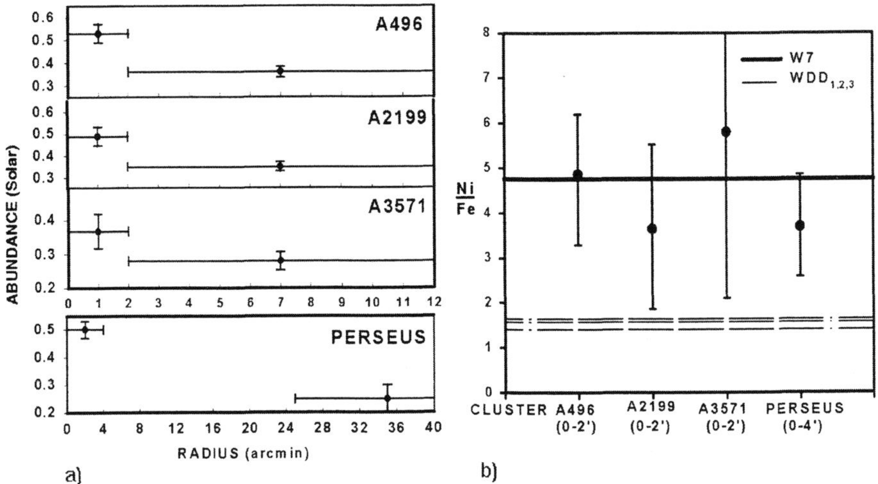

FIGURE 1. (a) - Radial Abundance Distributions. The central bin of all four clusters are spatially similar ($\approx 100h_{50}^{-1}$ kpc). Errors are 90 % confidence. (b) - Ni/Fe Abundance Ratio for the Central Clusters Regions. Predictions for SN Ia enrichment based on the W7 (solid thick) and 3 different WDD (dotted thin) models. SN II models predict Ni/Fe ≈ 1.9.

As seen in Table 1, a variety of abundance ratios provide mutually consistent estimates of the central SN Ia Fe mass fraction. This mutual consistency encourages us to test alternative theoretical models for SNe Ia. In particular, we compare the W7 model adopted above to delayed detonation models. We consider the delayed detonation models of Nomoto et al. [15] , who calculate yields for a sequence of models distinguished by a variety of densities ahead of the deflagration front, which modulates the onset of detonation. The results are also shown in Table 1. We find that the elemental yields of their models WDD1, WDD2 & WDD3 do not provide the same consistency as found with model W7 in estimating the SN Ia/II fraction.

It can be seen that W7 model provides more satisfactory results in three different ways: 1) more abundance ratios fall within the theoretical bounds of SN Ia & II, as they should, when W7 is adopted; 2) the dispersion in the average of the (remaining) individual SN Ia Fe mass fraction estimates tends to be smaller for the W7 model; 3) the mutual consistency of the SN Ia Fe mass fraction estimates is best for W7 model.

Ni is the most discriminatory element between WDD & W7 (displaying a large range of variation of mass yields). The Ni abundance is reasonably well constrained in the X-ray spectra of clusters with ASCA. Fe has the most precisely determined abundance. Therefore, we use the Ni/Fe ratio as a SN Ia model discriminator. Fig. 1b compares the Ni/Fe ratio observed in the four clusters analyzed in this work to the Ni/Fe ratio predicted by the W7 deflagration model, as well as from the three delayed detonation models described at Nomoto et al. [15]. The observed Ni/Fe

Cluster	Element Ratio	SN Ia Iron Mass Fraction			
		W7	WDD1	WDD2	WDD3
A 496	O/Fe	$0.77^{+0.13}_{-0.13}$	$0.82^{+0.11}_{-0.10}$	$0.77^{+0.12}_{-0.11}$	$0.74^{+0.12}_{-0.12}$
"	Si/Fe	$0.65^{+0.07}_{-0.08}$	≥ 0.96	$0.79^{+0.08}_{-0.09}$	$0.65^{+0.08}_{-0.08}$
"	Ni/Fe	$1.00^{+0.00}_{-0.27}$	—	—	—
"	O/Si	$0.86^{+0.10}_{-0.86}$	$0.72^{+0.17}_{-0.72}$	$0.77^{+0.15}_{-0.77}$	$0.81^{+0.12}_{-0.81}$
"	Si/Ni	$0.74^{+0.07}_{-0.09}$	—	—	—
"	WA	0.73 ± 0.05	$0.82^{+0.11}_{-0.10}$	0.78 ± 0.07	0.68 ± 0.07
Perseus	O/Fe	$0.67^{+0.24}_{-0.27}$	$0.73^{+0.20}_{-0.26}$	$0.67^{+0.24}_{-0.26}$	$0.63^{+0.26}_{-0.27}$
"	Si/Fe	$0.70^{+0.10}_{-0.10}$	≥ 0.98	$0.84^{+0.11}_{-0.12}$	$0.69^{+0.11}_{-0.11}$
"	Ni/Fe	$0.66^{+0.34}_{-0.35}$	—	—	—
"	O/Si	$0.59^{+0.38}_{-0.59}$	$0.37^{+0.55}_{-0.37}$	$0.43^{+0.50}_{-0.43}$	$0.50^{+0.45}_{-0.45}$
"	Si/Ni	$0.69^{+0.12}_{-0.17}$	—	—	≥ 0.92
"	WA	0.69 ± 0.07	0.66 ± 0.21	0.79 ± 0.11	0.67 ± 0.10

ratios are inconsistent with any of the delayed detonation models at better than 90% confidence. The observed Ni/Fe ratio is fully consistent with the W7 model.

ACKNOWLEDGMENTS. This work was partially supported by the NSF and the State of Alabama through EPSCoR grant EHR-9108761. REW also acknowledges partial support from NASA grant NAG 5-2574 and a National Research Council Senior Research Associateship at NASA GSFC. RAD acknowledges partial support from NASA grant NAG 5-3247. This research made use of the HEASARC ASCA database and NED.

REFERENCES

1. Hoyle, F., & Fowler, W. A. *ApJ* **132**, 565 (1960).
2. Branch, D. *ARAA* **36**, 17 (1998).
3. Nomoto, K., Iwamoto, K., & Kishimoto, N. *SCIENCE* **276**, 1378 (1997).
4. Nomoto, K., Thielemann, F.-K., & Yokoi, K., *ApJ* **286**, 644 (1984).
5. Arnett, W. D. & Livne, E. *ApJ* **315**, 565 (1994a).
6. Arnett, W. D. & Livne, E. *ApJ* **330**, 565 (1994b).
7. Mushotzky, R. F., & Loewenstein, M. *ApJL* **481**, L63 (1997).
8. Mushotzky, R. F., et al. *PASJ* **49**, 1 (1997).
9. Ishimaru, Y., & Arimoto, N. *PASJ* **50**, 187 (1998).
10. Fukazawa, Y., et al. *ApJ* **132**, 565 (1960).
11. Nagataki, S., & Sato, K., *ApJ* **504**, 629 (1998).
12. Dupke, R. A., & White, R. E. III *ApJ* **astro-ph/9902112**, submitted (1999).
13. Dupke, R. A., & White, R. E. III *ApJ* **astro-ph/9907343**, inpress (2000).
14. Dupke, R. A., & Arnaud, K. A. *ApJ* submitted (1999).
15. Nomoto, K., et al. *Nuclear Physics A* **A621**, 467c (1997b).
16. Nomoto, K., et al. *Nuclear Physics A* **A616**, 79 (1997a).

The Lick Observatory Supernova Search

W. D. Li, A. V. Filippenko, R. R. Treffers, A. Friedman,
E. Halderson, R. A. Johnson, J. Y. King, M. Modjaz, M. Papenkova,
Y. Sato, and T. Shefler

Department of Astronomy, University of California, Berkeley, CA 94720-3411 USA
email: (wli, alex, rtreffers)@astro.berkeley.edu

Abstract. We report here the current status of the Lick Observatory Supernova Search (LOSS) with the Katzman Automatic Imaging Telescope (KAIT). The progress on both the hardware and the software of the system is described, and we present a list of recent discoveries. LOSS is the world's most successful search engine for nearby supernovae.

INTRODUCTION

Located at Lick Observatory atop Mount Hamilton east of San Jose, California, the 0.75-m Katzman Automatic Imaging Telescope (KAIT) is a robotic telescope dedicated to the Lick Observatory Supernova Search (LOSS) and the monitoring of variable celestial objects. It is equipped with a CCD camera and an automatic autoguider (that is, the autoguider is able to find its own guide stars).

KAIT is the third robotic telescope in the Berkeley Automatic Imaging Telescope (BAIT) program. The predecessors to KAIT were two telescopes developed at the Leuschner Observatory, which is located about 10 miles east of the campus of the University of California, Berkeley. KAIT inherits the operational concept and the majority of the software from its two predecessors. More thorough descriptions of the BAIT system can be found in references 1–4.

LOSS discovered its first supernova in 1997 (SN 1997bs in NGC 3627; Treffers et al. 1997 [5]). Its performance improved dramatically in 1998 and 19 supernovae (SNe) were discovered. In 1999, 35 SNe were discovered by mid-December.

Multicolor photometry of SNe is an important scientific goal of KAIT. Because of the early discoveries of most of the LOSS SNe, many good light curves have been obtained.

We report our hardware and software setups for LOSS in Section 2, the SN search in Section 3, and the discoveries and follow-up observations in Section 4.

CP522, *Cosmic Explosions: Tenth Astrophysical Conference,*
edited by Stephen S. Holt and William W. Zhang
© 2000 American Institute of Physics 1-56396-943-2/00/$17.00

THE HARDWARE AND SOFTWARE OF LOSS

KAIT has a 30-inch diameter primary with a Ritchey-Chretién mirror set. The focal ratio is $f/8.2$ which results in a plate scale of $33.2''$ mm^{-1} at the focal plane. The telescope has a very compact design; it is lightweight and slews fast. An off-axis guider designed by one of us (RRT) enables the telescope to obtain long exposures.

The CCD camera is an Apogee AP7 with a SITe 512×512 pixel back-illuminated chip. It is thermoelectrically cooled to about 60°C below the ambient temperature. The quantum efficiency (QE) is good (peak 60%) and flat from 3000 Å to 8000 Å. The field of view is $6.7' \times 6.7'$ with a scale of $0.8''$ pixel^{-1}.

Observations done by KAIT are fully robotic. All hardware (telescope, filters, autoguider, CCD camera, slit, dome, weather station, etc.) are automatically controlled by the software (see Richmond, Treffers, and Filippenko 1993 for details). Dark, bias, and twilight flatfield observations are also done automatically. A focusing routine finds a good focus for the telescope in twilight, then runs every 90 minutes during the night. The images are automatically transferred to the U.C. Berkeley campus to be processed. The appropriate template image is subtracted from each galaxy image, and new objects are detected in the resulting images. The most promising candidates are reobserved during the same night, while others require human evaluation before rescheduling.

THE SUPERNOVA SEARCH

The LOSS galaxy sample includes about 5,000 nearby galaxies. Mosaic images are taken for some large, nearby galaxies. An automatic scheduler selects targets to be observed during the night according to their observation history. Follow-up observations of SNe or routine monitoring of some other objects (active galactic nuclei, variable stars, etc.) are also scheduled at the same time.

We have optimized the system in every possible way to increase the observation efficiency. The search images are taken through a hole in the filter wheel (i.e., no filter is used). This greatly increases the observation efficiency compared to observations through an R-band filter. The exposure time for the search images is only 25 seconds, but because of the high QE of the CCD camera we still reach a limiting magnitude of ~ 19 (sometimes deeper). The order of galaxies to be observed is optimized so as to minimize the accumulated movement of the telescope and dome during the night. Currently the observing efficiency is about 75 images per hour. KAIT can obtain more than 1,000 images during a winter night.

Because of our high observing efficiency, all the sample galaxies are observed every 3 to 5 days in periods of good weather. This ensures that most of the LOSS SNe are discovered considerably before their maxima. Follow-up observations of SNe are usually done starting the night after their discovery. A detailed logging system is also designed to keep track of the observation history of every galaxy, which is very useful for statistical studies (e.g., SN rates).

THE LOSS DISCOVERIES IN 1998 AND 1999

1998 discoveries:
> Supernovae : 19
> Novae   : 4
> Dwarf novae : 2
> Comets   : 1

1999 discoveries (through mid-December):
> Supernovae : 35
> Novae   : 7
> Dwarf novae : 2
> Comets   : 1

For a detailed list of the LOSS discoveries, please visit the LOSS Web page at http://astron.berkeley.edu/~bait/kait.html.

Multicolor photometric observations of SNe are always emphasized in LOSS. Our goal is to build up a multicolor database for nearby SNe. So far light curves have been obtained for 11 SNe in 1998 and 12 SNe in 1999. Examples of the LOSS discoveries and their light curves are presented in Figure 1.

Our supernova research at UC Berkeley is supported by NSF grant AST-9417213 and NASA grant GO-7434.

REFERENCES

1. Richmond, M. W., Treffers, R. R., and Filippenko, A. V., "Progress Report on the Berkeley Automatic Imaging Telescope," in *Robotic Telescopes in the 1990s*, ed. A. V. Filippenko (San Francisco: ASP Conf. Ser. 34, 1992), p. 105.

2. Treffers, R. R., Richmond, M. W., and Filippenko, A. V., "Technical Description of the Berkeley Automatic Imaging Telescope," in *Robotic Telescopes in the 1990s*, ed. A. V. Filippenko (San Francisco: ASP Conf. Ser. 34, 1992), p. 115.

3. Richmond, M. W., Treffers, R. R., and Filippenko, A. V., *P.A.S.P.* **105**, 1164 (1993).

4. Treffers, R. R., Filippenko, A. V., Van Dyk, S. D., Paik, Y., and Richmond, M. W., "The Berkeley Automatic Imaging Telescope: An Update," in *Robotic Telescopes*, ed. G. W. Henry and J. A. Eaton (San Francisco: ASP Conf. Ser. 79, 1995), p. 86.

5. Treffers, R. R., Peng, C. Y., Filippenko, A. V., and Richmond, M. W., *IAU Circ.* No. **6627** (1997).

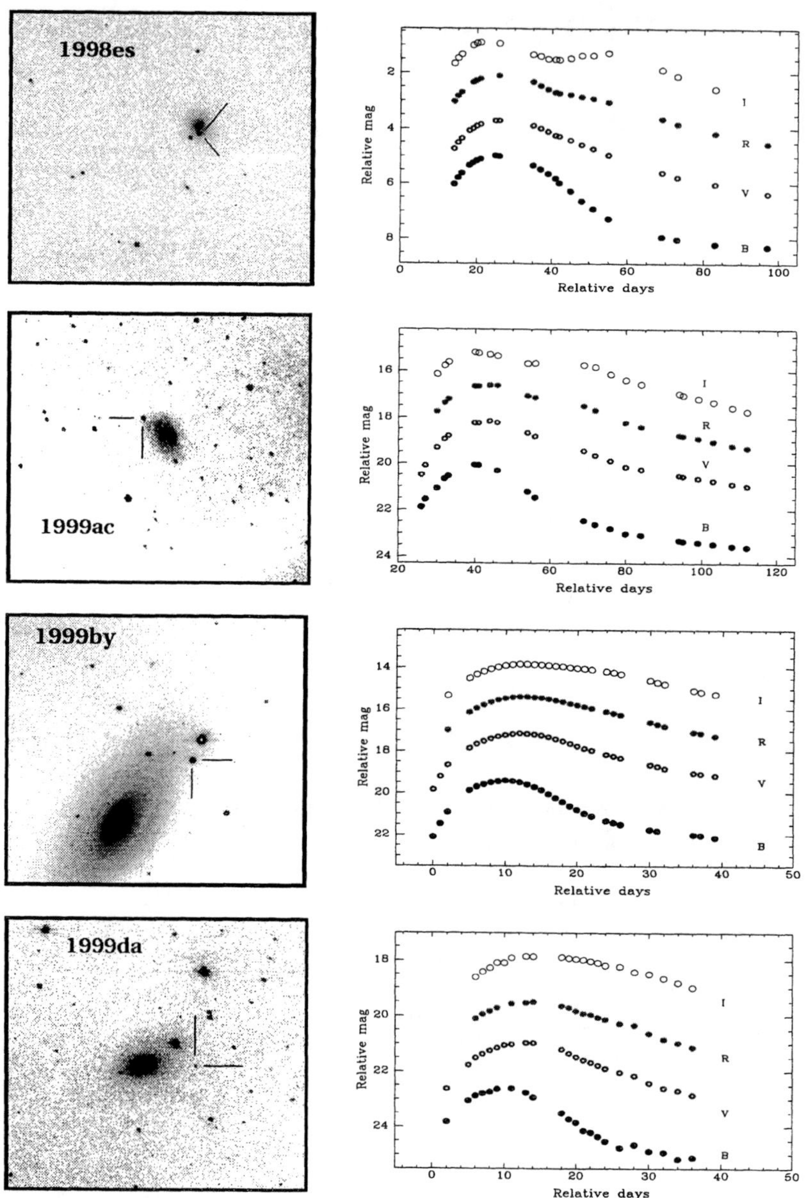

FIGURE 1. Some of the LOSS SN discoveries and their light curves. Examples shown (top to bottom) are the type Ia SNe 1998es, 1999ac, 1999by, and 1999da. The light curves have been arbitrarily shifted up and down for display purposes.

Supernova Rates in Abell Galaxy Clusters and Implications for Metallicity

Avishay Gal-Yam[1] and Dan Maoz[1]

[1]*School of Physics and Astronomy and Wise Observatory, Tel Aviv University, Tel Aviv 69978, Israel*

Abstract. Supernovae (SNe) play a critical role in the metal enrichment of the intracluster medium (ICM) in galaxy clusters. Not only are SNe the main source for metals, but they may also supply the energy to eject enriched gas from galaxies by winds. However, measurements of SN rates in galaxy clusters have not been published to date. We have initiated a program to find SNe in 163 medium-redshift ($0.06 < z < 0.2$) Abell clusters, using the Wise Observatory 1m telescope. We report here our on first results and describe our main scientific goals. Following the discovery of a SN in the remote periphery (78 Kpc) of the cD galaxy of Abell 403, we discuss a novel explanation for the centrally enhanced metal abundances indicated by X-ray observations of galaxy clusters.

THE PROJECT AND MAIN SCIENTIFIC OBJECTIVES

We have used the Wise Observatory 1m telescope to monitor monthly a sample of 163 rich (richness class $R > 0$) Abell galaxy clusters, with medium redshift ($0.06 < z < 0.2$) northern declination ($\delta > 0$) and small angular size ($r < 20'$). We have also observed "blank" flanking fields for a sub-sample of the clusters. These will be used to study the cluster vs. field SN rates, and to estimate the luminosity contributed by the cluster in each field. We have used unfiltered ("clear") observations to achieve maximum sensitivity, and have a characteristic limiting magnitude of $R \sim 22$. Variable objects are discovered by image subtraction. New subtraction methods have been developed for use in this project (see fig. 1).

Our main scientific goal is to derive from our data the SN rate as a function of various parameters such as host galaxy type and cluster environment: position within the cluster, cluster richness and cluster vs. field. SN rates can then be used to determine the current and past star formation rates in galaxy clusters [1]. Our measured SN rates can replace the assumed rates used so far in studies of metal abundances in the intracluster gas. We also intend to study the rate, distribution and properties of intergalactic SNe in galaxy clusters. A candidate intergalactic SN we have discovered, SN 1998fc (see fig. 3), will be discussed below. Our search is

CP522, *Cosmic Explosions: Tenth Astrophysical Conference,*
edited by Stephen S. Holt and William W. Zhang
© 2000 American Institute of Physics 1-56396-943-2/00/$17.00

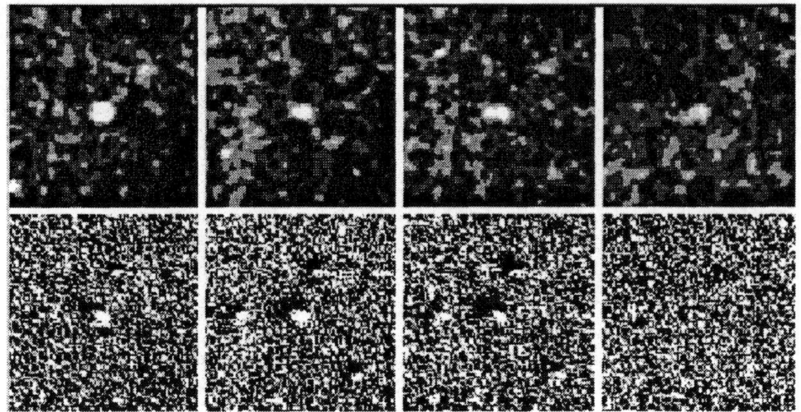

FIGURE 1. Difference images of SN1999ay 1,3,4 and 5 months after discovery produced with our new image subtraction method (upper panel) and with the standard one (lower panel). Note that with the new method this SN is still detectable after 5 months (upper right)

also sensitive to other optical transients, such as AGNs in the clusters and behind them, flares from tidal disruption of stars by dormant massive black holes in galactic nuclei and GRB afterglows. We may also detect the gravitational lensing effect of the clusters on background SNe [2].

FIRST RESULTS

Our program has already discovered 11 spectroscopically confirmed SNe at $z = 0.1 - 0.24$, (see table 1 and fig. 2) and several unconfirmed SNe. We have also detected variable stellar objects (some of which are AGN) and dozens of asteroids.

TABLE 1. Spectroscopically confirmed SNe discovered at the Wise Observatory

SN	Cluster	Cluster z	Type	SN z	Discovered	R mag	spectrum
1998cg	Abell 1514	0.199	Ia	0.12	1.5.98	18.5	ESO 3.6m
1998eu	Abell 125	0.188	Ia	0.181	14,11.98	19.7	AAT 4m
1998fc	Abell 403	0.103	Ia	0.10	20.12.98	20.5	ESO 3.6m
1998fd	Field		Ia	0.24	24.12.98	21.3	Keck II 10m
1999C	Abell 914	0.195	Ia	0.125	14.1.99	19.6	Keck II 10m
1999ax	Abell 1852	0.181	Ia	0.09	20.3.99	18.5	KPNO 4m
1999ay	Abell 1966	0.151	II?	0.15	21.3.99	18.0	KPNO 4m
1999cg	Abell 1607	0.136	Ia	0.14	15.4.99	19.2	Keck II 10m
1999ch	Abell 2235	0.151	Ia	0.15	13.5.99	19.8	KPNO 4m
1999ci	Abell 1984	0.124	Ia	0.12	15.5.99	19.3	KPNO 4m
1999ct	Abell 1697	0.183	Ia	0.18	13.6.99	20.8	Keck II 10m

FIGURE 2. A sample of SN images discovered at the Wise Observatory:
(From left to right and top to bottom) SN 1998cg, 1998eu, 1998fc, 1998fd, 1999C, 1999ax, 1999ay,
1999cg, 1999ch, 1999ci

INTERGALACTIC SNE AND ENHANCED CENTRAL METAL ABUNDANCES IN CLUSTERS

The existence of a diffuse population of intergalactic stars is supported by a growing body of observational evidence such as intergalactic planetary nebulae in the Fornax and Virgo clusters [3] [4] [5] [6], and intergalactic red giant stars in Virgo [7]. Recent imaging of the Coma cluster reveals low surface brightness emission from a diffuse population of stars [8], the origin of which is attributed to galaxy disruption [9] [10]. Since type Ia SNe are known to occur in all environments, there is no obvious reason to assume that such events do not happen within the intergalactic stellar population. SN 1998fc may be such an event. The intergalactic stellar population is centrally distributed [11]. Therefore, metals produced by intergalactic Ia SNe can provide an elegant explanation for the central enhancement of metal abundances with type Ia characteristics, recently detected in galaxy clusters [12].

SN 1998fc - An intergalactic SN candidate in Abell 403

SN 1998fc was detected near the cD galaxy of Abell 403 [13], and was spectroscopically confirmed as a type Ia SN at the cluster redshift [14] [15]. The most likely host for this SN, the cD galaxy, is very distant - at least 78 Kpc away. This may be an intergalactic SN whose progenitor star was a member of the diffuse intergalactic stellar population. Alternatively, the host may be a faint dwarf galaxy. The distribution of "hostless" SNe is expected to be different if the progenitors are members of the intergalactic population, centered near the cluster core [11], or members of dwarf galaxies, more abundant in the outskirts of galaxy clusters [16]. Therefore,

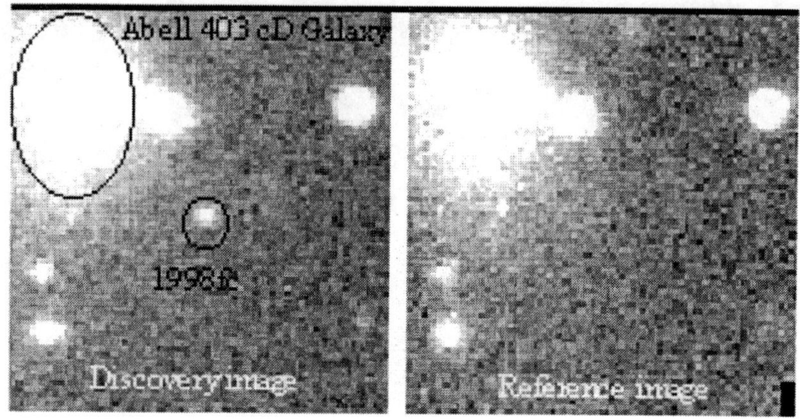

FIGURE 3. SN 1998fc - An intergalactic SN candidate in Abell 403.

the nature of such objects could be resolved with larger number statistics. In any event, the number of SNe with undetected hosts relative to the total number of cluster SNe can put an upper limit on the intergalactic stellar fraction.

REFERENCES

1. Madau, P., DellaValle, M., & Panagia, N. 1998, MNRAS, 297, L17
2. Kolatt, T., & Bartelmann, M. 1998, MNRAS, 296, 763
3. Theuns, T., & Warren, S.J. 1996, MNRAS, 284, L11
4. Arnaboldi, M. et al. 1996, ApJ, 472, 145
5. Ciardullo, R. et al. 1998, ApJ, 492, 62
6. Freeman, K. C. et al. 1999, astro-ph/9910057
7. Ferguson, H.C., Tanvir, N.R., & von Hippel, T. 1998, Nature, 391, 461
8. Gregg, M.D. & West, M.J., 1998, Nature, 396, 549
9. Dubinski, J., Mihos, J.C. & Hernquist, L., 1996, ApJ, 462, 576
10. Moore, B., et al. 1996, Nature, 379, 613
11. Dubinski, J., 1999, astro-ph/9902331
12. Dupke, R.A. & White, R.E.III, 1999, ApJ, submitted, astro-ph/9902112
13. Gal-Yam, A. & Maoz, D., 1999, IAUC 7082
14. Gal-Yam, A. & Maoz, D., 1999, IAUC 7093
15. Filippenko, A.V., et al. 1999, IAUC 7091
16. Phillipps, S., Driver, S., Couch, W. J. & Smith, R. M., 1998, ApJ, 498, 119

Supernovae II

What Powers Core-Collapse Supernovae?

Chris L. Fryer

UCO/Lick Observatory, UC Santa Cruz
Santa Cruz, CA, 95064

Abstract. Supernovae are one of nature's most energetic "cosmic explosions". But do we know how they work? Models based on the neutrino-driven mechanism have become increasingly sophisticated over the past 3 decades, and this mechanism still remains the leading explanation of the engine that drives supernovae. The leading *alternate* mechanism invokes magnetic fields. Unfortunately, energy constraints rule out this mechanism for the bulk of supernovae unless the current stellar progenitors are incorrect. Comparing these two mechanisms provides insight into the engines that power gamma-ray bursts (GRBs). The different geometry in black hole accretion disk GRB models allows many mechanisms which failed in core-collapse supernovae to work for GRBs.

INTRODUCTION

In 1934, Baade & Zwicky [1] proposed that one of nature's most energetic explosions (supernovae) is powered by the gravitational energy released ($\sim 10^{54}$ergs) as a massive stellar core collapses down to a neutron star. To explain the observed supernovae, theorists needed only come up with a mechanism to harness a mere 0.1% of this energy. Figuring out how one might convert this potential energy into kinetic energy has occupied theorists for the past 65 years. In this paper, I review the quest for an efficient energy converter in supernovae, concentrating primarily on the details of the neutrino-driven supernova paradigm. The leading alternative mechanism for supernovae invokes magnetic fields, and I also review the basic assumptions in this mechanism. The engine driving core-collapse supernovae is akin to the engine in many of the favorite gamma-ray burst models, and comparisons between these two engines will also be discussed.

NEUTRINOS AND CORE-COLLAPSE SUPERNOVAE

We start our review of core-collapse supernovae at the initial collapse of a massive star's iron core ($M_{\mathrm{star}} \gtrsim 8\,\mathrm{M_\odot}$). The collapse occurs when the densities and temperatures in the core are sufficiently high to cause both the dissociation of nuclei in the iron core (removing energy from the core) and the capture of electrons onto protons

CP522, *Cosmic Explosions: Tenth Astrophysical Conference,*
edited by Stephen S. Holt and William W. Zhang

(which further reduces the pressure in the core). This sudden decrease in pressure causes the core to collapse nearly at free-fall, and the collapse stops only when nuclear densities are reached and nucleon degeneracy pressure once again stabilizes the core and drives a bounce shock back out through the core. It was believed that this bounce shock (with energy equal to a few times 10^{51}ergs) might drive the supernova explosion [2]. However, in the 1960s, most of the relevant physics in the collapse of a massive star was still poorly understood. And, as often happens in theory, as our understanding of the relevant physics increased, this model became less and less likely. In most modern calculations of core-collapse, the bounce shock stalls at roughly 100-300 km because neutrino emission and iron dissociation sap its energy.

However, the shock itself uses only a small fraction of the total binding energy released as the core collapses to form a neutron star (Binding Energy= $GM_{\mathrm{NS}}^2/R_{\mathrm{NS}} \approx 10^{54}$ergs). Initially, the gravitational potential energy is converted into thermal energy. The matter is too dense for photon radiation to cool the star ($\rho > 10^{14}$gcm^{-3}) and the temperature gets sufficiently high to produce copious neutrino emission ($>10^{53}$ergs s^{-1}). Since the bulk of the potential energy released is being carried away by these neutrinos, theorists logically proposed that supernova explosions could be produced by absorbing a fraction of these neutrinos (only a few percent is necessary [3]). If you are comfortable with hand-waving that 1-3% of the neutrinos emitted in the first few seconds after core bounce are absorbed and can be used to power a supernova explosion, then the core-collapse supernova problem is solved!

Core-collapse theorists were *not* satisfied. Accurately calculating the actual efficiency at which neutrino energy is converted into explosion energy has occupied theorists for the past 3 decades! Over the years, simulations of core-collapse have become increasingly sophisticated, including detailed equations of state required to handle the dense, hot matter in the collapsing core, corrections for the effects of general relativity, and continually revised cross-sections for neutrino emission, scattering and absorption. Although at any point in time in the last 30 years, a single piece of physics has cornered center stage as *the solution* to the supernova problem, it is generally accepted that the problem of core-collapse supernovae is indeed one of details, where any quantitative estimates of explosion energies, remnant masses, etc. seem to depend sensitively on all of our above list of important physics (see [4,5] for reviews).

THE DELAYED NEUTRINO MECHANISM: CONVECTION AND ROTATION

The piece of physics that has played the starring role in the supernova problem over the past decade has been convection. As we discussed above, the bounce shock, which Colgate & Johnson [2] thought would drive a supernova explosion, stalls due to neutrino cooling. However, it leaves behind an entropy profile which is unstable to convection, and sets up a convective region from $\sim$50 km out to $\sim$300 km (Fig. 1)

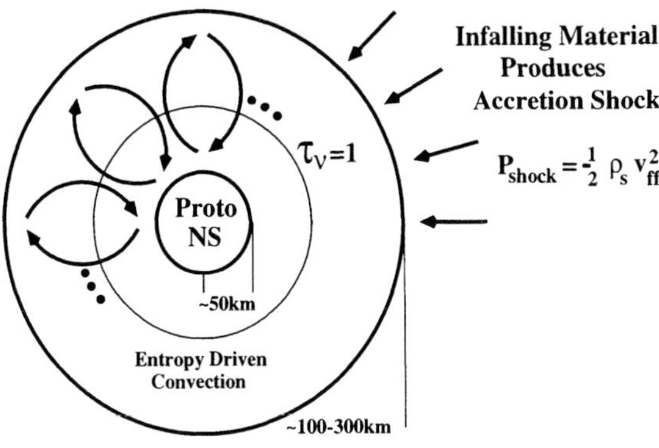

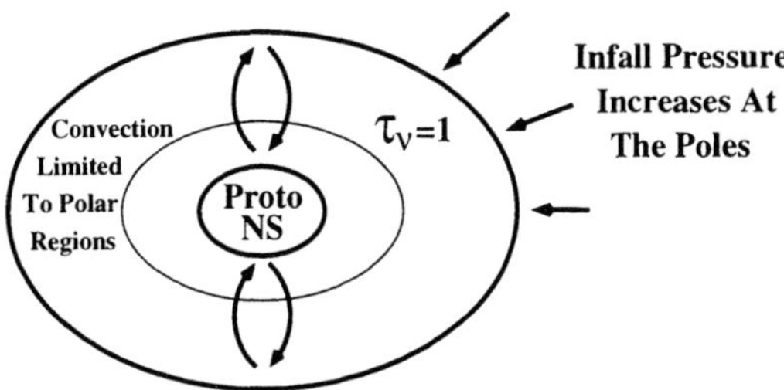

FIGURE 1. Entropy-driven convection has been found to enhance the energy deposition within the accretion shock. As the density of the infalling matter decreases, the convective region is able to push the shock outward, launching a supernova explosion. The angular momentum profile in rotating core-collapses stabilizes the convection along the equator.

capped by the accretion shock of infalling material. The convective region must overcome this cap to launch a supernova explosion.

Neutrino heating deposits considerable energy into the bottom layers of the convective region. If this material were not allowed to convect (which is the case for most of the 1-dimensional simulations), it would then re-emit this energy via neutrinos and it is difficult to make the matter gain much energy. In the meantime, the infall pressure increases as more material piles up at the accretion shock and the neutrino luminosity decreases as the proto-neutron star cools. Both make it increasingly difficult to launch an explosion. In the "delayed-neutrino" supernova mechanism [6–10], convection aids the explosion in two ways: a) as the lower layers of the convective region are heated, that material rises and cools adiabatically and converts the energy from neutrino deposition into kinetic and potential energy rather than re-radiating it as neutrinos, and b) the material does not simply pile up at the shock but instead convects down to the surface of the proto-neutron star

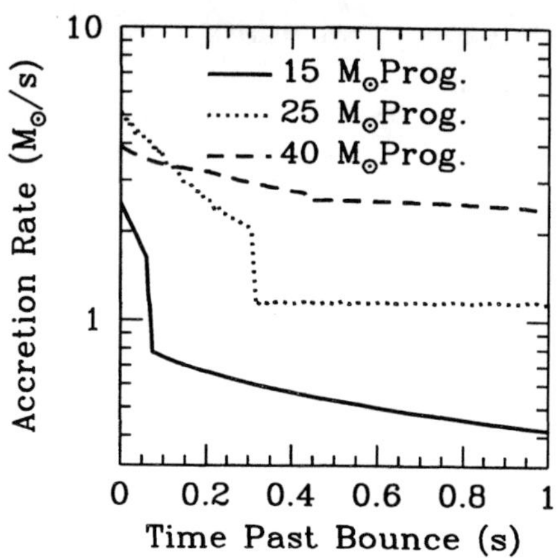

FIGURE 2. Mass infall rates for a 3 separate progenitor masses: $15,25,40\,M_\odot$. The mass infall rate for the $15\,M_\odot$ progenitor drops to 1/5th that of the 25 and $40\,M_\odot$ models in 100 ms. This allows it to explode sooner, leaving behind a smaller core. The infall rates of the 25 and $40\,M_\odot$ progenitors stay roughly the same for 300 ms past bounce, and hence their explosion energies are similar.

where it either accretes onto the proto-neutron star providing additional neutrino emission or is heated and rises back up. Thus, convection both increases the efficiency at which neutrino energy is deposited into the convective region and reduces the energy required to launch an explosion by reducing the pressure at the accretion shock. It appears that nature has conspired to make core-collapse supernovae straddle the line between success and failure, where convection, which for years was thought to be a mere detail, plays a crucial role in the explosion. It is not surprising that the outcome of the core collapse of massive stars depends upon numerical approximations of the physics (e.g.the algorithm for neutrino transport [11,12]).

Because the neutrino driven mechanism is sensitive to the details, it is also not surprising that the explosion energy depends upon the structure of the stellar core just before collapse, and hence, upon the progenitor evolution. Many uncertainties remain in stellar evolution (convection, opacities, mass loss from winds and binary effects). To make accurate estimates of supernova energies, remnant masses, and even kicks, we need to first understand the the progenitor models and their associated uncertainties. However, this strong dependence on the progenitor provides a natural explanation for the range observed in supernova energies. For instance, because the ram pressure at the top of the convective region depends strongly upon the collapsing core (Fig. 2), the structure of the core makes a big difference in the supernova produced by that core. As the initial progenitor mass increases, the ram pressure increases, and the explosion energy decreases [13]. In addition, if the core

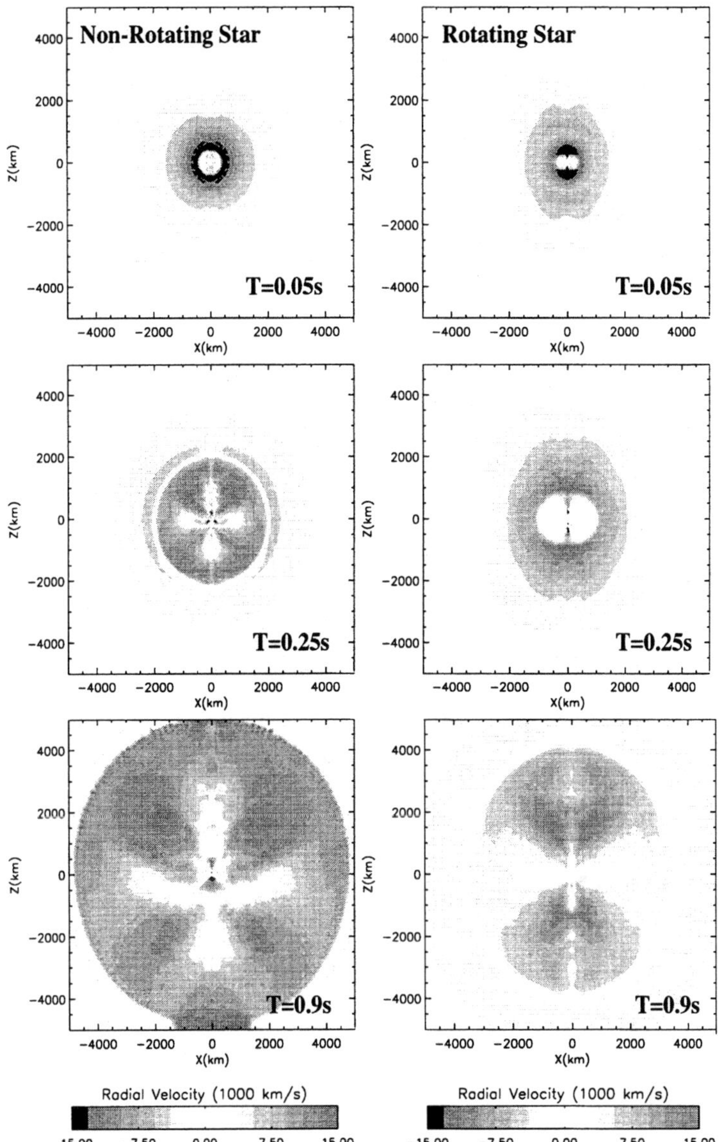

FIGURE 3. Radial velocity distribution of non-rotating and rotating models 0.05, 0.25, 0.5, and 0.9 s after bounce. At 0.9 s, the non-rotating model remains essentially spherical. The asymmetries in the velocities are caused by the buoyant convective bubbles which are driving the explosion. In contrast, the rotating model already shows strong asymmetries in the shock position and velocities.

(and hence the ram pressure) is affected by mass loss, then the explosion energy will also depend on winds (metallicity) and binary effects. Knowing the core structure is not enough to determine core-collapse explosion energies. Massive stars also have a range of spin periods, and rotation alters the supernova explosion energy. Hence, progenitor stars with the same core structure can have a range of explosion energies based on their spin alone.

Rotating stars have now been evolved to collapse [14] and these supernova progenitors have been modeled through bounce and to explosion [15]. Rotation limits the convection in core-collapse supernovae, both by weakening the shock and by constraining the convection to the polar regions. Because of these constraints, the convective region takes longer to overcome the accretion shock and the explosion occurs at later times and is less energetic [15]. However, by constraining the convection to the poles, the resultant supernova is stronger along the poles (Fig. 3). It is possible that these asymmetric explosions may help explain many as-yet unsolved mysteries in supernovae (e.g. polarization, extended mixing, etc.).

Although much more work is necessary to confirm many of these tantalizing results, it appears that the neutrino driven supernova mechanism may work and can explain most of the features in supernova observations.

MAGNETIC FIELDS

Not long after Colgate & White [3] proposed that neutrino absorption could drive supernova explosions, LeBlanc & Wilson [16] proposed that magnetic fields could be used to extract rotational energy to drive an explosion. The two difficulties in this mechanism are a) the energy available is limited to the rotational energy of the core and b) the collapse must generate extremely strong magnetic fields to extract the rotational energy.

To estimate the available rotation energy, we can simply use current rotating progenitors and allow them to collapse. Collapsing the inner $1.4\,M_\odot$ of the Heger et al. [14] rotating supernova progenitor down to a $10\,$km neutron star yields a total rotational energy of $\sim 2\times 10^{52}$ ergs. Thus, to drive a supernova explosion, the magnetic field mechanism need only extract 10% of its available energy resovoir. But this simple calculation overestimates the rotation energy during the first few seconds after bounce. Remember that as the star collapses, it converts a good deal of its potential energy into thermal energy ($\sim 10^{54}$ ergs) and this energy must be dissipated (mostly through neutrinos at a luminosity of 10^{53} ergs s^{-1}) before it can contract down to $10\,$km. In the first few seconds of the collapse, the core collapses down to a radius of roughly 50-100 km. Models of this core [15] yield rotation energies $\sim 8\times 10^{50}$ ergs and magnetic fields would have to extract >200% of the available rotational energy. This sort of efficiency is daunting even to the most ambitious theorists.

There are several options out of this predicament. One is to wait a few seconds for the core to cool and contract, thus increasing the rotational energy of the proto-neutron star. However, material continues to pile up on the proto-neutron star and by the time the core has collapsed enough to produce significant rotation energy, its mass will exceed $2\,M_\odot$, well above most observed neutron star masses. Hence, this is not a viable option when forming neutron star compact remnants. But if one only wants to produce black hole remnants, this mechanism (as we shall see in the next section) can produce strong explosions.

Alternatively, we need only assume that the core is rotating faster. If we raise the spin of the core by a factor of 5, then magnetic fields need only extract 10% of the rotation energy to drive an explosion. However, current progenitor models [14] are considered to be maximally rotating cores. Indeed, it is highly unlikely that any stars exist with spins more than a factor of a few faster than the current models [17]. This solution may work, but only if we assume from the start that the current progenitor models must be incorrect!

Even if such high rotation velocities are possible, we also need to produce a magnetic field strong enough to drive an explosion. But what constitutes "strong enough"? Certainly the magnetic energy density must be large enough to direct the hydrodynamics. Hence, the magnetic field energy must be greater than the energy density in the matter. As a simple estimate, let's assume that the magnetic energy density at the *accretion shock* must be greater than the energy density of the infalling matter. Assuming a dipole magnetic field arising from a 50 km proto-neutron star, the magnetic field energy is given by:

$$\epsilon_{\text{Mag.}} = \frac{B(r)^2}{8\pi} = \frac{B_{50\text{km}}^2}{8\pi}\left(\frac{50\,\text{km}}{R_{\text{Sho.}}}\right)^6 \tag{1}$$

where $B_{50\text{km}}$ is the magnetic field strength at 50 km, and $R_{\text{Sho.}}$ is the shock radius. The corresponding energy density of the infalling material is given by:

$$\epsilon_{\text{Inf.}} = \frac{1}{2}\rho v_{\text{Inf.}}^2 = \frac{(2GM_{\text{PNS}})^{0.5}\dot{M}}{8\pi R_{\text{Sho.}}^{2.5}} \tag{2}$$

where G is the gravitational constant, M_{PNS} is the mass within the shock radius and $\dot{M}$ is the accretion rate. Figure 4 lists the magnetic field ($B_{50\text{km}}$) strengths at which these two energy densities are equal for 3 accretion rates suitable for core-collapse of stars with masses between 15-40 $M_\odot$ as a function of the shock radius. Recall that the magnetic field in Figure 4 is the dipole field strength at 50 km. If the proto-neutron star retains its magnetic moment as it collapses, when these cores contract, their magnetic fields would exceed 10^{18}G! Although the favorite models of a subset of gamma-ray bursts requires magnetic fields as high as 10^{15}G [18], these fields are probably produced as the neutron star cools [19,15]. Again, this is too late to explain the bulk of supernovae which produce neutron stars.

So let's summarize what we currently know about magnetic fields. To use the LeBlanc & Wilson mechanism for all supernovae, we need to assume that stellar cores are rotating much faster than current progenitor models predict, and we have to somehow magically generate extremely strong magnetic fields and then somehow force them to dissipate as the proto-neutron star cools. Then we must also further wave our hands to argue that these two features will drive an explosion. On the other hand, with the current progenitors and current physics, we need only hand-wave a 1-3% efficiency. It is not surprising that most of the detailed work on core-collapse supernovae has concentrated on the neutrino mechanism. However, the behavior of magnetic fields in core-collapse is still poorly known, and these magical fields may well provide a viable supernova mechanism. Until a promising mechanism is proposed, however, it is unlikely any careful calculations using magnetic fields will be attempted. Until then, magnetic fields will remain as hand-wavy as the neutrino models were back in the 60s.

GAMMA-RAY BURSTS

It is interesting to compare the theoretical progress on gamma-ray burst mechanisms with that of supernovae. This is a field which remains strongly driven by the observations. The past decade has seen a flurry of observational "revelations" which have managed to push one class of models to the status of *standard* engine. The observed fact that GRBs are at high redshifts eliminated many models from the huge list (>200) offered by theorists. The current favorite is the class of models powered by dense accretion disks around black holes [20,21]. In these models, the fireball is produced either through neutrino annihilation above the disk (preferentially along the disk angular momentum axis) or through a magnetic field mechanism similar to the mechanism which produces jets in active galactic nuclei (see [20] for a review).

Both mechanisms send out warning signals to core-collapse theorists. Neutrino annihilation (an interaction where a neutrino and anti-neutrino annihilate) was also proposed as the means to convert neutrino energy into explosion energy [22]. Although core-collapse theorists quickly proved that neutrino annihilation is not likely in actual core-collapse models [23], claims that neutrino annihilation is the "solution" for core-collapse supernovae still persist. Knowing the history of core-collapse models, it is not surprising that core-collapse theorists approach neutrino-annihilation and black hole accretion disk GRB models with a degree of caution. However, these accretion disks have several advantages that make neutrino annihilation more plausible than in the core-collapse supernova model. The rate of neutrino annihilation depends sensitively on the neutrino flux density and the incident angle of the colliding neutrinos (it is much higher for head-on collisions). For core-collapse supernovae, by the time neutrino absorption and scattering become low, the neutrino flux density is also low and the neutrinos are traveling nearly radially outwards. Hence, the neutrino annihilation rate is also low. For black hole accretion disks, neutrino annihilation can become a dominate energy deposition source in a region (the disk rotation axis) where the flux density is still high and the neutrino collisions are still nearly head-on. These factors almost make the high

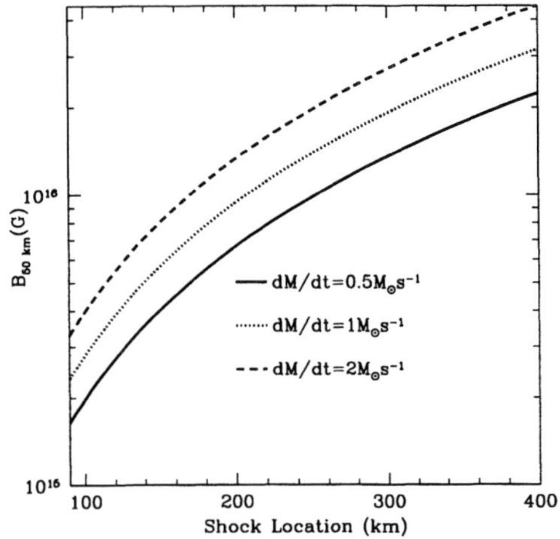

FIGURE 4. Magnetic field strength at which the magnetic energy density equals the infall density as a function of shock radius. The 3 curves correspond to 3 different accretion rates.

predicted energy depositions believable [20].

Magnetic field mechanisms are slightly more difficult to reconcile. Recall that in core-collapse supernovae, magnetic fields seem a hopeless mechanism for explosions because it is unlikely that we can produce strong enough magnetic fields to drive an explosion and, even if we could, there is not sufficient rotation energy available to explain the existing explosions. On the other hand, the beauty of the black hole accretion disk models is that they do have sufficient rotational energy to drive explosions. For instance, if we allow a massive star to collapse and form a $3M_\odot$ black hole with a disk around it, the rotational energy using the existing stellar models will exceed 10^{53}ergs! So the first stumbling block that magnetic fields encounter as explanations of supernovae is avoided in GRBs. However, no self-consistent magnetic field mechanism exists, so it is hard to compare the viability of such a mechanism with that of neutrino annihilation. If we are forced to invoke magnetic fields as the GRB mechanism, we will probably have to hand-wave conversion efficiencies for at least another decade or two.

The increasingly accepted connection between SN 1998bw and GRB 980425 has led to a frenzy of papers connecting gamma-ray bursts and supernovae. For those theorists actively modeling GRBs, this connection was not a surprise. Long duration GRBs are thought to be powered by collapsars or helium mergers [21], both of which should produce supernova-like outbursts. But they make up only a small fraction ($\sim$1%) of all supernovae. It is the formation process and aspects of the black hole accretion disk which make the GRB mechanisms (both neutrino annihilation and magnetic field jets) plausible. The proposed GRB mechanisms would not work in a standard core-collapse supernova to produces a neutron star remnant. It may be true that all gamma-ray bursts may produce supernovae, but it is unlikely that most supernovae produce gamma-ray bursts.

This research was funded by NASA (NAG5-8128) and the US DOE ASCI program (W-7405-ENG-48).

REFERENCES

1. Baade, W., & Zwicky, F. *Phys. Rev.*, 45, 138 (1934).
2. Colgate, S. A., & Johnson, H.J. *Phys. Rev. Letters*, 5, 573 (1960).
3. Colgate, S. A., & White, R.H. *ApJ*, 143, 626 (1966).
4. Burrows, A. On the Systematics of Core-Collapse Explosions, *to be published in the proceedings of the 9'th Workshop on Nuclear Astrophysics, held at the Ringberg Castle, Germany, March 23-29*, ed. E. Müller & W. Hillebrandt (1999).
5. Mezzacappa, A. *to appear in the Memoirs of the Italian Astronomical Society as the Proceedings of Future Directions of Supernova Research: progenitors to remnants, September 29-October 2, 1998, Gran Sasso* (1999).
6. Wilson, J. R., & Mayle, R. W. *Phys. Rep.*, 163, 63 (1988).
7. Miller, D. S., Wilson, J. R., & Mayle, R. W. *ApJ*, 415, 278 (1993).
8. Herant, M., Benz, W., Hix, W. R., Fryer, C. L., Colgate, S. A. *ApJ*, 435, 339 (1994).
9. Burrows, A., Hayes, J., & Fryxell, B. A. *ApJ*, 450, 830 (1995).
10. Janka, H.-T. & Müller, E. *A&A*, 306, 167 (1996).
11. Messer, O. E. B., Mezzacappa, A., Bruenn, S. W., & Guidry, M. W. *ApJ*, 507, 353 (1998).
12. Mezzacappa, A., Calder, A. C., Bruenn, S. W., Blondin, J. M., Guidry, M. W., Strayer, M. R., & Umar, A. S. *ApJ*, 495, 911 (1998).
13. Fryer, C. L. *ApJ*, 522, 413 (1999).
14. Heger, A., Langer, N., & Woosley, S. E. *ApJ*, in press (2000).
15. Fryer, C.L., & Heger, A. *submitted to ApJ* (1999).
16. LeBlanc, J.M., & Wilson, J.R. *ApJ*, 161, 541 (1970).
17. Heger, A. private communication
18. Duncan, R., & Thompson, C. 1992, ApJ, 392, L9
19. Thompson, C., & Duncan, R.C. *ApJ*, 408, 194 (1993).
20. Popham, R., Woosley, S.E., Fryer, C.L., *ApJ*, 518, 356 (1999).
21. Fryer, C.L., Woosley, S.E., & Hartmann, D.H. *ApJ*, 526, 152 (1999).
22. Goodman, J., Dar, A., & Nussinov, S. *ApJ*, 314, L7, (1987).
23. Cooperstein, J., van den Horn, L.J., & Baron, E. *ApJ*, 321, L129 (1987).

Optical Observations of Type II Supernovae

Alexei V. Filippenko

Department of Astronomy, University of California, Berkeley, CA 94720-3411

Abstract. I present an overview of optical observations (mostly spectra) of Type II supernovae. SNe II are defined by the presence of hydrogen, and exhibit a very wide variety of properties. SNe II-L tend to show evidence of late-time interaction with circumstellar material. SNe IIn are distinguished by relatively narrow emission lines with little or no P-Cygni absorption component and (quite often) slowly declining light curves; they probably have unusually dense circumstellar gas with which the ejecta interact. Some SNe IIn, however, might not be genuine SNe, but rather are super-outbursts of luminous blue variables. The progenitors of SNe IIb contain only a low-mass skin of hydrogen; their spectra gradually evolve to resemble those of SNe Ib. Limited spectropolarimetry thus far indicates large asymmetries in the ejecta of SNe IIn, but much smaller ones in SNe II-P. There is intriguing, but still inconclusive, evidence that some peculiar SNe IIn might be associated with gamma-ray bursts. SNe II-P are useful for cosmological distance determinations with the Expanding Photosphere Method, which is independent of the Cepheid distance scale.

INTRODUCTION

Supernovae (SNe) occur in several spectroscopically distinct varieties; see reference [1], for example. Type I SNe are defined by the absence of obvious hydrogen in their optical spectra, except for possible contamination from superposed H II regions. SNe II all prominently exhibit hydrogen in their spectra, yet the strength and profile of the Hα line vary widely among these objects.

The early-time ($t \approx 1$ week past maximum brightness) spectra of SNe are illustrated in Figure 1. [Unless otherwise noted, the optical spectra illustrated here were obtained by my group, primarily with the 3-m Shane reflector at Lick Observatory. When referring to phase of evolution, the variables t and τ denote time since *maximum brightness* (usually in the B passband) and time since *explosion*, respectively.] The lines are broad due to the high velocities of the ejecta, and most of them have P-Cygni profiles formed by resonant scattering above the photosphere. SNe Ia are characterized by a deep absorption trough around 6150 Å produced by blueshifted Si II λ6355. Members of the Ib and Ic subclasses do not show this

CP522, *Cosmic Explosions: Tenth Astrophysical Conference,*
edited by Stephen S. Holt and William W. Zhang
© 2000 American Institute of Physics 1-56396-943-2/00/$17.00

line. The presence of moderately strong optical He I lines, especially He I $\lambda 5876$, distinguishes SNe Ib from SNe Ic.

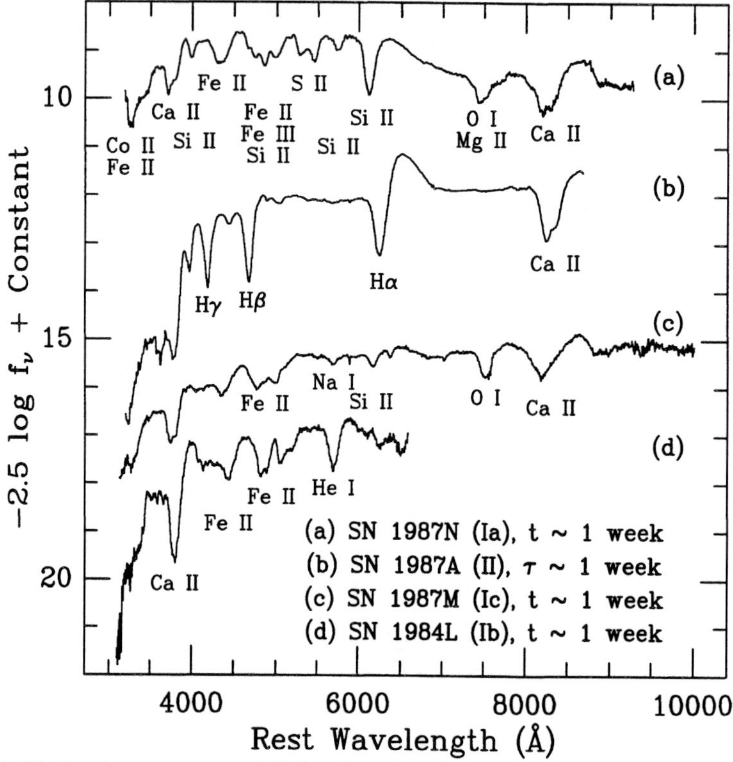

Figure 1: Early-time spectra of SNe, showing the main subtypes.

The late-time ($t \gtrsim 4$ months) optical spectra of SNe provide additional constraints on the classification scheme (Figure 2). SNe Ia show blends of dozens of Fe emission lines, mixed with some Co lines. SNe Ib and Ic, on the other hand, have relatively unblended emission lines of intermediate-mass elements such as O and Ca. At this phase, SNe II are dominated by the strong Hα emission line; in other respects, most of them spectroscopically resemble SNe Ib and Ic, but with narrower emission lines. The late-time spectra of SNe II show substantial heterogeneity, as do the early-time spectra.

To a first approximation, the light curves of SNe I are all broadly similar [2], while those of SNe II exhibit much dispersion [3]. It is useful to subdivide the majority of early-time light curves of SNe II into two relatively distinct subclasses [4,5]. The light curves of SNe II-L ("linear") generally resemble those of SNe I, with a steep decline after maximum brightness followed by a slower exponential tail. In contrast, SNe II-P ("plateau") remain within ~ 1 mag of maximum brightness for an extended period. The peak absolute magnitudes of SNe II-P show a very wide dispersion [6], almost certainly due to differences in the radii of the progenitor stars. The light curve of SN 1987A, albeit unusual, was generically related to those of SNe II-P; the initial peak was very low because the progenitor was a blue supergiant, much smaller than a red supergiant [7]. The remainder of this review concentrates on SNe II.

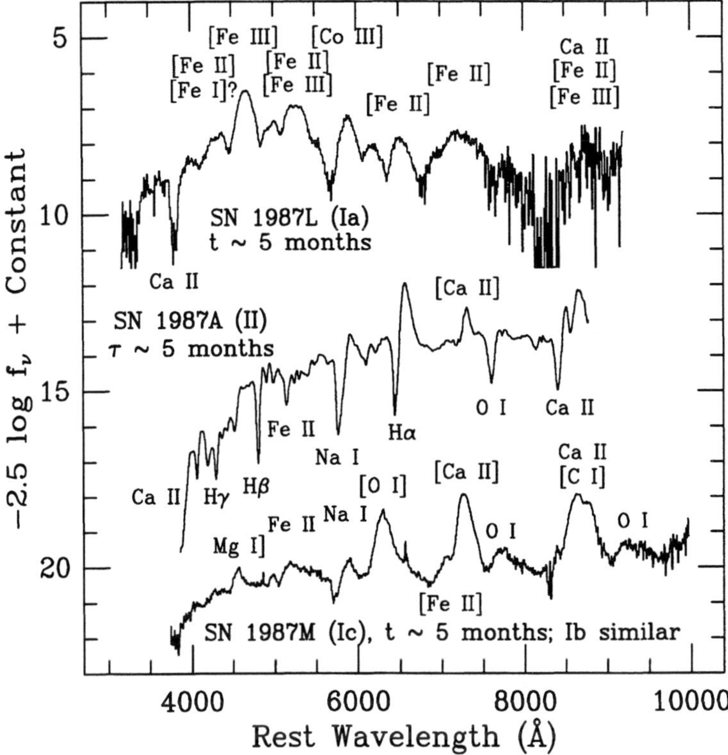

Figure 2: Late-time spectra of SNe. At even later phases, SN 1987A was dominated by strong emission lines of Hα, [O I], [Ca II], and the Ca II near-infrared triplet.

SUBCLASSES OF TYPE II SUPERNOVAE

Most SNe II-P seem to have a relatively well-defined spectral development, as shown in Figure 3 for SN 1992H (see also reference [8]). At early times the spectrum is nearly featureless and very blue, indicating a high color temperature ($\gtrsim$ 10,000 K). He I $\lambda5876$ with a P-Cygni profile is sometimes visible. The temperature rapidly decreases with time, reaching $\sim$ 5000 K after a few weeks, as expected from the adiabatic expansion and associated cooling of the ejecta. It remains roughly constant at this value during the plateau (the photospheric phase), while the hydrogen recombination wave moves through the massive ($\sim 10\ M_\odot$) hydrogen ejecta and releases the energy deposited by the shock. At this stage strong Balmer lines

and Ca II H&K with well-developed P-Cygni profiles appear, as do weaker lines
of Fe II, Sc II, and other iron-group elements. The spectrum gradually takes on
a nebular appearance as the light curve drops to the late-time tail; the continuum
fades, but Hα becomes very strong, and prominent emission lines of [O I], [Ca II],
and Ca II also appear.

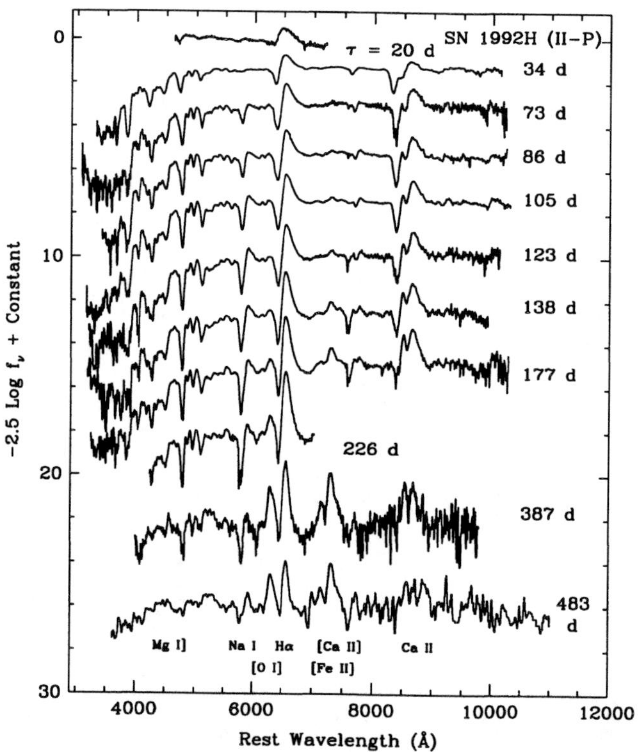

Figure 3: Montage of spectra of SN 1992H in NGC 5377. Epochs (days) are given
relative to the estimated time of explosion, February 8, 1992.

Few SNe II-L have been observed in as much detail as SNe II-P. Figure 4 shows
the spectral development of SN 1979C [9], an unusually luminous member of this
subclass. Near maximum brightness the spectrum is very blue and almost featureless, with a slight hint of Hα emission. A week later, Hα emission is more easily
discernible, and low-contrast P-Cygni profiles of Na I, Hβ, and Fe II have appeared.
By $t \approx 1$ month, the Hα emission line is very strong but still devoid of an absorption
component, while the other features clearly have P-Cygni profiles. Strong, broad
Hα emission dominates the spectrum at $t \approx 7$ months, and [O I] λλ6300, 6364
emission is also present. Several authors [10–12] have speculated that the absence

of Hα absorption spectroscopically differentiates SNe II-L from SNe II-P, but the small size of the sample of well-observed objects precluded definitive conclusions.

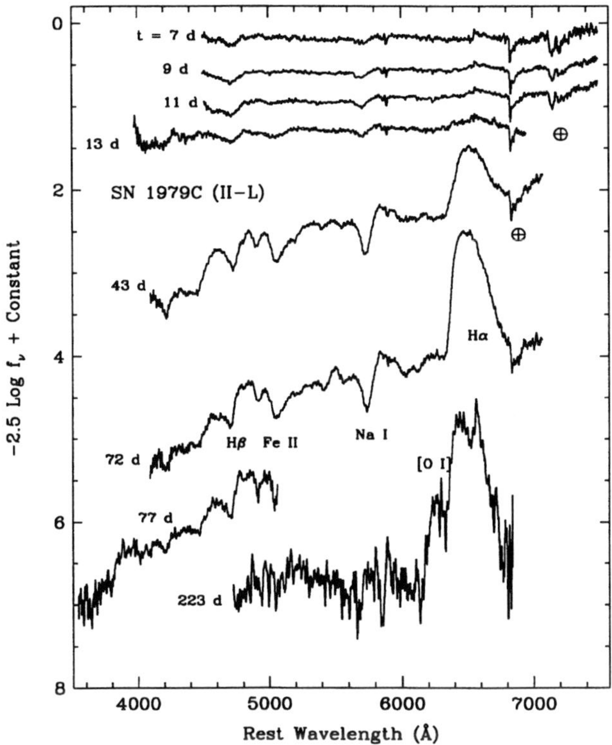

Figure 4: Montage of spectra of SN 1979C in NGC 4321, from reference [9]; reproduced with permission. Epochs (days) are given relative to the date of maximum brightness, April 15, 1979.

The progenitors of SNe II-L are generally believed to have relatively low-mass hydrogen envelopes (a few $M_\odot$); otherwise, they would exhibit distinct plateaus, as do SNe II-P. On the other hand, they may have more circumstellar gas than do SNe II-P, and this could give rise to the emission-line dominated spectra. They are often radio sources [13]; moreover, the ultraviolet excess (at $\lambda \lesssim 1600$ Å) seen in SNe 1979C and 1980K may be produced by inverse Compton scattering of photospheric radiation by high-speed electrons in shock-heated ($T \approx 10^9$ K) circumstellar material [14,15]. Finally, the light curves of some SNe II-L reveal an extra source of energy: after declining exponentially for several years, the Hα flux of SN 1980K reached a steady level, showing little if any decline thereafter [16,17]. The excess almost certainly comes from the kinetic energy of the ejecta being thermalized and radiated due to an interaction with circumstellar matter [18,19].

The very late-time optical recovery of SNe 1979C and 1980K [17,20,21] and other SNe II-L supports the idea of ejecta interacting with circumstellar material. The spectra consist of a few strong, broad emission lines such as Hα, [O I] $\lambda\lambda$6300, 6364, and [O III] $\lambda\lambda$4959, 5007. A *Hubble Space Telescope (HST)* ultraviolet spectrum of SN 1979C reveals some prominent, double-peaked emission lines with the blue peak substantially stronger than the red, suggesting dust extinction within the expanding ejecta [21]. The data show general agreement with the emission lines expected from circumstellar interaction [22], but the specific models that are available show several differences with the observations. For example, we find higher electron densities (10^5 to 10^7 cm^{-3}), resulting in stronger collisional de-excitation than assumed in the models. These differences can be used to further constrain the nature of the progenitor star. Note that based on photometry of the stellar populations in the environment of SN 1979C (from *HST* images), the progenitor of the SN was at most 10 million years years old, so its initial mass was probably 17–18 $M_\odot$ [23].

During the past decade, there has been the gradual emergence of a new, distinct subclass of SNe II [11,24,25,19] whose ejecta are believed to be *strongly* interacting with dense circumstellar gas, even at early times (unlike SNe II-L). The derived mass-loss rates for the progenitors can exceed $10^{-4}M_\odot$ yr^{-1} [26]. In these objects, the broad absorption components of all lines are weak or absent throughout their evolution. Instead, their spectra are dominated by strong emission lines, most notably Hα, having a complex but relatively narrow profile. Although the details differ among objects, Hα typically exhibits a very narrow component (FWHM $\lesssim$ 200 km s^{-1}) superposed on a base of intermediate width (FWHM $\approx$ 1000–2000 km s^{-1}; sometimes a very broad component (FWHM $\approx$ 5000–10,000 km s^{-1}) is also present. This subclass was christened "Type IIn" [25], the "n" denoting "narrow" to emphasize the presence of the intermediate-width or very narrow emission components. Representative spectra of five SNe IIn are shown in Figure 5, with two epochs for SN 1994Y.

The early-time continua of SNe IIn tend to be bluer than normal. Occasionally He I emission lines are present in the first few spectra (e.g., SN 1994Y in Figure 5). Very narrow Balmer absorption lines are visible in the early-time spectra of some of these objects, often with corresponding Fe II, Ca II, O I, or Na I absorption as well (e.g., SNe 1994W and 1994ak in Figure 5). Some of them are unusually luminous at maximum brightness, and they generally fade quite slowly, at least at early times. The equivalent width of the intermediate Hα component can grow to astoundingly high values at late times. The great diversity in the observed characteristics of SNe IIn provides clues to the various degrees and forms of mass loss late in the lives of massive stars.

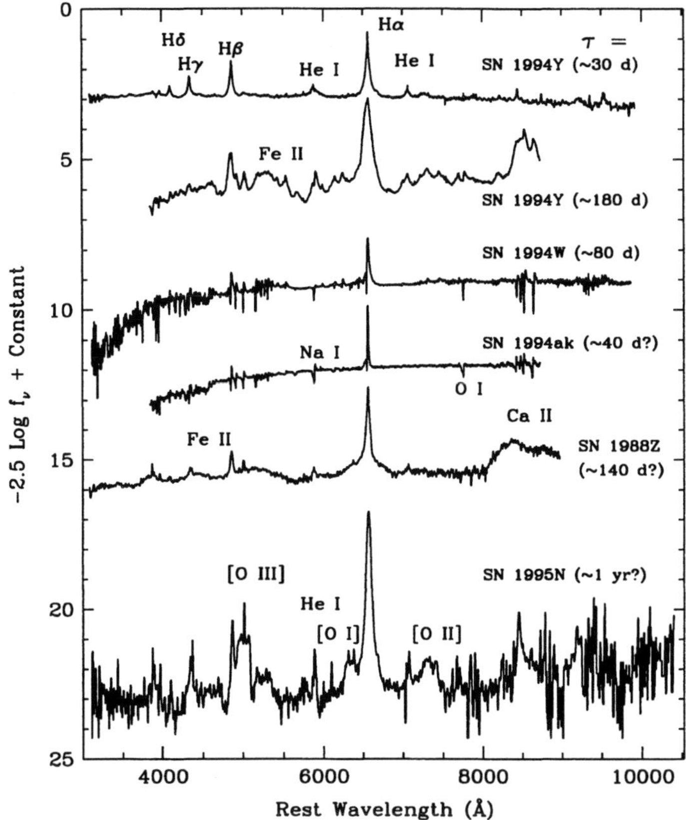

Figure 5: Montage of spectra of SNe IIn. Epochs are given relative to the estimated dates of explosion.

TYPE II SUPERNOVA IMPOSTORS?

The peculiar SN IIn 1961V ("Type V" according to Zwicky [27]) had probably the most bizarre light curve ever recorded. (SN 1954J, also known as "Variable 12" in NGC 2403, was similar [28].) Its progenitor was a very luminous star, visible in many photographs of the host galaxy (NGC 1058) prior to the explosion. Perhaps SN 1961V was not a genuine supernova (defined to be the violent destruction of a star at the end of its life), but rather the super-outburst of a luminous blue variable such as η Carinae [29,30].

A related object may have been SN IIn 1997bs, the first SN discovered in the
Lick Observatory Supernova Search (LOSS) that we are conducting with the 0.75-
m Katzman Automatic Imaging Telescope (KAIT) at Lick Observatory [31]. Its
spectrum was peculiar (Figure 6), consisting of narrow Balmer and Fe II emission
lines superposed on a featureless continuum. Its progenitor was discovered in an
HST archival image of the host galaxy [32]. It is a very luminous star ($M_V \approx -7.4$
mag), and it didn't brighten as much as expected for a SN explosion ($M_V \approx -13$ at
maximum). These data suggest that SN 1997bs may have been like SN 1961V —
that is, a supernova impostor. The real test will be whether the star is still visible
in future *HST* images obtained years after the outburst.

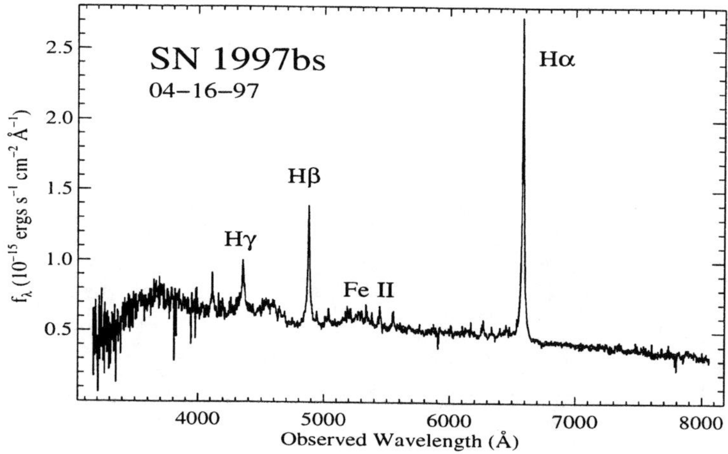

Figure 6: Spectrum of SN 1997bs, obtained on April 16, 1997 UT.

LINKS BETWEEN TYPE II AND TYPE Ib/Ic SUPERNOVAE

Filippenko [33] discussed the case of SN 1987K, which appeared to be a link
between SNe II and SNe Ib. Near maximum brightness, it was undoubtedly a
SN II, but with rather weak photospheric Balmer and Ca II lines. Many months
after maximum brightness, its spectrum was essentially that of a SN Ib. The
simplest interpretation is that SN 1987K had a meager hydrogen atmosphere at
the time it exploded; it would naturally masquerade as a SN II for a while, and
as the expanding ejecta thinned out the spectrum would become dominated by
emission from deeper and denser layers. The progenitor was probably a star that,
prior to exploding via iron core collapse, lost almost all of its hydrogen envelope

either through mass transfer onto a companion or as a result of stellar winds. Such SNe were dubbed "SNe IIb" by Woosley et al. [34], who had proposed a similar preliminary model for SN 1987A before it was known to have a massive hydrogen envelope.

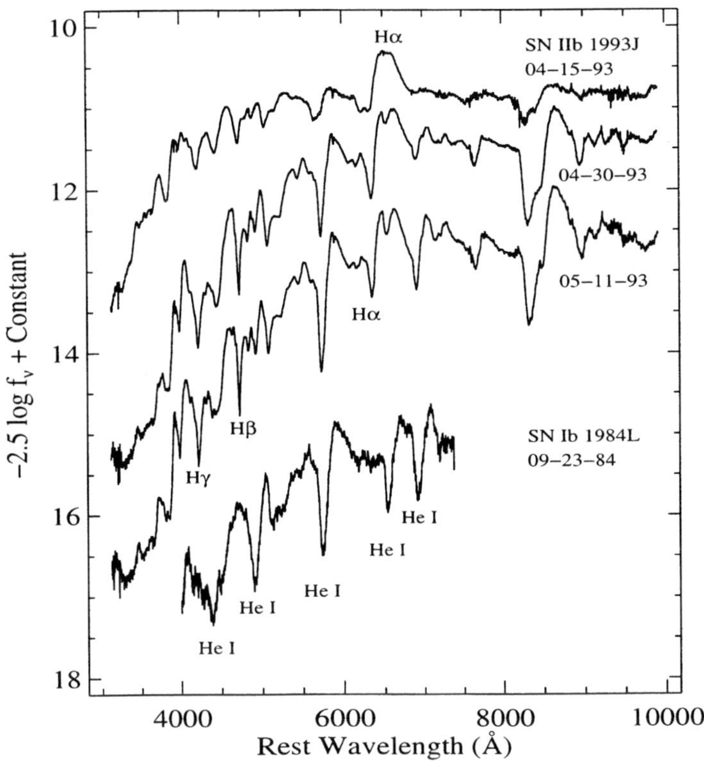

Figure 7: Early-time spectral evolution of SN 1993J. A comparison with the Type Ib SN 1984L is shown at bottom, demonstrating the presence of He I lines in SN 1993J. The explosion date was March 27.5, 1993.

The data for SN 1987K (especially its light curve) were rather sparse, making it difficult to model in detail. Fortunately, the Type II SN 1993J in NGC 3031 (M81) came to the rescue, and was studied in greater detail than any supernova since SN 1987A [35]. Its light curves [36] and spectra [37–39] amply supported the hypothesis that the progenitor of SN 1993J probably had a low-mass (0.1–0.6 $M_\odot$) hydrogen envelope above a $\sim 4\ M_\odot$ He core [40–42]. Figure 7 shows several early-time spectra of SN 1993J, showing the emergence of He I features typical of SNe Ib.

Considerably later (Figure 8), the Hα emission nearly disappeared, and the spectral resemblance to SNe Ib was strong. The general consensus is that its initial mass was $\sim 15\ M_\odot$. A star of such low mass cannot shed nearly its entire hydrogen envelope without the assistance of a companion star. Thus, the progenitor of SN 1993J probably lost most of its hydrogen through mass transfer to a bound companion 3–20 AU away. In addition, part of the gas may have been lost from the system. Had the progenitor lost essentially *all* of its hydrogen prior to exploding, it would have had the optical characteristics of SNe Ib. There is now little doubt that most SNe Ib, and probably SNe Ic as well, result from core collapse in stripped, massive stars, rather than from the thermonuclear runaway of white dwarfs.

SN 1993J held several more surprises. Observations at radio [43] and X-ray [44] wavelengths revealed that the ejecta are interacting with relatively dense circumstellar material [45], probably ejected from the system during the course of its pre-SN evolution. Optical evidence for this interaction also began emerging at $\tau \gtrsim 10$ months: the Hα emission line grew in relative prominence, and by $\tau \approx 14$ months it had become the dominant line in the spectrum [38,46,47], consistent with models [22]. Its profile was very broad (FWHM $\approx 17{,}000$ km s^{-1}; Figure 8) and had a relatively flat top, but with prominent peaks and valleys whose likely origin is Rayleigh-Taylor instabilities in the cool, dense shell of gas behind the reverse shock [48]. Radio VLBI measurements show that the ejecta are circularly symmetric, but with significant emission asymmetries [49], possibly consistent with the asymmetric Hα profile seen in some of the spectra [38].

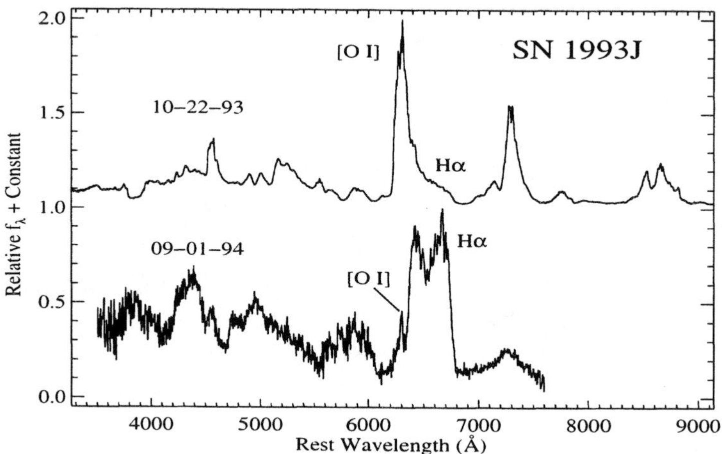

Figure 8: In the *top* spectrum, which shows SN 1993J about 7 months after the explosion, Hα emission is very weak; the resemblance to spectra of SNe Ib is striking. A year later (*bottom*), however, Hα was once again the dominant feature in the spectrum (which was scaled for display purposes).

SPECTROPOLARIMETRY OF TYPE II SUPERNOVAE

Spectropolarimetry of SNe can be used to probe their geometry [50]. The basic question is whether SNe are round. Such work is important for a full understanding of the physics of SN explosions and can provide information on the circumstellar environment of SNe. We have obtained spectropolarimetry of one object from each of the major SN types and subtypes, generally with the Keck-II 10-m telescope.

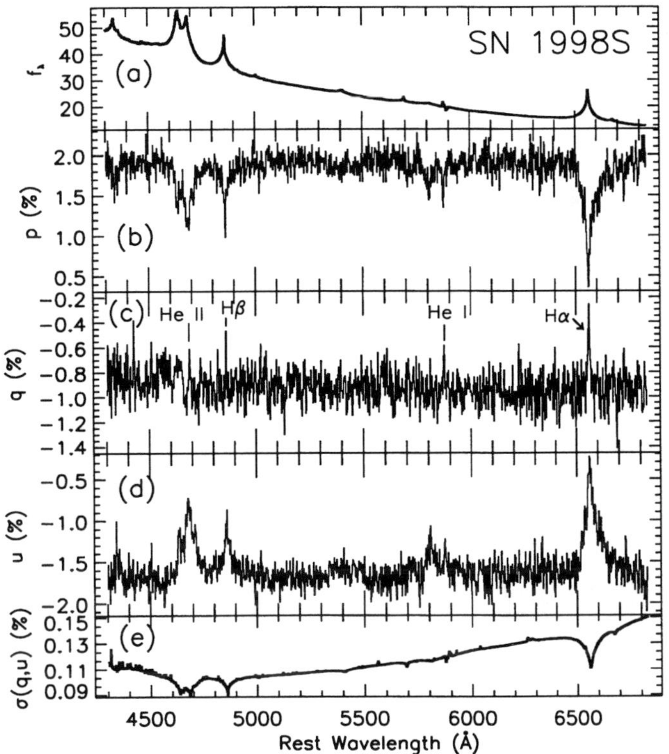

Figure 9: Polarization data for SN 1998S, obtained with Keck-II on March 7, 1998. *(a)* Total flux, in units of 10^{-15} ergs s^{-1} cm^{-2} Å^{-1}. *(b)* Observed degree of polarization. *(c,d)* The normalized q and u Stokes parameters, with prominent narrow-line features indicated. *(e)* Average of the (nearby identical) 1σ statistical uncertainties in the Stokes q and u parameters. See reference [51] for details.

We have completed our analysis of the peculiar Type IIn SN 1998S [51]. The data consist of one epoch of spectropolarimetry (5 days after discovery) and total flux spectra spanning the first 494 days after discovery. The SN is found to exhibit a high degree of linear polarization (Figure 9), implying significant asphericity for its continuum-scattering environment. Prior to removal of the interstellar polarization, the polarization spectrum is characterized by a flat continuum (at $p \approx 2\%$) with distinct changes in polarization associated with both the broad (symmetric, half width near-zero intensity $\gtrsim 10,000$ km s^{-1}) and narrow (unresolved, FWHM < 300 km s^{-1}) line emission seen in the total flux spectrum. When analyzed in terms of a polarized continuum with unpolarized broad-line recombination emission, however, an intrinsic continuum polarization of $p \approx 3\%$ results, suggesting a global asphericity of $\gtrsim 45\%$ from the oblate, electron-scattering dominated models of Höflich [52]. The smooth, blue continuum evident at early times is inconsistent with a reddened, single-temperature blackbody, instead having a color temperature that increases with decreasing wavelength. Broad emission-line profiles with distinct blue and red peaks are seen in the total flux spectra at later times, suggesting a disk-like or ring-like morphology for the dense ($n_e \approx 10^7$ cm^{-3}) circumstellar medium, generically similar to what is seen directly in SN 1987A, although much denser and closer to the progenitor in SN 1998S.

The Type IIn SN 1997eg also exhibits considerable polarization [50]; there are sharp polarization changes across its strong, multi-component emission lines, suggesting distinct scattering origins for the different components. Based on our rather small sample, it appears as though SNe II-P are considerably less polarized than SNe IIn, at least within the first month or two after the explosion. Leonard et al. [50] show some spectropolarimetric evidence of asphericity in the ejecta of SN II-P 1997ds, but it does not match the degree of polarization of SNe IIn 1998S and 1997eg. Moreover, SN II-P 1999em does not reveal significant polarization variation across the strong Balmer lines shortly after its explosion [53].

SUPERNOVAE ASSOCIATED WITH GAMMA-RAY BURSTS?

At least a small fraction of gamma-ray bursts (GRBs) may be associated with nearby SNe. Probably the most compelling example thus far is that of SN 1998bw and GRB 980425 [54–56], which were temporally and spatially coincident. SN 1998bw was, in many ways, an extraordinary SN; it was very luminous at optical and radio wavelengths, and it showed evidence for relativistic outflow. Its bizarre optical spectrum is often classified as that of a SN Ic, but the object should be called a "peculiar SN Ic" if not a subclass of its own; the spectrum was distinctly different from that of a normal SN Ic.

As discussed by several speakers at this meeting, models suggest that SNe associated with GRBs are highly asymmetric. Thus, spectropolarimetry should provide some useful tests. In particular, perhaps objects such as SN 1998S, discussed above, would have been seen as GRBs had their rotation axis been pointed in our direction. That of SN 1998S was almost certainly *not* aligned with us [51]; both the spectropolarimetry and the appearance of double-peaked Hα emission suggest an inclined view, rather than pole-on.

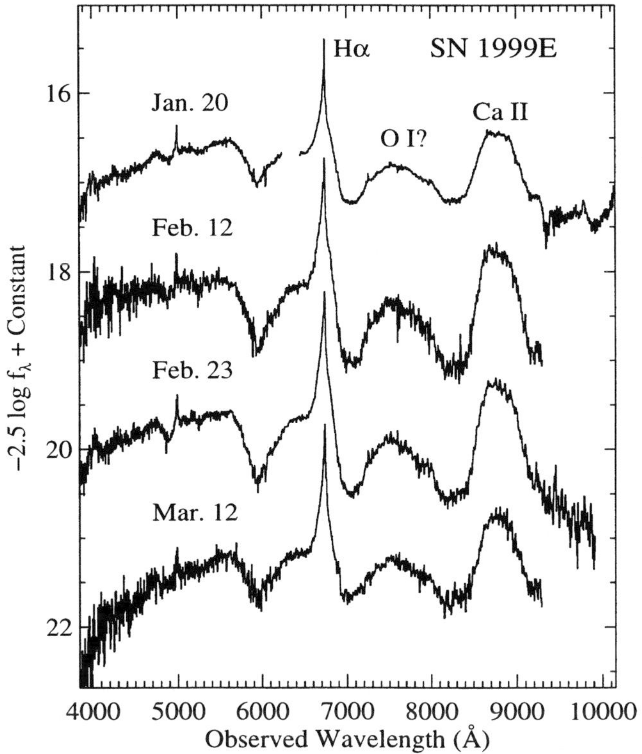

Figure 10: Spectral evolution of SN 1999E, which may have been associated with GRB 980910.

The case of GRB 970514 and the very luminous SN IIn 1997cy is also interesting [57,58]; there is a reasonable possibility that the two objects were associated. The optical spectrum of SN 1997cy was highly unusual, and bore some resemblance to that of SN 1998bw, though there were some differences as well. SN 1999E, which might be linked with GRB 980910 but with large uncertainties [59], also had an optical spectrum similar to that of SN 1997cy [60,61]; see Figure 10. The undulations are very broad, indicating high ejection velocities. Besides Hα, secure line identifications are difficult, though some of the emission features seem to be associated with oxygen and calcium. Perhaps SN 1999E was produced by the highly asymmetric collapse of a carbon-oxygen core.

Shortly before this meeting, SN IIn 1999eb was discovered with KAIT [62], and Terlevich et al. [63] pointed out that it might be associated with GRB 991002. However, KAIT data show that the optical SN was visible at least 10 days *before* the GRB occurred, making it very unlikely that the two were linked. If SN 1999eb ends up showing double-peaked Hα emission at late times, as did SN 1998S, it will be another argument against the SN/GRB association in this particular case, since our view will not have been pole-on.

THE EXPANDING PHOTOSPHERE METHOD

Despite *not* being anything like "standard candles," SNe II-P (and some SNe II-L) are very good distance indicators. They are, in fact, "custom yardsticks" when calibrated with the "Expanding Photosphere Method" (EPM); see [64]. A variant of the famous Baade-Wesselink method for determining the distances of pulsating variable stars, this technique relies on an accurate measurement of the photosphere's effective temperature and velocity during the plateau phase of SNe II-P.

Briefly, here is how EPM works. The radius (R) of the photosphere can be determined from its velocity (v) and time since explosion $(t - t_0)$ if the ejecta are freely expanding: $R = v(t - t_0) + R_0 \approx v(t - t_0)$, where we have assumed that the initial radius of the star $(R_0$ at $t = t_0)$ is negligible relative to R after a few days. The velocity of the photosphere is determined from measurements of the wavelengths of the absorption minima in P-Cygni profiles of weak lines such as those of Fe II or, better yet, Sc II. (The absorption minima of strong lines like Hα form far above the photosphere.) The angular size (θ) of the photosphere, on the other hand, is found from the measured, dereddened flux density (f_ν) at a given frequency. We have $4\pi D^2 f_\nu = 4\pi R^2 \zeta^2 \pi B_\nu(T)$, so

$$\theta = \frac{R}{D} = \left[\frac{f_\nu}{\zeta^2 \pi B_\nu(T)} \right]^{1/2} ,$$

where D is the distance to the supernova, $B_\nu(T)$ is the value of the Planck function at color temperature T (derived from broadband measurements of the supernova's brightness in at least two passbands), and ζ^2 is the flux dilution correction factor (basically a measure of how much the spectrum deviates from that of a blackbody, due primarily to the electron-scattering opacity).

The above two equations imply that $t = D(\theta/v) + t_0$. Thus, for a series of measurements of θ and v at various times t, a plot of θ/v versus t should yield a straight line of slope D and intercept t_0. This determination of the distance is independent of the various uncertain rungs in the cosmological distance ladder; it does not even depend on the calibration of the Cepheids. It is equally valid for nearby and distant SNe II-P.

An important check of EPM is that the derived distance be *constant* while the SN is on the plateau (before it has started to enter the nebular phase). This has been verified with SN 1987A [65] and a number of other SNe II-P [66,67]. Moreover, the EPM distance to SN 1987A agrees with that determined geometrically through measurements of the brightening and fading of emission lines from the inner circumstellar ring [68]. It is also noteworthy that EPM is relatively insensitive to reddening: an underestimate of the reddening leads to an underestimate of the color temperature T [and hence of $B_\nu(T)$ as well], but this is compensated by an underestimate of f_ν, yielding a nearly unchanged value of θ. Indeed, for errors in A_V [the visual extinction, or $\sim 3.1E(B-V)$] of 0–1 mag, one incurs an error in D of only $\sim$ 0–20% [66].

Of course, EPM has some caveats or potential limitations. A critical assumption is spherical symmetry for the expanding ejecta, yet polarimetry shows that SN 1987A was not spherical [69], as do direct *HST* measurements of the shape of the ejecta. As discussed above, the few other SNe II-P that have been studied don't show very much polarization, though it is possible that deviations from spherical symmetry could be a severe problem for some SNe II-P. On the other hand, the *average* distance derived with EPM for many SNe II-P might be almost unaffected, given random orientations to the line of sight. (Sometimes the cross-sectional area will be too large, and other times too small, relative to spherical ejecta.) Indeed, comparison of EPM and Cepheid distances to the same galaxies shows agreement to within the expected uncertainties for a sample of 6 objects ($D_{\mathrm{Cepheids}}/D_{\mathrm{EPM}} = 0.98 \pm 0.08$; [64]).

Another limitation of EPM is that one needs a well-observed SN II-P in a given galaxy in order to measure its distance; thus, the technique is most useful for aggregate studies of galaxies, rather than for distances of specific galaxies in a random sample. Finally, knowledge of the flux dilution correction factor, ζ^2, is critical to the success of EPM. Fortunately, an extensive grid of models [64] shows that the value of ζ^2 is mainly a function of T during the plateau phase of SNe II-P; it is relatively insensitive to other variables such as helium abundance, metallicity, density structure, and expansion rate. Also, it does not differ too greatly from unity during the plateau. There are, however, some differences of opinion regarding the treatment of radiative transfer and thermalization in expanding supernova atmospheres [70]. The calculations are difficult and various assumptions are made, possibly leading to significant systematic errors.

The most distant SN II-P to which EPM has successfully been applied is SN 1992am at $z = 0.0487$ [71]. The derived distance is $D = 180^{+30}_{-25}$ Mpc. This object, together with 15 other SNe II-P at smaller redshifts, yields a best fit value of $H_0 = 73 \pm 7$ km s^{-1} Mpc^{-1}, where the quoted uncertainty is purely statistical [67,64]. A systematic uncertainty of $\sim \pm 6$ km s^{-1} Mpc^{-1} should also be associated with the above result. The main source of statistical uncertainty is the relatively small number of SNe II-P in the EPM sample, and the low redshift of most of the objects (whose radial velocities are substantially affected by peculiar motions). My group is currently trying to remedy the situation with EPM measurements of additional nearby SNe II-P (at Lick Observatory), as well as with Keck spectra of SNe II-P in the redshift range 0.1–0.3.

ACKNOWLEDGMENTS

My recent research on SNe has been financed by NSF grant AST–9417213, as well as by NASA grants GO-6584, GO-7434, AR-6371, and AR-8006 from the Space Telescope Science Institute, which is operated by AURA, Inc., under NASA Contract NAS5-26555. The Committee on Research (U.C. Berkeley) provided partial travel support to attend this meeting. I am grateful to the students and postdocs who have worked with me on SNe over the past 14 years for their assistance and discussions. Tom Matheson and Doug Leonard were especially helpful with the figures for this review.

REFERENCES

1. Filippenko, A. V., *ARAA*, **35**, 309 (1997).
2. Leibundgut, B., Tammann, G. A., Cadonau, R., & Cerrito, D., *A&AS*, **89**, 537 (1991a).
3. Patat, F., Barbon, R., Cappellaro, E., & Turatto, M., *A&AS*, **98**, 443 (1993).
4. Barbon, R., Ciatti, F., & Rosino, L. *A&A*, **72**, 287 (1979).
5. Doggett, J. B., & Branch, D., *AJ*, **90**, 2303 (1985).
6. Young, T. R., & Branch, D., *ApJ*, **342**, L79 (1989).
7. Arnett, W. D., Bahcall, J. N., Kirshner, R. P., & Woosley, S. E., *ARAA*, **27**, 629 (1989).
8. Clocchiatti, A., et al., *AJ*, **111**, 1286 (1996).
9. Branch, D., Falk, S. W., McCall, M. L., Rybski, P., Uomoto, A. K., & Wills, B. J., *ApJ*, **224**, 780 (1981).
10. Wheeler, J. C., & Harkness, R. P., *Rep. Prog. Phys.*, **53**, 1467 (1990).
11. Filippenko, A. V., in *Supernovae*, ed. S. E. Woosley (New York: Springer), 467 (1991a).
12. Schlegel, E. M., *AJ*, **111**, 1660 (1996).
13. Sramek, R. A., & Weiler, K. W., in *Supernovae*, ed. A. G. Petschek (New York: Springer), p. 76 (1990).
14. Fransson, C., *A&A*, **111**, 140 (1982).
15. Fransson, C., *A&A*, **133**, 264 (1984).
16. Uomoto, A., & Kirshner, R. P., *ApJ*, **308**, 685 (1986).
17. Leibundgut, B., et al., *ApJ*, **372**, 531 (1991b).
18. Chevalier, R. A., in *Supernovae*, ed. A. G. Petschek (New York: Springer), p. 91 (1990).
19. Leibundgut, B., in *Circumstellar Media in the Late Stages of Stellar Evolution*, eds. R. E. S. Clegg, I. R. Stevens, & W. P. S. Meikle (Cambridge: Cambridge Univ. Press), p. 100 (1994).
20. Fesen, R. A., Hurford, A. P., & Matonick, D. M., *AJ*, **109**, 2608 (1995).
21. Fesen, R. A., et al., *AJ*, **117**, 725 (1999).
22. Chevalier, R. A., & Fransson, C., *ApJ*, **420**, 268 (1994).
23. Van Dyk, S. D., et al., *PASP*, **111**, 315 (1999a).

24. Filippenko, A. V., in *Supernova 1987A and Other Supernovae*, eds. I. J. Danziger & K. Kjär (Garching: ESO), p. 343 (1991b).

25. Schlegel, E. M., *MNRAS*, **244**, 269 (1990).

26. Chugai, N. N., in *Circumstellar Media in the Late Stages of Stellar Evolution*, eds. R. E. S. Clegg, I. R. Stevens, & W. P. S. Meikle (Cambridge: Cambridge Univ. Press), p. 148 (1994).

27. Zwicky, F, in *Stars and Stellar Systems*, Vol. 8, ed. L. H. Aller & D. B. McLaughlin, pp. 367. Chicago: Univ. Chicago Press (1965).

28. Humphreys, R. M., & Davidson, K., *PASP*, **106**, 1025 (1994).

29. Goodrich, R. W., Stringfellow, G. S., Penrod, G. D., & Filippenko, A. V., *ApJ*, **342**, 908 (1989).

30. Filippenko, A. V., et al., *AJ*, **110**, 2261 (1995) [Erratum: **112**, 806].

31. Li, W. D., et al., *These Proceedings* (2000).

32. Van Dyk, S. D., Peng, C. Y., Barth, A. J., & Filippenko, A. V., *AJ*, **118**, 2331 (1999b).

33. Filippenko, A. V., *AJ*, **96**, 1941 (1988).

34. Woosley, S. E., Pinto, P. A., Martin, P. G., & Weaver, T. A., *ApJ*, **318**, 664 (1987).

35. Wheeler, J. C., & Filippenko, A. V., in *Supernovae and Supernova Remnants*, eds. R. McCray & Z. Wang (Cambridge: Cambridge Univ. Press), p. 241 (1996).

36. Richmond, M. W., Treffers, R. R., Filippenko, A. V., & Paik, Y., *AJ*, **112**, 732 (1996).

37. Filippenko, A. V., Matheson, T., & Ho, L. C., *ApJ*, **415**, L103 (1993).

38. Filippenko, A. V., Matheson, T., & Barth, A. J., *AJ*, **108**, 2220 (1994).

39. Matheson, T., et al., in preparation (2000).

40. Nomoto, K., Suzuki, T., Shigeyama, T., Kumagai, S., Yamaoka, H., & Saio, H., *Nature*, **364**, 507 (1993).

41. Podsiadlowski, P., Hsu, J. J. L., Joss, P. C., & Ross, R. R., *Nature*, **364**, 509 (1993).

42. Woosley, S. E., Eastman, R. G., Weaver, T. A., & Pinto, P. A., *ApJ*, **429**, 300 (1994).

43. Van Dyk, S. D., Weiler, K. W., Sramek, R. A., Rupen, M. P., & Panagia, N., *ApJ*, **432**, L115 (1994).

44. Suzuki, T., & Nomoto, K., *ApJ*, **455**, 658 (1995).

45. Fransson, C., Lundqvist, P., & Chevalier, R. A., *ApJ*, **461**, 993 (1996).

46. Patat, F., Chugai, N., & Mazzali, P. A. *A&A*, **299**, 715 (1995).

47. Finn, R. A., Fesen, R. A., Darling, G. W., & Thorstensen, J. R., *AJ*, **110**, 300 (1995).

48. Chevalier, R. A., Blondin, J. M., & Emmering, R. T., *ApJ*, **392**, 118 (1992).

49. Marcaide, J. M., et al., *Science*, **270**, 1475 (1995).

50. Leonard, D. C., Filippenko, A. V., & Matheson, T., *These Proceedings* (2000a).

51. Leonard, D. C., Filippenko, A. V., Barth, A. J., & Matheson, T., *ApJ*, in press, astro-ph/9908040 (2000b).

52. Höflich, P., *A&A*, **246**, 481 (1991).

53. Leonard, D. C., Filippenko, A. V., & Chornock, R. T., *IAU Circ.* 7305 (1999).

54. Galama, T. J., et al., *Nature*, **395**, 670 (1998).

55. Iwamoto, K., et al., *Nature*, **395**, 672 (1998).

56. Woosley, S. E., Eastman, R. G., & Schmidt, B. P., *ApJ*, **516**, 788 (1999).

57. Germany, L. M., et al., *ApJ*, in press, astro-ph/9906096 (2000).
58. Turatto, M., et al., astro-ph/9910324 (2000).
59. Thorsett, S. E., & Hogg, D. W., *GCN Circ.* 197 (1999).
60. Filippenko, A. V., Leonard, D. C., & Riess, A. G., *IAU Circ.* 7091 (1999).
61. Cappellaro, E., Turatto, M., & Mazzali, P., *IAU Circ.* 7091 (1999).
62. Modjaz, M., et al., *IAU Circ.* 7268 (1999).
63. Terlevich, R., Fabian, A., & Turatto, M., *IAU Circ.* 7269 (1999).
64. Eastman, R. G., Schmidt, B. P., & Kirshner, R. P., *ApJ*, **466**, 911 (1996).
65. Eastman, R. G., & Kirshner, R. P., *ApJ*, **347**, 771 (1989).
66. Schmidt, B. P., Kirshner, R. P., Eastman, R. G., *ApJ*, **395**, 366 (1992).
67. Schmidt, B. P., *et al.*, *ApJ*, **432**, 42 (1994a).
68. Panagia, N., *et al.*, *ApJ*, **380**, L23 (1991).
69. Jeffery, D. J., *ApJ*, **375**, 264 (1991).
70. Baron, E., Hauschildt, P. H., & Mezzacappa, A., *MNRAS*, **278**, 763 (1996).
71. Schmidt, B. P., *et al.*, *AJ*, **107**, 1444 (1994b).

On the Nature of the X-Ray Point Source in Cassiopeia A

Hideyuki Umeda[1], Ken'ichi Nomoto[1], Sachiko Tsuruta[2], and Shin Mineshige[3]

[1] *Department of Astronomy and Research Center for the Early Universe, University of Tokyo, Bunkyo-ku, Tokyo 113-0033, Japan*
e-mail: umeda@astron.s.u-tokyo.ac.jp, nomoto@astron.s.u-tokyo.ac.jp
[2] *Department of Physics, Montana State University, Bozeman, Montana 59717, USA*
e-mail: uphst@gemini.oscs.montana.edu
[3] *Department of Astronomy, Kyoto University, Sakyo-ku, Kyoto 606-8502, Japan*
e-mail: minesige@kusastro.kyoto-u.ac.jp

Abstract.
The *Chandra* First Light Observation discovered a point-like source in Cassiopeia A supernova remnant. This detection was subsequently confirmed by the analyses of the archival data from ROSAT observations. Here we compare the results from these observations with the scenarios involving both black holes (BH) and neutron stars (NS). Although the current data of Cas A point-source can be consistent with both BH and NS scenarios, future observations by the *Chandra*, XMM, and other satellite missions should be able to distinguish between these cases. If distinct periodicity is found, the point source definitely should be a NS. From superior spectral data to be available from these observations, the actual temperature and luminosity of the cooler component may be measured. If the measurement gives $L^\infty \lesssim 10^{33}$ erg s^{-1}, this will become the first concrete evidence for NS non-standard cooling. If the observed radiation does not show precise periodicity but smaller scale burst-like variability is detected, the compact source can be a BH. If superior spectral data are best fit by a two-component model with lower energy blackbody (~ 0.1 keV) + steep power law, our proposed disk-corona BH accretion model will be well supported. If the source is found to be a BH, the implication is significant in the sense that this will offer the first observational evidence for BH formation through a SN explosion.

Cassiopeia A (Cas A) is an interesting supernova (SN) remnant in various aspects. The remnant is very young, about 320 years old. This ring-shaped (e.g. Holt et al. 1994) remnant is associated with jet like structures (Fesen et al. 1987). The observed abundances of heavy elements are in good agreement with the yields of a massive star (e.g., Hughes et al. 2000). The overabundance of nitrogen found in some knots (Fesen et al. 1987) implies that the progenitor was a massive Wolf-Rayet star (WN type).

Recently the ACIS on board the *Chandra* X-ray satellite observed Cas A and

CP522, *Cosmic Explosions: Tenth Astrophysical Conference,*
edited by Stephen S. Holt and William W. Zhang

found a point-like source (Tananbaum et al. 1999). Subsequently, Aschenbach et al. (1999) reported that the ROSAT/HRI image of Cas A also shows the point-like source at the similar location. They claim that if the entire observed radiation comes from a 10 km radius neutron star emitting blackbody, the temperature T^∞ is 1.6 million degrees (MK) which corresponds to luminosity $L^\infty \sim 8 \times 10^{33}$ erg s^{-1}. Very recently Pavlov et al. (1999) and Chakrabarty et al. (2000) reported the results of their analyses of the point-source data from the *Chandra* observation. They concluded that the data can be fitted both by power-law or thermal models.

In this paper, we discuss both neutron star (NS) and black hole(BH) models are consistent with current observational data, and describe how these models can be distinguished by the forthcoming long *Chandra* and XMM observations. Our earlier discussion is given in Umeda et al. (1999).

1. ACCRETING BLACK HOLE

If the Cas A progenitor is more massive than $\sim 25 M_\odot$ a BH may be formed in the explosion (e.g., Ergma & van den Heuvel 1998). After formation the inner part of the ejected matter may fall back onto the BH. The property of an accreting BH depends strongly on whether or not an accretion disk is formed. Here we discuss possible BH scenarios based either on the spherical or disk accretion models under the following observational constraints (see, e.g., Pavlov et al. 1999): (1) the single power-law X-ray luminosity of intermediate brightness, $L_{\rm x}(0.1 - 5.0 \text{ keV})= (2-60) \times 10^{34}$ erg s^{-1}, which is much lower than the Eddington luminosity, $L_{\rm EDD} \sim 7.5 \times 10^{38}(M_{\rm BH}/3M_\odot)$ erg s^{-1} for H-free matter. (2) no significant variability being detected between the *Einstein* and *Chandra* observations, and (3) large power-law photon index, $\Gamma \sim 2.6 - 4.1$.

Spherical accretion occurs if the fallback matter has too small angular momentum to form a disk. In this case the mass-flow rate decays in a power-law fashion (Colpi et al. 1996) as $\dot{M} \propto t^{-5/3}$. Accordingly, L also decays in a power fashion as $L \sim 7.5 \times 10^{33}(M_{\rm BH}/3M_\odot)^{2/3}(\rho_0/0.1 \text{ g cm}^{-3})^{5/6}(t_0/1 \text{ hr})^{20/9}(t/320 \text{ yr})^{-25/18}$ (Zampieri et al. 1998).

Next consider the case of disk accretion. If the fallback material has sufficently large angular momentum, a fallback disk will be formed. There will be no efficient mechanism for angular-momentum removal since the Cas A compact remnant is unlikely to have a binary companion (§3). Then the disk evolution most likely obeys the self-similar solution in which the total angular momentum within the disk is kept constant (e.g., Mineshige et al. 1993), and mass accretion rate will decay in a power-law fashion after the disk is formed (Mineshige et al. 1997) as $\dot{m} \equiv \dot{M}c^2/L_{\rm EDD} \sim 10^2(M_{\rm fallback}/0.1M_\odot)(\alpha/0.1)^{-1.3}(t/320{\rm yr})^{-1.3}(M_{\rm BH}/3M_\odot)^{-1.15}$, where $M_{\rm fallback}$ is the amount of fallback material and $\alpha \simeq 0.1 - 1.0$ is the viscosity parameter. According to the standard-disk relation, we estimate that $l \sim \eta\dot{m}$, where η (~ 0.1) is the energy-conversion efficiency. In order for the black-hole accretion scenario to be consistent with the observed $l \sim 10^{-4}$, therefore, we have $\dot{m} \sim 10^{-3}$ at $t = 320$ yr and hence $M_{\rm fallback} \sim 10^{-6}M_\odot$. The implication is that the amount of the fallback material should indeed be very small. We note that the luminosity decrease expected in this model is consistent with the luminosity variation between *Einstein* (in 1979) and the *Chandra* (in 1999) observations.

Next consider the important constraint (3), a large photon index. In the spherical accretion model, we assume that the initial thermal photon energy is small ($\ll 1$ keV) and photons acquire energy from high energy electrons by the inverse Compton scattering as in the model by Colpi (1988). The index can be significantly large if the Compton y parameter is sufficiently small, i.e., if the scattering is not strong. However, the problem here is how small y can be satisfied for spherical accretion. In the case of disk accretion, luminosity of $\sim 10^{33}$ erg s^{-1} is typical to quiescent Galactic BH candidates (GBHC) and its power-law photon indices should be as small as $\Gamma \sim 1.7$ (Tanaka & Lewin 1995), in conflict with the Cas A observation. However, GBHC generally exhibit two states, soft and hard. Quiescent GBHC are in the hard state and their accretion flow is advection dominated (ADAF, Narayan, McClintock, & Yi 1996). On the other hand, a large Γ is a characteristics of the soft-state emission which exhibits soft blackbody spectra with $kT \sim 1$ keV. The radiation from thermal photons with ~ 1 keV times the area of the emission region around a typical black hole of $3 - 10$ M$_\odot$ produces higher luminosity, $L_{\mathrm{x}} \sim 10^{36-38}$ erg s^{-1}, than observed from Cas A. Unlike GBHC, however, no further mass input is available in our Cas A BH accretion case (§3). Then the accretion rate should monotonically decrease, and so does the maximum blackbody temperature, as $T_{\mathrm{max}} \sim 0.1(M/3M_\odot)^{1/4}(\dot{m}/10^{-3})^{1/4}$ keV. Therefore, we get $T_{\mathrm{max}} \sim 0.1$ keV for $\dot{m} \sim 10^{-3}$ instead. The conclusion is that with the low accretion rate and lower soft photon temperature our Compton model with a disk-corona configuration naturally yields small y and large Γ with the observed luminosity.

2. COOLING NEUTRON STAR

Here let us assume that the Cas A point source is a NS, and compares the observational data with the NS cooling theory. In this case, the observed data should be dominated by the thermal emission from the NS. The best fit of the *Chandra* data for a one component blackbody model yields with the temperature $T^\infty = 6 - 8$ MK and the effective radius $R_e = 0.20 - 0.45$ km (Pavlov et al. 1999), which is considered to be the emission from hot spots. Pavlov et al. (1999) also constrained the temperature of the 10 km NS by a two temperature model. However, total number of photons ROSAT collected is more than those *Chandra* could collect in the low energy region, where cooler componet should dominate. Therefore, we take the luminosity observed by ROSAT. $L^\infty \sim 8 \times 10^{33}$ erg s^{-1}, as the upper limit to the NS surface emission outside the hot spots.

The neutron star thermal evolution is calculated with a 1D general relativistic evolutionary code (e.g., Umeda et al. 1994). Our results are summarized in Figure 1. The observed data is marginally consistent with the 'standard' cooling. However, it is still only an upper limit, and if the actual temperature of the cooler component turns out to be $\lesssim 10^{33}$erg s^{-1}, the result will be very interesting. This is because then the observed value will be considered the evidence for non-standard cooling scenarios such as those involving pion and/or kaon condensates.

For all the curves discussed above a representative equation of state (EOS) of medium stiffness is adopted. To show the effect of the stiffness of the EOS, the dashed curve labeled (PS) shows the direct URCA cooling with the very stiff PS EOS (Pandharipande, Pines & Smith 1976). The stiffer the EOS the more extended the crust will become. Consequently, the age at which the sudden drop occurs becomes larger for a stiffer EOS. Hence, if the Cas A source has a luminosity $L^{\infty} \ll 10^{34}$ erg s^{-1}, the EOS must be softer than the PS model.

3. CONSTRAINTS FROM PROGENITOR SCENARIOS

Here we discuss whether the formation of a NS or BH is consistent with the current models of stellar evolution and SN models. The overabundance of nitrogen in Cas A implies that the progenitor was a massive WN star which lost most of its hydrogen envelope before the SN explosion. Here we describe two possible evolutionary paths to form such a pre-SN WN star.

One path is the mass loss of a very massive *single* star. A star with the initial

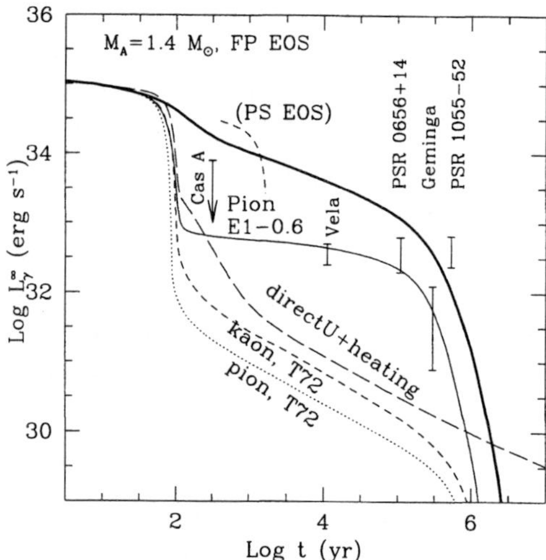

FIGURE 1. Various neutron star thermal evolution curves are compared with the observational data of the Cas A point source and several cooling NS candidates (e.g., Tsuruta 1998 for reference). The heavy solid curve refers to 'standard-cooling', the dashed and dotted curves show the 'non-standard' kaon and pion cooling scenarios when the superfluid effect is negligible, while the thin sold curve refers to the pion cooling with significant superfluid effect (E1 – 0.6 model, Umeda et al. 1994). The long dashed curve shows the effect of strong heating with a strength parameter $K = 10^{37}$ erg m$^{-3/2}$ s^2 on the direct URCA cooling.

144

mass M_{MS} larger than ~ 40 M$_\odot$ can lose its hydrogen-rich envelope via mass loss due to strong winds and become a Wolf-Rayet star. This WN star progenitor is massive enough to form a BH. The other evolutionary path to form a pre-SN WN star is mass loss due to binary interaction. If the progenitor is in a close binary system with a less massive companion star, the star loses most of its H-rich envelope through Roche lobe overflow. In this case, the WN progenitor can form from a star of $M_{\mathrm{MS}} \lesssim 40$ M$_\odot$. Its SN explosion of type Ib/c would leave either a BH (if $M_{\mathrm{MS}} \sim 25 - 40$ M$_\odot$) or a NS (if $M_{\mathrm{MS}} \lesssim 25$ M$_\odot$). If the compact remnant in Cas A turns out to be a NS, therefore, the progenitor must have been in a close binary system.

In the binary scenario, the companion to the Cas A progenitor cannot be more massive than a red dwarf, as constrained from the R & I band magnitude limit (van den Bergh & Pritchet 1986). When the companion star is such a small mass star, i.e., the mass ratio between the stars is large, the mass transfer is inevitably non-conservative (e.g., Nomoto, Iwamoto, & Suzuki 1995), and the companion star will spiral-in into the envelope of the Cas A progenitor.

If we take the model of $M_{\mathrm{MS}} = 25$ M$_\odot$, as an example, the star at the WN stage has 8 M$_\odot$. Since the explosion ejects Si and Fe from the deep layers, the mass of the compact remnants could not exceed 2 - 3 M$_\odot$. Then the binary system is very likely to be disrupted at the explosion. Then the compact star in the Cas A remnant does not have a companion star, and so no mass transfer can be postulated. The implication is also that the accretion onto the compact remnant can occur only as a result of fallback of the ejected matter, and so the composition of the fall back matter is mostly heavy elements with no hydrogen.

In either the single or binary scenario, the WN star blows a fast wind which collides with the red-giant wind material to form a dense shell (Chevalier & Liang 1989). If the red-giant wind formed a ring-like shell (due possibly to the spiral-in of the companion), the collision between the supernova ejecta and the shell could explain the observed ring-like structure of Cas A.

4. DISCUSSION

The investigation outlined in §1 suggests that if the source is an accreting BH the disk accretion is more promising than the spherical accretion. Accreting NS models are also possible (Chakrabarty et al. 2000). Here we emphasize that we can still, without difficulty, distinguish between the BH and NS accretion models because, for example, the radiation from an accreting NS is dominated by thermal emission from the stellar surface (Rutledge et al. 1999), which is absent if a BH is involved.

If the point source is a NS, the dominant radiation observed by *Chandra* must come from small regions. Pavlov et al. (1999) offered a two-component model with hydrogen polar caps and the rest of the cooling NS surface composed of Fe. The hotter polar caps are the result of higher conductivity of hydrogen as compared with Fe. In their model non-standard cooling is not allowed, since the temperature difference between the polar caps and the rest of the surface should be small, less than a factor of 2. Here we offer a promising alternative NS model.

There is some evidence for significant magnetospheric activities (which can be responsible for polar cap heating) in some NS where no radio pulsar has been found, such as Geminga. Therefore, the apparent absence of a radio pulsar and/or a plerion should not be used as evidence against polar cap heating (Paccini 2000). Also accreting NS model can explain small hot region (Chakrabarty et al. 2000). With these additional heat source for the hotter component, larger temperature difference between the hotter and cooler components is expected, and hence there is no conflict with the faster non-standard cooling.

REFERENCES

1. Aschenbach, B. et al. 1999, IAUC No. 7249
2. Chakrabarty, D., Pivovaroff, M.J., Hernquist, L.E., Heyl, J.S., & Narayan, R. 2000, ApJ, submitted (astro-ph/0001026)
3. Chevalier, R.A., & Lian, E. 1989, ApJ, 346, 847
4. Colpi, M. 1988, ApJ, 326, 223
5. Colpi, M., Shapiro, S.L., & Wasserman, I. 1996, ApJ, 470, 1075
6. Ergma, E., & van den Heuvel, E.P.J. 1998, A&A, 331, L29
7. Fesen, R.A., Becker, R.H., & Blair, R.H. 1987, ApJ, 313, 378
8. Holt, S.S., Gotthelf, E.V., Tsunemi, H., & Negoro, H. 1994, PASJ, 46, L151
9. Hughes, J.P., Rakowski, C.E., Burrows, D.N., & Slane, P.O. 2000, ApJ, in press (astro-ph/99104741)
10. Mineshige,S., Nomura,H., Hirose,M., Nomoto,K., & Suzuki,T. 1997, ApJ, 489, 22
11. Mineshige, S., Nomoto, K., & Shigeyama, T. 1993, A&A, 267, 95
12. Narayan, R., McClintock, J.E., & Yi, I. 1996, ApJ, 457, 821
13. Nomoto, K., Iwamoto, K., & Suzuki, T. 1995, Phys. Rep., 256, 173
14. Paccini, F. 2000, Proceedings of IAU Symposium 195, in press
15. Pandharipande, V. R., Pines, D., & Smith, R. A. 1976, ApJ, 208, 550
16. Pavlov, G.G., Shibanov, Yu.A., Ventura,J. & Zavlin, V.E., 1994, A&A, 289, 837
17. Pavlov, G.G., Zavlin, V.E., Aschenbach, B., Trumper, J., & Sanwal, D. 1999, ApJ, submitted (astro-ph/9912024)
18. Rutledge, R.E., Bildsten, L., Brown, E.F., Pavlov, G.G., & Zavlin, V.E. 1999, ApJ, in press (astro-ph/9909319)
19. Tanaka, Y., & Lewin, W. H. G., 1995, in X-ray binaries, ed. W.H.G. Lewin, J. van Paradijs, E.P.J. van den Heuvel (Cambridge U.P., Cambridge), 126
20. Tananbaum, H. et al. 1999, IAUC No. 7246
21. Tsuruta, S. 1998, Phys. Rep., 292, 1
22. Umeda, H., Tsuruta, S., & Nomoto, K. 1994, ApJ, 433, 256
23. Umeda, H., Nomoto, K., Tsuruta, S., & Mineshige, S. 1999, ApJ, submitted (astro-ph/9910113)
24. van den Bergh, S., & Pritchet, C.J. 1986, ApJ, 307, 723
25. Zampieri, L., Shapiro, S.L., & Colpi, M. 1998, ApJ, 502, L149

Black Hole Emergence in Supernovae

Shmuel Balberg[1], Stuart L. Shapiro [1,2] and Luca Zampieri[1,3]

[1]*Department of Physics, University of Illinois at Urbana–Champaign, Urbana, IL*
[2]*Department of Astronomy and National Center for Supercomputing Applications University of Illinois at Urbana–Champaign, Urbana, IL*
[3]*Department of Physics, University of Padova, Padova, Italy*

Abstract. If a black hole formed in a core-collapse supernova is accreting material from the base of the envelope, the accretion luminosity could be observable in the supernova light curve. We present results of a fully relativistic numerical investigation of the fallback of matter onto a black hole in a supernova and examine conditions which would be favorable for detection of the black hole. In general, heating by radioactive decays is likely to prevent practical detection of the black hole, but we show that low energy explosions of more massive stars may provide an important exception. We emphasize the particular case of SN1997D in NGC1536, for which we predict that the presence of a black hole could be inferred observationally within the next year.

INTRODUCTION

Theory suggests that the compact object formed in a core-collapse supernova can be either a neutron star or a black hole, depending on the character of the progenitor and the details of the explosion [1]. The presence of several radio pulsars in sites of known supernovae provides substantial observational evidence that neutron stars are indeed created in supernovae, but similar evidence for a black hole - supernova connection is still mostly unavailable (see [2] for recent indirect evidence).

A newly formed black hole in a supernova can be identified directly if it imposes an observable effect on the continuous emission of light that follows the explosion - the *light curve*. In particular, if some material from the bottom of the expanding envelope remains gravitationally bound to the black hole, it will gradually fall back onto it, generating an accretion luminosity [3]. The black hole can be said to "emerge" in the supernova light curve if and when this luminosity becomes comparable to the other sources that power the light curve.

BLACK HOLE EMERGENCE IN THE LIGHT CURVE

Since the material which remains bound to the black hole following a supernova is outflowing in an overall expansion, the accretion rate must decrease in time. The expansion will also cause pressure forces to become unimportant eventually, and the accretion will proceed as dust-like, following a power-law decline in time according to $\dot{M} \propto t^{-5/3}$ [4]. As shown in [3], the accretion flow and the radiation field proceed as a sequence quasi-steady-states, and the accretion luminosity can therefore be estimated according to the formula of Blondin [5] for stationary, spherical, hypercritical accretion onto a black hole ($L \propto \dot{M}^{5/6}$). The accretion luminosity then takes the form [3,6,7]:

$$L_{acc}(t) \propto L_{acc,0} t^{-25/18} \,, \tag{1}$$

where $L_{acc,0}$ depends on the kinetic energy, density and composition of the accreting material at the onset of dust-like flow.

Heating by decays of radioactive elements synthesized in the explosion may provide a significant source of luminosity in the late-time light curve. The time dependence of radioactive heating rate for an isotope X may be estimated as [8]

$$Q_X(t) = M_X \varepsilon_X f_{X,\gamma}(t) e^{-t/\tau_X} \,, \tag{2}$$

where M_X is the total initial mass of the isotope X in the envelope, τ_X is the isotope's life time, and ε_X is the energy generation rate per unit mass. The factor $f_{X,\gamma}(t)$ reflects that not all $\gamma-$rays emitted in the decays are efficiently trapped in the envelope (and so do not contribute to the UVOIR luminosity).

Since accretion luminosity decreases as a power law in time while radioactive heating declines exponentially, then - assuming that spherical accretion persists - the accretion luminosity must eventually become the dominant source in the light curve. Furthermore, the non-exponential character of the accretion luminosity should be readily distinguishable in observations, announcing that the black hole has "emerged" in the light curve.

REALISTIC SUPERNOVAE

The typical amount of radioactive elements observed in type II supernovae suggests that an observation of black hole emergence in the light curve will usually be impractical. For example, luminosity due to accretion onto a hypothetical black hole in SN1987A would become comparable to the heating rate due to positron emission in ^{44}Ti decays only ~ 900 years after the explosion. At this time the luminosity will have dropped to only $\sim 10^{32}$ ergs s^{-1} [3].

An important exception is expected in the case of higher mass progenitors, $M_* = 25 - 40 \, M_\odot$. Explosions of such stars are likely to involve significant early fallback even while the explosion is still proceeding. The survey of Woosley and Weaver [9] suggests that, in general, larger mass stars leave behind larger remnants and expel a smaller amount of radioactive isotopes (since these are synthesized in the deepest layers of the envelope, and a significant fraction is advected back onto the collapsed core). Clearly, for such an explosion, there is likely to be a larger reservoir of bound material for late time accretion, so that combined with the low background of radioactive isotopes, an actual detection of black hole emergence may become feasible. We have recently conducted a numerical investigation of the expected emergence of a black hole in such supernovae [7]. This investigation was carried out with the spherical, fully relativistic radiation-hydrodynamics code described in [3], modified to include a variable chemical composition with a detailed photon opacity table, and to account for radioactive heating.

Black Hole Emergence in "Radioactive-Free" Supernovae

The most favorable case for identifying black hole emergence in supernova would be a low-to-medium energy ($\leq 1.3 \times 10^{51}$ ergs) explosion of a progenitor star with a mass of $35 - 40 \ M_\odot$, where the ejected envelope is expected to be practically free of radioactive isotopes [9]. For such supernovae, the black hole should emerge within a few tens of days after the explosion. As an example, we show in Fig. 1 the calculated light curve of such an explosion, based on the theoretical model S35A of [9] ($M_* = 35 \ M_\odot$, $M_{BH} = 7.5 \ M_\odot$). The luminosity at emergence is $\gtrsim 10^{37}$ ergs s^{-1}, after which the light curve clearly follows a power law decline in time. If such a supernova were observed, it would offer an explicit opportunity to confirm the presence of a newly formed black hole [3].

SN1997D

While such an ideal candidate is not available at present, SN1997D may provide a marginally *observable* case for idenfying the emergence of a black hole. Discovered on January 14, 1997 in the galaxy NGC 1536, SN1997D is the most sub-luminous type II supernova ever recorded. Through an analysis of the light curve and spectra, Turatto et al. [10] suggested that the supernova was a low energy explosion, $\sim 4 \times 10^{50}$ ergs, of a 26 $M_\odot$ star. The observed late-time light curve (up to 416 days after the explosion) is consistent with only $\sim 0.002 \ M_\odot$ of ^{56}Co in the ejected envelope, much lower than the $\sim 0.1 \ M_\odot$ typical of most type II supernova (0.075 $M_\odot$ in SN1987A). In a preliminary investigation, Zampieri et al. [6] pointed out that the 3 $M_\odot$ black hole (prdecited for this model) may emerge in SN1997D as early as ~ 3 years after the explosion, with an accretion luminosity ranging between 10^{35} to as much as 5×10^{36} ergs s^{-1}.

Our calculated light curve for SN1997D based on the best-fit post-explosion model of [10] is shown in Fig. 1. The earlier part of the light curve is in good agreement with the observed data, while at a later time we find that the black hole emerges about 1050 days after the explosion - which corresponds to late 1999 - *NOW!*. Figure 2 compares the heating due to the isotopes ^{56}Co, ^{57}Co and ^{44}Ti to the accretion luminosity. Note that radioactive heating (especially ^{44}Ti) is never negligible with respect to the accretion luminosity, so the total luminosity does not fall off as an exact power law. Nonetheless, the presence of the black hole could still be inferred by attempting to decompose the total light curve.

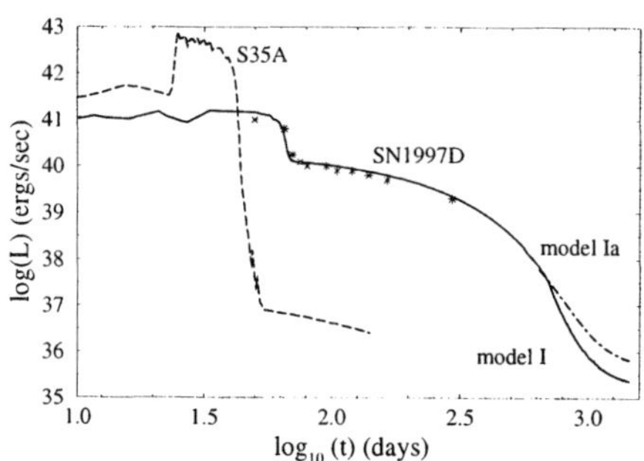

FIGURE 1. Light curves including accretion lumninosity for a 35 $M_\odot$ progenitor (model S35A of [9]), the best fit model of [10] for SN1997D (model I), and a variant of SN1997D where the initial conditions allow for a larger late time luminosity (model Ia).

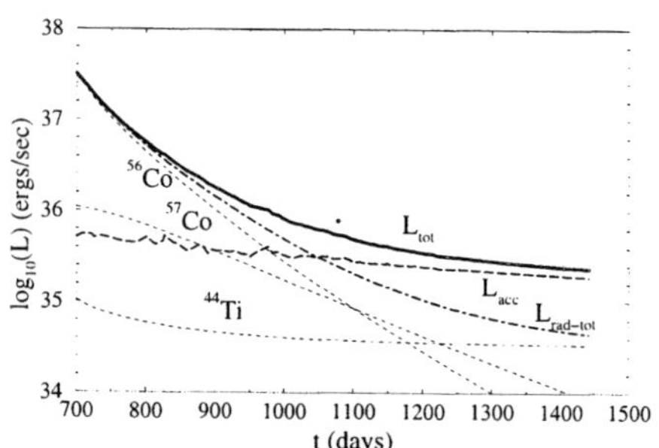

FIGURE 2. The total luminosity of model I for SN1997D and the contributions to the light curve by radioactive heating of ^{56}Co, ^{57}Co and ^{44}Ti, their total ($L_{rad-tot}$), and the accretion luminosity. The arrow marks the time at which $L_{acc} = \frac{1}{2}L_{tot}$.

In this calculation, the total luminosity at emergence is about 7×10^{35}ergs s^{-1}. However, this luminosity is dependent on the finer details of the initial profile, which are difficult to constrain from the early light curve. Considering these uncertainties and those regarding the abundances of ^{57}Co and ^{44}Ti, we find through a revised analytical estimates (which account for γ-ray transparency and Eddington-rate limits on the accretion flow), that while the time of emergence is fairly well determined, the plausible range for the luminosity at emergence is $0.5 - 2 \times 10^{36}$ ergs s^{-1}. One example, where the luminosity at emergence is $\sim 1.4 \times 10^{36}$ ergs s^{-1} is also shown in Fig. 1. In this case, the accretion luminosity is sufficiently high that contribution of radioactive heating does not cause any significant deviation from a power-law decay. We estimate that such a luminosity is still marginally detectable ($m_v \approx 29$) with the HST STIS camera, so that if observed, SN1997D could provide first direct observational evidence of black hole formation in supernova within the next year.

REFERENCES

1. Fryer, C. L., *in these proceedings* (1999).
2. Israelian, G. et al., Nature, 401, 142-144 (1999).
3. Zampieri, L., Colpi, M., Shapiro, S. L., and Wasserman, I., ApJ, 505, 876 (1998).
4. Colpi, M., Shapiro, S. L., and Wasserman, I., ApJ, 470, 1075 (1996).
5. Blondin, J. M., ApJ, 308, 755 (1986).
6. Zampieri, L., Shapiro, S. L., and Colpi, M., ApJL, 502, L149 (1998).
7. Balberg, S. Shapiro, S. L., & Zampieri, L., ApJ, *submitted* (1999)
8. Woosley, S. E., Pinto, P. A., and Hartmann, D., ApJ, 346, 395 (1989).
9. Woosley, S. E., and Weaver, T. A., ApJS, 101, 181 (1995).
10. Turatto, M. et al., ApJL , 498, L129 (1998).

Supernova Environments in Hubble Space Telescope Images

Schuyler D. Van Dyk[*], Chien Y. Peng[†],
Aaron J. Barth[‡], and Alexei V. Filippenko[††]

[*]*IPAC/Caltech, Pasadena, CA 91125*
[†]*Steward Observatory, University of Arizona, Tucson, AZ 85721*
[‡]*Harvard-Smithsonian Center for Astrophysics, Cambridge, MA 02138*
[††]*Dept. of Astronomy, University of California, Berkeley, CA 94720*

Abstract. The locations of supernovae in the local stellar and gaseous environment in galaxies contain important clues to their progenitor stars. Access to this information, however, has been hampered by the limited resolution achieved by ground-based observations. High spatial resolution *Hubble Space Telescope (HST)* images of galaxy fields in which supernovae had been observed can improve the situation considerably. We have examined the immediate environments of a few dozen supernovae using archival post-refurbishment *HST* images. Although our analysis is limited due to signal-to-noise ratio and filter bandpass considerations, the images allow us for the first time to resolve individual stars in, and to derive detailed color-magnitude diagrams for, several environments. We are able to place more rigorous constraints on the masses of these supernovae. A search was made for late-time emission from supernovae in the archival images, and for the progenitor stars in presupernova images of the host galaxies. In particular, we highlight the results for the Type II SN 1979C in M100. In addition, we have identified the progenitor of the Type IIn SN 1997bs in NGC 3627. We also add to the statistical inferences that can be made from studying the association of SNe with recent star-forming regions.

INTRODUCTION

A primary goal of supernova (SN) research is an understanding of the progenitor stars and explosion mechanisms of the different SN types. Unfortunately, a SN leaves few traces of the star that underwent the catastrophic event. Unambiguous information can be derived if the progenitors can be directly identified in pre-explosion images, but to date this has been possible only for SNe 1987A, 1961V, 1978K, and 1993J. In the absence of direct information about the progenitors, scrutiny of the host galaxies and local stellar and gaseous SN environments, in favorable cases, yields useful constraints on progenitor ages and masses. However, previous studies have been hampered by the limited spatial resolution of ground-

CP522, *Cosmic Explosions: Tenth Astrophysical Conference,*
edited by Stephen S. Holt and William W. Zhang

based observations. The superior angular resolution of the *Hubble Space Telescope* (*HST*) offers the potential for greater understanding of SN environments.

ARCHIVAL DATA AND PHOTOMETRY

An investigation using *HST* archival data was begun, to (1) study the stellar populations in the immediate SN environments in cases where individual stars or clusters are resolved, and to use this information to help constrain the progenitor star age and mass; (2) search for progenitor stars in images taken prior to SN explosions; (3) search for old visible SNe; (4) augment ground-based data on the statistical association of the different SN types with star-forming regions; and (5) measure magnitudes for SNe observed "accidentally" while still bright.

Van Dyk et al. [1,2] report on WFPC2 results for the environments of 12 SNe II, 5 SNe Ib/c, and 16 SNe Ia. Although *HST* images offer the tremendous advantage of higher angular resolution over similar ground-based data, the images were almost exclusively obtained for other purposes. Limitations include small field of view, images underexposed at the SN sites, and images made through only one filter.

When possible, we employed PSF-fitting DAOPHOT photometry, using Tiny Tim PSFs [3]. To determine uncertainties, we tested with artificial stars of known brightnesses. Count rates were converted to magnitudes using either the synthetic photometric zeropoints from [4] or [5]. We analyze resulting color-magnitude diagrams with theoretical isochrones for solar metallicity from [6]. Where no resolved stars are apparent, we performed aperture photometry, providing a larger-scale measurement of magnitude, and where possible, color.

The problem of overall uncertainty in SN positions is particularly severe for *HST* images. We performed careful verification of the image astrometry, which, in "coarse" or "gyro" mode, can be grossly in error. Despite $\leq 1''$ SN position accuracy, *HST* astrometry is good to $\lesssim 1''.5-2''$. We assign an estimated error centered on the nominal SN position and regard the error circle as the SN environment. Even with the relative lack of precision, the fine image detail afforded by *HST* in many SN environments provides unprecedented information about SN progenitors.

RESULTS AND CONCLUSIONS

For the first time, through deep multicolor imaging, undertaken, in particular, by the *HST* Cepheid projects, we have been able to produce color-color and color-magnitude diagrams for several SN environments, providing constraints on the ages and masses of stellar populations associated with the SNe, and allowing us to indirectly infer the ages and masses of the SN progenitors, for, e.g., SNe 1940B, 1983V, 1987K, and, 1988M. From archival and GO images of the Type II SN 1979C environment in M100 (Figure 1), Van Dyk et al. [2] were able to rigorously constrain the SN progenitor initial mass, to $M \approx 17-18 \ M_\odot$.

We may have identified the stellar progenitor of SN 1997bs, clearly visible on a pre-SN *HST* image of NGC 3627 (Figure 2), as a luminous supergiant star with $M_V \simeq -7.4$ mag. (However, SN 1997bs may be an η Car-like superoutburst, like SN 1961V.) This would be only the fifth SN progenitor to ever be identified in pre-explosion images.

We have recovered the old Type II SNe 1979C, 1986J, and possibly 1981K, as expected, since these have shown evidence of interaction with circumstellar gas. From GO multi-band imaging obtained 17 years after explosion, we recover SN 1979C at, e.g., $m_{F439W} = 23.37$ ($m_B[\mathrm{max}] = 11.6$ mag). Additionally, the Type Ia SN 1994D was caught "by accident" in *HST* images, adding to its existing ground-based light curves.

The colors of the diffuse emission in Type Ia SNe environments indicate that the stellar populations are generally old and red, consistent with red giants, although four SN environments contained bright clusters and H II regions.

The five Type Ib/c SNe appear more closely associated with brighter, more massive stellar regions than do the Type II SNe, contrary to the ground-based results [7], suggesting that the Type Ib/c SN progenitors may be more massive, in general, than Type II SN progenitors.

We will continue to acquire images of SN environments and will build on the current study in future papers.

REFERENCES

1. Van Dyk, S. D., et al., *AJ*, **118**, 2331 (1999a).
2. Van Dyk, S. D., et al., *PASP*, **111**, 313 (1999b).
3. Krist, J., in Calibrating *Hubble Space Telescope*: Post Servicing Mission (Baltimore: STScI), 311 (1995).
4. Holtzman, J. A., et al., *PASP*, **107**, 1065 (1995).
5. Hill, R. J., et al., *ApJ*, **496**, 648 (1998).
6. Bertelli, G., et al., *A&AS* **106**, 275 (1994).
7. Van Dyk, S. D., Hamuy, M., & Filippenko, A. V., *AJ*, **111**, 2017 (1996).

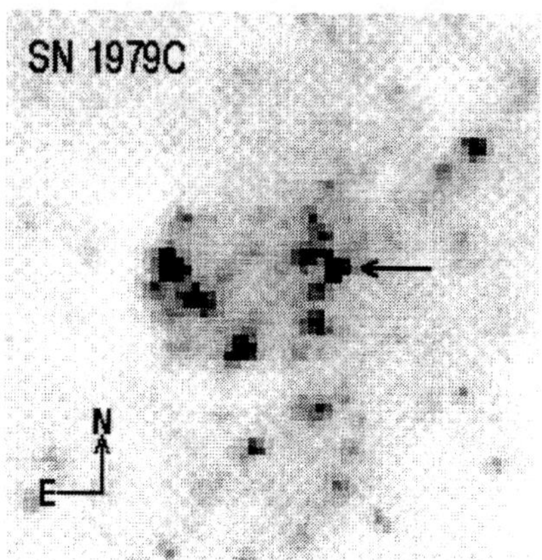

FIGURE 1. *HST* WFPC2 F555W image of the stellar environment of SN 1979C in M100. The arrow points to the SN.

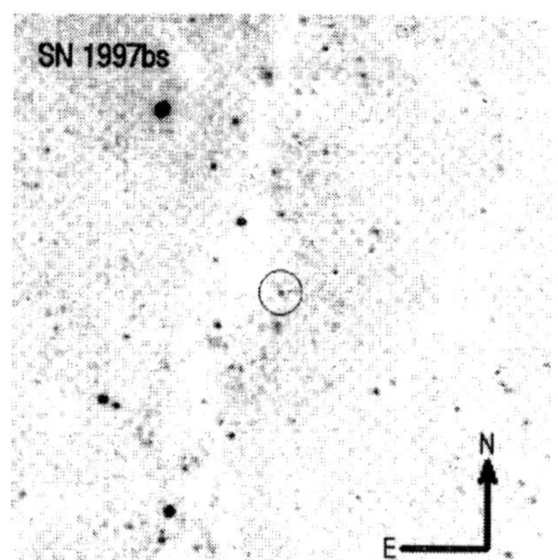

FIGURE 2. *HST* WFPC2 F606W image of the stellar progenitor of SN 1997bs in NGC 3627 (within circle).

Properties of SN1978K from Multi-wavelength Observations

Eric M. Schlegel[*], Stuart Ryder[†], L. Staveley-Smith[‡], E. Colbert[♯],
R. Petre[♭], M. Dopita[◇], & D. Campbell-Wilson[△]

[*]*Harvard-Smithsonian Center for Astrophysics, Cambridge, MA USA,*
[†]*Anglo-Australian Observatory, Epping, Australia,*
[‡]*Australia National Telescope Facility, Epping, Australia,*
[♯]*University of Maryland, College Park, MD USA,*
[♭]*NASA-Goddard Space Flight Center, Greenbelt, MD USA,*
[◇]*Mount Stromlo and Siding Spring Observatories, Weston Creek, Australia,*
[△]*University of Sydney, Sydney, Australia*

Abstract.
We update the light curves from the X-ray, optical, and radio bandpasses which we have assembled over the past decade, and present two observations in the ultraviolet using the *Hubble Space Telescope* Faint Object Spectrograph. The HRI X-ray light curve is constant within the errors over the entire observation period which is confirmed by *ASCA* GIS data obtained in 1993 and 1995. In the UV, we detected the Mg II doublet at 2800 Å and a line at $\sim$3190 Å attributed to He I 3187 at SN1978K's position. The optical light curve is formally constant within the errors, although a slight upward trend may be present. The radio light curve continues its steep decline.

The longer time span of our radio observations compared to previous studies shows that SN1978K belongs in the class of highly X-ray and radio-luminous supernovae. The Mg II doublet flux ratio implies the quantity of line optical depth times density is $\sim 10^{14}$ cm^{-3}. The emission site must lie in the shocked gas.

SN 1978K is one of a handful of extremely luminous, X-ray emitting supernovae [1]. X-ray observations are difficult to obtain, so to date, few observations exist. Currently, eleven supernovae are known to emit, or have emitted, X-rays sometime in the months and years following their outbursts (SN 1978K, SN 1979C, SN 1980K, SN 1986J, SN 1987A, SN 1988Z, SN 1993J, SN 1994I, SN 1994W, SN 1995N, and SN 1998bw (references: see [2]). Of these, the X-rays from SN 1987A largely resulted from Compton scattered γ-rays from the radioactive decay of ^{56}Co. SN 1987A is starting to undergo a circumstellar interaction [3]; it is expected to become a very bright X-ray source [4–6]. SN 1998bw may be associated with the GRB 980425, so its relationship to the other X-ray emitting SNe is uncertain [7].

The remaining nine supernovae are all believed to emit X-rays solely because

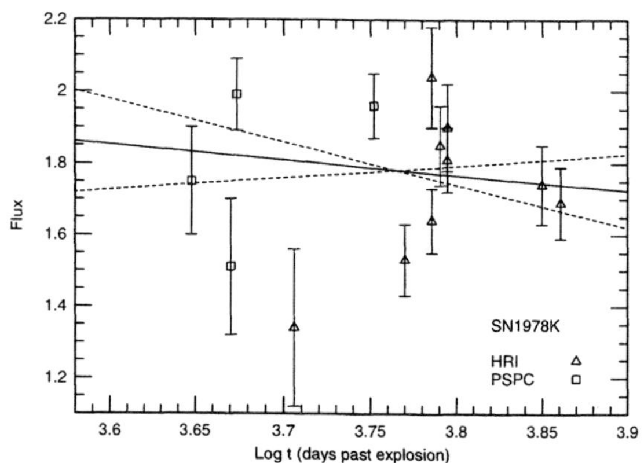

FIGURE 1. The *ROSAT* X-ray light curve in the 0.2-2.0 keV band for SN 1978K. The dashed lines show the ±90% range of the slope. The vertical axis is in units of 10^{-12} ergs s^{-1} cm^{-2}.

of a circumstellar interaction. About half are SN IIn [8,9]. Of the nine, six have one or two observations, leaving an "observable" three: SN 1978K, SN 1986J, and SN 1993J. SN 1986J and SN 1993J have been reported to be fading [10,11]; the X-ray light curve of SN 1978K appeared to be constant as of 1995 [12].

Since that paper, we have accumulated another six observations, doubling the data and increasing the coverage to 7 years, about one third of the life of SN 1978K. The light curve shows all of the data of SN1978K obtained by *ROSAT*. The X-ray emission is constant; the range on the power law index of the density profile is ∼4-12. *ASCA* has observed SN 1978K twice (1993, 1995) with no change detectable between the observations, supporting the flat *ROSAT* light curve. The luminosities of the *ASCA* (1-10 keV) and *ROSAT* (0.1-2.4 keV) bands provide an estimate of the temperature of the reverse shock [13,14] of T_e ∼6×10^6 K.

We detected the Mg II 2800 doublet and He I 3187 Å at the location of SN 1978K using the Faint Object Spectograph on *Hubble Space Telescope*. None of the lines are resolved which implies that the shock velocities are low; the instrumental resolution is of the order of 10-12Å which corresponds to ∼500-600 km s^{-1}. The Mg II doublet implies the product of line optical depth times number density is ∼10^{14} cm^{-3}. This is high for the intercloud wind [15] even if the line optical depth is 10^{2-3}.

SN 1978K has been monitored in the optical by photometry and spectroscopy;

FIGURE 2. *HST* FOS spectra at SN1978K. The top plot shows the 1996 red spectrum at the position of SN 1978K; the middle plot shows the spectrum obtained 1".5 away to the NW. The spectra have not been corrected for reddening nor for the redshift of SN 1978K (439 km s^{-1}). Note that both the top and middle spectra contain a small flat excision, just blueward of the Mg II line, where a noisy pixel was located. The bottom spectra expand the regions around the Mg II line at 2800Å (left) and the line at 3190Å (right).

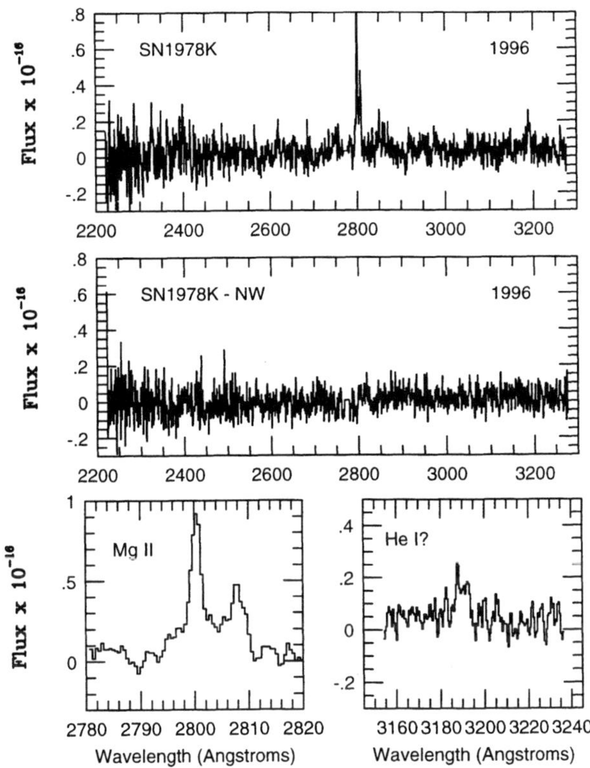

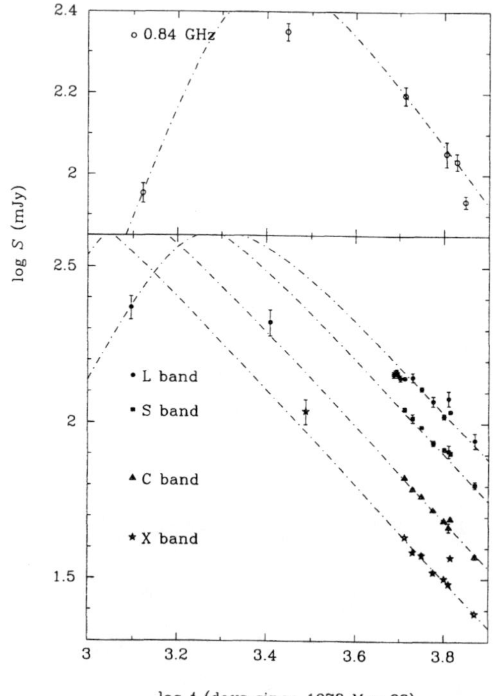

space precludes including the data. The B band light curve is flat within the errors, although a slight rise may be present. An AAO spectrum obtained in October 1996 shows emission lines of H, He, Mg, [O I], [O III], and [Fe II].

SN 1978K has also been monitored at irregular intervals using the 6 antennas of the Australia Telescope Compact Array (ATCA) since its discovery [16]. Deconvolved images of SN 1978K show the source still to be unresolved at all frequencies. No combination of parameters can make the L-band model fit both of the early [17] points, and fit the data at late epochs.

A detailed picture is emerging for the X-ray luminous supernovae, particularly the SN IIn variety. The supernova explodes into a very dense ($\sim 10^{6-8}$ cm^{-3}) circumstellar medium. In such a medium, the evolution of the supernova is accelerated, so the evolving remnant goes directly from the explosive phase to the radiative phase [21–24]. The object then radiates large quantities of energy in the X-ray, ultraviolet, and radio bands. The velocities of the emission lines are low

($\lesssim 10^3$ km s^{-1}) representing the extreme deceleration that has occurred. SN1978K clearly belongs to this category of behavior.

In contrast, the late-time observations of the Type II-L supernovae SN 1979C and SN 1980K [25] show line widths of several 1000 km s^{-1} and emission line strengths well-described by the shell model [14]. Details are in [2].

REFERENCES

1. Schlegel, Eric M. 1995, RepProgPhys, 58, 1375
2. Schlegel, Eric M. et al. 1999, AJ, in press (December)
3. Hasinger, G., Aschenbach, B., & Trümper, J. 1996, A&A, 312, L9
4. Chevalier, R. & Dwarkadas, V. V. 1995, ApJ, 452, L45
5. Masai, K. & Nomoto, K. 1994, ApJ, 424, 924
6. Suzuki, T., Shigeyama, T., & Nomoto, K. 1993, AJ, 274, 883
7. Pian, E. et al. 1999, ApJ, in press
8. Schlegel, Eric M. 1990, MNRAS, 244, 269
9. Filippenko, A. 1989, AJ, 97, 726
10. Zimmermann, H.-U. et al. 1994, Nature, 367, 621
11. Houck, J. C. et al. 1998, ApJ, 493, 431
12. Schlegel, Eric M., Petre, R., & Colbert, E. J. 1996, ApJ, 456, 187
13. Fransson, C., Lundqvist, P., & Chevalier, R. 1996, ApJ, 461, 993
14. Chevalier, R. & Fransson, C. 1994, ApJ, 420, 268
15. Chugai, N. N., Danziger, I. J., & Della Valle, M. 1995, MNRAS, 276, 530
16. Ryder, S. et al. 1993, ApJ, 416, 167
17. Peters, W. L. et al. 1994, MNRAS, 269, 1025
18. Weiler, K. W., Panagia, N., & Sramek, R. 1990, ApJ, 364, 611
19. Weiler, K. W. et al. 1986, ApJ, 301, 790
20. Montes, M. J., Weiler, K. W., & Panagia, N. 1997, ApJ, 488, 792
21. Chevalier, R. 1974, ApJ, 188, 501
22. Shull, J. M. 1980, ApJ, 237, 769
23. Terlevich, R., Tenorio-Tagle, G., Franco, J., & Melnick, J. 1992, MNRAS, 255, 713
24. Wheeler, J. C., Mazurek, T., & Sivaramakrishnan, A. 1980, ApJ, 237, 781
25. Fesen, R. et al. 1999, AJ, 117, 725

Radio Emission from Supernova 1993 J

W. K. Rose

Department of Astronomy, University of Maryland, College Park, MD 20742

Abstract.
Supernova 1993 J exploded in the nearby spiral galaxy M81. Its initial spectrum contained hydrogen lines and therefore it was classified as a Type II supernova. However, subsequent disappearance of hydrogen lines and appearance of helium lines made it similar to a type Ib supernova. Radio emission was detected from SN 1993 J shortly after the first peak in optical brightness. In this paper we present a model for the observed radio radiation that involves the interaction of a collisionless shock front generated by the supernova ejecta and a stellar wind produced by the progenitor supergiant. We discuss electron acceleration, magnetic field generation and interpret an observed break in the radio spectrum.

INTRODUCTION

Supernova 1993 J was a remarkable supernova from several points of view. It was discovered on March 28 at magnitude V = 11.8 in M81 [1] and on March 30 it reached maximum brightness of V = 10.7 magnitude, which made it the brightest supernova in 60 years in the northern hemisphere. SN 1993 J is a transition object because it initially appeared with Type II supernova spectral characteristics and then evolved into a Type Ib object [2,3]. Its light curve is also unusual [4] and it can be modeled as caused by core collapse and subsequent explosion in a reddish yellow supergiant that had lost most of its hydrogen-rich envelope. This indicates that at the time of outburst the progenitor supergiant was close to evolving into a hydrogen deficient Wolf-Rayet star. Such stars are known to have stellar winds with mass loss rates of 10^{-5} - 10^{-4} $M\odot$/yr and it has been suggested that they are progenitors of Type Ib supernovae.

Radio emission was first observed from SN 1993 J on April 2, 1993 at 1.3 cm [5]. Its radio structure has been measured with VLBI observations [6,7]. These observations showed that strong radio emission came from an approximately spherically symmetric region of high ($\simeq$ 15,000 - 18,000 kms^{-1}) velocity. It has been suggested that the radio emission is produced in a shock front formed between the supernova ejecta and circumstellar gas emitted from the progenitor supergiant as a stellar wind. In addition the radio observers proposed that the relativistic electrons responsible for the radio synchrotron were accelerated to relativistic energies by the

CP522, *Cosmic Explosions: Tenth Astrophysical Conference,*
edited by Stephen S. Holt and William W. Zhang
© 2000 American Institute of Physics 1-56396-943-2/00/$17.00

supernova shock front.

Under most physical conditions the thickness of a shock front is approximately a collisional mean free path. However, it has been recognized for some time that a supernova shock front is collisionless (i.e. the formation of the shock is the result of collisionless plasma processes). In previous publications we have discussed the collisionless interaction between plasmas in the context of extragalactic jets [8,9] and galactic X-ray sources [10,11] where it is known that relativistic velocities are relevant. These calculations were based on a model for plasma turbulence described most completely in references 8 and 10. Here we suggest that our calculations may also be useful in providing a physical explanation for how electrons are accelerated to relativistic energies in SN 1993 J and other supernova shocks propagating into a stellar wind. Because of the high velocity ($\simeq$ 18,000 kms^{-1}) between the shock front and ionized circumstellar gas excitation of the well known plasma electron two stream instability is predicted to occur. This instability leads directly to the generation of resonant plasma wave energy. As wave energy builds up it reaches the necessary threshold for plasma wave-wave instabilities such as the oscillating two stream instability and other parametric instabilities to occur. These latter instabilities generate nonresonant plasma and ion acoustic waves. The resultant plasma turbulence is believed to lead to the formation of Langmuir solitons [12], which are localized regions of electric field that oscillate at the plasma frequency (ω_p = 5.6 x 10^4 $\sqrt{n_e}$ with n_e equal to the electron number density). Because they are oscillatory Langmuir solitons preferentially accelerate high speed plasma electrons to relativistic energies. Low speed electrons do not experience significant acceleration because the sign of the electric field changes a number of times during their traversal of a soliton.

RESULTS

In this paper we interpret VLBI measurements [6] made at 30.2, 50.2 and 91.0 days after supernova outburst at four frequencies namely 22.2 GHz, 14.9 GHz, 8.4 GHz and 4.86 GHz. On the first of these dates radio emission was only detected at 22.2 GHz because free-free absorption made the signal too weak for detection at lower frequencies. At 91 days after outburst the radio source became detectable at 4.86 GHz and free-free absorption does not appear to appreciably affect observed intensities at higher frequencies. Assuming the low frequency decline in intensity was caused by free-free absorption we calculate that the progenitor supergiant had a stellar wind with mass loss rate and outflow velocity equal to 10^{-4} M$\odot$/yr and 10 kms^{-1} respectively. In making this estimate we have assumed that gas flowing into the shock has a temperature of 10^5 K. Since this estimate is only sensitive to the density and temperature of circumstellar gas a higher (lower) mass loss rate and correspondingly lower (higher) outflow velocity are also consistent with observations. This estimate gives the gas density into which the shock propagates at particular times after outburst. Following [6] we take the shock outflow velocity during the first 91 days after outburst to be 18,000 kms^{-1} and the distance to M81 where the supernova took place to be 4 Mpc.

Although the presence of stellar wind magnetic fields sufficiently strong to produce the observed synchrotron radiation can not be ruled out, it is unlikely that their spatial distribution would be approximately spherical, and therefore even if the shock were spherical the observed synchrotron radiation would be asymmetric whereas it is observed to be approximately spherical [6,7]. Faraday's law of induction implies that if the electric field required to decelerate stellar wind protons (and ions) approaching the collisionless shock is not describable by an electrostatic scalar potential then a magnetic field can be generated by the shock. The shock front electric field formed by decelerated electrons must slow 2MeV protons and therefore if the stopping distance of the plasma moving into the shock can be estimated we can also determine approximately the magnetic field generated in the shock from Faraday's law of induction. From references 9 and 10 we find that an order of magnitude estimate for the stopping time of incoming plasma is

$$t_s \sim 3 \times 10^7/\omega_p \tag{1}$$

where ω_p is the plasma frequency. It follows that the stopping distance is $L_p = vt_s = 3 \times 10^7 \, v/\omega_p$ and from Faraday's law the induced magnetic field is

$$\Delta B \sim \frac{c \, Et_s}{L_p} = \frac{c}{v} \, E \tag{2}$$

Since eEL$_p$ = 2 MeV we find

$$\Delta B \sim 7 - 2 \times 10^{-4} gauss \tag{3}$$

at times 30-91 days after outburst. The above estimate for the magnetic field requires that electron Lorentz factors are $\sim 10^4$ in the radio emitting region in order to explain observed radio frequencies. Such high values for characteristic electron γ's would imply that electron inverse Compton losses caused by radiation from the supernova (L $\sim 10^9$ L$\odot$) would be much greater than synchrotron losses and therefore required relativistic electron energies would become too high. Moreover, the observed break in the radio spectrum at 91 days after outburst would not be explained and electron gyroradii would probably be too large. We take the characteristic magnetic field strength in the radio emitting region 30- 91 days after outburst to be ~ 1 gauss and propose that it is generated by plasma turbulence behind the shock wave. With this choice for magnetic field strength inverse Compton losses do not exceed synchrotron losses during the period of observation, gyroradii are not too large and the required Lorentz factors are $\gamma \simeq 10^2$.

If electrons are accelerated continuously to a power law spectrum for some time t and energy losses are from synchrotron radiation then electrons above a certain energy will lose all their energy in time t whereas lower energy electrons will not lose much energy. Therefore we only observe high energy electrons produced during the previous synchrotron lifetime and the energy spectrum of these electrons is one power of E steeper than lower energy electrons [13]. The radio flux observations at 91 days after outburst [6] gives a spectral index between 22.2 GHz and 14.8 GHz of α = .86 whereas between 14.8 GHz and 8.4 GHz it is .487. This change of $\simeq$ .4 in spectral index can be interpreted as a change of one in electron power law distribution (i.e. $N(E) \alpha E^{-p}$ to αE^{-p-1} with $p \simeq 2$) at $\simeq$ 14.8 GHz. If we interpret this break in spectral index [$\alpha = (p-1)/2$] to be caused by continuous acceleration of relativistic electrons and take $\gamma = 10^2$ and B = 1 gauss then the timescale of synchrotron losses becomes approximately one month, which is consistent with observations.

Finally we estimate the number of relativistic electrons required to explain the observed radiation of approximately 50 - 100 mJy assuming the distance to M81 is 4 Mpc. We find that the number of relativistic electrons is about 1.5 x 10^{48}. These electrons must be accelerated to high ($\gamma \simeq 10^2$) energy in less than a month. Assuming the energy source for relativistic electrons is the kinetic energy of electrons passing through the shock then we find that the available energy is about ten times greater than necessary to explain observed fluxes.

REFERENCES

1. Ripero, J. 1993, IAU Circ., No. 5731.
2. Filippenko, A. V., Matheson, T. and Ho, L. C. 1993, ApJ, 415, L103.
3. Swartz, D. A., Clocchiatti, A., Benjamin, R., Lester, D. F., and Wheeler, J. C. 1993, Nature, 365, 232.
4. Schmidt, B. P., et al. 1993, Nature, 364, 600.
5. Van Dyk, S. C., Weiler, K. W., Sramek, R. A., Rupen, M. P., and Panagia, N. 1994, ApJ, 432, L115.
6. Bartel, N. et al. 1994, Nature, 368, 610.
7. Marcaide, J. M. et al. 1995, Science, 270, 1475.
8. Rose, W. K., Guillory, J., Beall, J. H., and Kainer, S. 1984, ApJ, 280, 550.
9. Rose, W. K., Beall, J. H., Guillory, J., and Kainer, S. 1987, ApJ, 314, 94.
10. Rose, W. K. 1994 in Evolution of X-ray Binaries, eds. S. S. Holt and C. S. Day (AIP: New York), p. 499.
11. Rose, W. K. 1995, MNRAS, 276, 1191.
12. Nicholson, D. R. 1983, Introduction to Plasma Theory (John Wiley: New York).
13. Longair, M. S. 1994, High Energy Astrophysics, Volume 2 (Cambridge Univ. Press: Cambridge).

Probing the Geometry of Supernovae with Spectropolarimetry

Douglas C. Leonard, Alexei V. Filippenko, and Thomas Matheson

Department of Astronomy, University of California at Berkeley, Berkeley, CA 94720-3411

Abstract. We present results from a spectropolarimetric survey of young supernovae completed at the Keck Observatory, including at least one example from each of the major supernova types: Ia (1997dt), Ib (1998T, 1997dq), Ib/c-pec (1997ef), IIn (1997eg), and II-P (1997ds). All objects show evidence for intrinsic polarization, suggesting that asphericity may be a common feature in young supernova atmospheres.

INTRODUCTION

Are supernovae (SNe) round? This simple question belies a menacing observational challenge, since all extragalactic SNe remain unresolvable point sources throughout the crucial early phases of their evolution. Since a hot young supernova (SN) atmosphere is dominated by electron scattering, which, by its nature, is highly polarizing, a powerful tool for investigating SN geometry is spectropolarimetry of the expanding fireball shortly after the explosion (Fig. 1).

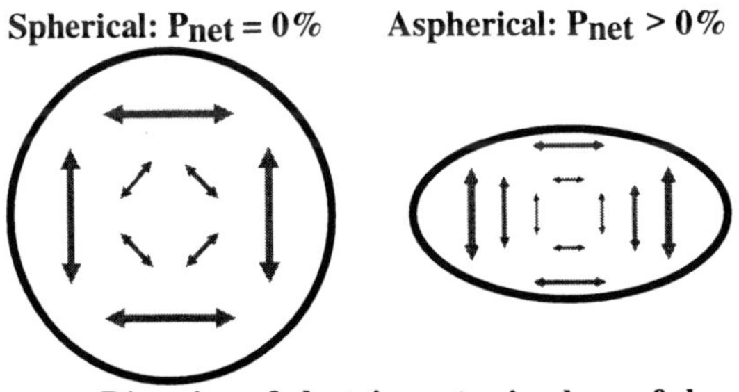

FIGURE 1. Polarization magnitude and direction in the plane of the sky for resolved electron-scattering atmospheres; for an unresolved source (i.e., a SN), only the *net* magnitude and direction can be measured. An aspherical source produces a net linear polarization due to incomplete cancellation of the directional components of the electric vectors. Note that the more highly polarized light (longer arrows) comes from the limb regions in this simple model.

CP522, *Cosmic Explosions: Tenth Astrophysical Conference,*
edited by Stephen S. Holt and William W. Zhang
© 2000 American Institute of Physics 1-56396-943-2/00/$17.00

Typical polarizations of $\sim 1\%$ are expected for moderate ($\sim 20\%$) SN asphericity [1]. Detecting such low polarization requires a very high signal-to-noise ratio, which has limited previous detailed spectropolarimetric studies to only the two brightest recent events, SN 1987A [2] and SN 1993J [3,4]. We thus began a program to obtain spectropolarimetry of nearby SNe using the 10-m Keck telescopes.

A complication in the interpretation of all polarization measurements is disentangling the polarization intrinsic to the object from interstellar polarization (ISP) produced by dust along the line-of-sight. Fortunately, the ISP is constant with time and a smoothly varying function of wavelength. Therefore, we consider distinct spectral polarization features, temporal changes in the overall polarization level, or continuum polarization characteristics differing from the known form produced by interstellar dust as evidence for intrinsic SN polarization.

RESULTS AND DISCUSSION

Single-epoch polarization data for six SNe of various types are shown in Fig. 2. Since determining the intrinsic SN polarization level requires knowledge of the (unknown) ISP, we focus instead on the sharp changes seen in the polarization at the location of strong features in the total flux spectra; these features remain, regardless of the ISP contribution. Since all the objects studied possess spectropolarimetric line features, we conclude that *all* types of SNe show evidence for intrinsic polarization at early times, suggesting that asphericity may be a ubiquitous SN characteristic.

The fact that the strongest spectropolarimetric features are often seen in the troughs of strong P-Cygni lines is not surprising. A simple explanation may be that P-Cygni absorption selectively blocks photons coming from the central, more forward-scattered (and thus less polarized) regions, thereby enhancing the relative contribution of the more highly polarized photons from the limb regions (c.f., Fig. 1). Unfortunately, since different (allowable) choices for the ISP can make inferred intrinsic polarization dips become peaks and vice-versa [5], we cannot say for certain whether the changes seen here in P-Cygni troughs represent increases or decreases in the intrinsic polarization level. We do note, however, that trough polarization increases are seen in the ISP-corrected data of both SN 1987A [2] and SN 1993J [4].

A total flux spectrum dominated by strong line emission without P-Cygni absorption is a distinguishing characteristic of SNe IIn [6], likely resulting from an intense interaction between the SN and a dense circumstellar environment (CSM). SN 1997eg (Fig. 2) shows sharp polarization changes across its strong, multi-component emission lines, suggesting distinct scattering origins for the intermediate (full width at half maximum (FWHM) $\approx$ 2000 km/s) and broad (FWHM $\approx$ 15000 km/s) components. Two additional spectropolarimetric epochs (not shown) revealed a change in continuum polarization level of $\sim 1\%$ over 78 days, further confirming the presence of intrinsic polarization. A detailed analysis combining

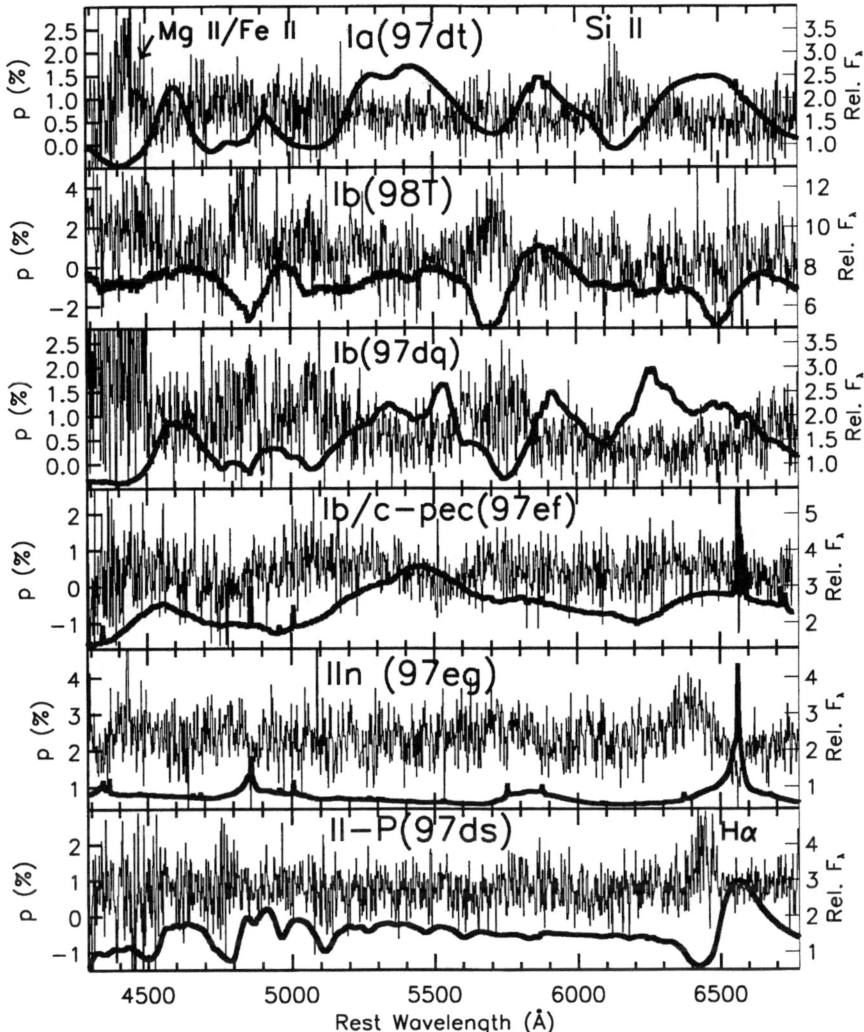

FIGURE 2. Spectropolarimetry (thin lines) and total flux spectra (thick lines) of nearby SNe, all within two months of the explosion. The recession velocities of the host galaxies have been removed for all objects. Note that arbitrary, but allowable (from reddening considerations), estimates of the ISP have been removed from the polarizations of SNe 1997dt, 1998T, 1997dq, and 1997ds in order to better highlight the line features. Since the ISP is in general not known, the intrinsic continuum polarization levels are uncertain; spectropolarimetric line features, however, exist regardless of the ISP, indicating polarization intrinsic to all of these SNe.

spectropolarimetry and total flux spectra of another IIn event, SN 1998S, also found evidence for a highly aspherical ($\gtrsim 45\%$) continuum scattering region, with the CSM likely distributed in a disk-like or ring-like morphology, quite similar to what is seen directly in SN 1987A [7], except much closer to the progenitor in the case of 1998S [5].

CONCLUSION

The number of SNe studied spectropolarimetrically is still very small, but early indications are that all types reveal intrinsic polarization if examined in sufficient detail. In addition to the implications of spectropolarimetry on the core-collapse mechanism, the mass-loss history of evolved stars, and the spatial distribution of SN ejecta, this work has direct consequences on the use of SNe as cosmological distance indicators. Although the empirically-based, standard-candle technique used to measure SN Ia distances does not rely on spherical symmetry, distances derived to SNe II-P through the "expanding photosphere method" [8] would need to be corrected for directionally-dependent flux if asphericity is found to be common among this SN class. We note that SN 1999em, a type II-P event discovered shortly after this conference, showed no evidence for intrinsic polarization when it was observed less than two weeks after the explosion [9]. It will be interesting to see if it remains unpolarized at an age comparable to the II-P observation (SN 1997ds, observed ~ 50 days after explosion) presented here.

ACKNOWLEDGMENTS

We thank Aaron Barth for useful discussions and assistance with the observations and data reduction. Supernova research is supported at UC Berkeley through NSF grant AST-9417213 and NASA grant GO-7434.

REFERENCES

1. Höflich, P. A., *Astron. Astrophys.* **246**, 481 (1991).
2. Jeffery, D. J., *Ap. J.* **375**, 264 (1991).
3. Trammell, S. R., Hines, D. C., and Wheeler, J. C., *Ap. J.* **414**, 21 (1993).
4. Tran, H. D., Filippenko, A. V., Schmidt, G. D., Bjorkman, K. S., Jannuzi, B. T., and Smith, P. S., *P.A.S.P.* **109**, 489 (1997).
5. Leonard, D. C., Filippenko, A. V., Barth, A. J., and Matheson, T., *Ap. J.* submitted (astro-ph/9908040).
6. Schlegel, E. M., *M.N.R.A.S.* **244**, 269 (1990).
7. Crotts, A. P. S., and Heathcote, S. R., *Ap. J.* in press (astro-ph/9907367).
8. Kirshner, R. P., and Kwan, J., *Ap. J.* **193**, 27 (1974).
9. Leonard, D. C., Filippenko, A. V., and Chornock, R. T., *IAU Circ.*, No. 7305 (1999).

SNEWS: A Neutrino Early Warning System for Galactic SN II

Alec Habig* for the SNEWS collaboration

*Boston University Physics Dept., Boston, MA 02215

Abstract. The detection of neutrinos from SN1987A confirmed the core-collapse nature of SN II, but the neutrinos were not noticed until after the optical discovery. The current generation of neutrino experiments are both much larger and actively looking for SN neutrinos in real time. Since neutrinos escape a new SN promptly while the first photons are not produced until the photospheric shock breakout hours later, these experiments can provide an early warning of a coming galactic SN II. A coincidence network between neutrino experiments has been established to minimize response time, eliminate experimental false alarms, and possibly provide some pointing to the impending event from neutrino wave-front timing.

INTRODUCTION

In a supernova (SN) driven by the gravitational collapse of a massive stellar core into a neutron star (*e.g.* Type II or Type Ib SNe), the resulting huge pulse of neutrinos escapes the star promptly after the collapse (for a detailed review and discussion of SN neutrinos, see [1]). In the case of the nearby SN1987A, these neutrinos were observed by the Kamiokande [2] and IMB [3] neutrino detectors in offline analyses performed after the optical discovery of the SN. Although the neutrinos escape the star promptly, photons do not get out until the shock wave travels from the core through the stellar envelope and breaks out of the stellar photosphere – hours later, depending upon the size of the envelope (ref. [4] contains a simple model of this delay). Thus, if the neutrinos could be detected in real-time, they would provide advance warning of the coming photons from the SN. Such a warning would allow preparation of observations over the whole electromagnetic spectrum, maximizing the information gathered during the previously unobserved first few hours of light from a new SN.

The Supernova Early Warning System (SNEWS) [5] is a coincidence trigger between the world's neutrino telescopes. While the individual neutrino detectors all have near real-time SN monitors which are sensitive to SNe in the Milky Way galaxy and its satellites (*e.g.* [6–8]), those monitors could be fooled by instrumental effects. To avoid releasing such a false alarm, any single experiment would want

CP522, *Cosmic Explosions: Tenth Astrophysical Conference,*
edited by Stephen S. Holt and William W. Zhang
© 2000 American Institute of Physics 1-56396-943-2/00/$17.00

its SN monitor output to be carefully scrutinized by a human. Unfortunately, the time scale on which humans operate is of the same order as the lead time gained by the neutrinos over the photons, which would waste much of the advance notice. Those extra hours would be better spent by observers preparing to observe the impending photons. However, the likelihood of two independent experiments experiencing a false alarm in coincidence is very small, therefore an automated alert can be issued with confidence. If each input experiment has a false alarm rate of $< 1/\text{week}$, the false coincidence rate will be $\ll 1/\text{century}$. Thus, the SNEWS network can eliminate the need for active human supervision and provide an alert to the astronomical community as promptly as possible by providing an automated alarm when multiple experiments simultaneously see a SN-like neutrino signal.

THE SNEWS NETWORK

There are a number of experiments capable of detecting neutrinos from a galactic SN (Tab. 1) either active now or coming online within the next few years. Experiments participating in the SNEWS network send a standard UDP packet via the internet to a remote server when their local automated SN monitor detects a SN-like signal in their detector. This remote server forms a blind coincidence trigger between incoming alert packets and issues an alarm to interested observers should it record multiple experimental SN triggers time-stamped within 10 seconds (in UTC) of each other.

When the SNEWS server sees such a coincidence between experiments, it will issue an alert via email to all interested parties. Email to pager gateways will provide the fastest initial notification. Currently the MACRO, Super-Kamiokande, and LVD experiments are providing automated inputs to this coincidence trigger. It is anticipated that the SNO and AMANDA experiments will begin providing inputs early in the year 2000. To verify the prediction of a vanishingly small rate of false alarms, the coincidence server is presently running in a test mode that would not send out an automated alarm. Since no false coincidences have occurred, SNEWS will be switched over to its final, fully automated configuration in the year 2000.

WHAT INFORMATION CAN SNEWS PROVIDE?

The single most important information provided by SNEWS would be simply the fact that within hours, a nearby SN will soon be visible. If nothing else, this alarm will allow observers get set up so as to be able to start observing as soon as more information becomes available.

Naturally more information would be desirable, particularly that which tells observers where to look. There are two methods which could provide pointing information from the neutrino signal. The first takes advantage of the directionality of some neutrino interaction channels in individual detectors, most notably electron scattering ($\nu_x + e^- \rightarrow \nu_x + e^-$) in the large water Cherenkov detectors. For a SN

Detector	Type	Mass (kton)	Location	# events @10 kpc	Status
Super-K	water Cherenkov	32	Japan	4400	**signaling SNEWS since May 1998**
MACRO	scint.	0.6	Italy	150	**signaling SNEWS since March 1998**
LVD	scint.	0.7	Italy	170	**signaling SNEWS since Feb. 1999**
SNO	H_2O, D_2O	1.7 1	Canada	350 430	running
AMANDA	long string	$M_{eff} \sim$ 2/pmt	Antarctica	N/A	running
Baksan	scint.	0.33	Russia	70	running
Borexino	scint.	1.3	Italy	$\sim$200	2000
Kamland	scint.	1	Japan	300	2001
OMNIS	high Z	10 kT Fe, 4 kT-Pb	USA	2000	2000+
LAND	high Z	1	Canada	450	2000+
Icanoe	liquid argon	9	Italy		2000+

TABLE 1. Current and near-future neutrino detectors capable of detecting the neutrino signal from a galactic core-collapse SN. **Bold** entries are currently providing input to the SNEWS network.

at 10 kpc, it is estimated that Super-K could point to a $\sim 5°$ cone on the sky, and SNO a $\sim 20°$ cone [9]. While hardly precise by photon astronomy standards, these solid angles are easily covered by large field of view instruments. This directional information is not a product of the SNEWS network but that of the individual experiments' analyses, but SNEWS could play an important role in disseminating and correlating such information as it becomes available.

The second pointing method is unique to SNEWS. The precise timing of the neutrino wavefront arrival at the different detectors' widely separated locations on the earth could be used to reconstruct where the neutrinos were coming from. Unfortunately, a detailed analysis [9] of the statistics available to the current detectors suggests that this "triangulation" approach would be substantially less precise than the $\nu + e$ scattering, being mostly valuable as a confirmation rather than as a position refinement – but this itself is both a helpful and important cross-check.

PLANS FOR USE OF AN EARLY WARNING

Should a core-collapse SN occur in the range of the neutrino detectors (within the Milky Way galaxy or its satellites), an alert would be issued hours before light is produced by the new SN. This short detection range ensures that any SN signal triggering SNEWS will be very nearby and thus a potential scientific bonanza. Conversely, this short range translates to a probe of a very limited volume, making for a low observable SN rate [10] of the order of several SNe per century. In order to make the most of this once in a lifetime opportunity, good plans should be developed in advance, to be put into action when a SN does occur.

Two examples of such planning are currently in place. The first is from the editors of *Sky & Telescope* magazine, who are preparing to both provide a means to disseminate the alarm information to the amateur astronomer community, and to manage the resulting flow of observations such an alarm would generate [11]. The celebrated accidental observations of SN1987A by the amateur astronomer Albert Jones were instrumental in studying the early stages of this important event. An organized effort by the many similarly skilled observers available would not only provide many more valuable early observations, but also aid greatly in precisely locating the new SN, given both the wide angle instruments commonly used by amateurs and their sheer numbers, experience with the sky, and enthusiasm.

The second example is a Hubble Space Telescope Target of Opportunity proposal from J. Bahcall *et al* [12]. Given the long response time of the HST and the precise positional information needed to point it, these observations would not occur until days after the initial alarm. However, the HST is the only facility available to take the high resolution UV spectra needed to understand the nearby environment of the new SN, and having the detailed observation plans on file and ready to be used would not only save time but provide better results when they are needed.

Any astronomer who might want to observe such an important event is encouraged to make similar plans. Plans and preparations now not only will make good observations easier when the exciting but hectic time comes to observe a galactic SN, but there are undoubtably many good ideas as yet unborn. Some good thinking now in the idle years while waiting for such an event will allow these ideas to be fleshed out and properly put into practice when it counts.

REFERENCES

1. A. Burrows, D. Klein and R. Gandhi, Phys. Rev. **D45**, 3361 (1992).
2. K. Hirata *et al.*, Phys. Rev. Lett. **58**, 1490 (1987).
3. R.M. Bionta *et al.*, Phys. Rev. Lett. **58**, 1494 (1987).
4. T. Shigeyama, K. Nomoto, M. Hashimoto, D. Sugimoto, *Nature* **328**, 320 (1987).
5. K. Scholberg, Proceedings of the 3rd Amaldi Conference on Gravitational Waves, astro-ph/9911359; Also see http://hep.bu.edu/~snnet/ for more information.
6. M. Aglietta *et al.*, Nuovo Cim. **105A**, 1793 (1992).
7. Y. Oyama, M. Yamada, T. Ishida, T. Yamaguchi and H. Yokoyama, Nucl. Instrum. Meth. **A340**, 612 (1994).
8. M. Ambrosio *et al.*, Astropart. Phys. **8**, 123 (1998).
9. J.F. Beacom and P. Vogel, Phys. Rev. **D60**, 033007 (1999).
10. G.A. Tammann *et al*, Astrophys. J. Suppl. **92**, 487 (1194).
11. L. Robinson, Sky & Telescope **98-2**, 30 (August, 1999).
12. J. Bahcall *et al*, HST proposal #8404, cycles 8 and 9.

Remnants of Core-Collapse SNe: Hydrodynamical Simulations with Fe-Ni Bubbles

Kazimierz J. Borkowski, John M. Blondin, William J. Lyerly, and Stephen P. Reynolds

North Carolina State University
Department of Physics
Raleigh, North Carolina 27695-8202

Abstract. Observations of core-collapse supernovae (SNe) have revealed the presence of extensive mixing of radioactive material in SN ejecta. The mixing of radioactive material, mostly freshly synthesized Ni, is not complete, which leads to a two-phase SN ejecta structure. The low-density phase consists of Fe bubbles, created by the energy input from radioactive Co and Ni, surrounded by compressed high-density metal-rich ejecta. During the interaction of SN ejecta with the ambient medium, the ejecta are heated to X-ray emitting temperatures. With the powerful new X-ray satellites, both Fe bubbles and ambient ejecta should be detectable in X-ray images and X-ray spectra of young supernova remnants (SNRs).

We report on the theoretical investigation of SNR dynamics with the two-phase SN ejecta. We first study a single Fe bubble immersed in an outer ejecta envelope. Next, instead of a single bubble, we consider randomly distributed Fe bubbles. The SNR dynamics was simulated with the VH-1 hydrocode in 2 dimensions. Our main finding is that the presence of Fe bubbles leads to vigorous turbulence and mixing of Fe with other heavy elements and with the ambient normal-abundance gas.

INTRODUCTION

Large-scale mixing in core-collapse SNe is now known to be nearly universal. The early emergence of X-ray and γ-ray radiation produced by the radioactive decay in SN 1987A demonstrated this mixing dramatically. This was followed by intensive observations of Type II and Type Ib, c SN light curves and spectra, which clearly showed that large-scale mixing occurs in these SNe. But this mixing appears to be macroscopic, so that the chemical elements are not fully mixed. The evidence for this is provided by the meteoritic data on dust grains formed in SN ejecta, by presence of inhomogeneous ejecta with very different abundances of heavy-elements in young SNRs such as Cas A, and by detailed studies of the SN 1987A ejecta at its late SN stage. In particular, there is evidence for Fe-Ni bubbles in SN 1987A

[1], based on observations of Fe, Co, and Ni lines. Ni-rich material is heated by its own radioactive energy input, and expands in the ambient substrate of other heavy elements, forming low-density Fe bubbles. The SN structure is then expected to be like a Swiss cheese, with Fe bubbles occupying a substantial (~ 0.5) fraction of the ejecta volume. We report on our investigation of how this two-phase ejecta structure affects dynamics of young SNR.

DYNAMICS OF SNRS WITH FE BUBBLES

We use the Virginia Hydrodynamics (VH-1) hydrocode to study the complex dynamics of SNRs with Fe bubbles. The two-dimensional simulations reported here were performed at the North Carolina Supercomputing Center.

Hydrodynamical simulations with a single Fe bubble are shown in Figure 1. The density of unshocked SN ejecta outside the bubble is modeled as a power law in radius, and the ambient medium is uniform. Without the bubble, the interaction is described by a self-similar solution [2,3]. This self-similar structure has the following components: the outer blast wave, the wavy contact discontinuity between the shocked ambient medium and the shocked SN ejecta, and the inner reverse shock.

The bubble, with a fractional radius of 0.4, is 100 times less dense than the ambient SN ejecta. When the reverse shock reaches the outer edge of the bubble, it encounters a much reduced ram pressure because of the low bubble density. The reverse shock accelerates quickly into the bubble's interior, the gas pressure drops and a rarefaction wave propagates back into the shocked SN ejecta. The shocked ejecta is then pushed into the bubble's interior by a more tenuous shocked ambient gas. This situation is strongly Raleigh-Taylor unstable, which leads to mixing of dense shocked ejecta with Fe in the bubble. This mixing is enhanced when a reverse shock reflects from the bottom of the bubble (bottom panels of Figure 1). The reverse shock is also transmitted into dense ejecta, generating vorticity at the bubble's boundary, and the shocked ejecta at the bubble's bottom quickly become unstable. Transverse flows are present along the bubble's bottom, which generate prominent vortices at the bubble's edge. The final result is mixing of Fe with the dense ejecta and with the shocked ambient gas. The shock reflected from the bubble's bottom eventually overtakes the blast wave. This marks the transition to the original self-similar stage, and the end of the transient stage associated with the bubble's presence.

A more realistic case of multiple bubbles is shown in Figure 2. In this example, bubbles with a fractional diameter of 0.2 and with the density 100 times smaller than the ejecta density are distributed randomly in the ejecta, with the average filling fraction of 0.5. The presence of multiple bubbles makes the interaction region very turbulent and inhomogeneous, leading to vigorous mixing of shocked ejecta, Fe in the bubbles, and the ambient gas. The structure of the turbulent shock interaction is self-similar, with the same power law for the expansion of the blast wave as in the standard self-similar solution [2,3].

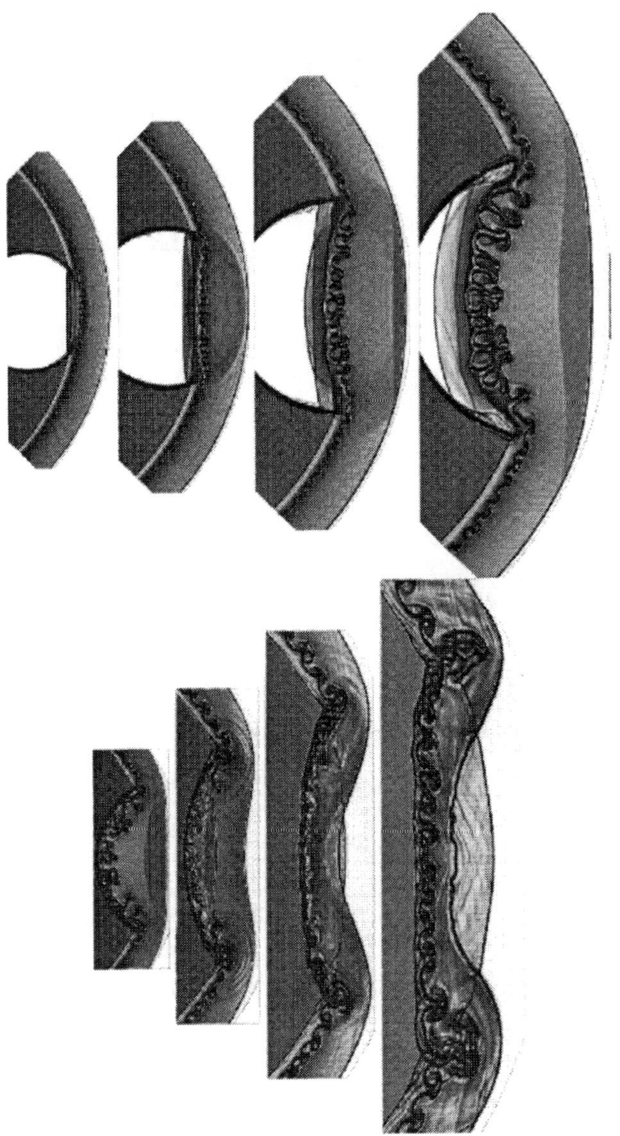

FIGURE 1. Dynamics of a single Fe bubble, shown as a shaded-surface representation of density, with panels arranged in a time sequence from left to right and from top to bottom. Only the interaction region near the bubble is shown, with the explosion center located to the left of each panel. The spatial scale of bottom panels is 2.5 times larger than that of top panels.

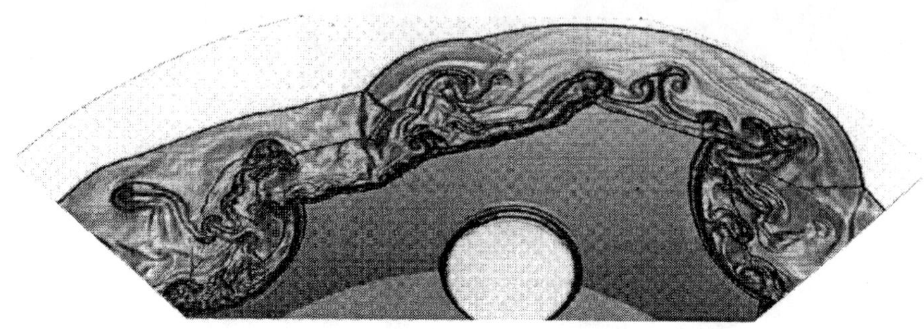

FIGURE 2. A snapshot of an SNR with multiple Fe bubbles, again shown as a shaded-surface representation of density. The shocked ambient gas is bounded on the outside by an irregular blast wave, while an even more ragged reverse shock propagates into the bubbly ejecta. One bubble in the unshocked ejecta and outlines of two recently shocked bubbles can be clearly seen. The highly turbulent nature of the interaction is apparent.

CONCLUSIONS

Large-scale inhomogeneities associated with the presence of Fe bubbles in heavy-element ejecta lead to vigorous turbulence and mixing in young SNRs. This is inevitable, based on our understanding of the core-collapse SNe, where large-scale mixing is expected and even required for a successful explosion. The famous core-collapse SNR, Cassiopeia A, is indeed very turbulent. In view of the turbulent nature of young SNRs, one-dimensional models are clearly not adequate, multi-dimensional hydrodynamical modeling is essential for understanding dynamics of heavy-element ejecta in young SNRs.

The presence of the two-phase heavy-element ejecta in young remnants should affect their X-ray spectra. For example, before Fe from low-density bubbles is mixed with other heavy elements, X-ray lines from freshly synthesized Fe should be weaker than lines from other abundant heavy elements. Fe lines indeed appear to be weak in Cas A [4,5]. Turbulent mixing will lead to a gradual strengthening of Fe lines with respect to other heavy elements as remnants become older.

REFERENCES

1. Li, H., McCray, R., & Sunyaev, R. A. 1993, ApJ, 419, 824
2. Chevalier, R. A. 1982, ApJ, 258, 790
3. Chevalier, R. A., Blondin, J. M., & Emmering, R. T. 1992, ApJ, 392, 118
4. Borkowski, K. J., Szymkowiak, A. E., Blondin, J. M., & Sarazin, C. L. 1996, ApJ, 466, 866
5. Vink, J., Kaastra, J. S., & Bleeker, J. A. M. 1996, A&A, 307, L41

Gamma Ray Bursts

Gamma-Ray Burst Observations

Gerald J. Fishman

Space Science Department, Code SD 50
NASA-Marshall Space Flight Center
Huntsville, AL 35812 USA

Abstract. Gamma-ray bursts (GRBs) are the most luminous known objects in the Universe. Their brief, random appearance in the gamma-ray region had made their study difficult since their discovery, over thirty years ago. There is a rich diversity in the duration and morphology of GRB time profiles. The spectra are characterized by a smooth continuum, usually peaking in the range from ~ 0.1 MeV to 1 MeV. The recent discovery of counterparts to gamma-ray bursts and afterglow radiation in other wavelengths has provided the long-sought breakthrough in the direct determination of their distance and luminosity. Delayed gamma-ray burst photons extending to GeV energies have been detected.

I INTRODUCTION

While gamma-ray bursts have been observed for over thirty years by a wide variety of instruments on many spacecraft, their detailed study was elusive because of their random and transient nature. Furthermore, for the first twenty years or so, GRB observations were often a secondary objective of other experiments. The detectors and their associated instrumentation did not have the needed sensitivity, processing capability or telemetry rates needed to significantly advance the field.

The Burst and Transient Source Experiment (BATSE) on the Compton Gamma-Ray Observatory (CGRO), launched in 1991, provided the necessary large area and versatile data system required to detect a large number of GRBs and study their temporal and spectral properties in detail. It also allowed statistical correlation studies of large numbers of bursts together with a measure of the intensity and sky distributions from a large sample of GRBs derived over many years of operation. Although the locational accuracy derived by BATSE was not sufficient for follow-up observations by narrow-field instruments, it was good enough to derive measurements of the GRB sky distribution and to be of use for wide-field observations by robotic telescopes for counterpart observations.

It was not until the BeppoSAX spacecraft came into operation in 1996 that rapid and precise GRB locations could be provided. This soon led to immediate follow-up observations of GRBs by telescopes in other wavelength regions and a quick

CP522, *Cosmic Explosions: Tenth Astrophysical Conference,*
edited by Stephen S. Holt and William W. Zhang

2609 BATSE Gamma-Ray Bursts

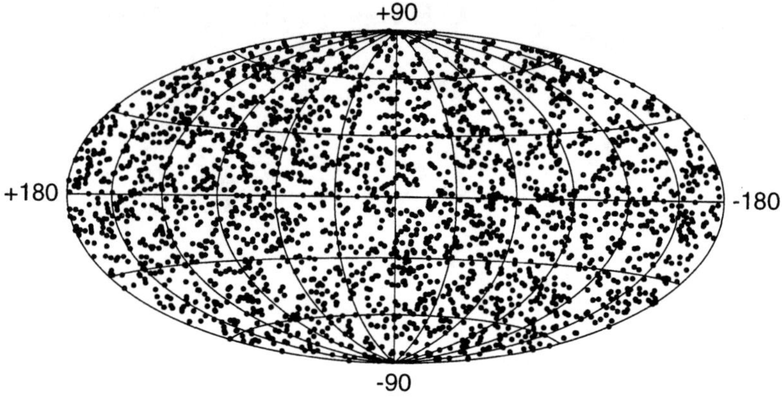

FIGURE 1. The sky distribution of over 2600 gamma-ray bursts observed with BATSE in 8.8 years of operation.

succession of breakthrough observations since 1997. The resulting x-ray, optical and radio discoveries of afterglow emission, the identification of host galaxies, and the measurements of the redshift of GRBs, have now opened the field to detailed study and provided fundamentally new GRB data for GRB models.

II   DISCOVERY AND EARLY OBSERVATIONS

It is quite likely that the problem of gamma-ray bursts will be considered as one of the greatest mysteries in astronomy of the twentieth century. The gamma-ray bursts were discovered by accident in the late 1960's by the Vela series of spacecraft. These spacecraft were developed by the US Department of Defense, designed to detect clandestine nuclear test detonations in space that could be detonated by other countries. From the spacecraft data, it was noticed that widely separated spacecraft were simultaneously detecting bursts of gamma rays that were not coming from the direction of the earth or any other body in the solar system. At first these observations were classified, but after several years, they were published in The Astrophysical Journal in 1973 [1]. Although gamma-ray bursts are now being studied by hundreds of scientists in many countries of the world, the observations of them have been limited to relatively few satellite experiments and research groups.

Early BATSE observations showed an isotropic GRB distribution [2], even for those weak GRBs that showed an inhomogeneity (or a confinement), based on the GRB frequency vs. intensity ($\log N(> P) - \log P$) distribution. (The latest sky map of the sky distribution is shown in Figure 1.) Some researchers reconciled these GRB distribution observations by placing the GRB sources in a very large,

isotropic and uniform galactic halo. However, such a distribution was not observed
to exist with any other galactic component, but it was given some credence by the
observation of high-velocity neutron stars, presumably ejected with a high velocity
at their birth. The supporters of this view held on to the galactic neutron star
hypothesis, which had been modeled in great detail in numerous papers in the
1980s. This group had several strong and vocal supporters. Another group shifted
to a view that the new data indicated a cosmological distance scale for the bursts.
Thus, the GRB community was largely split into two separate camps. Many were
not convinced of either interpretation, and waited until more or new data were
available.

In 1997, BeppoSAX was able, for the first time, to quickly and accurately locate
the position of a GRB so that rapid follow-up observations could be made [3]. Soon
afterward, a fading optical counterpart was discovered [4], a radio counterpart was
discovered [5], and a redshift was measured [6]. A new and exciting era had begun
in GRB research.

III   TEMPORAL PROPERTIES

Perhaps the most striking features of the time profiles of gamma-ray bursts are
their morphological diversity and the large range of burst durations (Figures 2-
4). Coupled with this diversity is the general inability to place many gamma-ray
bursts into well-defined classifications. The durations of gamma-ray bursts range
from about 10 ms to over 1000 s in the energy range in which most bursts are
observed. However, EGRET observations show high energy (> 100 MeV) emission
over 90 minutes after the burst trigger [7], [8]. Sub-millisecond structure has been
detected within a burst [9]. A cursory examination of burst profiles indicates that
some are chaotic and spiky with large fluctuations on all timescales, while others
show rather simple structures with few peaks. However, some bursts are seen with
both characteristics present within the same burst. The longest GRB ever observed
by BATSE is shown in Figure 5.

The duration of a gamma-ray burst is difficult to quantify since this quantity
is dependent on the intensity of the burst and the sensitivity of the experiment.
The BATSE group has settled on a T_{90} measure, the time over which 90% of the
burst fluence is detected. A recent compilation of the distribution of durations is
shown in Figure 6 (left) [10]. A bimodality is seen in the logarithmic distribution
with broad, un-resolved peaks at about 0.3 s and 20 s and a minimum at around
2 s. The fall-off at short durations is partially due to an instrumental bias, since
the minimum time scale over which the BATSE experiment can trigger on bursts
is 64 ms. The shorter bursts are also seen to have harder spectra, as measured by
a hardness ratio [11]. The clustering of two distinct classes of GRBs is evident in
the hardness-duration diagram in Figure 6 (right). Recently, renewed evidence for
a hard x-ray afterglow, decaying as a power-law, has been seen that overlaps with
the main part of the GRB emission [12], as seen in Figure 7.

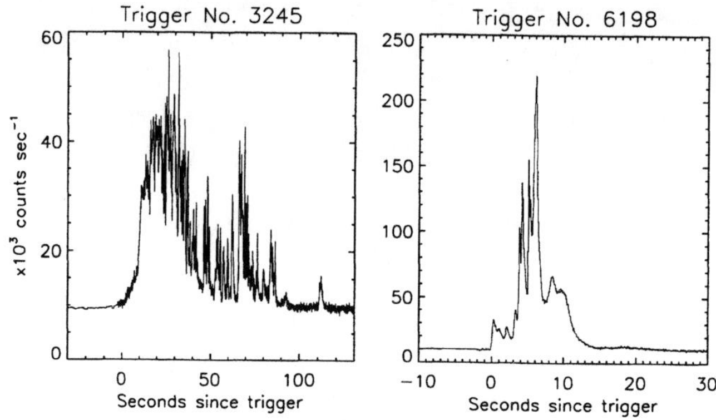

FIGURE 2. Examples of GRBs with fast variability and complex time histories.

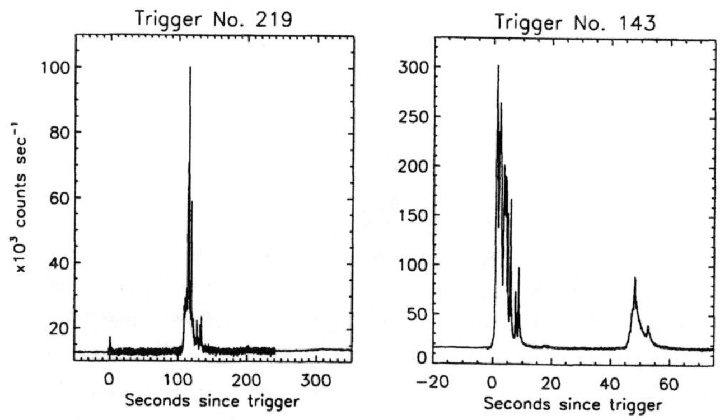

FIGURE 3. Examples of GRBs with separated emission episodes.

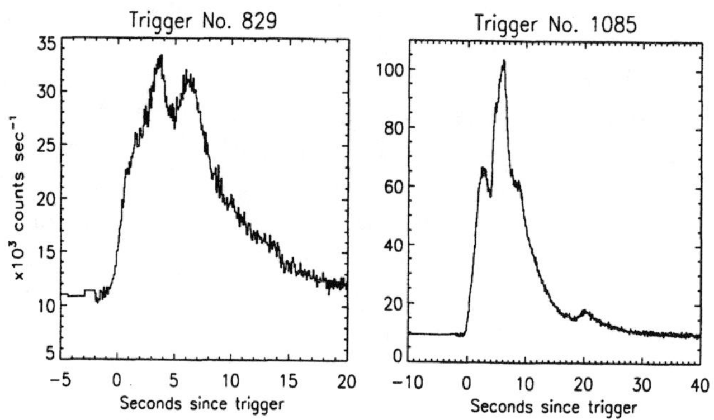

FIGURE 4. Examples of GRBs with smooth time histories.

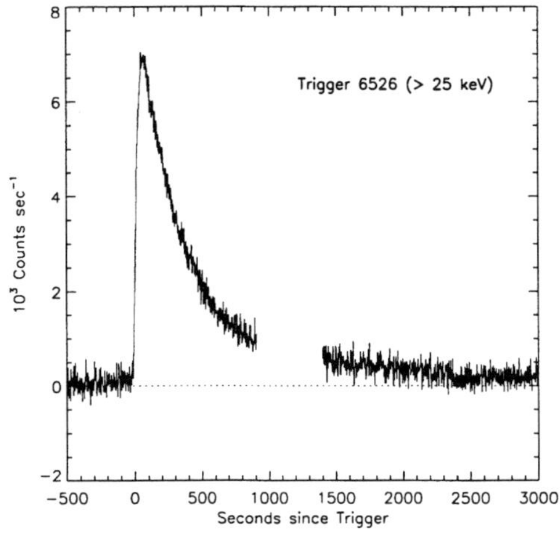

FIGURE 5. Time history of the longest GRB observed by BATSE.

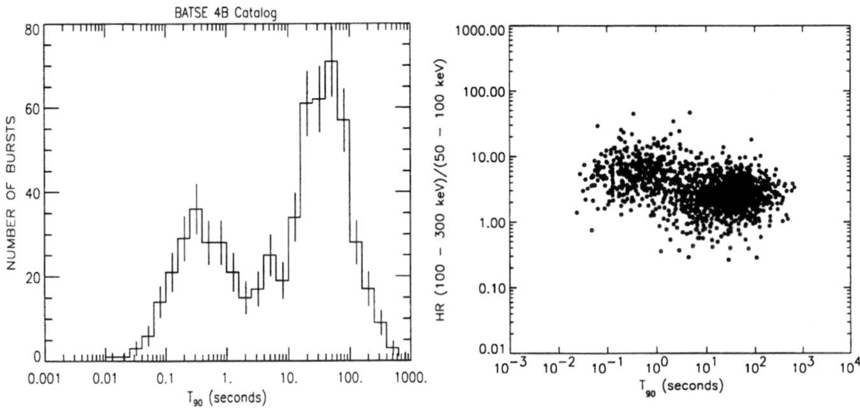

FIGURE 6. The BATSE T_{90} distribution (left) and hardness-duration diagram (right).

No periodic structures have been seen in gamma-ray bursts. However, there are many bursts that have similar shaped sub-pulses within the burst. There is another general characteristic of the time profiles. At higher energies, the overall burst durations are shorter and sub-pulses within a burst tend to have shorter rise-times and fall-times (sharper spikes). This is shown in Figure 8. Most bursts also show an asymmetry, with shorter leading edges than trailing edges. This has been quantified by Link, Epstein and Priedhorsky [13] and by Nemiroff et al. [14].

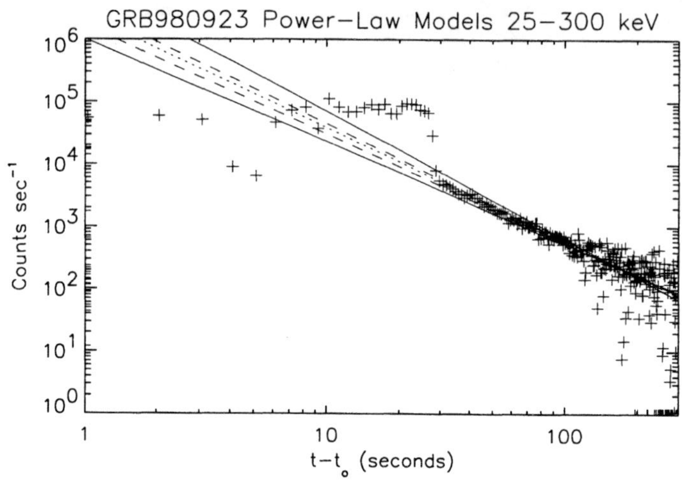

FIGURE 7. Early high-energy afterglow observed from BATSE GRB 980923.

IV SPECTRAL PROPERTIES

The unique feature of gamma-ray bursts is their high-energy emission: most of the power is emitted above 50 keV. Some bursts show emission as low as 1 keV, but the power at low energies is only a few percent of the total power [15]. Most bursts show rather simple continua spectra which appear similar in shape when integrated over the entire burst and when sampled on various timescales within a burst. Most spectra are well fit by a relatively simple analytical function, which has become known as the Band expression [16]. Many GRB spectra are seen to have cut-offs, in which there is no detectable emission above $\sim$ 300 keV [17]. This phenomenon is seen in $\sim$ 25% of all GRBs. It is also within some pulses in individual GRBs (e.g., see Figure 8).

Spectra within the BATSE range from 20 keV to $\sim$ 2 MeV have been fit to numerous GRBs [18]. In these spectra, both the low and high-energy ends of the broad spectral distribution are fit by a power-law within the BATSE bandpass. The observed distributions of the low-energy and high-energy power law indices (α and β, respectively) are shown in Figure 9. The peak of the νF_ν spectra is referred to as the E_{peak} energy. The distribution of E_{peak} for an ensemble of bursts is shown in Figure 10. The high-energy spectra of GRBs that are seen to extend above 10 MeV are best fit by a power-law with a spectral index of $\sim$ 2, but the number of GRBs for which these data are available are limited. The particularly intense GRB of 23 January 1999, has been well fit with the Band function using data from all four CGRO experiments, over a very large energy range, as shown in Figure 11 [19].

Liang and Kargatis [20] have noted a correlation between the spectral evolution

184

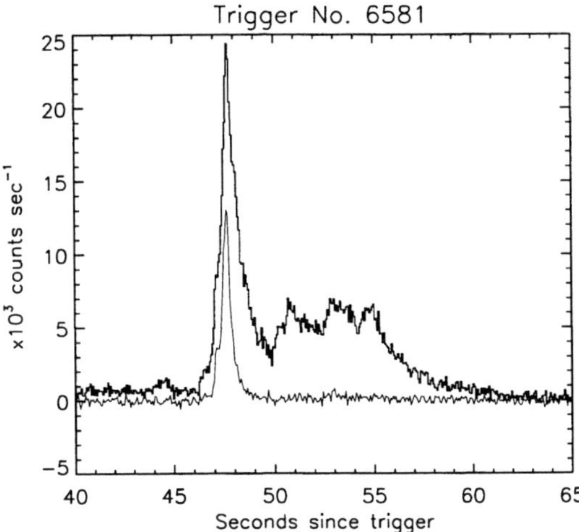

FIGURE 8. Example of a GRB with a high-energy (> 300 keV) pulse, followed by extended emission without any detectable photons above 300 keV. In a significant number of GRBs, this high energy radiation is never present.

of sub-peaks within a given GRB and the intensity of these sub-peaks. Spectral evolution within GRBs is another property that must be considered in burst emission models but is usually not, in contrast to the well-modeled spectral evolution of the GRB afterglow emission. Crider et al. [21] find that the spectral luminosity peak energy decays linearly with energy fluence in most GRBs. They also show that the low-energy spectral form is in disagreement with synchrotron shock models of GRB emission. Other relatively simple correlations between spectral properties, pulse decay time and burst intensity have also been found by recent analyses [22].

The reported observation of spectral line features in gamma-ray bursts, and their interpretation as cyclotron lines produced in the intense magnetic fields of neutron stars had been a primary reason for associating gamma-ray bursts with neutron stars in the 1980s. A search for line features (either absorption or emission features) with the detectors of BATSE has thus far been unable to confirm the earlier reports of spectral line features from gamma-ray bursts [23], [24].

V   A NEW ERA IN GAMMA RAY BURST RESEARCH

Ever since the initial discovery of gamma-ray bursts, there has been a quest to discover a counterpart to a gamma-ray burst in any other wavelength region before, during, or after the gamma-ray event. These searches have taken many forms, including searches for statistical associations of known objects with bursts that have poorly known locations as well as searches of archival plates and other data bases

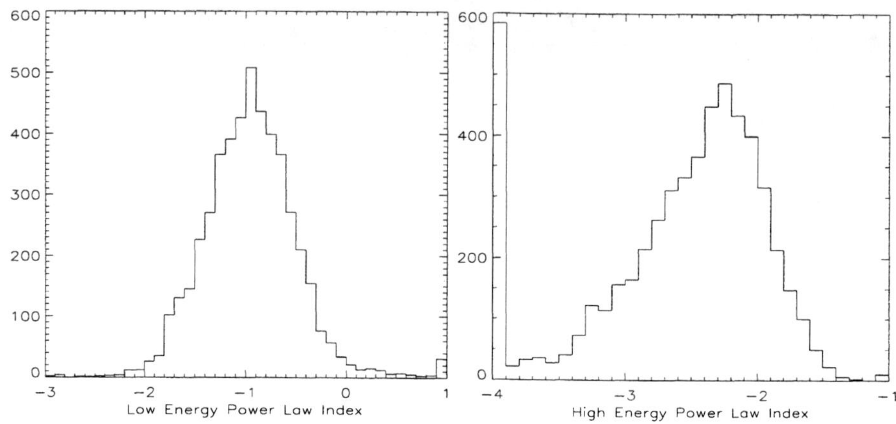

FIGURE 9. The α distribution (left) and β distribution (right).

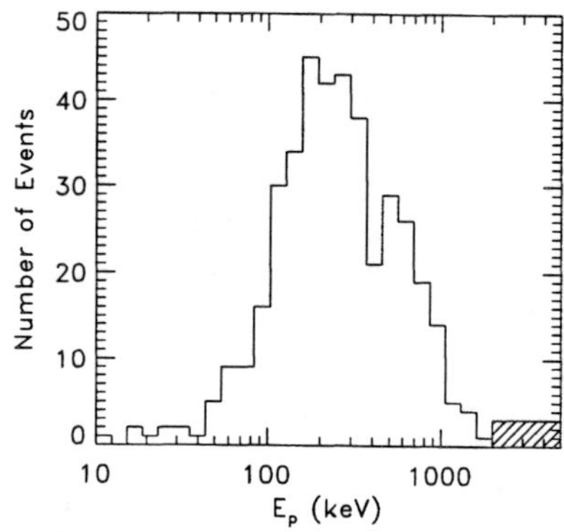

FIGURE 10. The BATSE E_{peak} distribution.

for transient or unusual objects within the error boxes of well-determined burst locations. The arc-minute locations available from BeppoSAX within hours of a GRB proved to be the decisive factor leading to counterpart identifications, afterglow observations, host galaxy identifications, and redshift measurements. Precise locations and the identification of afterglows has led to an identification of the host galaxies of GRBs in almost all cases. These host galaxies are observed to be very faint; most are in the magnitude range (R band) ~ 22 to ~ 26 [25].

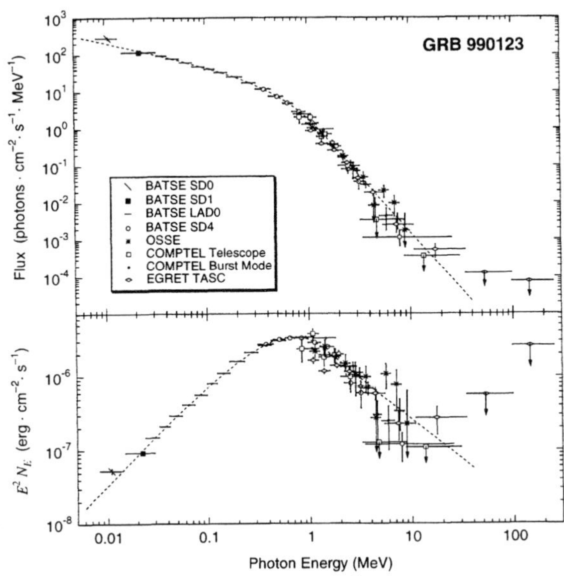

FIGURE 11. Broad band gamma-ray spectrum of GRB 990123.

VI   GRB 980425 (=SN 1998BW ?)

There has been considerable discussion regarding the association of GRB 980425 with the unusual supernova SN 1998bw. If the physical association were real, it would provide an example if a very under-luminous (or mis-aligned) gamma-ray burst, as well as a better understanding of the relationship of these two explosive phenomena. Several papers at this conference have argued for the connection of these two explosive events. Others have argued against the association based upon the improbable observation of such a nearby GRB, assuming that the space density and frequency of GRB sources is nearly homogeneous out to large distances. Kippen et al. [26] find no other associations of GRBs with locations of known GRBs and conclude that the fraction of high-fluence GRBs from known supernovae is less than 0.2%. Thus the association remains problematic and unresolvable based upon this one occurrence. There has also remained some controversy regarding the variable nature, if any, of the associated x-ray object.

In the BATSE gamma-ray data, GRB 980425 does not appear to be unusual in any sense. The observed peak photon flux, 1.25 ph cm^{-2} s^{-1} (50 to 300 keV), is $\sim$ 30x less than the largest peak flux seen with BATSE from a GRB. The burst duration, as measured by the usual BATSE T_{90} parameter, is 23 s. The morphology of this burst is a single smooth peak; this shape has been compared to many others observed with BATSE [27]. GRB 980425 has a soft spectrum, with very little emission above 300 keV and an $E_{peak} \sim 200$ keV. This type of spectrum is observed in $\sim 20\%$ of GRBs.

VII   CURRENT DATA AND FUTURE OBSERVATIONS

The most comprehensive set of BATSE GRB data is in the BATSE catalog now available on the Web, and continually updated. After eight years of operation, the BATSE GRB data set continues to provide the major fraction of GRB data in the energy range from ~ 20 keV to ~ 2 MeV. Most data, including time profiles, are available within several days of their occurrence. The future of BATSE, the Compton Observatory, and of BeppoSAX appears to be somewhat uncertain at this time.

Two missions that are expected to result in many additional GRB counterpart and afterglow observations are HETE-2, planned for launch in May 2000 and Swift, with a planned launch date of 2003. Swift has a primary detector that is much more sensitive than BATSE, although with a limited field-of-view of ~ 3 steradians. This unique spacecraft will be capable of rapid slewing in order to rapidly observe afterglow emission with x-ray, UV and optical telescopes.

It is now recognized that as the most luminous known objects in the Universe and observable to high redshifts, GRBs can become an important probe of star-forming regions in the early Universe [28]. Future GRB observations will not only be of interest in their own right, but they will also become important tools for the study of the early Universe, just as supernovae have become in recent years.

REFERENCES

1. Klebesadel, R., Strong, I., and Olsen, R., *ApJ* **182**, L85 (1973).
2. Meegan, C., Fishman, G., Wilson, R., Paciesas, W., Pendleton, G., et al., *Nature* **355**, 143 (1992).
3. Costa, E., et al., *Nature* **387**, 783 (1997).
4. van Paradijs, J., et al., *Nature* **386**, 686 (1997).
5. Frail, D., et al., *Nature* **389**, 261 (1997).
6. Metzger, M., et al., *Nature* **387**, 878 (1997).
7. Hurley, K. et al., *Nature* **372**, 652 (1994).
8. Dingus, B.L. et al., *AIP Conf. Proc.* **307**, 22 (1994).
9. Bhat, P.N. et al., *Nature* **359**, 219 (1992).
10. Paciesas, W. S. et al., *ApJS* **122**, 465 (1999).
11. Kouveliotou, C. et al., *ApJ* **413**, L101 (1993).
12. Giblin, T. et al., 1999, *ApJ* **524**, L47 (1999).
13. Link, B., Epstein, R.I. and Priedhorsky, W.C., *ApJ* **408**, L81 (1993).
14. Nemiroff, R.J. et al., *ApJ* **423**, 432 (1994).
15. Yoshida, A. et al., *Pub. Astron. Soc. Japan* **41**, 509 (1989).
16. Band, D., Matteson, J., Ford, L., et al., *ApJ* **413**, 281 (1993).
17. Pendleton, G.N. et al., *ApJ* **489**, 175 (1997).

18. Preece, R. D., et al., *ApJS* **125**, in press (2000).
19. Briggs, M., Pendleton, G., Kippen, M., et al., *ApJ* **524**, 82 (1999).
20. Liang, E. and Kargatis, V., *Nature* **381**, 49 (1996).
21. Crider, T., et al., *ApJ* **519**, 206 (1999).
22. Ryde, F. and Svensson, R., *ApJ* **529**, L13 (2000).
23. Band, D. et al., *AIP Conf. Proc.* **428**, 329 (1998).
24. Briggs, M., et al., *PASP Conf. Proc.* **190**, 133 (1999).
25. Hogg, D.W. and Fruchter, A.S., *ApJ* **520**, 54 (1999).
26. Kippen, R.M., et al., *ApJ* **506**, L27 (1998).
27. Bonnell, J., et. al 1999, *Astron. & Astrophys.* **138**, 471 (1999).
28. Lamb, D.Q. and Reichart, D.E., *ApJ*, in press [astro-ph/9909002] (1999).

The Afterglows of Gamma-Ray Bursts

S. R. Kulkarni*, E. Berger*, J. S. Bloom*, F. Chaffee¶,
A. Diercks*, S. G. Djorgovski*, D. A. Frail†, T. J. Galama*,
R. W. Goodrich¶, F. A. Harrison*, R. Sari*, and S. A. Yost*

*California Institute of Technology, Pasadena, CA 91125, USA
†National Radio Astronomy Observatory, Socorro, NM 87801, USA
¶W. M. Keck Observatory, Kamuela, HI 96743, USA

Abstract. Gamma-ray burst astronomy has undergone a revolution in the last three years, spurred by the discovery of fading long-wavelength counterparts. We now know that at least the long duration GRBs lie at cosmological distances with estimated electromagnetic energy release of $10^{51} - 10^{53}$ erg, making these the brightest explosions in the Universe. In this article we review the current observational state, beginning with the statistics of X-ray, optical, and radio afterglow detections. We then discuss the insights these observations have given to the progenitor population, the energetics of the GRB events, and the physics of the afterglow emission. We focus particular attention on the evidence linking GRBs to the explosion of massive stars. Throughout, we identify remaining puzzles and uncertainties, and emphasize promising observational tools for addressing them. The imminent launch of *HETE-2* and the increasingly sophisticated and coordinated ground-based and space-based observations have primed this field for fantastic growth.

I INTRODUCTION

GRBs have mystified and fascinated astronomers since their discovery. Their brilliance and their short time variability clearly suggest a compact object (black hole or neutron star) origin. Three decades of high-energy observations, culminating in the definitive measurements of CGRO/BATSE, determined the spatial distribution to be isotropic yet inhomogeneous, suggestive of an extragalactic population (see [14] for a review of the situation prior to the launch of the BeppoSAX mission). Further progress had to await the availability of GRB positions adequate for identification of counterparts at other wavelengths.

In the cosmological scenario, GRBs would have energy releases comparable to that of supernovae (SNe). Based on this analogy, Paczyńsk & Rhoads [65] and Katz [44] predicted that the gamma-ray burst would be followed by long-lived but fading emission. These papers motivated systematic searches for radio afterglow,

CP522, *Cosmic Explosions: Tenth Astrophysical Conference,*
edited by Stephen S. Holt and William W. Zhang
© 2000 American Institute of Physics 1-56396-943-2/00/$17.00

including our effort at the VLA [15]. The broad-band nature of this "afterglow" and its detectability was underscored in later work [59,78].

Ultimately, the detection of the predicted afterglow had to await localizations provided by the Italian-Dutch satellite, BeppoSAX [6]. The BeppoSAX Wide Field Camera (WFC) observes about 3% of the sky, triggering on the low-energy (2 – 30 keV) portion of the GRB spectrum, localizing events to $\sim 5 - 10$ arcminutes. X-ray afterglow was first discovered by BeppoSAX in GRB 970228, after the satellite was re-oriented (within about 8 hours) to study the error circle of a WFC detection with the 2 – 10 keV X-ray concentrators. The detection of fading X-ray emission, combined with the high sensitivity and the ability of the concentrators to refine the position to the arcminute level, led to the subsequent discovery of long lived emission at lower frequencies [10,77,16].

Optical spectroscopy of the afterglow of GRB 970508 led to the definitive demonstration of the extragalactic nature of this GRB [60]. The precise positions provided by radio and/or optical afterglow observations have allowed for the identification of host galaxies, found in almost every case. Not only has this provided further redshift determinations, but it has been useful in tying GRBs to star formation through measurements of the host star formation rate (e.g. [46,11]). HST with its exquisite resolution has been critical in localizing GRBs within their host galaxies and thereby shed light on their progenitors (e.g. [29,41,4]). Observations of the radio afterglow have directly established the relativistic nature of the GRB explosions [16] and provided evidence linking GRBs to dusty star-forming regions. Radio observations are excellent probes of the circumburst medium and the current evidence suggests that the progenitors are massive stars with copious stellar winds. The latest twist is an apparent connection of GRBs with SNe [5]. Separately, an important development is the possible association of a GRB with a nearby (40 Mpc) peculiar SN [30,47].

In this paper we review the primary advances resulting from afterglow studies. §II discusses the statistics of detections to-date, including possible causes for the lack of radio and optical afterglows from some GRBs. In §III we review constraints on the nature of the progenitor population(s), in particular evidence linking some classes of GRBs to SNe. §IV describes the status of current understanding of the physics of the afterglow emission. Here we compare observations to predictions of the basic spherically-symmetric model, and describe complications arising from deviations from spherical symmetry and non-uniform distribution of the circumburst medium. We conclude with speculations of the near and long-term advances in this field (§V).

We point out that this review has two biases. First, given the concentration of previous review articles on optical and X-ray observations, we emphasize the unique contributions of radio afterglow measurements. Second, this article is intended to also provide a summary of the efforts of the Caltech-NRAO-CARA GRB collaboration, and therefore details our work in particular. This review is in response to review talks given at the 1999 Maryland October meeting (SRK) and the 5th Huntsville GRB meeting (DAF and SRK).

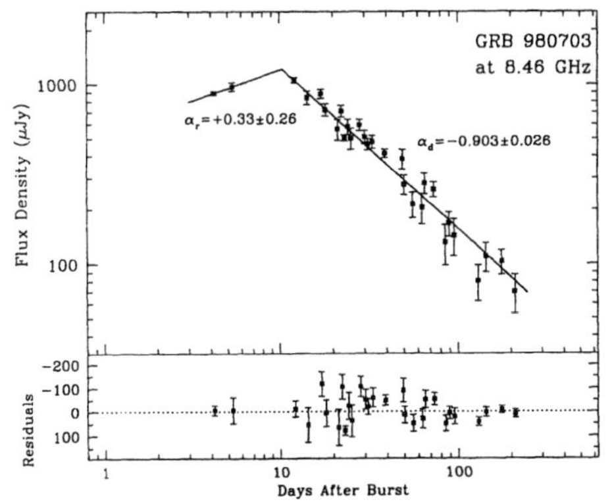
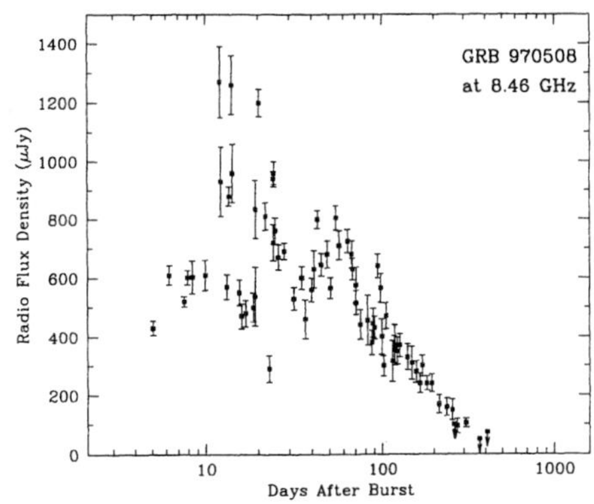

FIGURE 1. *Left: The radio light curve of GRB 980703. This is a typical afterglow, a rise to a peak followed by a power law decay. The longer lifetime of the radio afterglow allows us to see both the rise and the fall of the afterglow emission. In contrast, at optical and X-ray emission, most of the times we see only the decaying portion of the light curve. Right: The radio light curve of GRB970508 [21]. The wild fluctuations of the light curve in the first three weeks are chromatic. At later times, the fluctuations become broad-band and subdued. These fluctuations are a result of multi-path propagation of the radio waves in the Galactic interstellar medium. As the source expands (at superluminal speeds) the scintillation changes from diffractive to refractive scintillation. This is analogous to why stars twinkle but planets do not.*

II STATISTICS OF AFTERGLOW DETECTIONS

Afterglow emission was first detected from GRB 970228, both at X-ray [10] and optical frequencies [77], but not at radio wavelengths [17]. The first radio afterglow detection came following the localization of GRB 970508 [16]. Figure 1 shows two examples of radio lightcurves. The radio afterglow of GRB 970508 is famous for several reasons: it was the first radio detection, it gave the first direct demonstration of relativistic expansion, and it remains the longest-lived afterglow [21].

Afterglow emission is now routinely detected across the electromagnetic spectrum. BeppoSAX has been joined in studying the X-ray afterglows by the All Sky Monitor (ASM) aboard the X-ray Timing Explorer (XTE), the Japanese ASCA mission, and recently the Chandra X-ray observatory (CXO). A veritable armada of optical facilities (ranging from 1-m class telescopes to the 10-m Keck telescopes) routinely discover and study optical afterglows. The HST has been primarily used to make exquisite images of the host galaxies (see above) but in the near future we expect other uses such as UV spectroscopy and identification of underlying SNe. The VLA has led the detection in radio. However, other centimeter-wavelength facilities (the Australia Telescope National Facility, Westerbork Synthesis Radio Telescope, the Ryle Telescope) and millimeter wavelengths (James Clerk Maxwell Telescope, the Owens Valley Millimeter Array, IRAM and the Plateau de Bure Interferometer) are now regularly contributing to afterglow studies.

Figure 2 summarizes the statistics of afterglow detections. In almost all cases, X-ray emission has been detected, establishing the critical importance of prompt X-ray observations. Optical afterglow appears to be detected in about 2/3 of all well-localized events if sufficiently deep optical images are taken rapidly (i.e. within a day or so of the burst). Radio afterglows are detected in 40% of the cases – far more often than usually assumed. We refer the reader to the Frail *et al.* [22] for a comprehensive summary of the X-ray/optical/radio afterglow detection statistics. The non-detections are, as discussed below, as interesting as the detections.

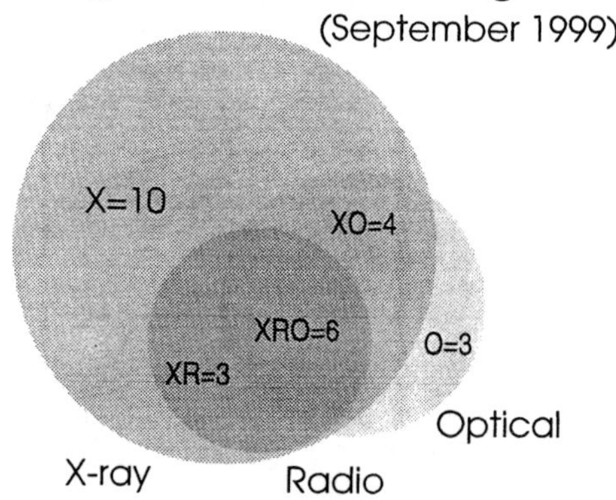

FIGURE 2. *A Venn diagram showing the detection statistics for 26 well-localized GRBs in the Northern and Southern hemispheres. Of the 23 GRBs for which X-ray afterglows have been detected to date, 10 have optical afterglows (XO + XOR) and 9 have radio afterglows (XR + XOR). In total there are 13 optical and/or radio afterglows with corresponding X-ray afterglows.*

Radio Non-detection. The failure to find radio afterglow is most likely due to lack of sensitivity. The brightest radio afterglow to date is that from GRB 991208 (Frail GCN[1] 451) with a peak flux of 2 mJy, a 60-σ detection (at centimeter wavelengths) whereas the weakest afterglow is typically around 5σ. In contrast, at optical and X-ray wavelengths, afterglow emission is routinely detected at hundreds of sigma. If the VLA were to be upgraded by a factor of 10 in sensitivity, then we

---

194

predict that radio afterglow emission would, like X-ray emission, be detected from most GRBs.

Optical Non-detection. Non-detection at optical wavelengths is more interesting, as it may result in some cases from extinction along the line of sight or within the source. Bad weather as well as rapid fading of the afterglow has certainly hindered some optical searches, which, due to notification delays, typically begin some hours after the event. Furthermore, low Galactic latitude events may be obscured, or hidden in crowded foregrounds. However, in some cases deep searches have been performed with no success. Here, non-detection likely results from extinction by dust in the burst host galaxy and/or absorption by the intergalactic medium. GRB 970828 [38] is one example, as is the more dramatic case of GRB 980329. This burst was one of the brightest events in the WFC [42]. Searches for optical afterglow emission failed to identify any counterpart. VLA observations identified an unusual radio variable in the field [76]. Soon thereafter, a red afterglow and a bright IR afterglow were identified (Klose GCN 43, Larkin et al. GCN 44). Taylor et al. [76] suggest that the GRB arose in a region with high extinction. Further optical and IR work on this interesting afterglow can be found in [34], [64], and [69].

Optically dim "red" but bright IR afterglows can also result from the GRB being located at high redshift. Intergalactic HI absorption will result in a wavelength cutoff below the Lyman limit, $< 912(1 + z)$ Å, where z is the redshift of the source. This effect was originally invoked to explain the faint R-band but bright IR emission from GRB 980329 [27]. We now know, based on recent Keck observations, that the GRB host is blue, incompatible with a high-z origin. Rather, it is more tenable that the host is a typical star-forming galaxy with dusty star-forming regions, and that the GRB occurred in one such region [76]. We are presently carrying out IR spectroscopy of this host to determine the redshift and the star formation rate (SFR). While searching for "R dropouts" may in the future provide an effective method for finding high-redshift events, it is clear that cross-calibrated multi-band photometry of higher quality than currently exists will be required to make this useful.

X-ray Non-detection. The spectra of most GRB events clearly extend into the X-ray band, as established by *GINGA* observations [75]. How the X-ray emission observed during the burst connects to the X-ray afterglow is uncertain, due to sensitivity limitations of wide-field monitors. X-ray afterglow emission appears to be ubiquitous. Observations of the X-ray afterglow are important for two reasons: (i) the observations of the X-ray afterglow by sensitive imaging instruments (e.g. the concentrators aboard BeppoSAX) result in sufficiently precise (arcminute) localization and (ii) a significant (perhaps even a dominant) fraction of the explosion energy appears to be radiated in this band. Of all the SAX bursts, GRB 970111 is peculiar for the absence of X-ray afterglow (admittedly the data were obtained about 17 hours after the burst) [25]. In view of the critical role played by X-ray afterglow in localization of GRBs we regard this non-detection to be worthy of

TABLE 1. Basic Properties of Selected GRBs

GRB	α(J2000) (h m s)	δ(J2000) ($^\circ$ ' ")	R_{host} (mag)	$S \times 10^{-6}$ (erg cm^{-2})	z	References[a]
970228	05 01 47	+11 46.9	25.2[b]	1.7	0.695	Djorgovski et al. GCN 289
970508	06 53 49	+79 16.3	25.7	3.1	0.835	[60,2]
970828	18 08 32	+59 18 52	TBD	74	0.957	[12]
971214	11 56 26	+65 12.0	25.6	11	3.418	[46]
980326	08 36 34	-18 51.4	$\gtrsim 27.3$	1	$\ldots$	
980329	07 02 38	+38 50.7	25.4	50	$\ldots$	
980519	21 22 21	+77 15.7	26.2	25	$\ldots$	
980613	10 17 58	+71 27.4	24.5	1.7	1.096	Djorgovski et al. GCN 189
980703	23 59 07	+08 35.1	22.6	37	0.966	[11]
981226	23 29 37	-23 55 54	$\gtrsim 22$	N.A.	$\ldots$	
990123	15 25 31	+44 46 00	24.4	265	1.600	[48]
990510	13 38 07	-80 29 49	$\gtrsim 28$	23	1.619	Vreeswijk et al. GCN 324
990712	22 31 53	-73 24 29	21.78	N.A.	0.430	Galama et al. GCN 388
991208	16 33 54	+46 27 21	$\gtrsim 25$	100	0.706	Dodonov et al. GCN 475
991216	05 09 31	+11 17 07	24.5	256	1.020	Vreeswijk et al. GCN 496

[a] References to redshift determination.

[b] V-band magnitude from HST. All others are R magnitude in the Johnson system.

further investigation.

III   THE NATURE OF THE PROGENITORS

In almost all cases, a host galaxy has been identified at the location of the fading afterglow. GRB redshifts can be obtained either via absorption spectroscopy (when the transient is bright) or by emission spectroscopy of the host galaxy. In Figure 3 and Table 1 we summarize the measured redshifts and host galaxy magnitudes. While the distance scale debate is settled (at least for the class of long duration GRBs, see below) we remain relatively ignorant of the nature of the central engine. Currently popular GRB models fall into two categories: (i) the coalescence of compact objects (neutron stars, black holes and white dwarfs [13,54,61,63]) and (ii) the collapse of the central iron core of a massive star to a spinning black hole, a "collapsar" [85,57]. We now summarize the light shed on the progenitor problem by afterglow studies.

The Location of GRBs Within Hosts. A fundamental insight into the nature of SNe came from their location with respect to other objects within the host galaxy (specifically HII regions and spiral arms) and the morphology of the host galaxy itself (elliptical versus spiral). In a similar manner, we are now making progress in understanding GRB progenitors by measuring offsets with respect to other objects in the host galaxies. The rather good coincidence of GRBs with host galaxies already suggests that they are unlikely to be a halo population (as would be expected in the coalescence scenario [3]). On the other hand, with the possible

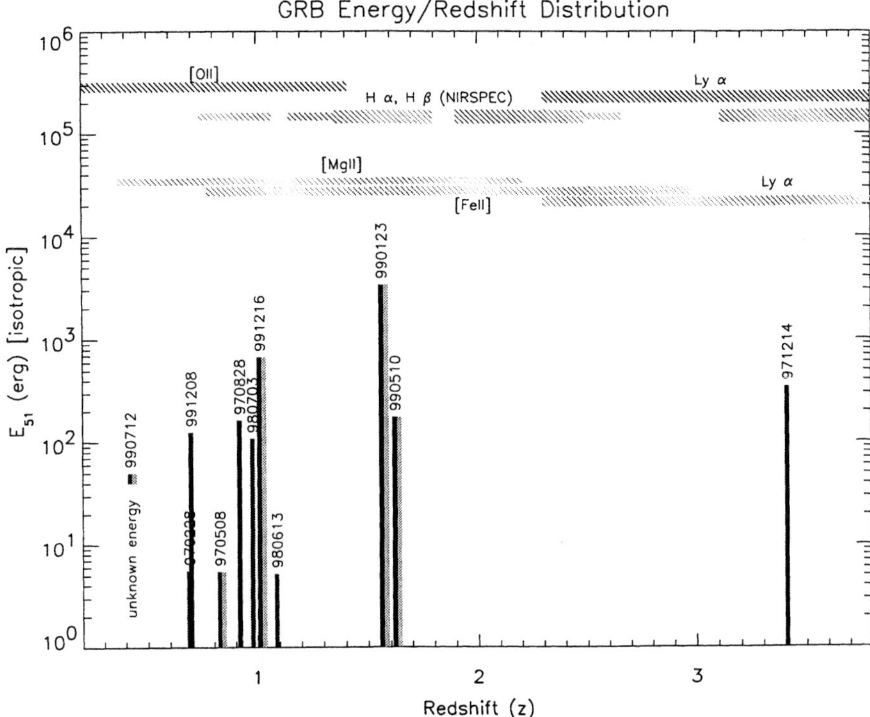

FIGURE 3. *The isotropic gamma-ray energy distribution of GRBs with confirmed redshifts. Bursts indicated in black are those with spectroscopically confirmed emission lines from the host galaxies; bursts indicated by a shaded column (e.g. 990123) are those with absorption line redshifts. The relevant key absorption or emission features are noted at the top of the figure.*

exception of GRB 970508 [66], they are clearly not associated with galactic nuclei (i.e. massive central black holes). Typical offsets of GRBs from the centroid of their host galaxies are comparable to the half-light radii of field galaxies at comparable magnitudes, suggesting that GRBs originate from stellar populations.

Host Galaxies. Demonstrating a direct link between GRBs and (massive) star formation is more difficult. On the whole, the population of identified hosts seems typical in comparison to field galaxies in the same redshift and magnitude range. The hosts have average luminosities for field galaxies, modulo corrections due to evolution. Their emission line fluxes and equivalent widths are also statistically indistinguishable from the normal field galaxy population. The observed star formation rates, derived from recombination line fluxes (mostly the [O II] 3727 $\overset{\circ}{A}$ line) and from the UV continuum flux range from less than $1\,M_\odot\,\mathrm{yr}^{-1}$ to several tens of $M_\odot$ yr^{-1} – typical of normal galaxies at comparable redshifts (extinction corrections can increase these numbers by a factor of a few, but similar corrections apply to the comparison field galaxy population as well). It will probably be necessary to have a sample of several tens of GRB hosts before a correlation of GRBs with the (massive) star formation rate can be tested statistically. However, below we point to several specific examples which are suggestive of a link between GRBs and star-forming regions.

Association with Starforming Regions. There is evidence showing that GRBs arise from dusty regions within their host galaxies. In this respect, radio observations provide a unique tool for detecting events in regions of high ambient density (as was the case for GRB 980329). An even more extreme example is GRB 970828, where the host was identified based *solely* on the VLA discovery of a radio flare [12]. Interestingly enough, this is the dustiest galaxy in the sample of GRB hosts to-date.

Second, some GRBs appear to be located within identifiable star-forming regions. An example is GRB 990123 [4,28,41]. VLA observations of GRB 980703 [19] are perhaps more convincing. The radio observations can be sensibly interpreted by appealing to free-free absorption from a foreground HII region (which would dwarf the Orion complex). If this interpretation is correct then this would be strong evidence for a GRB being located within a starburst region.

The GRB–SN link. If GRBs arise from the collapse of a massive star, it is an unavoidable consequence that emission from the underlying supernova should be superimposed on the afterglow. Bloom et al. [5] may have made the first detection of a possible SN component in the GRB 980326 lightcurve (Fig. 4). These authors noted that SNe, in contrast to afterglows, have distinctive temporal and spectral signatures: rising to a maximum at $\sim 20(1+z)$ days, with little emission blueward of about 4000 Å in the restframe (and certainly blueward of 3000 Å) owing to a multitude of resonance absorption lines. This discovery has led to other possible SN detections, most notably GRB 970228 [31,68].

The suggestion of a GRB–SN connection is an intriguing one but it has yet to be placed on a firm footing. Important questions are: (i) are all long-duration GRBs accompanied by SNe? (ii) if so, are these SNe of type Ib/c? Ground-based observations are possible in those cases where the afterglow decays rapidly (e.g. GRB 980326) or if high quality optical and IR observations exist (e.g. GRB 970228).

We need more examples to test the GRB–SN link. Future progress will depend on a combination of ground and HST observations. For relatively nearby GRBs especially those with a rapidly decaying optical afterglow it would be attractive and feasible to obtain the spectrum of the SN around the time when the flux from the SN peaks. A moderate quality spectrum with SN-like features would have the singular advantage of definitively confirming the SN interpretation (as opposed to alternatives involving re-radiation by dust [80]). However, for most GRBs, we expect HST observations to play a critical role. HST's widely recognized strengths in

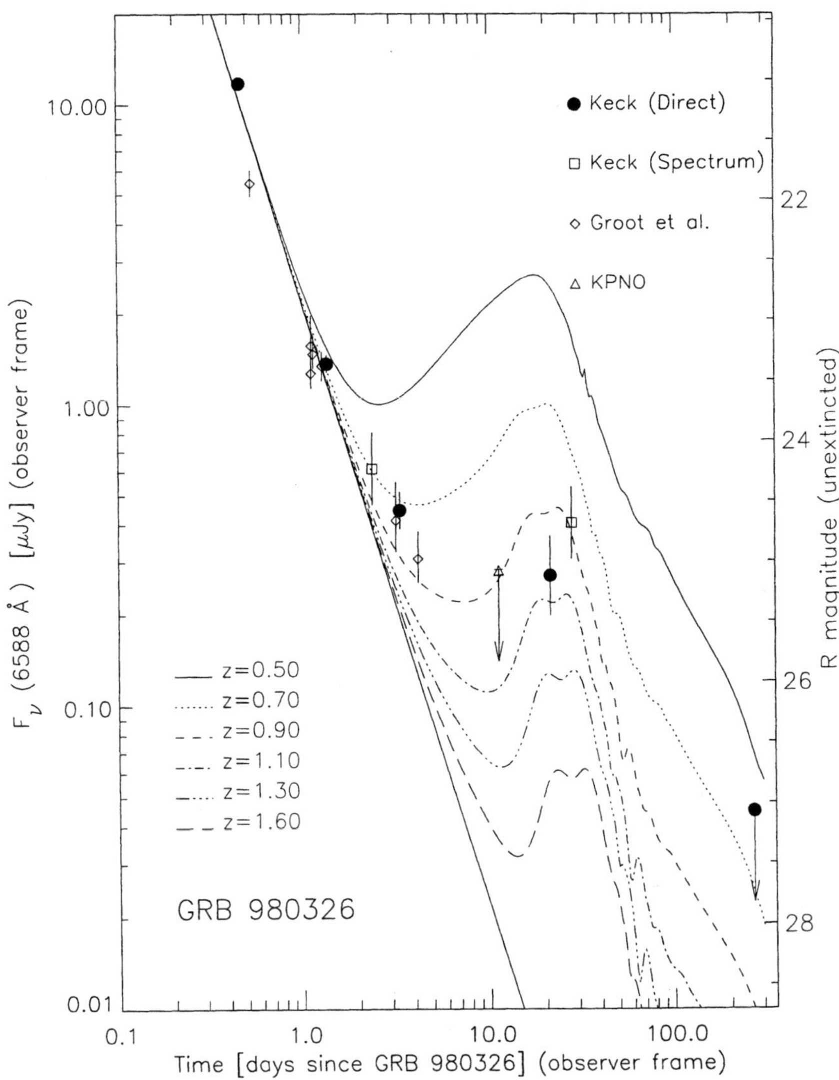

FIGURE 4. *R-band light curve of GRB 980326 and the sum of an initial power-law decay plus Ic supernova light curve for redshifts ranging from $z = 0.50$ to $z = 1.60$; from [5].*

accurate photometry of sources embedded in galaxies [32] and photometric stability make the detection of a faint SN against the optical afterglow and the host galaxy possible.

Diversity of the Progenitor Population. As was the case with SNe, it is likely naive to think of a single progenitor population. Below, we discuss the two additional classes which show some promise: the mysterious short duration GRBs and a possible class of low luminosity GRBs associated with SNe.

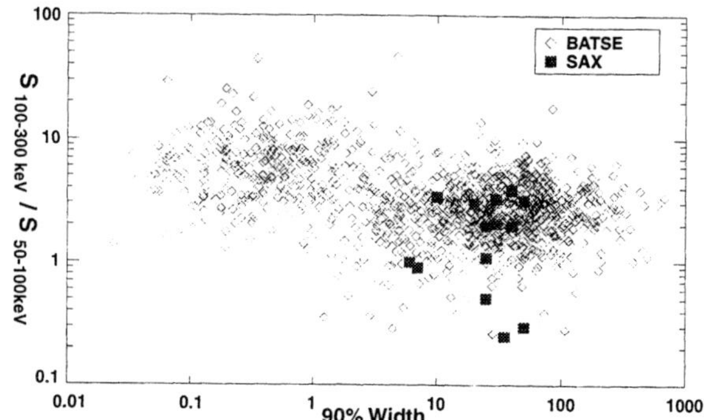

FIGURE 5. *Distribution of duration (T_{90}) vs. spectral hardness for BATSE bursts (diamonds) from the 4B catalogue. There is a clear suggestion of two groups of GRBs: short/hard and long/soft events. Events localized by BeppoSAX (solid squares) appear to belong to the long duration class.*

Short Events. It has been known for some time that the distribution of the duration of GRBs appears to be bimodal [14]; see Figure 5. Furthermore, these two groups may have different spatial distributions [45], with the short bursts being detected out to smaller limiting redshifts. However, we know very little about this class of GRBs since, as noted earlier, all bursts localized by BeppoSAX and RXTE thus far are of long duration (Figure 5). Fortunately, improvements in BeppoSAX and the imminent launch of HETE-2 provide for the first time the opportunity to follow-up short GRBs.

The short duration bursts are difficult to accommodate in the collapsar model, given the long collapse time of the core. However, they find a natural explanation in the coalescence models. How would these bursts manifest themselves? Li & Paczyński [56] speculate that if the short-duration bursts result from NS–NS mergers then they may leave a bright, but short-lived ($\lesssim 1$ day) optical transient. Radio observations provide a complementary tool for determining the nature of the short duration bursts. The low ambient density would result in weak afterglows (since flux $\propto \rho^{1/2}$) which are potentially detectable. Radio observations have additional advantages of a longer lived afterglow, immunity from weather and freedom from the diurnal cycle.

Gamma-ray Bursts Associated with Supernovae. Observers and theorists alike have been intrigued by the possibility that the bright supernova, SN 1998bw, discovered by Galama et al. [30] in the error circle of GRB 980425 [67], is associated with the gamma-ray event (Figure 6). Kulkarni et al. [47] discovered that the SN had

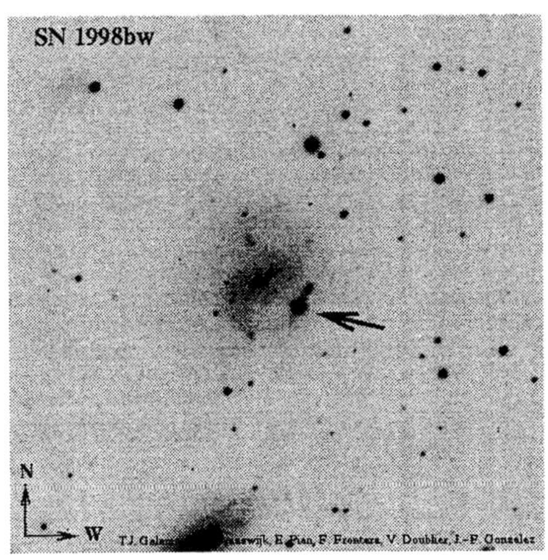

FIGURE 6. *Discovery image of SN 1998bw [30]. The SN is the bright object (marked with an arrow) SW of the nucleus. Relative to typical SNe, this SN is more energetic and appears to have synthesized ten times more Nickel.*

an extremely bright radio counterpart; see Figure 7. We noted that the inferred brightness temperature exceeded the inverse Compton catastrophe limit of 5×10^{11} K and to avoid rapid cooling we postulated the existence of a relativistically expanding blastwave ($\Gamma \gtrsim 2$). This relativistic shock is, of course, in addition to the usual sub-relativistic SN shock. This relativistic shock may have produced the GRB at early times. (We note here that we disagree with the much lower energy estimates of [81]; our recent calculations using the same assumptions as those made in [81] result in an energy estimate similar to that obtained earlier [47] from minimum-energy formulation). The optical modeling of the lightcurve and the spectra show that GRB 980425 was especially energetic [43,86] with an energy release of 3×10^{52} erg and Nickel production of nearly nearly a solar mass.

If GRB 980425 is associated with 1998bw, then this type of event is rare among the SAX localizations. GRB 980425 is most certainly not a typical GRB: the redshift of SN 1998bw is 0.0085 and the γ-ray energy release in GRB 980425 is at least four orders of magnitude less than in other cosmologically located GRBs. For this reason, most astronomers (especially those in the GRB field; see Wheeler's foray in experimental sociology [82]) do not believe the association between GRB 980425 and SN 1998bw. On the other hand, as evidenced by the intense interest in and modeling of the radio and optical data of SN 1998bw, this object is of considerable interest to the SN community. Indeed, we believe that the proposed GRB–SN association controversy has muddied the main issue: SN 1998bw is an interesting SN in its own right.

What is the true distinguishing feature of SN 1998bw that may connect it to a GRB event? Is it the large energy release, as suggested by several authors [43,33]?

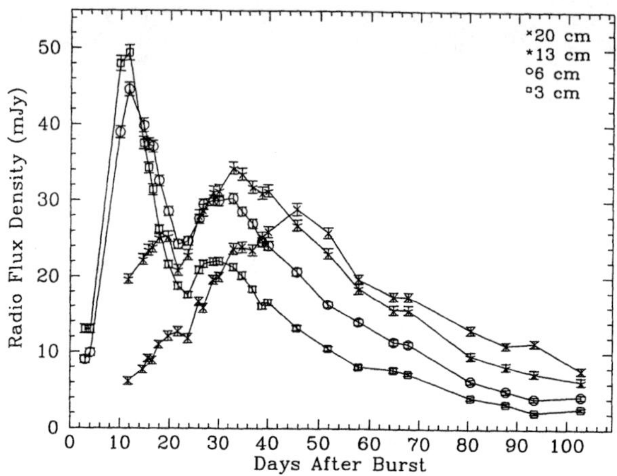

FIGURE 7. *The radio light curve of SN 1998bw at four wavelengths [47]. The peak brightness temperature of SN 1998bw at early times is 10^{13} K, well in excess of the inverse Compton limit of 5×10^{11} K, and can be best understood if the radio emission originates from a relativistic shock ($\Gamma \gtrsim 2$).*

We argue that in fact it is the energy *coupled into relativistic ejecta* that most closely connects SN 1998bw to a GRB. In a typical SN, about 10^{51} erg is coupled to the envelope of the star (a small fraction of the total SN energy release of 10^{53} erg). In a GRB, a similar amount of energy (10^{51}–10^{52} erg depending on the event) is coupled to a much smaller ejecta mass, resulting in relativistic outflow. For SN 1998bw, applying the minimum energy formulation to the radio observations we infer the relativistic shell to contain $\sim 10^{50}$ erg. Not only is this uncharacteristic of a typical SN (there exists no evidence for relativistic ejecta in ordinary SN), but it is not dissimilar from the energy implied for GRB outflows. One could therefore envisage a continuum of physical phenomenon between SN 1998bw and cosmological GRBs provided we use the energy in the relativistic ejecta as the basic underlying parameter and not the isotropic gamma-ray release.

IV AFTERGLOW: THE PHYSICS AND ENERGETICS OF THE FIREBALL

One can consider a GRB to be like a SN explosion with a central source releasing energy E_0 (comparable to the mechanical release of energy in an SN). This is the so-called fireball model. The difference between an SN and a GRB is primarily in ejecta mass: 1–10 $M_\odot$ for SNe whereas only 10^{-5} $M_\odot$ for GRBs. The evolution of a GRB is much faster than that of a SN due to two factors: the ejecta expand relativistically

202

and, thanks to the smaller ejecta mass, the optical depth is considerably smaller.

As the ejecta encounter ambient gas, two shocks are produced: a short-lived reverse shock (traveling through the ejecta) and a long-lived forward shock (propagating into the swept-up ambient gas). Afterglow emission is identified with emission from the forward shock. In order to obtain significant afterglow emission, several conditions are necessary. (1) Rapid equipartition of electrons with the shocked protons (which hold most of the energy). (2) Acceleration of electrons to a power law spectrum (particle Lorentz factor distribution, $dN/d\gamma \propto \gamma^{-p}$). (3) Rapid growth of the magnetic field with energy density in the range of 10^{-2} of that of the protons. Under these circumstances, afterglow emission is dominated by synchrotron emission of the accelerated particles; see [71,79]. The weakness of this model is the assumption of growth in the magnetic field strength to the high values noted above (R. Blandford, pers. comm.).

The theoretically expected afterglow spectrum is shown in Figure 8. Three key frequencies can be identified: ν_a, the synchrotron self-absorption frequency; ν_m, the frequency of the electron with a minimum Lorentz factor (corresponding to the thermal energy behind the shock) and ν_c, the cooling frequency. Electrons which radiate above ν_c cool on timescales equal to the age of the shock. The evolution of these three frequencies is determined by the hydrodynamical evolution of the shock which in turn is affected by two principal factors: the environment of the GRB and the geometry of the explosion.

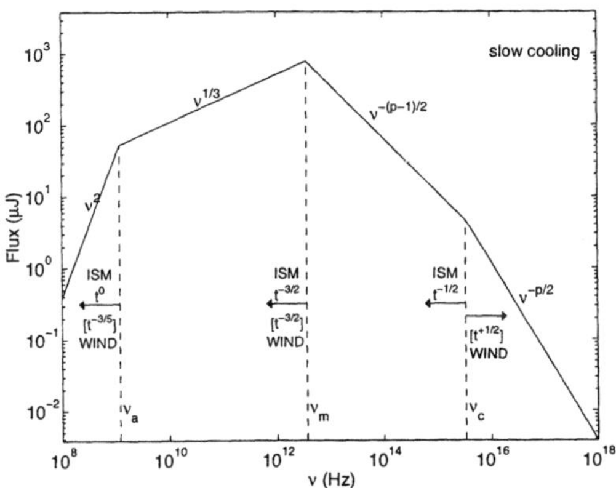

FIGURE 8. *Broad-band spectrum (f_ν) of the afterglow from a spherical fireball with constant density ("ISM" model; see text) and $\rho \propto r^{-2}$ medium ("wind" model; see text). This is representative of the observed spectrum few days after the burst. Note the distinct evolution of ν_a and ν_c in the two models.*

The GRB environment. The earliest afterglow models made the simplifying assumption of expansion into a constant density medium. This is an appropriate assumption should the GRB progenitor explode into a typical location of the host galaxy. However, there is increasing evidence tying GRBs to massive stars (see §III). It is well known that massive stars lose matter throughout their lifetime and thus one expects the circumburst medium to exhibit a density profile, $\rho \propto r^{-2}$ where r is the distance from the progenitor. Chevalier & Li [8] refer to these two models as the ISM (interstellar medium) and the wind model respectively. As can be seen from Figure 8 these two models give rise to rather different evolution of the three critical frequencies.

Geometry: Jets versus Spheres. The hydrodynamics is also affected by the geometry of the explosion. Many powerful astrophysical sources have jet-like structure. There is evidence (from polarization observations) indicating asymmetric expansion in SNe [82], so it is only reasonable to assume that GRB afterglows also have jet-like geometry as well. A clear determination of the geometry is essential in order to infer the true energy of the explosion. This is especially important for energetic bursts such as GRB 990123 whose isotropic energy release approaches $M_\odot c^2$.

Let the opening angle of the jet be θ_0. As long as the bulk Lorentz factor, Γ, is larger than θ_0^{-1}, the evolution of the jet is exactly the same as that of a sphere (for an observer situated on the jet axis). However, once Γ falls below θ_0^{-1} then two effects become important. First, for a well defined jet, the on-axis observer sees an edge and thus one expects to see a break in the afterglow emission. Second, the lateral expansion of the jet (due to heated and shocked particles) will start affecting the hydrodynamical explosion.

Wind or ISM? The two key diagnostics to distinguish these two models are the evolution of the cooling frequency (see Figure 8) and the early behavior of the radio emission. In the wind model, the radio emission rises rapidly (relative to the ISM model) and the synchrotron self-absorption frequency falls rapidly with time. Both these result from the fact that the ambient density decreases with radius (and hence in time) in the wind model.

Unfortunately, in general, the current data are not of sufficient quality to firmly distinguish the two models. For example in GRB 980519, the same optical and X-ray data appear to be adequately explained by the jet+ISM model [73] and the sphere+wind model [8]. Including the radio data tips the balance, but only slightly, in favor of the wind model [23]. In our opinion, the best example for the wind model is that of GRB 980329 [20]; see Figure 9. This afterglow exhibits the two unique signatures of the wind model: high ν_a and a rapid rise. Given the importance of making the distinction between the wind and the ISM model we urge early wide band radio observations (especially at high frequencies).

Energetics. Of all the physical parameters of the fireball, the most eagerly sought parameter is the total energy E_0. By analogy with supernovae, it is E_0 which sets the GRB phenomenon apart from other astrophysical phenomena. Classes of GRBs may eventually be distinguished and ranked by their energy budget; for example, long-duration events, short duration events and supernova-GRBs (see §III).

One approach has been to use the isotropic γ-ray energy as a measure of E_0; see Figure 3. There are three well known problems with such estimates. First, collimation of the ejecta (jets) will result in overestimation of the total energy release. For GRB 990510 where a good case for a jet has been established (Figure 9), the standard isotropic energy estimate is probably a factor of 300 more than the true energy [39]. Second, even after accounting for a possible jet geometry, the efficiency of converting the shock energy into gamma-ray emission is very uncertain. For example, some authors [51] advocate low efficiency ($\sim 1\%$) which would result in an enormous upward correction to the usual isotropic estimates. Third, the bulk Lorentz factor is extremely high during the emission of γ-rays and thus the estimates critically depend on assumption of the geometry and granularity [52] of the emitting region. In particular, if the emission is from small blobs [52] then the inferred estimates are grossly in error.

In contrast to this highly uncertain situation, afterglows offer (in principle) more robust methods to evaluate E_0. In view of the importance of determining E_0 we summarize the different methods of determining E_0 from afterglow observations. One approach is to fit a "snapshot" broad-band afterglow spectrum (from radio to X-rays) to an afterglow model; this approach was pioneered by Wijers & Galama [83]. The strength of this method is that the estimated E_0 is, in principle, robust. Specifically, the estimate does not depend on the usually unknown environmental factors (run of density). However, in practice, this method is very sensitive to the values of the critical frequencies (Figure 8) which are usually not well determined.

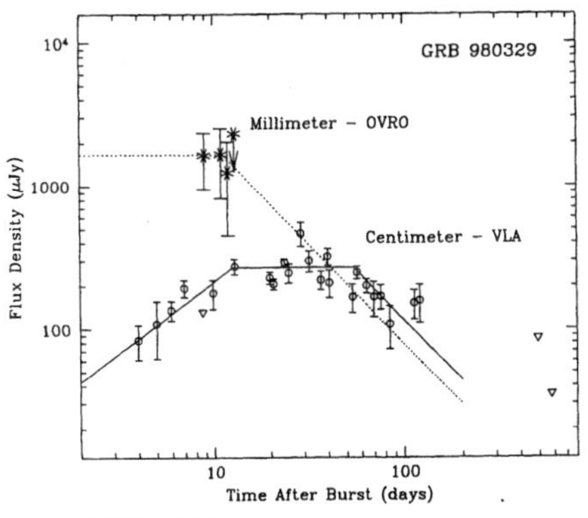

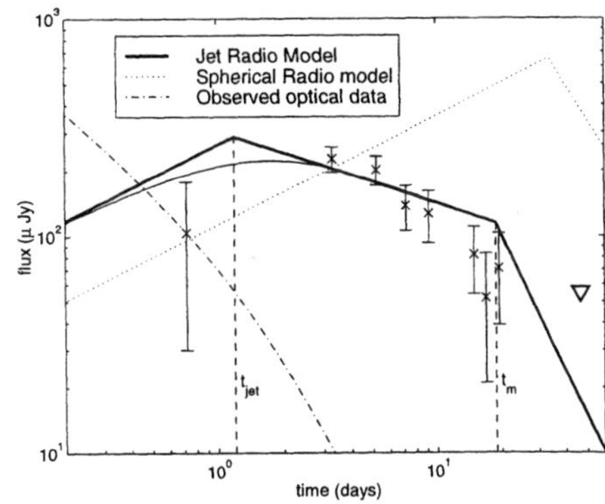

FIGURE 9. *Left: Radio afterglow of GRB 980329 [20]. The rapid rise of the centimeter flux and the high absorption frequency (signified by the considerable strength of the millimeter emission) offer good support for GRB 980329 expanding into a circumburst medium with density falling as inverse square distance. The lines represents a wind model based on X-ray, optical, IR, mm and cm data. Right: Observed and model radio light curves of GRB 990510 [39]. The model predictions for the radio afterglow emission are displayed by the solid line (jet fireball model) and dotted line (spherical fireball model). The observed optical afterglow emission is displayed by the dotted-dashed line; see text for more details.*

This difficulty explains the wildly differing estimates of E_0 for GRB 970508 [83,35]. Furthermore, this method uses measurements obtained at early times (when the afterglow at high frequencies is bright) with the result that the true source geometry is hidden by relativistic beaming.

A second approach is to model the light curves of the afterglow in a given band, specifically a radio band. The advantages of this method are the photometric stability of radio interferometers and the low Lorentz factor at the epoch of the peak of the radio emission. The disadvantages are two-fold: the sensitivity to the environmental parameters (density) and the assumption of the constancy of the microphysics parameters (electron and magnetic field equipartition factors). Application of this approach to GRB 980703 has resulted in seemingly accurate measures of the fireball parameters [19].

Freedman & Waxman [24] take yet another approach, and estimate the energy release from late time X-ray observations. They show that the X-ray flux is insensitive to the GRB environment, and obtain robust estimates of the fireball energy per unit solid angle: from 3×10^{51} erg to 3×10^{53} erg.

With all the above approaches, however, the possible collimation of the ejecta in jets is still a major uncertainty. This can be addressed by observing the evolution of the afterglow as the "edge" of the jet becomes visible. In most cases no evidence for jets has been seen, with the notable exceptions of GRB 990510 and possibly GRB 990123. In addition, a variety of statistical arguments (the absence of copious numbers of "orphan afterglows") [37,36,70] suggests that, on average, the collimation cannot be extreme, and that for most bursts the opening angle is not less than 0.1 radian. Thus the total energy for most bursts may be reduced to the range of 10^{50} erg to 3×10^{51} erg, but could easily be much higher in at least some cases.

Possibly the best approach to determining the energetics, which minimizes uncertainties due both to collimation (jets) and to the environment is to model the afterglow after it becomes non-relativistic. This method builds on the well established minimum energy formulation and the self-similarity of the Sedov solution. Not only are the ejecta truly non-relativistic, but they are also essentially spherical, as by this time jets will have had sufficient time to have undergone significant lateral expansion. Indeed, we can justifiably call this "fireball calorimetry" [21]. Applying this technique to the long-lived afterglow of GRB 970508 (Figure 1) led to the surprising result that $E_0 \sim 5 \times 10^{50}$ erg – weaker than a standard SN! This is an astonishing result. If true, this result would suggest that it is not E_0 which is the prime distinction between GRBs and SNe but the ejecta mass. However, Chevalier & Li [9] interpret the same data in the wind framework and derive much larger E_0. Clearly, we need more well studied afterglows with sufficient observations to first distinguish the circumburst environment (wind versus ISM) and then radio observations over a sufficiently long baseline to undertake calorimetry. Nonetheless, one should bear in mind that the current evidence for large energy release in GRBs is not as strong as is usually assumed.

V   EPILOGUE AND FUTURE

Clearly, the GRB field is evolving rapidly. Along what direction[s] will this field proceed in the coming years? One way to anticipate the future is by considering analogies from the past.

In §III we already discussed the parallels between the SN field and the GRB field. Here we discuss the numerous parallels with quasar astronomy. First discovered at radio wavelengths, we now study quasars across the electromagnetic spectrum. Although still identified by their gamma-ray properties, we now recognize the tremendous value of pan-chromatic GRB and afterglow studies. In both cases, there was considerable controversy about the distance scale. However, once this issue was settled, it became clear that quasars are the most energetic objects (sustained power) whereas GRBs are the most brilliant. For both, the ultimate energy appears to be related to black holes (albeit of different masses).

The raging issues in GRB astronomy today are the same that fueled quasar studies in the 60's: the spatial distribution, the extraction of energy from the central engine, the transfer of energy from stellar scales to parsec scales, and the geometry of the relativistic outflow (sphere or jet). Astronomers took decades to unify the seemingly diverse types of quasars, and to conclude that there are two types of central engines: radio loud and radio quiet. Likewise, there may well be two types of GRB engines: rapidly and slowly spinning black holes emerging respectively from collapse of a rotating core of a massive star or coalescence of compact objects and the collapse of a massive star. This picture could potentially explain both the cosmologically located GRBs and SN 1998bw. Finally, we can project that in the future, GRBs may be used to probe distant galaxies, just as quasars are used today to study the IGM.

There is a feeling in the astronomical community (outside the GRB community) that the GRB problem is "solved". The truth is that the GRB problem is now getting defined! We now summarize our view of the major issues and anticipated near term advances. In our opinion the major issues are Diversity, Progenitors and Energy Generation.

As discussed earlier, high energy observations suggest the existence of two classes: short and long duration bursts. It is possible that afterglow observations may demarcate additional classes. If so, one can contemplate that within a year (assuming abundant localizations by HETE-2) that we will have new GRB designations such as sGRBs (GRBs with late time bump indicative of an underlying SN), wGRBs (GRBs whose afterglow clearly indicates a wind circumburst medium shaped by stellar winds), iGRBs (GRBs which explode in the interstellar medium) and so on.

The broad indications are that GRBs are associated with stars and most likely massive stars. However, we know little beyond this. Comparing the unbeamed GRB event rate of 1.8×10^{-10} yr^{-1} Mpc^{-3} [74] with 3×10^{-5} Type Ibc SN yr^{-1} Mpc^{-3} and 10^{-6} yr^{-1} NS–NS merger Mpc^{-3} [53] shows that GRBs events are extremely rare; here we note that the present data do not support a collimation correction in excess of 100. It will be quite some time before we will be in a position to identify the conditions necessary for a star to die as a GRB.

It is our opinion that SN 1998bw is a major development in the field of stellar collapse. The association (or lack) with GRB 980425 unfortunately has distracted our attention of this important development. The existence of a significant amount of mildly relativistic material, $\sim 10^{50}$ erg [47], is fascinating and it is ironic that none of the models can account for this inferred value whereas most of the theoretical effort has gone into explaining the gamma-ray burst itself (especially considering the uncertain association of GRB 980425 with SN 1998bw). Clearly, SN 1998bw is a rare event but we are convinced that more such events will be found and accordingly have mounted a major campaign to identify these SNe. The robust signatures of this class are high T_B and prompt X-ray emission since these are necessary consequences of a relativistic ejecta. We note that if these future events are as bright as SN 1998bw then the energy in the relativistic ejecta can be directly measured by VLBI observations of the expanding radio shell.

It is vitally important to make quantitative progress in determining the energy release in GRBs. As discussed in §IV, firm estimates of the energy release require well sampled broad-band data at early times and densely sampled radio light curves out to late times. This will require a *coordinated* approach and necessarily involve many observatories around the world and in space. The same datasets will also help us understand a profound puzzle: if GRBs indeed arise from the death of massive stars then why do we not see signatures for a circumburst medium shaped by stellar winds in *all* long duration GRBs? Even ardent supporters of the wind model [8,9] concede that some GRBs (e.g. GRB 990123, 990510) are due to a jet expanding into a constant density medium.

We now discuss the anticipated returns. True to our tradition as observers, we order the discussion by wavelength regimes!

Radio Observations: Dusty galaxies, Circumstellar Edges and Reverse Shocks. Perhaps the most exciting use of radio afterglow is in identifying dusty star-forming host galaxies. Such host galaxies are not readily seen at optical wavelengths. Currently, such galaxies are eagerly sought and studied at sub-millimeter wavelengths. However, the sensitivity and localization of such galaxies by sub-millimeter telescopes is poor. In contrast, GRB host galaxies are identified at the sub-arcsecond level. The present radio afterglow detection rate of 40% already places an upper limit on the amount of star-formation in dusty regions, viz. this rate is not larger than that measured from optical observations. This result is entirely independent of the conclusion based on studies in the sub-millimeter regime, or the diffuse cosmic FIR background found in the COBE data. However, the result does rely on two assumptions: (i) GRBs trace star formation and (ii) the GRB explosion and its aftermath does not radically alter the ambient medium (i.e., with a prompt and complete destruction of dust grains along the line of sight).

Radio observations of SNe offer a probe of the distribution of the circumstellar matter. A spectacular example is SN 1980K whose radio flux dropped 14 yrs after the explosion [58]. A progenitor star which suffered mass loss with variation in the wind speed could explain the observations. Indeed, one *expects* significant radial structure in the circumburst medium as the progenitor evolves from a blue star to a red supergiant and thence to possibly a blue supergiant etc. If GRBs come from binary stars which undergo a phase of common envelope envolution [7] then the structure would be even more complicated. Thus radio observations have the potential (in fortunate circumstances) to give us insight into the mass loss history of the progenitor star[s].

The prompt optical emission from GRB 990123 [1] has been interpreted to arise from the reverse shock [72]. Far less discussed is the prompt radio emission – a radio flare – also seen from this burst [49]. Sari & Piran [72] suggest that the radio emission also originates from the reverse shock as the electrons cool. Observations related to the reverse shocks are important since it is only through these observations that we have a chance of studying the elusive ejecta. We now have four such examples of radio flares [50] and this represents an order of magnitude better success rate than ROTSE+LOTIS. We urge theorists to pay attention to these new findings. More to the point, radio observations appear to be fruitful for the study of reverse shocks, especially when combined with observations of the prompt optical emission. This bodes well for the coming years given the efforts underway to increase the sensitivity of ROTSE [1].

X-ray Observations: Diversity & Progenitors. GINGA identified a number of X-ray rich GRBs. BeppoSAX has found several such examples with some bursts lacking significant gamma-ray emission – the so-called X-ray flashes [40]. We know very little about these X-ray transients. Could they be GRBs in a very dense environment (with red giant progenitors)? We need to take such transients more seriously and intensively followup on such bursts.

Another interesting finding from *GINGA* was the discovery of precursor soft X-ray emission [62]. There is no simple explanation for this phenomenon in the current internal-external shock model. We suggest that the soft X-ray emission precursor is similar to the UV breakout of ordinary SNe. This hypothesis can be confirmed or rejected by obtaining the redshift to such bursts.

The X-ray rich GRB 981226 [26,18] was marked with two additional peculiarities: a precursor emission and afterglow emission which is seemingly undetectable after about 12 hours but then rises rapidly before commencing decay. Above we alluded to the fact that massive stars do not have a single phase of mass loss but instead have a veritable history of mass loss (from birth to death). The X-ray observations of GRB 981226 could be accounted for in a model in which the progenitor has first a red supergiant wind followed by a blue supergiant wind.

Optical Observations: SN link, Short bursts & Geometry. The GRB–SN connection is best probed by optical observations. The value of optical observations has already been demonstrated by the current observations of GRB 980326 and 970228. Clearly, more observations are needed to establish this link. Once this link is established then one can undertake detailed spectroscopic studies of the SN with large ground-based telescopes and photometric studies with HST.

Offsets of GRBs and the morphology of the host galaxies will continue to be of great interest. Such observations will help us differentiate whether some GRBs come from nuclear regions or always from star-forming regions. Under the current paradigm, the discovery of GRBs coincident with elliptical galaxies would be a major surprise. On the other hand, one expects short bursts to arise in the halo of their galaxies and thus in this case no coincidence is expected. We expect HETE-2 to contribute significantly to these issues. Finally, polarization measurements offer a very convenient way to probe the geometry of the emitting region as has already been demonstrated from the discovery of polarization in GRB 990510 (e.g. [55,84]).

Acknowledgments. Our research is supported by NASA and NSF. JSB holds a Fannie & John Hertz Foundation Fellowship, AD holds a Millikan Postdoctoral Fellowship in Experimental Physics, TJG holds a Fairchild Foundation Postdoctoral Fellowship in Observational Astronomy and RS holds Fairchild Foundation Senior Fellowship in Theoretical Astrophysics. The VLA is a facility of the National Science Foundation operated under cooperative agreement by Associated Universities, Inc. The W. M. Keck Observatory is operated by the California Association for Research in Astronomy, a scientific partnership among California Institute of Technology, the University of California and the National Aeronautics and Space Administration. It was made possible by the generous financial support of the W. M. Keck Foundation.

REFERENCES

1. Akerlof, C., Balsano, R., Barthelmy, S. et al., *Nature* **398**, 400 (1999).
2. Bloom, J. S., Djorgovski, S. G., Kulkarni, S. R. & Frail, D. A., *ApJ* **508**, L17 (1998).
3. Bloom, J. S., Sigurdsson, S. & Pols, O. R., *MNRAS* **305**, 763 (1999).
4. Bloom, J. S. et al., *ApJ* **518**, L1, (1999).
5. Bloom, J. S. et al., *Nature* **401**, 453 (1999).
6. Boella, G. et al., *A.&A. Suppl. Ser.* **122**, 298 (1997).
7. Brown, G. E., Lee, C. -H., Lee, H. K. & Bethe, H. A., astro-ph/9911458 (1999).
8. Chevalier, R. A. & Li, Z.-Y., *ApJ* **520**, L29, (1999).
9. Chevalier, R. A. & Li, Z. -Y., astro-ph/9908272 (1999).
10. Costa, E. et al., *Nature* **387**, 783 (1997).
11. Djorgovski, S. G., Kulkarni, S. R., Bloom, J. S., Goodrich, R., Frail, D. A., Piro, L & Palazzi, E., *ApJ* **508**, L17 (1998).
12. Djorgovski, S. G. et. al., in preparation, (2000).
13. Eichler, D., Livio, M., Piran, T., & Schramm, D. N., *Nature* **340**, 126 (1989).
14. Fishman, G. J. & Meegan, C. A., *Annu. Rev. Astron. Astrophys.* **33**, 415 (1995).
15. Frail, D. A. & Kulkarni, S. R., *Astrophys. Space Sci.* **231**, 277 (1995).
16. Frail, D. A., Kulkarni, S. R., Nicastro, S. R., Feroci, M., & Taylor, G. B., *Nature* **389**, 261 (1997).
17. Frail, D. A., Kulkarni, S. R., Shepherd, D. S., & Waxman, E., *ApJ* **502**, L119, (1998).
18. Frail, D. A., et al., *ApJ* **525**, L81 (1999).

19. Frail, D. A., Bloom, J. S., Kulkarni, S. R., Sari, R., & Taylor, G. B., in preparation, (2000).

20. Frail, D., Kulkarni, S., Sari, R., Taylor, G., Shepherd, D., Bloom, J., Young, C., Nicastro, L., & Masetti, N., *ApJ in press*, (1999).

21. Frail, D., Waxman, E. & Kulkarni, S. R., *ApJ in press*, (2000).

22. Frail, D. A., Kulkarni, S. R., Wieringa, M. H. et al., astro-ph/9912171, (1999).

23. Frail, D. A., Kulkarni, S. R., Sari, R. et al., astro-ph/9910060, (2000).

24. Freedman, D. L. & Waxman, E., astro-ph/9912214 (1999).

25. Frontera, F., Amati, L., Costa, E. et al., astro-ph/9911228, (1999).

26. Frontera, F., Antonelli, L. A., Amati, L. et al., *ApJ*, submitted (2000).

27. Fruchter, A. S., *ApJ* **512**, L1 (1999).

28. Fruchter, A. S., Thorsett, S., Metzger, M. R., *ApJ* **519**, L13 (1999).

29. Fruchter, A. S., Pian, E., Thorsett, S. E. et al., *ApJ* **516**, 683 (1999).

30. Galama, T. J. et al., *Nature* **395**, 670, (1998).

31. Galama, T. J. et al., *ApJ* in press, (1999).

32. Garnavich, P. M. et al., *ApJ* **493**, L53 (1998).

33. Germany, L., Reiss, D. J., Sadler, E. M., Schmidt, B. P. & Stubbs, C. W., astro-ph/9906096 (1999).

34. Gorosabel, J., Castro-Tirado, A. J., Pedrosa, A. et al., *A.&A.* **347**, L31 (1999).

35. Granot, J., Piran, T. & Sari, R., *ApJ* **527**, 236 (1999).

36. Greiner, J., *et al.*, *A& AS* **138**, 441 (1999)

37. Grindlay, J. E., *ApJ* **510**, 710 (1999)

38. Groot, P. J., Galama, T. J., van Paradijs, J. et al., *ApJ* **493**, L27 (1998).

39. Harrison, F. A., et al., *ApJ* **523**, L121 (1999).

40. Heise, J., talk at the 5th Hunstville GRB conference, (1999).

41. Holland, S. & Hjorth, J., *A&A* **344**, L67 (1999).

42. In't Zand, J. J. M., Amati, L., Antonelli, L. A. et al., *ApJ* **505**, 1191 (1998).

43. Iwamoto, K. et al., *Nature* **395**, 672, (1998).

44. Katz, J. I., *ApJ* **422**, 248 (1993).

45. Katz, J. I. & Canel, L. M., *ApJ* **471**, 915 (1996).

46. Kulkarni, S. R., Djorgovski, S. G., Ramaprakash, A. N. *et al.*, *Nature* **393**, 35 (1998).

47. Kulkarni, S. R., Frail, D. A., Wieringa, M. H. et al., *Nature* **395**, 663 (1998).

48. Kulkarni, S. R., Djorgovski, S. G., Odewahn, S. C. et al., *Nature* **398**, 389 (1999).

49. Kulkarni, S. R. et al., *ApJ* **522**, L97 (1999).

50. Kulkarni, S. R. & Frail, D. A., in preparation, (2000).

51. Kumar, P., astro-ph/9907096 (1999).

52. Kumar, P. & Piran, T., astro-ph/9909014 (1999).

53. Lamb, D. Q., astro-ph/9909026 (1999).

54. Lattimer, J. M. & Schramm, D. N., *ApJ* **192**, L145 (1974).

55. Lazzati, D., Covino, S. & Ghisellini, G., astro-ph/9912247 (1999).

56. Li, L.-X. & Paczynski, B., *ApJ* **507**, L59, (1998).

57. Macfadyen, A. I. & Woosley, S. E., *ApJ* **524**, 262, (1999).

58. Montes, M. J., van Dyk, S. D., Weiler, K. W., Sramek, R. A. & Panagia, N., *ApJ* **506**, 874, (1998).

59. Mészáros, P. & Rees, M. J., *ApJ* **476**, 232 (1997).

60. Metzger, M. R., Djorgovski, S. G., Kulkarni, S. R., Steidel, C. C., Adelberger, K. L., Frail, D. A., Costa, E., & Frontera, F., *Nature* **387**, 879 (1997).
61. Mochkovitch, R., Hernanz, M., Isern, J., & Martin, X., *Nature* **361**, 236–238, (1993).
62. Murakami, T. et al., *Nature* **350**, 592 (1991).
63. Narayan, R., Paczyński, B., & Piran, T., *ApJ* **395**, L83 (1992).
64. Palazzi, E. et al., *A.&A.* **336**, L95 (1998).
65. Paczyński, B. & Rhoads, J., *ApJ* **418**, L5 (1993).
66. Pian, E., Fruchter, A. S., Bergeron, L. E. et al., *ApJ* **492**, L103 (1999).
67. Pian, E. *et al.*, *A&AS* **138**, 463 (1999).
68. Reichart, D. E., *ApJ* **521**, L111 (1999).
69. Reichart, D. E., Lamb, D. Q., Metzger, M. R. et al., *ApJ* **517**, 692 (1999).
70. Rhoads, J. E., *ApJ* **487**, L1 (1997)
71. Sari, R., Piran, T. & Narayan, R., *ApJ* **497**, L17 (1998).
72. Sari, R. & Piran, T., *ApJ* **517**, L109 (1999).
73. Sari, R., Piran, T., & Halpern, J. P., *ApJ* **519**, L17 (1999).
74. Schmidt, M., *ApJ* **523**, L117 (1999).
75. Strohmeyer, T. E., Fenimore, E.E., Murakami, T. & Yoshida, A., *ApJ* **500**, 873 (1998).
76. Taylor, G. B., Frail, D. A., Kulkarni, S. R. et al., *ApJ* **502**, L115 (1998).
77. van Paradijs, J., Groot, P. J., Galama, T. et al., *Nature* **368**, 686 (1997).
78. Vietri, M., *ApJ* **488**, L105 (1997).
79. Waxman, E., *ApJ* **489**, L33 (1997).
80. Waxman, E., & Draine, B. T., astro-ph/9909020 (1999).
81. Waxman, E. & Loeb, A., *ApJ* **515**, 721 (1999).
82. Wheeler, J. C., astro-ph/9912403, (1999).
83. Wijers, R. A. M. J. & Galama, T. J., *ApJ* **523**, 177, (1999)
84. Wijers, R. A. M. J., Vreeswijk, P. M., Galama, T. J. et al., astro-ph/9906346 (1999).
85. Woosley, S. E., *ApJ* **405**, 273, (1993).
86. Woosley, S. E., Eastman, R. G. & Schmidt, B. P., *ApJ* **516**, 788 (1999).

The Fireball Shock Model of Gamma Ray Bursts

P. Mészáros [1,2,3]

[1] *Pennsylvania State University, 525 Davey, University Park, PA 16802*
[2] *California Institute of Technology, MS 105-24, Pasadena, CA 91125*
[3] *E-mail address: nnp@astro.psu.edu*

Abstract. Gamma-ray bursts are thought to be the outcome of a cataclysmic event leading to a relativistically expanding fireball, in which particles are accelerated at shocks and produce nonthermal radiation. We discuss the theoretical predictions of the fireball shock model and its general agreement with observations. Some of the recent work deals with the collimation of the outflow and its implications for the energetics, the production of prompt bright flashes at wavelenghts much longer than gamma-rays, the time structure of the afterglow, its dependence on the central engine or progenitor system behavior, and the role of the environment on the evolution of the afterglow.

I   INTRODUCTION

Gamma-ray bursts (GRB) have been studied in gamma-rays for over 25 years, but except for rare and fleeting X-ray detections, until a few years ago there existed no longer-lasting detections at softer wavelengths. However in early 1997 the Italian-Dutch satellite Beppo-SAX suceeded in providing accurate X-ray locations and images that allowed their follow-up with large ground-based optical and radio telescopes. The current interpretation of the gamma-ray and longer wavelength radiation is that the progenitor trigger produces an expanding relativistic fireball which can undergo both internal shocks leading to gamma-rays, and (as it decelerates on the external medium) an external blast wave and a reverse shock producing a broad-band spectrum lasting much longer.

A strong confirmation of the generic fireball shock model came from the correct prediction [43], in advance of the observations, of the quantitative nature of afterglows at longer wavelengths, in substantial agreement with the subsequent data [89,85,91,73,96]. The measured γ-ray fluences imply a total energy of order $10^{54}(\Omega_\gamma/4\pi)$ ergs, where $\Delta\Omega_\gamma$ is the solid angle into which the gamma-rays are beamed. Collimation may indeed be present, evidence having been recently reported for this [34,18,12]. In any case, such energies are possible [44] in the context of compact mergers involving neutron star-neutron star (NS-NS) or black hole-neutron star (BH-NS) binaries, or in hypernova/collapsar models involving a massive stellar progenitor [56,68]. In both cases, one is led to rely on MHD extraction of the spin energy of a disrupted torus and/or a central BH to power a relativistic outflow.

II THE GENERIC FIREBALL SHOCK SCENARIO

Whatever the GRB trigger is, the ultimate result must unavoidably be an $e^{\pm}, \gamma$ fireball, which is initially optically thick. The initial dimensions must be of order $r_{min} \lesssim ct_{var} \sim 10^7$ cm, since variability timescales are $t_{var} \lesssim 10^{-3}$ s. Most of the spectral energy is observed above 0.5 MeV, hence the $\gamma\gamma \to e^{\pm}$ mean free path is very short. Many bursts show spectra extending above 1 GeV, indicating the presence of a mechanism which avoids degrading these via photon-photon interactions to energies below the threshold $m_e c^2 = 0.511$ MeV. The inference is that the flow must be expanding with a very high Lorentz factor Γ, since then the relative angle at which the photons collide is less than Γ^{-1} and the threshold for the pair production is diminished [25]. However, the observed γ-ray spectrum is generally a broken power law, i.e., highly nonthermal. In addition, the expansion would lead to a conversion of internal into kinetic energy, so even after the fireball becomes optically thin, it would be radiatively inefficient, most of the energy being kinetic, rather than in photons.

The simplest way to achieve high efficiency and a nonthermal spectrum is by reconverting the kinetic energy of the flow into random energy via shocks after the flow has become optically thin [70]. Two different types of shocks may arise in this scenario. In the first case (a) the expanding fireball runs into an external medium (the ISM, or a pre-ejected stellar wind [70,39,30,78]. The second possibility (b) is that [71,54], even before external shocks occur, internal shocks develop in the relativistic wind itself, faster portions of the flow catching up with the slower portions. This is a generic model, which is independent of the specific nature of the progenitor.

External shocks will occur in an impulsive outflow of total energy E_o in an external medium of average particle density n_o at a radius $r_{dec} \sim (3E_o/4\pi n_o m_p c^2 \eta^2)^{1/3} \sim 10^{17} E_{53}^{1/3} n_o^{-1/3} \eta_2^{-2/3}$ cm , and on a timescale $t_{dec} \sim r_{dec}/(c\Gamma^2) \sim 3 \times 10^2 E_{53}^{1/3} n_o^{-1/3} \eta_2^{-8/3}$ s, where $\eta = \Gamma = 10^2 \eta_2$ is the final bulk Lorentz factor of the ejecta. Variability on timescales shorter than t_{dec} may occur on the cooling timescale or on the dynamic timescale for inhomogeneities in the external medium, but generally this is not ideal for reproducing highly variable profiles [79] (see however [15]). However, it can reproduce bursts with several peaks [58] and may therefore be applicable to the class of long, smooth bursts.

In a wind outflow [55], one assumes that a lab-frame luminosity L_o and mass outflow $\dot{M}_o$ are injected at $r \sim r_l$ and continuously maintained over a time t_w;

here $\eta = L_o/\dot{M}_o c^2$. In such wind model, internal shocks will occur at a radius [71] $r_{dis} \sim ct_{var}\eta^2 \sim 3 \times 10^{14} t_{var}\eta_2^2$ cm, on a timescale $t_w \gg t_{var} \sim r_{dis}/(c\eta^2)$ s, where shells of different energies $\Delta\eta \sim \eta$ initially separated by ct_v (where $t_v \leq t_w$ is the timescale of typical variations in the energy at r_l) catch up with each other. In order for internal shocks to occur above the wind photosphere $r_{ph} \sim \dot{M}\sigma_T/(4\pi m_p c\Gamma^2)$ $= 1.2 \times 10^{14} L_{53}\eta_2^{-3}$ cm, but also at radii greater than the saturation radius (so the bulk of the energy does not appear in the photospheric quasi-thermal component) one needs to have $7.5\times10^1 L_{51}^{1/5} t_{var}^{-1/5} \lesssim \eta 3\times10^2 L_{53}^{1/4} t_{var}^{-1/4}$. This type of models have the advantage [71] that they allow an arbitrarily complicated light curve, the shortest variation timescale $t_{var} \gtrsim 10^{-3}$ s being limited only by the dynamic timescale at r_l, where the energy input may be expected to vary chaotically. Such internal shocks have been shown explicitly to reproduce (and be required by) some of the more complicated light curves [79,33,61].

A potentially valuable diagnostic tool for the central engine of GRB is the power density spectrum (PDS). An analysis of BATSE light curves [3] indicates that the logarithmic slope of the PDS between 10^{-2} and 2 Hz is approximately -5/3, and there is a cutoff of the average PDS above 2 Hz. Using a simple kinematical

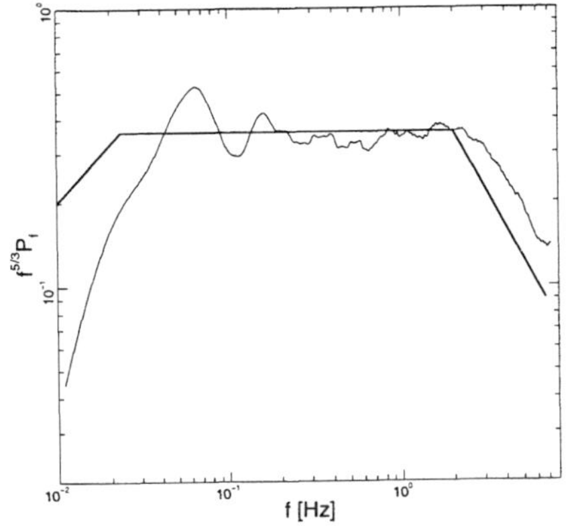

FIGURE 1. Average power density spectrum P_f of simulated bursts from internal shocks, compared with the observed PDS (thick line), using a square-sine modulated Lorentz factor and a cosmological distribution satisfying the observed logN-logP (Spada, Panaitescu & Mészáros 1999).

model for the ejection and collision of relativistic shells [63,83] have calculated the light curves and PDS expected for a range of total burst energies and for a total mass ejected and bulk Lorentz factor distribution compatible with the internal shock scenario (Figure 1). The redshift distribution also affects the PDS, and the observed logN-logP relation is used as a constraint. For optically thin winds, a slope approaching -5/3 requires a non-random Lorentz factor distribution, e.g. with an asymmetrical time modulation so as to produce a larger number of collisions at low frequencies (see also [4]). A cutoff at high frequencies (~ 2 Hz) can be understood in terms of shocks which increasingly occur below the scattering photosphere of the outflow, or a deficit of energy in short pulses due to the modulation of the Lorentz factors favoring shocks arising further out.

A significant fraction of bursts appear to have low energy spectral slopes steeper than 1/3 in energy [69,14]. This has motivated consideration of a thermal or non-thermal [35,36] comptonization mechanism, which can be put in the astrophysical context [23] of internal shocks leading to self-regulated pair formation. There is also

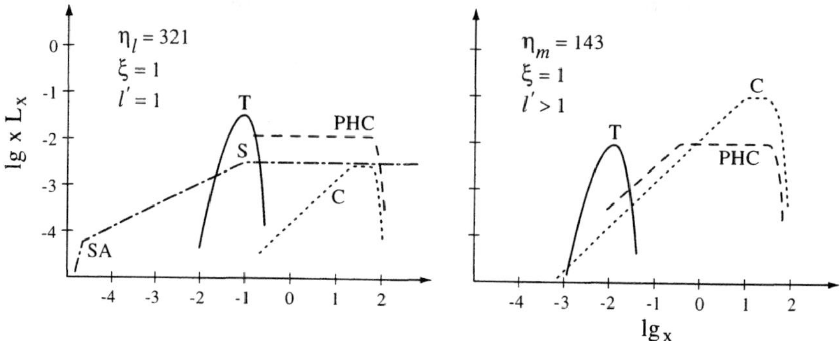

FIGURE 2. Luminosity per decade xL_x vs. $x = h\nu/m_ec^2$ for two values of $\eta = L/\dot{M}c^2$ and marginal (left) or large (rigt) pair compactness. T: thermal photosphere, PHC: photospheric comptonized component; S: shock synchrotron; C: shock pair dominated comptonized component (Mészáros & Rees, 1999b).

evidence that the clustering of the break energy of GRB spectra in the 50-500 keV range may not be due to observational selection [69,10,16]. Models using Compton attenuation [11] require reprocessing by an external medium whose column density adjusts itself to a few g cm^{-2}. More recently a preferred break has been attributed to a blackbody peak at the comoving pair recombination temperature in the fireball photosphere [17]. steep low energy spectral slope being due to the Rayleigh-Jeans part of the photosphere. In order for such photospheres to occur at the pair recombination temperature in the accelerating regime requires an extremely low baryon load. For very large baryon loads, a related explanation has been invoked [87], considering scattering of photospheric photons off MHD turbulence in the coasting portion of the outflow, which upscatters the adiabatically cooled photons up to the observed break energy. These ideas have been synthesized [50] to produce a generic scenario (see Figure 2) in which the presence of a photospheric component as well as shocks subject to pair breakdown can produce steep low energy spectra and preferred breaks.

III   SIMPLE STANDARD AFTERGLOW MODEL

The dynamics of GRB and their afterglows can be understood independently of any uncertainties about the progenitor systems, using a generalization of the method used to model supernova remnants. The simplest hypothesis is that the afterglow is due to a relativistic expanding blast wave, which decelerates as time goes on [43]. The complex time structure of some bursts suggests that the central trigger may continue for up to 100 seconds, the γ-rays possibly being due to internal shocks. However, at much later times all memory of the initial time structure would be lost: essentially all that matters is how much energy and momentum has been injected; the injection can be regarded as instantaneous in the context of the afterglow. As pointed in [70], the external shock bolometric luminosity builds up as $L \propto t^2$ and decays as $L \propto t^{-(1+q)}$. Beyond the deceleration radius the bulk Lorentz factor decreases as a power law in radius, $\Gamma \propto r^{-g} \propto t^{-g/(1+2g)}$, $r \propto t^{1/(1+2g)}$, with $g = (3, 3/2)$ for the radiative or adiabatic regime (in which $\rho r^3 \Gamma \sim$ constant or $\rho r^3 \Gamma^2 \sim$ constant).

The synchrotron peak frequency in the observer frame is $\nu_m \propto \Gamma B' \gamma^2$, and both the comoving field B' and electron Lorentz factor γ are expected to be proportional to Γ [39]. As Γ decreases, so will ν_m, and the radiation will move to longer wavelengths. For the forward blast wave, [53,31] discussed the possibility of detecting at late times a radio or optical afterglow of the GRB. A more detailed treatment of the fireball dynamics indicates that approximately equal amounts of energy are radiated by the forward blast wave, moving with $\sim \Gamma$ into the surrounding medium, and by a reverse shock propagating with $\Gamma_r - 1 \sim 1$ back into the ejecta [39]. The electrons in the forward shock are hotter by a factor Γ than in the reverse shock, producing two synchrotron peaks separated by Γ^2, one peak being initially in the optical (reverse) and the other in the γ/X band (forward) [41,42]. Detailed calculations and predictions of the time evolution of such a forward and reverse shock afterglow model ([43]) preceded the observations of the first afterglow GRB970228 ([13,88]), which was detected in γ-rays, X-rays and several optical bands, and was followed up for a number of months.

The simplest spherical afterglow model concentrates on the forward blast wave only. For this, the flux at a given frequency and the synchrotron peak frequency decay at a rate [43,49]

$$F_\nu \propto t^{[3-2g(1-2\beta)]/(1+2g)} \quad , \quad \nu_m \propto t^{-4g/(1+2g)}, \tag{1}$$

where g is the exponent of $\Gamma \propto r^{-g}$ and β is the photon spectral energy slope. The decay rate of the forward shock F_ν in equ.(1) is typically slower than that of the reverse shock [43], and the reason why the "simplest" model was stripped down to its forward shock component only is that, for the first two years 1997-1998, afterglows were followed in more detail only after the several hours needed by Beppo-SAX to acquire accurate positions, by which time both reverse external shock and internal shock components are expected to have become unobservable. This simple standard model has been remarkably successful at explaining the gross features and light curves of GRB 970228, GRB 970508 (after 2 days; for early rise, see §IV) e.g. [96,85,91,73].

This simplest afterglow model has a three-segment power law spectrum with two breaks. At low frequencies there is a steeply rising synchrotron self-absorbed spectrum up to a self-absorption break ν_a, followed by a $+1/3$ energy index spectrum up to the synchrotron break ν_m corresponding to the minimum energy γ_m of the power-law accelerated electrons, and then a $-(p-1)/2$ energy spectrum above this break, for electrons in the adiabatic regime (where γ^{-p} is the electron energy distribution above γ_m). A fourth segment and a third break is expected at energies where the electron cooling time becomes short compared to the expansion time, with a spectral slope $-p/2$ above that. With this third "cooling" break ν_b, first calculated in [47] and more explicitly detailed in [80], one has what has come to be called the simple "standard" model of GRB afterglows. One of the predictions of this model [43] is that the relation between the temporal decay index α, for $g = 3/2$ in $\Gamma \propto r^{-g}$, is related to the photon spectral energy index β through $F_\nu \propto t^\alpha \nu^\beta$, with $\alpha = (3/2)\beta$. This relationship appears to be valid in many (although not all) cases, especially after the first few days, and is compatible with an electron spectral index $p \sim 2.2 - 2.5$ which is typical of shock acceleration, e.g. [91,80,95], etc. As the remnant expands the photon spectrum moves to lower frequencies, and the flux in a given band decays as a power law in time, whose index can change as breaks move through it. For the simple standard model, snapshot overall spectra have been deduced by extrapolating spectra at different wavebands and times using assumed simple time dependences [92,95]. These can be used to derive rough fits for the different physical parameters of the burst and environment, e.g. the total energy E, the magnetic and electron-proton coupling parameters ϵ_B and ϵ_e and the external density n_o.

IV   "POST-STANDARD" AFTERGLOW MODELS

The most obvious departure from the simplest standard model occurs if the external medium is inhomogeneous: for instance, for $n_{ext} \propto r^{-d}$, the energy conservation condition is $\Gamma^2 r^{3-d} \sim$ constant, which changes significantly the temporal decay rates [47]. Such a power law dependence is expected if the external medium is a wind, say from an evolved progenitor star as implied in the hypernova scenario (such winds are generally used to fit supernova remnant models). Another departure from a simple impulsive injection approximation is obtained if the mass and energy injected during the burst duration t_w (say tens of seconds) obeys $M(> \Gamma) \propto \Gamma^{-s}$, $E(> \Gamma) \propto \Gamma^{1-s}$, i.e. more energy emitted with lower Lorentz factors at later times (but still shorter than the gamma-ray pulse duration). This would drastically change the temporal decay rate and extend the afterglow lifetime in the relativistic regime, providing a late "energy refreshment" to the blast wave on time scales comparable to the afterglow time scale [72]. These two cases lead to a decay rate

$$\Gamma \propto r^{-g} \propto \begin{cases} r^{-(3-d)/2} & ; \ n_{ext} \propto r^{-d}; \\ r^{-3/(1+s)} & ; \ E(> \Gamma) \propto \Gamma^{1-s}. \end{cases} \tag{2}$$

Expressions for the temporal decay index $\alpha(\beta, s, d)$ in $F_\nu \propto t^\alpha$ are given by [47,72], which now depend also on s and/or d. The result is that the decay can be flatter (or

218

steeper, depending on s and d) than the simple standard $\alpha = (3/2)\beta$. A third non-standard effect, which is entirely natural, occurs when the energy and/or the bulk Lorentz factor injected are some function of the angle. A simple case is $E_o \propto \theta^{-j}$, $\Gamma_o \propto \theta^{-k}$ within a range of angles; this leads to the outflow at different angles shocking at different radii and its radiation arriving at the observed at different delayed times, and it has a marked effect on the time dependence of the afterglow [47], with $\alpha = \alpha(\beta, j, k)$ flatter or steeper than the standard value, depending on j, k. Thus in general, a temporal decay index which is a function of more than one parameter

$$F_\nu \propto t^\alpha \nu^\beta \ \ , \text{with} \ \ \alpha = \alpha(\beta, d, s, j, k, \cdots) \ , \tag{3}$$

is not surprising; what is more remarkable is that, often, the simple relation $\alpha = (3/2)\beta$ is sufficient to describe the overall behavior at late times.

Evidence for departures from the simple standard model is provided by, e.g., sharp rises or humps in the light curves followed by a renewed decay, as in GRB 970508 ([64,66]). Detailed time-dependent model fits [62] to the X-ray, optical and radio light curves of GRB 970228 and GRB 970508 show that, in order to explain the humps, a *non-uniform* injection or an *anisotropic* outflow is required. These fits indicate that the shock physics may be a function of the shock strength (e.g. the electron index p, injection fraction ζ and/or ϵ_b, ϵ_e change in time), and also indicate that dust absorption is needed to simultaneously fit the X-ray and optical fluxes. The effects of beaming (outflow within a limited range of solid angles) can be significant [60], but are coupled with other effects, and a careful analysis is needed to disentangle them.

Prompt optical, X-ray and GeV flashes from reverse and forward shocks, as well as from internal shocks, have been calculated in theoretical fireball shock models for a number of years [41,42,57,43,81], as have been jets (e.g. [38,40,42], and in more detail [76,62,60,77]). However, observational evidence for these effects were largely lacking, until the detection of a prompt (within 22 s) optical flash from GRB 990123 with ROTSE by [2], together with X-ray, optical and radio follow-ups [34,21,18,1,12,27]. GRB 990123 is so far unique not only for its prompt optical detection, but also by the fact that if it were emitting isotropically, based on its redshift $z = 1.6$ [34,1] its energy would be the largest of any GRB so far, 4×10^{54} ergs. It is, however, also the first (tentative) case in which there is evidence for jet-like emission [34,18,12]. An additional, uncommon feature is that a radio afterglow appeared after only one day, only to disappear the next [21,34].

The prompt optical light curve of GRB 990123 decays initially as $\propto t^{-2.5}$ to $\propto t^{-1.6}$ [2], much steeper than the typical $\propto t^{-1.1}$ of previous optical afterglows detected after several hours. However, after about 10 minutes its decay rate moderates, and appears to join smoothly onto a slower decay rate $\propto t^{-1.1}$ measured with large telescopes [21,34,18,12] after hours and days. The prompt optical flash peaked at 9-th magnitude after 55 s [2], and in fact a 9-th magnitude prompt flash with a steeper decay rate had been predicted more than two years ago [43], from the synchrotron radiation of the reverse shock in GRB afterglows at cosmological redshifts (see also [81,41,42]). An origin of the optical prompt flash in internal shocks [43,49] cannot be ruled out, but is less likely since the optical light curve and the γ-rays appear not to correlate well [82,21]. The subsequent slower decay agrees with predictions for the forward external shock [43,82,49].

The evidence for a jet is based on an apparent steepening of the light curve after about three days [34,18,12]. If real, this steepening is probably due to the transition between early relativistic expansion, when the light-cone is narrower than the jet opening, and the late expansion, when the light-cone has become wider than the jet, leading to a drop in the effective flux [76,34,49,77]. A rough estimate leads to a jet opening angle of 3-5 degrees, which would reduce the total energy requirements to about 4×10^{52} ergs. This is about two order of magnitude less than the binding energy of a few solar rest masses, which, even allowing for substantial inefficiencies, is compatible with currently favored scenarios (e.g. [68,37]) based on a stellar collapse or a compact merger.

V LOCATION AND ENVIRONMENTAL EFFECTS

The location of the afterglow relative to the host galaxy center can provide clues both for the nature of the progenitor and for the external density encountered by the fireball. A hypernova model would be expected to occur inside a galaxy in a high density environment $n_o > 10^3 - 10^5$ cm^{-3}. Most of the detected and well identified afterglows are inside the projected image of the host galaxy [6], and some also show evidence for a dense medium at least in front of the afterglow ([52]).

In NS-NS mergers one would expect a BH plus debris torus system and roughly the same total energy as in a hypernova model, but the mean distance traveled from birth is of order several Kpc [8], leading to a burst presumably in a less dense environment. The fits of [95] to the observational data on GRB 970508 and GRB 971214 in fact suggest external densities in the range of $n_o = 0.04$–0.4 cm^{-3}, which would be more typical of a tenuous interstellar medium. These could be within the volume of the galaxy, but for NS-NS on average one would expect as many GRB inside as outside. This is based on an estimate of the mean NS-NS merger time of 10^8 years. BH-NS mergers would occur in timescales $\sim 10^7$ years, and would be expected to give bursts inside the host galaxy ([8]; see however [19]). In at least one "snapshot" standard afterglow spectral fit for GRB 980329 [75] the deduced external density is $n_o \sim 10^3$ cm^{-3}. In some of the other detected afterglows there is other evidence for a relatively dense gaseous environments, as suggested, e.g. by evidence for dust [74] in GRB970508, the absence of an optical afterglow and presence of strong soft X-ray absorption [24,51] in GRB 970828, the lack an an optical afterglow in the (radio-detected) afterglow ([86]) of GRB980329, and spectral fits to the low energy portion of the X-ray afterglow of several bursts [52]. One important caveat is that all afterglows found so far are based on Beppo-SAX positions, which is sensitive only to long bursts $t_b \gtrsim 20$ s [28]. This is significant, since it appears likely that NS-NS mergers lead [37] to short bursts with $t_b \lesssim 10$ s. To make sure that a population of short GRB afterglows is not being missed will probably need to await results from HETE [26] and from the planned Swift [84] mission, which is designed to accurately locate 300 GRB/yr.

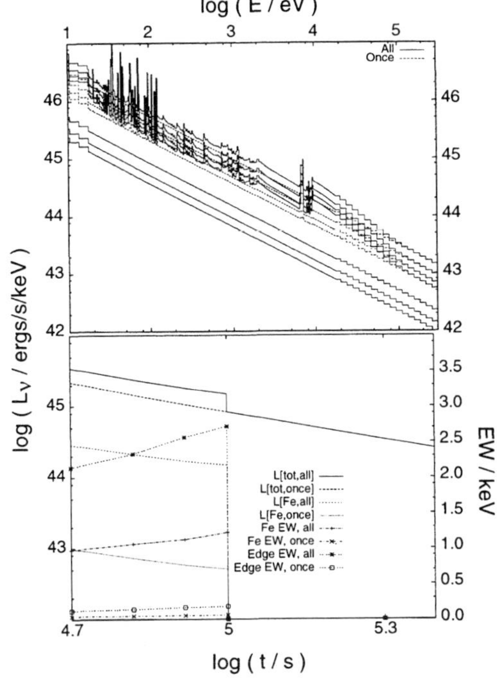

FIGURE 3. Spectrum (top) of a hypernova funnel model for various observer times, showing (bottom) the total and Fe light curves and equivalent widths (Weth, Mészáros , Kallman & Rees 1999), for with $R = 1.5 \times 10^{16}$ cm, $n = 10^{10}$ cm^{-3}, and Fe abundance 10^2 times solar.

The environment in which a GRB occurs may also lead to specific spectral signatures from the external medium imprinted in the continuum, such as atomic edges and lines [5,65,46]. These may be used both to diagnose the chemical abundances and the ionization state (or local separation from the burst), as well as serving as potential alternative redshift indicators. (In addition, the outflowing ejecta itself may also contribute blue-shifted edge and line features, especially if metal-rich blobs or filaments are entrained in the flow from the disrupted progenitor debris [45], which could serve as diagnostic for the progenitor composition and outflow Lorentz factor). To distinguish between progenitors, an interesting prediction ([46]; see also [22,9]) is that the presence of a measurable Fe K-α X-ray *emission* line could be a diagnostic of a hypernova, since in this case one may expect a massive envelope at a radius comparable to a light-day where $\tau_T \lesssim 1$, capable of reprocessing the X-ray continuum by recombination and fluorescence. Detailed radiative transfer calculations have been performed to simulate the time-dependent X/UV line spectra of massive progenitor (hypernova) remnants [93], see Figure 3. Two groups [67,98] have in fact recently reported the possible detection of Fe emission lines in GRB 970508 and GRB 970828.

An interesting case is the apparent coincidence of GRB 980425 with the unusual SN Ib/Ic 1998bw [20], which may represent a new class of SN [29,7]. If true, this could imply that some or perhaps all GRB could be associated with SN Ib/Ic [90], differring only in their viewing angles relative to a very narrow jet. Alternatively, the GRB could be (e.g. [97]) a new subclass of GRB with lower energy $E_\gamma \sim 10^{48}(\Omega_j/4\pi)$ erg, only rarely observable, while the great majority of the observed GRB would have the energies $E_\gamma \sim 10^{54}(\Omega_j/4\pi)$ ergs as inferred from high redshift observations. The difficulties are that it would require extreme collimations by factors $10^{-3} - 10^{-4}$, and the statistical association is so far not significant [32]. However, two more GRB light curves may have been affected by an anomalous SNR (see, e.g. the review of [94]).

VI CONCLUSIONS

The fireball shock model of gamma-ray bursts has proved quite robust in providing a consistent overall interpretation of the major features of these objects at various frequencies and over timescales ranging from the short initial burst to afterglows extending over many months. Significant progress has been made in understanding both the phenomenology and the physics of these obejcts, which may the most widely studied type of black holes sources. There still remain a number of mysteries, especially concerning the identity of their progenitors, the nature of the triggering mechanism, the transport of the energy, the time scales involved, and the nature and effects of beaming. However, the collective theoretical and observational understanding is vigorously advancing, and with dedicated new and planned observational missions under way, further significant progress may be expected in the near future.

I thank M.J. Rees, A. Panaitescu, M. Spada and C. Weth for stimulating collaborations, NASA NAG-5 2857, the Guggenheim Foundation, and the Division of Physics, Math & Astronomy, Astronomy Visitor and Merle Kingsley funds at Caltech.

REFERENCES

1. Andersen et al. 1999, Science, 283, 2073
2. Akerlof, C., *et al.*, 1999, Nature, 398, 389
3. Beloborodov, A, Stern, B & Svensson, R (1998) ApJ, 508, L25.
4. Beloborodov, A, astro-ph/9911122
5. Bisnovatyi-Kogan, G & Timokhin, A, 1997, Astr. Rep. 41, 423
6. Bloom, J., etal, 1998, A& A Supp.,in press (Procs. Rome Conference on GRB)
7. Bloom, J, *et al.*, 1998, ApJ 506, L105
8. Bloom, J, Sigurdsson, S & Pols, O, 1999, MNRAS (astro-ph/9805222)
9. Böttcher, M, *et al.*, 1998, astro-ph/9809156

10. Brainerd, J et al.(1999) in *Abstr 19th Texas Symp*, Paris (astro-ph/9904039).

11. Brainerd, J. et al.(1998), ApJ, 501:325.

12. Castro-Tirado et al., 1999, Science, 283, 2069

13. Costa, E., et al., 1997, Nature, 387, 783

14. Crider, A. et al.(1997), ApJ, 479:L39.

15. Dermer, C & Mitman, K, 1998, astro-ph/9809411

16. Dermer, C.D., et al., 1999, Ap.J., 515, L49.

17. Eichler, D & Levinson, A (1999), ApJ, subm(astro-ph/9903103).

18. Fruchter, A. etal, Ap.J. subm (astro-ph/9902236)

19. Fryer, C & Woosley, S, 1998, ApJ(Lett) subm (astro-ph/9804167

20. Galama, T. et al., 1998, Nature, 395, 670

21. Galama, T. et al., 1999, Nature, 398, 394

22. Ghisellini, G, et al., 1998, astro-ph/9808156

23. Ghisellini, G. and Celotti, A. (1999), ApJ, 511, L93.

24. Groot, P. et al., 1997; in *Gamma-Ray Bursts*, Meegan, C., Preece, R & Koshut, T, eds., 1997 (AIP: New York), p. 557

25. Harding, A.K. and Baring, M.G., 1994, in *Gamma-ray Bursts*, ed. G. Fishman, *et al.*, p. 520 (AIP 307, NY)

26. HETE, http://space.mit.edu/HETE/

27. Hjorth, etal, 1999, Science, 283, 2073

28. Hurley, K., A& A Supp.,in press (Procs. Rome Conference on GRB)

29. Iwamoto, K, et al., 1998, Nature 395, 672

30. Katz, J., 1994a, ApJ, 422, 248

31. Katz, J., 1994b, ApJ, 432, L107

32. Kippen, R.M. et al., 1998, ApJ subm (astro-ph/9806364)

33. Kobayashi, S, Piran, T & Sari, R, 1998, ApJ, 490, 92

34. Kulkarni, S., et al., 1999, Nature, 398, 389

35. Liang, E et al.(1997), Ap.J., 491, L15.

36. Liang, E. et al.(1999), Ap.J., 519, L21.

37. Macfadyen, A & Woosley, S, 1999, ApJ in press (astro-ph/9810274)

38. Mészáros , P & Rees, M.J., 1992, ApJ, 397, 570

39. Mészáros , P. & Rees, M.J., 1993a, ApJ, 405, 278

40. Mészáros , P., Laguna, P & Rees, M.J., 1993, ApJ, 415, 181

41. Mészáros , P. and Rees, M.J., 1993b, Ap.J., 418, L59

42. Mészáros , P., Rees, M.J. & Papathanassiou, H, 1994, Ap.J., 432, 181

43. Mészáros , P & Rees, M.J., 1997a, ApJ, 476, 232

44. Mészáros , P & Rees, M.J., 1997b, ApJ, 482, L29

45. Mészáros , P & Rees, M.J., 1998a, ApJ, 502, L105

46. Mészáros , P & Rees, M.J., 1998b, MNRAS, 299, L10

47. Mészáros , P, Rees, M.J. & Wijers, R, 1998, Ap.J., 499, 301 (astro-ph/9709273)

48. Mészáros , P, Rees, M.J & Wijers, R,1999,New Ast, 4, 313(astro-ph/9808106)

49. Mészáros , P & Rees, M.J., MNRAS, in press (astro-ph/9902367)

50. Mészáros , P. & Rees, M.J., ApJ, in press (1999) (astro-ph/9908126).

51. Murakami, T. et al., 1997, in *Gamma-Ray Bursts*, Meegan, C., Preece, R & Koshut, T, eds., 1997 (AIP: New York), p. 435

52. Owen, A., et al, 1998, Astron.&Astrophys. in press (astro-ph/9809356),

53. Paczyński , B. & Rhoads, J, 1993, Ap.J., 418, L5

54. Paczyński , B. & Xu, G., 1994, ApJ, 427, 708

55. Paczyński , B., 1990, Ap.J., 363, 218

56. Paczyński , B., 1998, ApJ, 494, L45

57. Papathanassiou, H & Mészáros , P, 1996, ApJ, 471, L91

58. Panaitescu, A & Mészáros , P, 1998a, ApJ, 492, 683

59. Panaitescu, A. & Mészáros , P., 1998b, ApJ, 501, 772

60. Panaitescu, A. & Mészáros , P., 1999, ApJ, in press (astro-ph/9806016)

61. Panaitescu, A. & Mészáros , P., 1998d, ApJ, subm (astro-ph/9810258)

62. Panaitescu, A, Mészáros , P & Rees, MJ, 1998, ApJ, 503, 314

63. Panaitescu, A., Spada, M., and Mészáros , P. (1999), ApJ, 522:L105.

64. Pedersen, H. *et al.*(1998), ApJ 496, 311

65. Perna, R. & Loeb, A., 1998, ApJ, 503, L135

66. Piro, L, *et al.*, 1998, A & A, 331, L41

67. Piro, L, *et al.*, 1998b, A& A Supp.,in press (Procs. Rome Conference on GRB)

68. Popham, R., Woosley, S & Fryer, C., 1998, ApJ n press (astro-ph/9807028)

69. Preece, R. *et al.*(1998), ApJ, 496, 849.

70. Rees, M.J. & Mészáros , P., 1992, MNRAS, 258, P41

71. Rees, M.J. & Mészáros , P., 1994, ApJ, 430, L93

72. Rees, M.J. & Mészáros , P., 1998, ApJ, 496, L1

73. Reichart, D., 1997, ApJ, 485, L57

74. Reichart, D., 1998, ApJ, 495, L99

75. Reichart, D & Lamb, D.Q, 1998, A&A Supp,in press (Procs. Rome Conf)

76. Rhoads, J, 1997, Ap.J., 487, L1

77. Rhoads, J, 1999, ApJ subm, astro-ph/9903383

78. Sari, R. & Piran, T., 1995, ApJ, 455, L143

79. Sari, R & Piran, T, 1998, ApJ, 485, 270

80. Sari, R , Piran, T & Narayan, R, 1998, ApJ, 497, L17 (astro-ph/9712005)

81. Sari, R & Piran, T., 1999a, A&A subm (astro-ph/9901105)

82. Sari, R & Piran, T, 1999b, Ap.J. subm (astro-ph/9902009)

83. Spada, M, Panaitescu, A & Mészáros , P, 1999 ApJ subm(astro-ph/9908097)

84. Swift, http://swift.gsfc.nasa.gov/

85. Tavani, M., 1997, ApJ, 483, L87

86. Taylor, G.B., et al., 1997, Nature, 389, 263

87. Thompson, C., 1994, MNRAS, 270, 480

88. van Paradijs, J, *et al.*, 1997, Nature,386, 686

89. Vietri, M., 1997a, ApJ, 478, L9

90. Wang, L. & Wheeler, J.C., 1998, ApJ subm (astro-ph/9806212)

91. Waxman, E., 1997, ApJ, 485, L5

92. Waxman, E., 1997b, ApJ, 489, L33

93. Weth, C, Mészáros , P, Kallman, T & Rees, M.J, 1999 ApJ in press(astro-ph/9908243)

94. Wheeler, J.C. (1999), in *Supernovae & Gamma Ray Bursts*, eds. M Livio *et al.*(Cambridge U.P.)(astro-ph/9909096).

95. Wijers, R.A.M.J. & Galama, T., 1998, ApJ, in press (astro-ph/9805341)
96. Wijers, R.A.M.J., Rees, M.J. & Mészáros , P., 1997, MNRAS, 288, L51
97. Woosley, S., Eastman, R. & Schmidt, B., 1998, ApJ, subm (astro-ph/9806299)
98. Yoshida, A, *et al.*, 1998, A& A Supp.,in press (Procs. Rome Conf on GRB)

Gamma Ray Bursts: The Future

N. Gehrels[*] and D. Macomb[†]

[*]*Laboratory for High Energy Astrophysics Goddard Space Flight Center*
[†]*Astronomy Program, Universities Space Research Association*

Abstract. Gamma-ray bursts are the most dramatic and powerful cosmic explosions known. They also continue to be the most puzzling. Thanks to breakthrough observations over the last decade, however, a picture has emerged of gamma-ray bursts being at cosmological distances and capable of releasing more than 10^{51} ergs of energy within seconds. Despite the emergence of this picture, the physical origin of bursts is still unknown and the classification of different types of bursts is still in its infancy. Further understanding of gamma-ray bursts requires the wise use of our current resources and the development of new observational capabilities. We outline the current state of our knowledge of bursts and describe the present and future instrumentation which will enable us to understand these baffling blasts.

INTRODUCTION

Cosmic explosions are at their most remarkable in the form of gamma-ray bursts (GRBs). Since their discovery in the early 1970's by the Vela satellites [1], understanding these intense yet brief beacons of gamma-ray emission has been one of the outstanding problems in astrophysics. Two decades of post-discovery work led most researchers to believe that GRBs were associated with galactic neutron stars. The last decade changed all that through a series of groundbreaking observations.

The transformation in our understanding of GRBs came with the measurement of burst isotropy and inhomogeneity using data compiled by the BATSE instrument on the Compton Gamma-Ray Observatory (CGRO) [2]. This measurement established GRBs as either cosmological or in an extended galactic halo. As the years went by, BATSE continued to shrink the phase-space which could be occupied by a putative halo and established a phenomenal database of over 2000 bursts.

The next major advance in GRB studies came with the detection of X-ray, optical, and radio counterparts and the firm establishment of a cosmological distance scale. The good imaging ability of the X-ray instruments aboard *BeppoSAX* provided the position of GRB 970228 to within a few arcminutes during the actual burst. Follow-up observations revealed the existence of an X-ray afterglow [3] and a fading optical counterpart [4]. This was the first of ≈ 15 precise GRB localizations which in several cases provided a determination of the host galaxy and a redshift

CP522, *Cosmic Explosions: Tenth Astrophysical Conference,*
edited by Stephen S. Holt and William W. Zhang
2000 American Institute of Physics 1-56396-943-2

[5–9]. Subsequently, one of the most dramatic developments was the detection by the ROTSE instrument [10] of bright optical emission (V=9) from GRB 970123 during the burst .

These measurements verified what BATSE had determined was highly probable, that GRBs are at cosmological distances ($z \approx 1$)and that the energy release exceeds 10^{51} ergs (ignoring beaming). Further progress in solving the GRB mystery depends on increasing the number of finely localized bursts along with reliable distance measurements.

GRB ASTROPHYSICS

The current understanding of GRBs still does not provide us with a basic insight into the ultimate source of the burst. While evidence for the so-called fireball model involving a relativistic outflow and emission from either internal or external shocks is successful at describing the burst and afterglow (for instance see [11] for a good review), the underlying source class is still unkown. Even the basic phenomenology of gamma-ray bursts is not understood since there is evidence for different classes (see [12]). For example, there is a clear bi-modality in the duration distribution of bursts (e.g. [13]) To date, counterparts have been found for only the long bursts due to the lack of trigger capability for short bursts by *BeppoSAX*. Also, there is a possible subclass of GRBs associated with supernovae such as GRB 980425 [14]. How these classes arise and their relationship to possibly different burst mechanisms and progenitors will likely be a very fruitful area of investigation.

There is no shortage of speculation on the fundamental source of GRBs. Some common theories of burst formation involve merging neutron stars [15], neutron star-black hole mergers [16], or hypernova or collapsar models [17]. Determining which models are applicable will have a deep impact on questions about the birth of black holes, the history of massive star formation, and studies of the high-z universe.

Before this impact is felt, however, there are key questions remaining to be answered by future observations including:

- What are the progenitors of GRB's?

- Are there different classes of bursts with different physical processes at work?

- How does the blast wave evolve and interact with its surroundings?

- Are GRBs occurring in star-forming regions?

- Are GRBs related to black hole formation?

- What is the distribution of z? Do GRBs occur at high z ($z > 10$)?

- What are the counterparts of short bursts?

- What is the GRB-supernova connection?

- How does the blast wave evolve at early times?

- What is the intrinsic GRB luminosity function?

- Are GRBs beamed? If so, is the beaming different at different energies?

- What is the role of internal, external, and reverse shocks?

- How common are bright optical events like GRB 990123?

- What is the cause and occurrence rate of GeV afterglows?

- Are there undiscovered classes of GRBs? Dim, soft, ultra-short, ultra-long . .

Answering these and other questions requires a new era in GRB detection and study. Primary GRB detectors need to provide sensitive observations of bursts independent of duration. Follow-up observations must build on this to observe the development of the initial burst and afterglow over a broad energy range. In the following sections, we describe the current capabilities in terms of primary burst detection, rapid follow-up capabilities, and sensitive deep follow-up observations. We also describe future instrumentation which will provide the improvements needed to answer the questions posed above. Many of these instruments and their expected dates of operation are shown in Figure 1. The most important requirements for future progress are as follows:

- Provide optical and X-ray observations during and immediately after burst

- Detect 100's of GRBs with redshifts and hosts

- Enable the study of short burst counterparts

- Provide gamma-ray observations at 1/10 BATSE sensitivity

- Determine sub-arcsec positions for > 100 GRBs

- Enable deep IR spectroscopy to find high-z bursts

- Enable radio observations at 1/10 VLA sensitivity

- Provide high energy gamma-ray coverage for > 100 GRBs

PRIMARY GAMMA-RAY BURST DETECTORS

Current Status

Until the new generation of gamma-ray burst detectors become available, the workhorse instruments which have been responsible for the breakthroughs of the last decade will continue to contribute. These instruments - primarily BATSE,

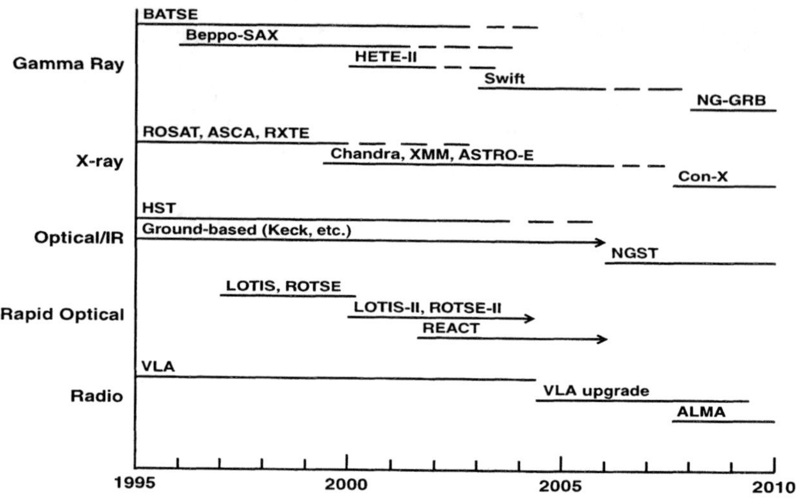

FIGURE 1. GRB experiment timeline for various wavelengths.

BeppoSAX, and the Interplanetary Network - are listed in Table 1 along with the future gamma-ray burst missions.

The BATSE instrument on the Compton Gamma-ray Observatory continues to detect nearly a burst per day [18]. As the primary component of the GCN network (see below), BATSE provides a unique database for both class studies of GRBs and for triggering follow-up observations. Also onboard CGRO, the imaging Compton Telescope, COMPTEL, detects a few bursts per year in the energy range from 1-30 MeV [19].

The *BeppoSAX* satellite is expected to operate through at least 2001 detecting a much smaller rate of bursts, but providing excellent positional information for those it does detect. *BeppoSAX* features a gamma-ray burst monitor (GRBM) which can detect about 60 GRBs/year [20], and an X-ray Wide Field Camera (WFC). The WFC detects 10 bursts per year with 10 arcmin localizations [21]. The spacecraft can be maneuvered via ground command in a few (> 6) hours to point on-board X-ray telescopes at the burst.

The Interplanetary Network (IPN) continues to be an important means for excellent GRB source localization [22]. This long baseline network includes burst detectors on NEAR, Ulysses, Konus/Wind, CGRO and RXTE. Time delays between the various detectors are used to triangulate the position of about one burst per week to an accuracy of several arcminutes.

Future Observatories

HETE II

An exciting series of GRB experiments will be launched in the next five years. The first major step is the January 2000 launch of the HETE II satellite (Figure 2). HETE II will combine a moderate burst detection rate of about 50/yr with excellent localizations of 10 arcsec to 5 arcmin during its 1-2 year mission lifetime [23]. The instruments on HETE II are provided by an international collaboration. They include a NaI gamma-ray burst detector with 2π sr field-of-view operating from 6 keV to 1 MeV (FREnch GAmma TElescope - FREGATE), an X-ray monitor (WXM) with 2 sr field of view operating from 2 - 25 keV and 4 arcminute PSF, and a soft X-ray camera (SXC) operating from 0.5 - 14 keV with a 1 sr field of view and 1.3 arcminute PSF. Locations will be sent to observers within 5 seconds of the burst onset. Breakthroughs for this mission include the ability to perform soft X-ray observations during a burst, improved coverage of short bursts, and rapid notification of multi-wavelength observers.

Swift

The approach of having a single satellite provide realtime multiwavelength observations reaches its pinnacle with Swift. Swift, shown in Figure 3, is a MIDEX mission selected by NASA for launch in 2003 [24]. Swift is an international with three onboard instruments: a 2 sr field-of-view Burst Alert Telescope (BAT) sensitive from 10 - 150 keV with 5000 cm^2 of CdZnTe detectors, an X-Ray Telescope (XRT) operating from 0.2 - 10 keV with 15 arcsec PSF and 110 cm^2 of effective area, and an UV/Optical Telescope (UVOT) with 0.3 arcsec PSF capable of 24th

TABLE 1. Current and future GRB instrumentation.

Name	Energy Range[a]	Dates	Rate[b]	Localization[c]	Delay
BATSE [18]	25-2000 keV	1991-2002	300	100 - 240	seconds
COMPTEL [19]	1-30 MeV	1991-2002	7	75	hours
IPN [22]	—	till 2003	≈ 70	$\approx 1 - 5$	day
BeppoSax WFC [21]	1.8-28 keV	1996-2001	10	$\approx 1 - 5$	hours
RXTE ASM [31]	1.5-12 keV	1996-2002	5	≈ 10	hours
HETE II [23]	6-400 keV	2000-2002	50	0.17-10	seconds
CATSAT [25]	200-5000 keV	2001	10	> 60	hours
INTEGRAL [27]	15-10000 keV	2001-2003	20	≈ 1	12-55 s
AGILE [28]	0.03-50 GeV	2002-2005	10	≈ 5	
Swift [24]	10-150 keV	2003-06	300	0.01	seconds
GLAST [29]	0.02-300 GeV	2005-2010	100	≈ 5	

[a] range of the primary burst detector

[b] units are bursts/year

[c] typical error radius, units are arcminutes

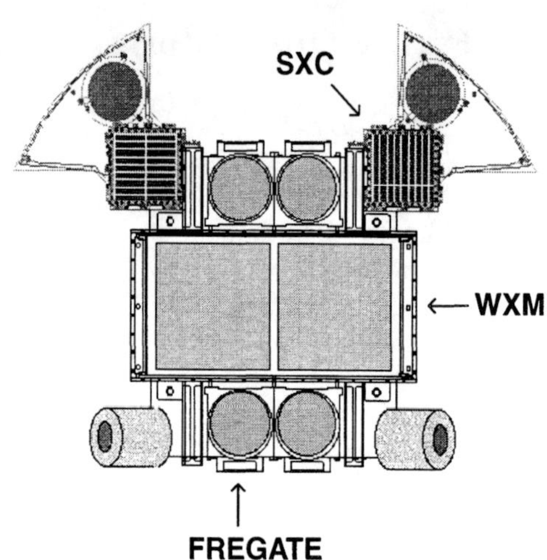

FIGURE 2. Drawing of the HETE II spacecraft showing the instrument positions.

magnitude sensitivity in a 1000 second observation. Swift is expected to detect approximately 300 bursts/year. A primary capability is to rapidly and autonomously re-point the spacecraft in 20-70 seconds to orient the XRT and UVOT to burst locations.

With this approach, Swift will be capable of providing multiwavelength observations from optical to gamma-ray energies. Within 10 seconds of a burst onset, a BAT position is calculated and sent to the ground at the same time that the spacecraft slew begins. An arcsecond position is available within about 65 seconds once the XRT has acquired the source. Then within about 160 of the burst onset, a UVOT finding chart is available for distribution to other experiments. Thus within minutes of a burst, a broad-band energy measurement and sub-arcsecond position is available for further follow-up.

Other Future Instruments

Beyond the major advances presented by HETE II and Swift, a number of missions will provide important capabilities. The most immediate is the Cooperative Astrophysical and Technology Satellite (CATSAT) which is part of NASA's STEDI program [25]. CATSAT features a soft X-ray spectrometer to detect photoelectric absorption features and an Earth albedo polarimeter which will measure the X-ray polarization of bursts. CATSAT will be launched in early 2000 for a nominal one year mission and measure $\approx$ 50 GRBs/year. Another small mission called Ballerina is being studied in Denmark for all-sky GRB coverage and X-ray follow-up [26].

The launch of the INTErnational Gamma-Ray Astrophysical Laboratory (IN-

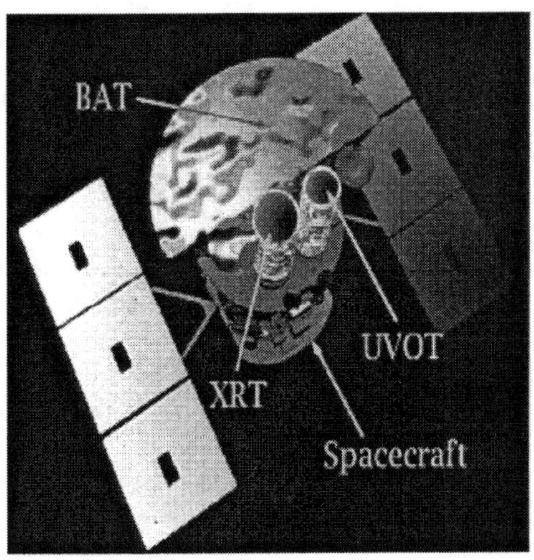

FIGURE 3. The Swift satellite showing the layout of the science instruments.

TEGRAL) in 2001 will provide other GRB observations. INTEGRAL is expected to detect about 20 bursts/year with a coded mask CdTe array operating from 15 keV - 10 MeV, and make arcminute localizations available after 5-100 seconds [27].

At still higher gamma-ray energies, two planned missions will contribute to understanding GRBs. The AGILE (Astro-rivelatore Gamma a Immagini LEggero) mission will operate in the 30 MeV - 50 GeV band with a large field of view and low deadtime. AGILE will detect 10 GRBs/year over the expected 2002-2005 lifetime and plans are being made for rapid distribution of locations [28]. The GLAST mission will also operate at high energies, from 30 MeV - 300 GeV, and is expected to launch in 2005 with a 5 year mission [29]. GLAST, with a large 2 sr field of view, large area (8000 cm^2), and low deadtime, should detect about 100 bursts per year and localize them to within a few arcminutes.

Beyond these planned missions, longer term mission concepts are in the works. The Gamma-ray Astronomy Programs Working Group (GRAPWG), an ad-hoc NASA panel which gives NASA recommendations about the future of gamma-ray astronomy, suggests a Next Generation GRB mission (NG-GRB), beyond Swift to continue to provide a platform for GRB detection, observation and triggering. The future is quite bright for detecting gamma-ray bursts.

GCN

The principal tool for enabling rapid follow-up observations is provided by the Gamma-Ray Burst Coordinates Network or GCN [30]. This service distributes 300 positions/year primarily using BATSE triggers with delays of seconds and 4 degree localizations. In addition to BATSE, locations based on *BeppoSAX*, COMPTEL, RXTE, and the IPN are available through GCN. These alerts are made via email, socket connections, the web and beepers. One such alert allowed the ROTSE instrument to detect the optical transient associated with GRB 990123 while the burst was in progress.

CURRENT OPTICAL INSTRUMENTS

Most of the opportunities for economical and rapid follow-up are for telescopes operating in the optical range. The ideal instrument has a field-of-view matched to the GRB positional error, can rapidly slew to the target, and has a good limiting sensitivity. Given the constraints on optical observations, it is valuable to have a large number of such instruments operating to achieve full sky coverage. These telescopes typically trigger off GCN alerts and can usually acquire a target within 10's of seconds following the trigger. In the last few years, the limiting sensitivities of these telescopes has continued to improve. Some of the current rapid follow-up experiments are listed in Table 2, along with the limiting sensitivities and fields-of-view. These telescopes are valuable for long-term GRB monitoring and providing improved source localizations for the more powerful instruments described below.

HIGH SENSITIVITY FOLLOW-UP

Once a GRB has been detected and finely localized, the entire electromagnetic spectrum is available for sensitive observations to constrain physical models of GRB emission. Deep observations of GRBS at X-ray, optical and radio wavelengths, GRB observations are beginning to contribute a great deal to our understanding.

Bursts have been studied most extensively in the hard X-ray and low-energy gamma-ray ranges. At lower energies, future X-ray observations are likely to take us well beyond the capabilities of BeppoSAX and RXTE. For instance, the ACIS instrument on the Chandra Observatory will be able to measure X-ray afterglows

TABLE 2. Rapid GRB follow-up instrumentation.

Name	field-of-view[a]	sensitivity	Acquisition time[b]	Dates
ROTSE [33]	16.4 × 16.4	$V \approx 14$ (5 s)	10 s	1998+
ROTSE II [33]	1.9 × 1.9	$V \approx 16$ (10 s)	20 s	1999+
LOTIS [34]	17.4 × 17.4	$R \approx 11$ (10 s)	10 s	1997+
Super-LOTIS [35]	0.84 × 0.84	$V \approx 19$ (30 s)	< 300 s	1999+
BOOTES WFC [36]	16 × 11	$V \approx 12$ (30 s)	0-60 s	1999+
BOOTES Robotic [36]	0.82 × 0.55	$V \approx 17$ (60 s)	0-60 s	1999+
TAROT [37]	2 × 2	$V \approx 17$ (10 s)	3 s	1998+

[a] in deg × deg

[b] time to acquire target after trigger notification

for bright bursts for several months for bursts with localizations less than 5 arcminutes [32]. In addition, the XMM and Astro-E satellites will add powerful new spectroscopy capabilities to GRB studies.

One of the most important future aspects of GRB research will be distance determinations. Sensitive IR spectroscopy is crucial in this case for reliable distance measurements and for studies of high-z bursts. Satellite observatories such as HST and NGST as well as well as a large contingent of ground-based observatories are required to achieve the necessary coverage.

At still lower energies, VLA observations of GRB radio emission are opening up an entire new window on bursts. The importance of radio observations to any source with relativistic outflows of particles is manifest. Future observatories such as the VLA upgrade and the proposed Atacama Large Millimeter Array (ALMA) will provide the more sensitive observations that are now lacking.

SUMMARY

While we are still far from understanding these tremendous explosions, the last decade has witnessed encouraging progress in unravelling the gamma-ray burst mystery. Gamma-ray bursts are not unlike other astrophysical sources in that they are best studied using the most sensitive possible instruments over the broadest energy range possible. Unfortunately, in the past their transience and poor localization conspired to made it impossible to apply this simple prescription. These handicaps will be overcome by a combination of new gamma-ray burst observatories which can perform multi-wavelength observations from a single platform and a network of instruments capable of rapidly following up on the primary burst measurements.

REFERENCES

1. Klebesadel, R.W., Strong, I.B. and Olson, R.A., 1973, ApJ, 182, L85
2. Meegan, C.A., et al., 1992, Nature, 355, 143
3. Costa, E., et al., 1997, Nature, 387, 783
4. van Paradijs, J., et al., 1997, Nature, 386, 686
5. Metzger, R.M., Cohen, J.G., Chaffee, F.H., & Blandford, R.D., 1997, IAU Circ., 6676
6. Kulkarni, S.R., et al., 1998, Nature, 393, 35
7. Djorgovski, S.G., et al., 1999, GCN Circ., 189
8. Djorgovski, S.G., et al., 1998, ApJ, 508, L17
9. Fruchter, A.S., 1999, ApJ, 512, L1
10. Akerlof, C.W. et al., 1999, Nature, 398, 400
11. Piran, T., 1999, Physics Reports, 314, 575
12. Lamb, D.Q., 1999, A&AS, 138, 607
13. Fishman, G., et al., 1994, ApJS, 92, 229
14. Pian, E., et al., 1999, A&A, in press

15. Eichler, D., et al., 1989, Nature, 340, 126

16. Paczynski, B., 1991, Acta Astr., 41, 257

17. MacFadyen A.I., and Woosley S.E., 1999, ApJ, 524, 262

18. Pacieasas, W.S., et al., 1999, ApJS, 122, 465

19. Kippen, R.M., et al., 1998, ApJ, 492, 246

20. Feroci, M., 1998, in AIP Conf. Proc. 428, *Gamma-Ray Bursts : 4th Huntsville Symposium*, Eds. C.A. Meegan, R.D. Preece, and T.M. Koshut. (AIP: New York), 451

21. Costa, E., et al., 1999, A&AS, 138, 425

22. Cline, T.L., et al., 1999, A&AS, 138, 557

23. Ricker, G. R., 1997, in *All-Sky X-Ray Observations in the Next Decade*, RIKEN, Japan, eds. M. Matsuoka and N. Kawai, p. 366

24. Gehrels, N., et al., American Astronomical Society, HEAD meeting, 31, 12.05 (1999)

25. Forrest, D.J., Vestrand, W.T., McConnell, M., Ryan, J.M., Owens, A., 1995, Ap&SS, 231, 459

26. Brand, S. & Lund, N., in *Small Missions for Energetic Astrophysics*, ed. S.P. Brunby (AIP:New York), p. 33 (1999)

27. Kretschmar, B., 1999, A&AS, 138, 571

28. Tavani, M., et al., 1999, A&AS, 138, 569

29. Gehrels, N. & Michelson, P., 1999, Astroparticle Physics, 11, 277

30. Barthelmy, S. D., et al., 1998, in AIP Conf. Proc 428, *Gamma-Ray Bursts : 4th Huntsville Symposium*, Eds. C.A. Meegan, R.D. Preece, & T.M. Koshut. (AIP:New York), 99

31. Smith, D.A., 1999, ApJ, 526, 683

32. Garmire, G.P., 1999, A&AS, 138, 567

33. Kehoe, R., et al., 1999, astro-ph/9909219

34. Williams, G.G., 1998, in AIP Conf. Proc. 428, *Gamma-Ray Bursts: 4th Huntsville Symposium*, Eds. C.A. Meegan, R.D. Preece, and T.M. Koshut (AIP:New York), 837

35. Park, H.S., 1998, in AIP Conf. Proc. 428, *Gamma-Ray Bursts: 4th Huntsville Symposium*, Eds. C.A. Meegan, R.D. Preece, and T.M. Koshut (AIP:New York), 842

36. Castro-Tirado, A.J., 1999, A&AS, 138, 583

37. Boer, M., 1999, A&AS, 138, 579

Stellar Explosions Powered by Black Hole Accretion

A. I. MacFadyen

Astronomy Department
UC Santa Cruz
Santa Cruz, CA 95064

Abstract. The fate of massive ($M_{ms} \gtrsim 25$ $M_\odot$) rotating stars for which the Type-II supernova mechanism fails to launch a strong explosion is studied by means of 1D and 2D hydrodynamical simulations. In the extreme case ($M_{ms} \gtrsim 35$ $M_\odot$) a collapsar - a rapidly accreting ($\dot{M} \approx 0.1$ $M_\odot$ s^{-1}) stellar mass black hole at the center of a collapsing star - forms promptly. Here we study the alternative case of weak to moderately energetic supernova explosions in rotating stars. For explosion energies (at infinity) in the range 0.2 - 1.7 $\times 10^{51}$ erg, part of the exploding stellar mantle (0.1 - 5 $M_\odot$) fails to reach terminal escape velocity and accretes onto the central compact object. Accretion rates of $10^{-2} - 10^{-6}$ $M_\odot$ s^{-1} result over time scales of 100 - 100,000 s. Since the accreting gas has sufficient angular momentum to form an accretion disk around the central black hole, a jet engine similar to currently favored GRB engine models forms within an extended stellar envelope. 2D modeling demonstrates the propagation of the ensuing jet-driven explosion through the stellar envelope. The resulting asymmetric explosions have isotropic equivalent energies ($4\pi dE/d\Omega$) at break out ranging from 10^{54} ergs at the pole to 10^{51} ergs near the equator. "Cold" jets with less internal pressure than the exploding star through which it propagates can be compressed by the star and become extremely well focussed to less than 0.1% of the sky. Were the stars stripped of their envelopes by mass loss due to a wind or accretion onto a companion, they would be capable of producing long, t ≈ 1000 s, GRBs.

INTRODUCTION

Current modeling of the Type II supernova explosion mechanism is inconclusive on the robustness of the neutrino mechanism in launching successful explosions [1] (Fryer, this meeting). Especially for high masses, $M_{\mathrm{ZAMS}} \gtrsim 25$ $M_\odot$, the shallow density gradient exterior to the iron core in the presupernova star leads to high ram pressure of the imploding core which may be detrimental to launching a strong explosion. MacFadyen and Woosley [2] (henceforth MW99) studied the fate of a 35 $M_\odot$ for which the the iron core collapsed promptly to a black hole. Here we discuss an alternative model [3] of a 25 $M_\odot$ star in which relatively weak explosions are launched. While these explosion energies are sufficient to expel the hydrogen

CP522, *Cosmic Explosions: Tenth Astrophysical Conference,*
edited by Stephen S. Holt and William W. Zhang

237

envelope of the star, large amounts of the stellar mantle fail to achieve escape velocity and fall back onto the central neutron star. In the case of of a 2.55×10^{50} erg explosion (final energy at infinity) 3.63 $M_\odot$ falls back onto the core over a time scale of 100 seconds, while for 1.01×10^{51} erg explosion, 0.83 $M_\odot$ falls back over 900 seconds. In both these cases (certainly the first) the central object quickly exceeds the maximum mass for a neutron star and must collapse to a black hole. The 25 $M_\odot$ progenitor star was computed from the main sequence with angular momentum transport and has enough angular momentum at collapse for the accreting gas to form a disk around the black hole (note, however, that magnetic fields were not considered in these presupernova calculations). The resulting configuration - up to several solar masses accreting through a disk into a rotating black hole - is potentially a powerful source of energy for producing an explosion. [2] calculated the efficiency of neutrinos at transporting some of the binding energy of the accreting gas to low density regions above the disk. Though neutrinos cannot be a viable energy transport mechanism at the low accretion rates relevant here [4], the same electro-magnetic mechanisms for energy extraction thought to be important in active galactic nuclei and gamma-ray burst engines may operate here as well.

MW99 also calculated that about one half of the gas accreting at large radii would be ejected in a disk wind (see also [5]). This wind had total energy in excess of 10^{51} erg and could power a violent stellar explosion, independent of any energy in the jet involving, e.g. the rotational energy of the black hole. The wind also originates in regions of the disk hotter than 5×10^9 K where the disk gas has photodisintegrated to free nucleons. The nucleons will recombine as the wind cools, forming ^{56}Ni which could power a radioactive tail to the light curve of the explosion, as well as other unusual nucleosynthetic products.

I   THE MODELS

We have studied the response of a 25 $M_\odot$ red supergiant star to jets assumed to form from the accretion of gas into a recently formed central black hole at its center [3]. Though the jet mechanism is unspecified, we tie the jet energy directly to the accretion rate through the inner boundary at 10^9 cm and parameterize the jet efficiency as a fraction, ϵ, of the rest mass accretion rate $\dot{E}_{jet} = \epsilon \dot{M}_{accrete} c^2$. The jet is injected with an opening angle of $10°$ and a velocity of 10^{10}cm s^{-1} through the inner boundary at $r_{in} = 10^9$ cm. The thermal pressure of the jet was parametrized by its ratio to the ram pressure, $f_P \equiv P_{jet}/(1/2\rho v_{jet}^2)$, a quantity closely related to the Mach number. The value of f_P depends on the nature of the mechanism responsible for launching the jet. For jets driven purely by energy deposition (e.g. neutrino annihilation) in a stationary gas, f_P is initially infinite, but rapidly decreases as the gas expands due to its pressure and converts its thermal energy into kinetic energy. Jets driven by MHD processes on the other hand may be relatively cold i.e. $f_P \ll 1$. In this case the jet can be confined by the toroidal density structure of the collapsed star due to equatorial rotational support and

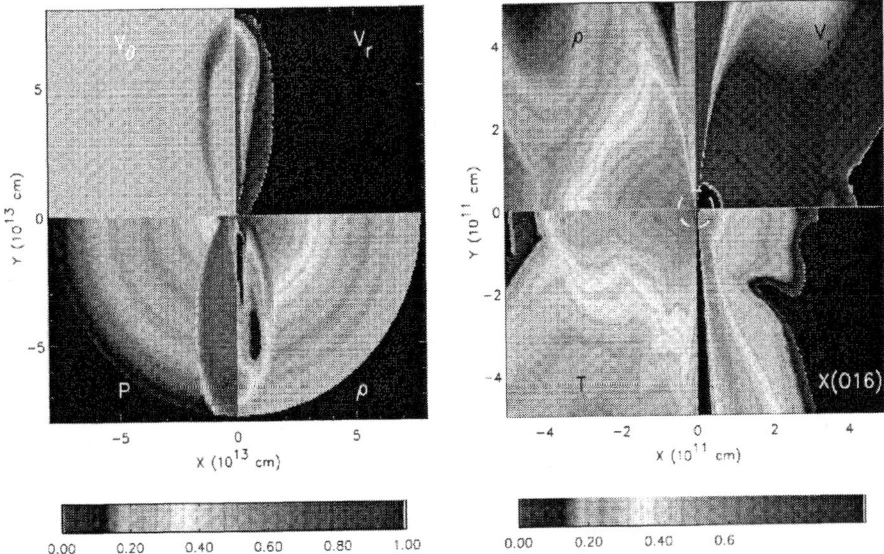

FIGURE 1. Left: Radial velocity, theta velocity, pressure, and mass density are plotted counterclockwise from the upper right as the jet reaches the surface of red supergiant. The minimum and maximum values for v_θ are -1.5×10^9 and 1.5×10^9, for radial velocity, v_r, -6.7×10^7 and 1.0×10^{10}, for density, $\log_{10}\rho$, -9 and -6.8, and for pressure, $\log_{10} P$, 2.5 and 10.9, all in cgs units. Right: Radial velocity, mass density, temperature and oxygen abundance plotted counterclockwise from upper right for exploding helium star after it has expanded by a factor of ten. The minimum and maximum values for radial velocity are v_r, 0 and 5.0×10^9 (though saturated region near pole is moving at 1.0×10^{10}), for density, $\log_{10}\rho$, -6 and 2, for temperature, $\log_{10} T$, 6.2 and 7.9, and for oxygen mass fraction, $\log_{10} X(016)$, -2 and 0. The dashed circle represents the initial radius of the helium core. In both figures the color bar indicates the linear scale between these minimum and maximum value given for each variable. For example, the medium gray region of the oxygen mass fraction plot represents $\log X(016) = -2 + 0.60 * (0 - (-2)) = -0.8$.

emerge from the star with extremely small ($\lesssim 5°$) opening angles ([6], [3]).

II ASYMMETRIC EXPLOSIONS IN HELIUM STARS

The right panel of Figure 1 shows the velocity, density temperature and oxygen abundance structure of a jet-driven explosion of a helium core with an initial spherically symmetric supernova explosion of 2.55×10^{50} erg about 500 seconds after core collapse. An additional 1.5×10^{51} erg has been deposited in a jet assuming 0.1% efficiency of converting accretion rest mass into a jet [3]. The star was exploded as described above and is shown after expanding about a factor ten in radius. The jet itself has escaped the helium core and is traveling at nearly relativistic speeds ($v_r = 10^{10}$ cm s^{-1}). What is left behind is a flattened distribution of the core expanding with at several thousand km s^{-1}. The oxygen has two components, the flattened equatorial component moving at $\approx 5,000$ km s^{-1}, and a high velocity component moving with the jet with speeds in excess of $50,0000$ km s^{-1}. There is also some oxygen near the inner boundary, within the dashed circle representing the original helium core radius of $r_{HE} = 4.5 \times 10^9$ cm, which is accreting.

REFERENCES

1. Fryer, C. L. 1999, talk presented at this meeting and astro-ph/9902315
2. MacFadyen, A., & Woosley, S. E. 1999, ApJ, 524, 262 (MW99)
3. MacFadyen, A., Woosley, S. E., & Heger, A. 1999, submitted to ApJ, astro-ph/9910034)
4. Popham, R., Woosley, S. E., & Fryer, C. 1999, ApJ, 518, 356, astro-ph/9807028
5. Stone, J. M., Pringle, J. E., Begelman, M. C. 1999, MNRAS, in press, astro-PM/908185
6. Aloy, M. A., Ibanez, J. M., Marti, J. M., Müller, E. & MacFadyen, A. I. 1999, ApJL submitted, astro-ph/9911098

On Jets and Luminosity Function of GRBs Associated with Supernovae

H. Che

Department of Physics, Michigan Technological University, 1400 Townsend Dr., Houghton, MI, 49931, hche@mtu.edu

Abstract. If Gamma-ray Bursts (GRBs) are generally associated with supernovae, a relatively wide intrinsic luminosity function is implied. If it is assumed that the intrinsic luminosity function of GRBs is a power-law: $\phi(L) \propto L^{-\beta}$, data from t he BATSE 4B catalog can be used to constrain slope index β and the dynamic range width Log $\frac{L_{max}}{L_{min}}$. Using a K-S test comparison with the observational Log N - Log P, we find constraints on the GRB fireball model, GRB jets. We find the acceptable dynamic range for $10^2 < L_{max}/L_{min} < 10^7$. Our results show that jet model is more likely to be related more highly energetic explosion than fireball model. Our studies also show that the luminosity function provided by a pure ly special relativistic effect on a jet would not dominate.

INTRODUCTION

In this paper, we compare GRB rates derived from supernova rates and density evolution to that of the measured rates from the Fourth BATSE Catalog of GRBs to investigate possible constraints on the GRB luminosity function and jet parameters. We also discuss if the special relativistic effects of a jet can provide by itself an acceptable luminosity function for GRBs? This question provides an alternative way to test if the gamma-ray burst emission is isotropic or random in the jet's comoving frame.

MODEL

If supernovae are the mysterious sources of GRBs, then the GRB "equivalent isotropic" explosion energy could be within a relatively large range: from $\sim 10^{54}$ erg/s, deduced for GRB 990123 to $\sim 10^{44}$ erg/s estimated from SN Ib/c - GRBs associations [1]. We assume the unknown intrinsic luminosity distribution follows a power-law:

$$\phi(L) = A \left(\frac{L}{L_{max}}\right)^{-\beta}, \qquad L_{min} \leq L \leq L_{max}, \qquad \beta > 0, \qquad (1)$$

CP522, *Cosmic Explosions: Tenth Astrophysical Conference*,
edited by Stephen S. Holt and William W. Zhang
© 2000 American Institute of Physics 1-56396-943-2/00/$17.00

where A is the normalization factor. We fix $L_{max} = 10^{53}$ erg/s. L_{min} was allowed to vary in the range 10^{44} - L_{max}. Observed spectra of GRBs can be approximated as a power-law: $dL(E)/dE \propto E^{-\alpha}$. For source density, w e adopt the broken power-law function we have used before as a rough description for SN Ib/c evolution [2]:

$$n(z) = n_0 \left(\frac{1+z}{1+z_0}\right)^\eta \begin{cases} \eta = \eta_1 > 0, z < z_0 \\ \eta = \eta_2 < 0, z > z_0 \end{cases}$$

where n_0 is the comoving GRB density at redshift $z = z_0$ per Mpc^3 yr^{-1}. We set $z_0 = 1$, $\eta_1 = 4$, and $\eta_2 = 0$.

We assume a flat Friedmann universe with cosmological constant $\Lambda = 0$. the number of observed gamma-ray bursts peak flux brighter than P is:

$$N(> P) = \int_{L_{min}}^{L_{max}} \phi(L)dL \int_0^{z_{max}(L,P)} \frac{4\pi n(z)}{1+z} r^2(z) \frac{dr(z)}{dz} dz \qquad (2)$$

The observed peak flux of a GRB at redshift z with intrinsic luminosity L is:

$$P(L,z) = L(1+z)^{-\alpha}/4\pi r^2(z) \qquad (3)$$

where we fixed the spectral index to be $\alpha = 1.1$ [3], and $P_{min} \sim 2 \times 10^{-7}$ erg cm^{-2} s$^{-1} \sim 1$ photon cm^{-2} s^{-1}.

$$L = \int_{50}^{300} \frac{dL(E)}{dE} dE \qquad (4)$$

where E is measured in KeV and $r(z)$ is proper motion distance. The total number of bursts we observe every year is $N_{total}(> P_{min})$ as derived from Eqn. 2.

RESULTS

A   Luminosity Function Constraints from Log N - Log P

We selected 939 bursts with peak flux greater than $P_{min} = 2 \times 10^{-7} erg/cm^2 s$ from the BATSE 4B catalog to get the measured Log N - Log P relation that will be used for comparison. We compared this Log N - Log P to GRBs following the evolving supernova rate, finding the best fits to the luminosity function range parameter L_{min} and slope parameter β. The rate was normalized by N_{total} as given in Eq. 5. We searched the acceptable range of L_{min} and β with the K-S confidence level 1%. The luminosity function width was allowed to vary from $1 < L_{max}/L_{min} < 10^9$, while the luminosity function slope was allowed to vary from $0 < \beta < 6$. The search result is shown in Figure 1: the acceptable range of luminosity function widths were $L_{max}/L_{min} > 100$, while the acceptable range for the luminosity function slope was $1.8 < \beta < 2.6$.

Constraints on GRB Fireballs and Jets

Setting the $N_{total} = 800$ yr^{-1}, which is the BATSE yearly rate corrected for the duty cycle [4], and constraining Log $\frac{L_{max}}{L_{min}}$ and β from equation (5), we can find lines of constant rate (n_0) in the β - Log $\frac{L_{max}}{L_{min}}$ frame. If we further suppose that GRBs are connected with supernovae, then for the fireball model, the GRB event rate should follow the supernovae rate $\sim 10^{-4}$ h^3 Mpc^{-3} yr^{-1}. For a jet with a Lorentz facto r $\gamma = 100$,the GRB rate will decrease to $\sim 10^{-8}$ h^3 Mpc^{-3} yr^{-1}.

Two lines of constant GRB rate are plotted in Figure 1 and labeled as lines II and III, representing isotropically emitting fireball GRBs and GRBs beamed into a $\gamma = 100$ jets respectively. For the fireball model, line II traverses the K-S test accepta ble region at around $L_{min} \sim 10^{46}$ erg/s, which is enough to produce a gamma-ray burst by shock, but is quite close to the hypothesized theoretical minimum (Meszaros,private communicatipon).

Line III shows that the minimum isotropic explosion energy of GRBs from a jet is $\sim 10^{49}$ erg/s. From lines II and III, possibly indicating that jet models are more

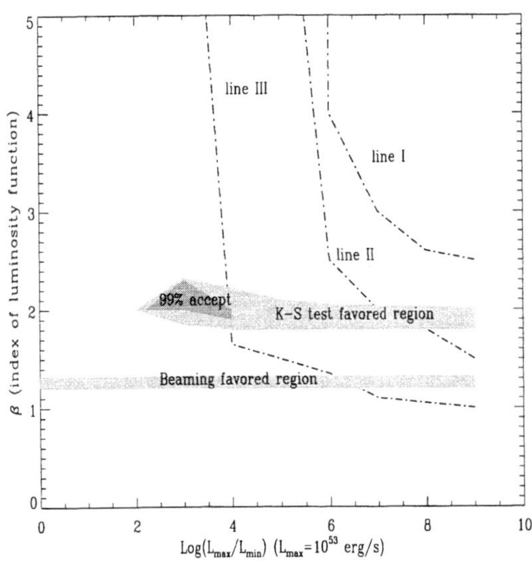

FIGURE 1. The shaded area indicates the allowed region of beaming induced GRB luminosity functions to the BATSE Log N - Log P. The boundary of the shaded region corresponds to a K-S confidence level of 1%. Line I satisfies the relation between the index of Power-Law Luminosity function & the width $Log\frac{L_{max}}{L_{min}}$ required by the rate producing a comparable observational diffuse gamma-ray background between 50-300 KeV.Line II and Line III satisfie the relation between the index of Power-Law Luminosity function & the width $Log\frac{L_{max}}{L_{min}}$ required by the supernova rate 10^{-4} h^3 Mpc^{-3} yr^{-1} and by the rate 10^{-8} h^3 Mpc^{-3}yr^{-1}

likely to be related to massive stars.

Beaming-Induced Luminosity Function

A relativistic jet or fireball has been considered as a way to solve the enormous energy release of GRBs. The simple assumption that the GRB radiation direction is random in the rest frame of the shocks cannot be excluded absolutely. We test this assumpti on by checking if the GRB luminosity function can be induced by a purely special relativistic effect.

We assume the viewing angle is θ, the velocity of beaming is v and L_{in} is the intrinsic luminosity of GRB in the beaming comoving frame, then the observational flux come from a source at redshift z is [5]:

$$P(z, \theta) = \Gamma^{2+\alpha} \frac{L_{in}}{4\pi(1+z)^2 r(z)^2} \tag{5}$$

$$\Gamma = \frac{1}{\gamma(1 - \frac{v}{c}\cos\theta)} \tag{6}$$

where $\gamma = 1/\sqrt{1 - v^2/c^2}$. We assume the beaming-induced $(L_{in})_{min} = \Gamma_{min}^{2+\alpha} L_{in}$, here Γ_{min} corresponds to the biggest viewing angle θ_{max} we can see a burst. $(L_{in})_{max} = L_{in}\Gamma^{2+\alpha}(\theta = 0)$, seen by the observer within the beam, $\frac{L}{L_{in}} = \Gamma^{2+\alpha}$. The induced luminosity function $\phi(\frac{L}{L_{in}})$ is:

$$\phi(\frac{L}{L_{in}}) = \frac{(L/L_{in})^{-(\frac{3+\alpha}{2+\alpha})}}{(2+\alpha)\gamma\frac{v}{c}} \tag{7}$$

With $\alpha = 1.1\pm0.3$, the induced luminosity function index is constrained to be $1.3 < \beta < 1.36$, far below the required index for $\text{Log } N$ - $\text{Log } P$. This contradiction implies that the GRB luminosity function is not created by purely special relativistic bulk motion. From inspection of Figure 1, we see if beaming creates $L_{max}/L_{min} < 100$, it will not dominate the wider intrinsic luminosity function and from Equ. 7, we see the $\theta_{max} \leq 1/\gamma$, the beaming effect could be hidden by the intrinsic luminosity function of GRBs.

REFERENCES

1. Bloom, J.S. et al. *ApJ*,**506**, L105 (1998)
2. Che, H., Yang, Y & Nemiroff, R. J., *ApJ*,**516**, 559 (1999)
3. Mallozzi,R.S. et al., *ApJ*, **471**, 636 (1996)
4. Meegan, C. A., et al., *Nature*, 353, 143 (1992)
5. Blandford, R. D. & Konigl, A., *ApJ*, **232**, 34 (1979)

An Instability-driven Dynamo for Gamma-ray Burst Black Holes

Rafael Ángel Araya-Góchez[1]

Laboratorio de Investigaciones Astrofísicas, Escuela de Física
Universidad de Costa Rica, San José, Costa Rica.

Abstract. We show that an MHD-instability driven dynamo (IDD) operating in a hot accretion disk is capable of generating energetically adequate magnetic flux deposition rates above and below a mildly advective accretion disk structure. The dynamo is driven by the magnetorotational instability (MRI) of a toroidal field in a shear flow and is limited by the buoyancy of 'horizontal' flux and by reconnection in the turbulent medium. The efficiency of magnetic energy deposition is estimated to be comparable to the neutrino losses although an MHD collimation mechanism may deem this process a more viable alternative to neutrino-burst–driven models of γ-ray bursts.

I INTRODUCTION

The combined redshift and fluence measurements of at least five γ-ray burst sources plus very large photon energy detections in certain bursts and tight size constraints derived from the rapid risetimes of burst triggers strongly suggest that the release of energy is highly focused by the central engine that propels a γ-ray burst. In spite of the very large γ-ray energy requirements, the efficiency of energy deposition into electromagnetic channels is likely to be very poor if the burst is driven by a neutrino burst in analogy to the processes thought to give birth to supernovæ (MacFadyen & Woosley 1998, 1999) or if the burst involves major energy losses to gravitational radiation such as might be the case in compact object merger scenarios (Rasio and Shapiro 1994, Davies *et al.* 1994, Ruffert & Janka 1998). Indeed, these measurements pose a serious energy budget problem for arguably all gravitational collapse powered models of γ-ray bursts if the energy release is not moderately collimated.

An attractive solution to this problem starts with a (directional) Poynting-flux dominated outflow (Thompson 1994, Mészáros & Rees 1997) under the premise that such a flow may carry very little baryonic contamination if deposited along a centrifugally (or gravitationally) evacuated funnel such as the angular momentum

---

[1] e-mail: araya@twinkie.gsfc.nasa.gov

CP522, *Cosmic Explosions: Tenth Astrophysical Conference,*
edited by Stephen S. Holt and William W. Zhang
© 2000 American Institute of Physics 1-56396-943-2/00/$17.00

axis of a black hole-accretion disk system. Yet, formal motivation for an external field of the desired strength and topology has yet to be investigated in this setting.

We motivate a reasonable set of heuristic two dimensional dynamo equations for magnetic field components in the comoving frame of a mildly advective disk under the premise of negligible generation of meridional field (invoking Parker's undulate instability to promote the growth (loss) of vertical (horizontal) field is questionable in turbulent disks where the turbulence is fed by MRI's). A self-consistent turbulent steady state is achieved when the non-linear damping rate of the turbulent cascade equals the inverse of the linear growth timescale (Zhang, Diamond & Vishniac 1994). Accretion disk models are far from being self-consistent in this respect.

The heuristic dynamo equations account for field generation by the shear flow and by the non-axisymmetric MRI in Paczyński's pseudo-potential well $\Phi \propto (r - r_g)^{-1}$ where $r_g = 2GM/c^2$; with flux buoyancy and turbulent reconnection providing for field loss terms. The model predicts *azimuthally averaged* field components and depends explicitly on the magnetic Mach number of the turbulence (assumed Alfvenic), on local pressure ratios, and on the relativistic generalization of the shear parameter ($A_{\text{Oort}}^{\text{Rel}} = $ Oort's A constant).

II DYNAMO EQUATIONS AND SCALINGS

In Lagrangian coordinates

$$\partial_t B_\varphi = \frac{B_r}{\tau_s} - \frac{B_\varphi}{\tau_B} - \frac{B_\varphi}{\tau_r} \quad \text{and} \quad \partial_t B_r = \frac{B_\varphi}{\tau_M} - \frac{B_r}{\tau_B} - \frac{B_r}{\tau_r}. \tag{1}$$

The shear flow is parameterized linearly by a relativistic generalization of Oort's first constant (Novikov & Thorne 1974), $\tau_s^{-1} = 2A_{\text{Oort}} = \gamma^2 d_{\ln r}\Omega$, where γ is the bulk Lorentz factor of the flow. Shear forces the radial wavenumber of perturbations to evolve according to $k_r(t) = k_r^0 - 2Ak_\varphi t$.

We use exact analytical MRI scalings from Foglizzo & Tagger (1995). With $\alpha_\varphi \equiv B_\varphi^2/(8\pi p)$ and $\hat{A} \equiv A/\Omega < 0$, the maximum growth of non-axisymmetric MRI modes occurs for wavenumber

$$k_\varphi \equiv \eta \frac{\Omega}{v_\varphi^{\text{Alf}}} \xrightarrow{\alpha_\varphi \ll 1} \{2\hat{A} + (1 + \alpha_\varphi)\hat{A}^2\} \frac{\Omega}{v_\varphi^{\text{Alf}}} \tag{2}$$

at a rate, τ_M^{-1}, that obeys

$$|\hat{A}|^2 \tau_M^2 \xrightarrow{\alpha_\varphi \ll 1} 1 + \alpha_\varphi(2 + \hat{A}) \tag{3}$$

as long as $k_\varphi \gtrsim 1/r$. On the other hand, in a very strongly sheared flow

$$-\hat{A} < 1 + (2\alpha_\varphi + 1)^{-1}, \tag{4}$$

the slow branch of MHD propagation (to which the toroidal MRI belongs) is destabilized into a radial interchange mode (T. Foglizzo, Priv. Comm.) in accordance with the Rayleigh criterium.

A plausible mechanism that promotes baryon unloading from field lines is turbulent pumping (Vishniac 1995a) by the MRI which must favorably compete with turbulent diffusion of matter back onto flux ropes. Under this assumption, the stretch, twist and fold of field lines by (enthalpy-weighted) sub-Alfvenic turbulence augments the field energy density and releases matter from field lines that otherwise would be "frozen-in". For the marginal case of Alfvenic turbulence, nearly empty $B_\perp$ flux ropes (i.e. flux residing on surfaces perpendicular to the local meridian) in a gas pressure dominated disk acquire a drag limited buoyant velocity $v_b \propto (v_{\perp\theta}^{\mathrm{Alf}})^2/c_s$. Moreover, assuming efficient diffusion of radiation and $e^\pm$ pairs into the flux tubes, in this picture the buoyancy loss rate, $\tau_B^{-1} = v_b/\mathcal{H}_\theta$, is enhanced by a factor $\lesssim p/p_{\mathrm{gas}} \equiv \xi$ (Vishniac 1995b).

Reconnection of the field at sub-MRI optimal lengthscales (where the field lines are only weakly stochastic) occurs at the Alfvén speed for Alfvenic turbulence (Lazarian & Vishniac 1999). The rates may be written as inverse Alfvén transit times calculated from the component of the field that undergoes reconnection, $\tau_{\mathrm{rec}} \approx l_{\mathrm{rec}}^{\perp i}/v_i^{\mathrm{Alf}} = \sqrt{2/\Gamma}\,(l_{\mathrm{rec}}^{\perp i}/\mathcal{H}_\Theta)\,(c_s/v_i^{\mathrm{Alf}})\,\Omega^{-1}$, where Γ is the adiabatic index of the fluid.

The perpendicular lengthscales, $l_{\mathrm{rec}}^{\perp i}$, associated with the mean distance for field reversal are derived from the fundamental linear lengthscales for coherent field pumping. These are supplied by half of the toroidal MRI's (wave)length scale. Azimuthal, radial field reversal, Y_r, directly involves the optimal wavenumber of the non-axisymmetric MRI. In addition, following Tout & Pringle (1992), an estimate of the radial, azimuthal field reversal lengthscale, X_φ follows by noting that the time evolution of wavenumbers implied by shear during one MRI timescale couples the azimuthal lengthscale to the radial lengthscale, i.e. $l_y^\perp = (\tau_\mathcal{M}/\tau_s) \times l_x^\perp$. Thus $Y_r = \pi/k_\mathcal{M}$ and $X_\varphi = (\tau_s/\tau_\mathcal{M}) \times (\pi/k_\mathcal{M})$.

III   EQUILIBRIUM SOLUTIONS

Scaling wavenumbers to the inverse pressure scale height $k \to \hat{k}/\mathcal{H}_\Theta$, and *redefining* $\hat{\mathrm{A}} \equiv |A|/\Omega > 0$; a set of normalized dynamo equations follows by representing fields in (Alfvén) velocity units and normalized to the soundspeed ($B' = B \times \sqrt{4\pi\varrho}\,c_s$), and the time normalized to the inverse of the Keplerian frequency $t' = \Omega t$.

In a steady state, these equations must satisfy

$$B_r' = \frac{B_\varphi'}{2\hat{\mathrm{A}}}\left\{\frac{1}{\xi}B_\perp'^2 + \frac{\sqrt{2}}{\pi}\frac{\tau_\mathcal{M}'}{\tau_s'}\eta\right\}, \quad \text{and} \quad B_\varphi' = \frac{B_r'}{\hat{\mathrm{A}}}\left\{\frac{1}{\xi}B_\perp'^2 + \frac{\sqrt{2}}{\pi}\frac{B_r'}{B_\varphi'}\eta\right\} \qquad (5)$$

which we solve numerically.

IV THE ENERGY DEPOSITION RATE

The accretion disk setting is envisioned to follow the standard hyper-accreting black hole model of Popham, Woosley & Fryer (1999, hereafter PWF) where $M_{\mathrm{bh}} = 3\ \mathrm{M}_\odot$; $\alpha_{\mathrm{SS}} = 0.1$, and $\dot{M} = 0.1\ \mathrm{M}_\odot\ \mathrm{sec}^{-1}$. By adopting their published pressure ratios and assuming Keplerian rotation for $r \in [2.25,\ 20]r_g$, we "piggyback" the hydromagnetic energy conversion process on this model.

We find that the magnetic output rate from buoyancy, $\dot{E} \simeq 1.77_{+51}\ \mathrm{erg\ sec}^{-1}$, is comparable to the neutrino luminosity $L_\nu \simeq 3.3_{+51}\ \mathrm{erg\ sec}^{-1}$ (PWF). The 'half-luminosity' radius is located at $r_{\mathcal{L}} \approx 5.75\,r_g$ and the IDD becomes operational (against the radial interchange instability, c.f. Eq [4] and Araya-Gochez 1999) at $r_{\min} \simeq 2.55\,r_g$. Curiously, the derived value of the magnetic viscosity $\alpha_{\mathrm{SS}} = B'_r B'_\varphi$ hovers on 0.1 at the innermost radii and decreases to about .087 at $r_{\mathcal{L}}$ in good agreement with the adopted value. Lastly, we note that the MRI pumps the field preferably at the lowest wavenumbers for a near-equipartition field in a strongly sheared flow. Thus, most of this energy could in principle go into a Poynting jet with interesting consequences for γ-ray bursts.

REFERENCES

1. Araya-Gochez, R.A. 1999, M.N.R.A.S. (submitted) astro-ph/9912324
2. Davies *et al.* 1994, Ap.J., 431, 742
3. Foglizzo, T. & Tagger M. 1995, A. & A., 301, 293
4. Lazarian, A. & Vishniac, E.T. 1999, Ap.J. 517, 700 (astro-ph/9811037)
5. MacFadyen & Woosley 1998, Ap.J. 524, 262 (astro-ph/9810274)
6. Mészáros, P., & Rees, M.J. 1997, Ap.J.Lett, 482, L29
7. Mészáros, P., Rees, M.J. & Wijers, R.A. 1999, New Ast., 4, 303 (astro-ph/9808106)
8. Novikov, I. & Thorne, K.S. 1973, in Black Holes, eds. C. DeWitt & B.S. DeWitt (New York: Gordon & Breach), 343
9. Popham, R., Woosley, S., & Fryer, C. 1999, Ap.J., 518, 356 (astro-ph/9807028)
10. Rasio, F.A. & Shapiro S.L. 1994, Ap.J., 432, 242.
11. Ruffert & Janka 1999, A&A 344, 573 (astro-ph/9809280)
12. Thompson, C. 1994, M.N.R.A.S., 270, 480
13. Tout, C.A. & Pringle, J.E. 1992, M.N.R.A.S., 259, 604; 1995, M.N.R.A.S., 281, 219
14. Vishniac, E.T. 1995a, Ap.J., 446, 724; 1995b, Ap.J., 451, 816
15. Zhang, W., Diamond P.H. & Vishniac, E.T. 1994, Ap.J., 420, 705

The Detectability of Gamma-Ray Bursts and Their Afterglows at Very High Redshifts

Donald Q. Lamb and Daniel E. Reichart

Department of Astronomy & Astrophysics, University of Chicago, 5640 South Ellis Avenue, Chicago, IL 60637

Abstract. There is increasingly strong evidence that gamma-ray bursts (GRBs) are associated with star-forming galaxies, and occur near or in the star-forming regions of these galaxies. These associations provide indirect evidence that at least the long GRBs detected by BeppoSAX are a result of the collapse of massive stars. The recent evidence that the light curves and the spectra of the afterglows of GRB 970228 and GRB 980326 appear to contain a supernova component, in addition to a relativistic shock wave component, provide more direct clues that this is the case. Here we establish that GRBs and their afterglows are both detectable out to very high redshifts ($z \gtrsim 5$).

INTRODUCTION

We first show that the GRBs with well-established redshifts could have been detected out to very high redshifts (VHRs). Then, we show that their soft X-ray, optical, and infrared afterglows could also have been detected out to these redshifts.

DETECTABILITY OF GRBS

We first show that GRBs are detectable out to very high redshifts. The peak photon number luminosity is

$$L_P = \int_{\nu_l}^{\nu_u} \frac{dL_P}{d\nu} d\nu \, , \tag{1}$$

where $\nu_l < \nu < \nu_u$ is the band of observation. Typically, for BATSE, $\nu_l = 50$ keV and $\nu_u = 300$ keV. The corresponding peak photon number flux P is

$$P = \int_{\nu_l}^{\nu_u} \frac{dP}{d\nu} d\nu \, . \tag{2}$$

CP522, *Cosmic Explosions: Tenth Astrophysical Conference,*
edited by Stephen S. Holt and William W. Zhang

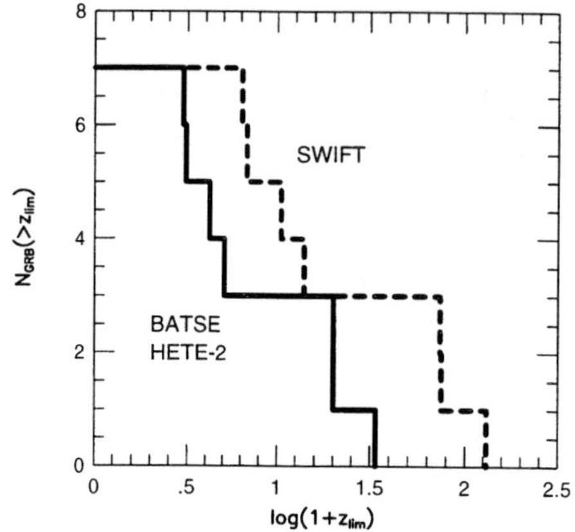

FIGURE 1. Cumulative distributions of the limiting redshifts at which the seven GRBs with well-determined redshifts and published peak photon number fluxes would be detectable by BATSE and HETE-2, and by *Swift*.

Assuming that GRBs have a photon number spectrum of the form $dL_P/d\nu \propto \nu^{-\alpha}$ and that L_P is independent of z, the observed peak photon number flux P for a burst occurring at a redshift z is given by

$$P = \frac{L_P}{4\pi D^2(z)(1+z)^\alpha} \, , \tag{3}$$

where $D(z)$ is the comoving distance to the GRB. Taking $\alpha = 1$, which is typical of GRBs [1], Equation (3) coincidentally reduces to the form that one gets when P and L_P are bolometric quantities.

Using these expressions, we have calculated the limiting redshifts detectable by BATSE and HETE-2, and by *Swift*, for the seven GRBs with well-established redshifts and published peak photon number fluxes. In doing so, we have used the peak photon number fluxes given in Table 1 of [2], taken a detection threshold of 0.2 ph s^{-1} for BATSE [3] and HETE-2 [4] and 0.04 ph s^{-1} for *Swift* [5], and set $H_0 = 65$ km s^{-1} Mpc^{-1}, $\Omega_m = 0.3$, and $\Omega_\Lambda = 0.7$ (other cosmologies give similar results).

Figure 1 displays the results. This figure shows that BATSE and HETE-2 would be able to detect four of these GRBs (GRBs 970228, 970508, 980613, and 980703) out to redshifts $2 \lesssim z \lesssim 4$, and three (GRBs 971214, 990123, and 990510) out to redshifts of $20 \lesssim z \lesssim 30$. *Swift* would be able to detect the former four out to redshifts of $5 \lesssim z \lesssim 15$, and the latter three out to redshifts in excess of $z \approx 70$, although it is unlikely that GRBs occur at such extreme redshifts (see §3 below). Consequently, if GRBs occur at VHRs, BATSE has probably already detected them, and future missions should detect them as well.

DETECTABILITY OF GRB AFTERGLOWS

The soft X-ray, optical and infrared afterglows of GRBs are also detectable out to VHRs. The effects of distance and redshift tend to reduce the spectral flux in

GRB afterglows in a given frequency band, but time dilation tends to increase it at a fixed time of observation after the GRB, since afterglow intensities tend to decrease with time. These effects combine to produce little or no decrease in the spectral energy flux F_ν of GRB afterglows in a given frequency band and at a fixed time of observation after the GRB with increasing redshift:

$$F_\nu(\nu, t) = \frac{L_\nu(\nu, t)}{4\pi D^2(z)(1+z)^{1-a+b}},$$ (4)

where $L_\nu \propto \nu^a t^b$ is the intrinsic spectral luminosity of the GRB afterglow, which we assume applies even at early times, and $D(z)$ is again the comoving distance to the burst. Many afterglows fade like $b \approx -4/3$, which implies that $F_\nu(\nu, t) \propto D(z)^{-2}(1+z)^{-5/9}$ in the simplest afterglow model where $a = 2b/3$ [6]. In addition, $D(z)$ increases very slowly with redshift at redshifts greater than a few. Consequently, there is little or no decrease in the spectral flux of GRB afterglows with increasing redshift beyond $z \approx 3$.

For example, [7] find in the case of GRB 980519 that $a = -1.05 \pm 0.10$ and $b = -2.05 \pm 0.04$ so that $1 - a + b = 0.00 \pm 0.11$, which implies no decrease in the spectral flux with increasing redshift, except for the effect of $D(z)$. In the simplest afterglow model where $a = 2b/3$, if the afterglow declines more rapidly than $b \approx 1.7$, the spectral flux actually *increases* as one moves the burst to higher redshifts!

As another example, we calculate the best-fit spectral flux distribution of the early afterglow of GRB 970228 from [8], as observed one day after the burst, transformed to various redshifts. The transformation involves (1) dimming the afterglow,[1] (2) redshifting its spectrum, (3) time dilating its light curve, and (4) extinguishing the spectrum using a model of the Lyα forest. For the model of the Lyα forest, we have adopted the best-fit flux deficit distribution to Sample 4 of [9] from [10]. At redshifts in excess of $z = 4.4$, this model is an extrapolation, but it is consistent with the results of theoretical calculations of the redshift evolution of Lyα absorbers [11]. Finally, we have convolved the transformed spectra with a top hat smearing function of width $\Delta\nu = 0.2\nu$. This models these spectra as they would be sampled photometrically, as opposed to spectroscopically; i.e., this transforms the model spectra into model spectral flux distributions.

Figure 2 shows the resulting K-band light curves. For a fixed band and time of observation, steps (1) and (2) above dim the afterglow and step (3) brightens it, as discussed above. Figure 2 shows that in the case of the early afterglow of GRB 970228, as in the case of GRB 980519, at redshifts greater than a few the three effects nearly cancel one another out. Thus the afterglow of a GRB occurring at a redshift slightly in excess of $z = 10$ would be detectable at K ≈ 16.2 mag one hour after the burst, and at K ≈ 21.6 mag one day after the burst, if its afterglow were similar to that of GRB 970228 (a relatively faint afterglow).

---

[1] Again, we have set $\Omega_m = 0.3$ and $\Omega_\Lambda = 0.7$; other cosmologies yield similar results.

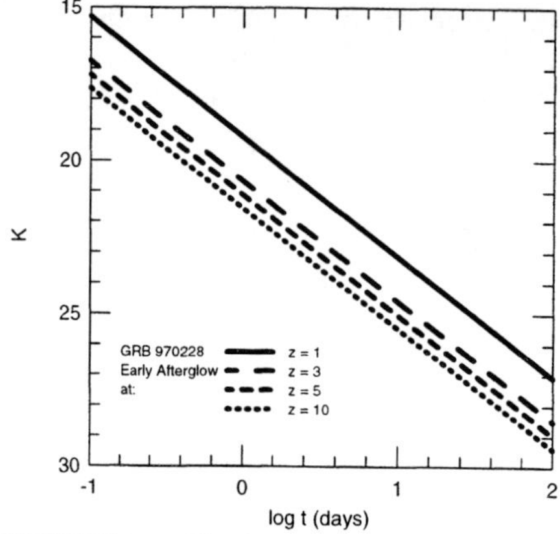
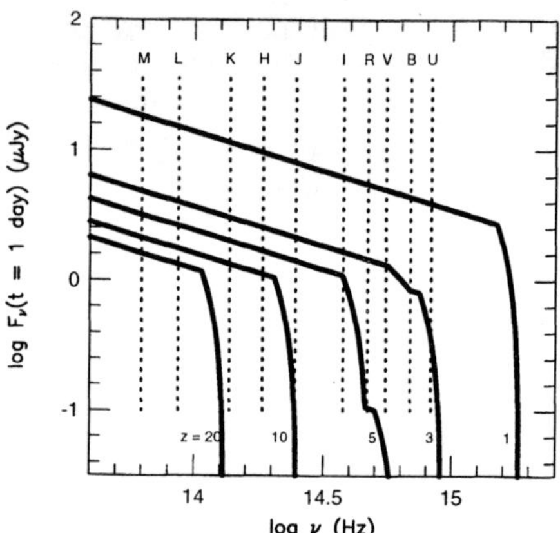

FIGURE 2. The best-fit light curve of the early afterglow of GRB 970228 from Reichart (1999), transformed to various redshifts.

FIGURE 3. The best-fit spectral flux distribution of the early afterglow of GRB 970228 from Reichart (1999), as observed one day after the burst, after transforming it to various redshifts, and extinguishing it with a model of the Lyα forest.

Figure 3 shows the resulting spectral flux distribution. The spectral flux distribution of the afterglow is cut off by the Lyα forest at progressively lower frequencies as one moves out in redshift. Thus high redshift ($1 \lesssim z \lesssim 5$) afterglows are characterized by an optical "dropout" [12], and very high redshift ($z \gtrsim 5$) afterglows by an infrared "dropout."

In conclusion, if GRBs occur at very high redshifts, both they and their afterglows would be detectable.

REFERENCES

1. Mazzoli, R. S., Pendleton, G. N., & Paciesas, W. S. 1996, ApJ, 471, 636
2. Lamb, D. Q., & Reichart, D. E., 2000, ApJ, in press (astro-ph/9909002)
3. Meegan, C. A., et al. 1993, Second BATSE Catalog
4. Ricker, G. 1998, BAAS, 30, Abstract 33.14
5. Gehrels, N. 1999, BAAS, 31, 993
6. Wijers, R. A. M. J., Rees, M. J., & Mészáros, P. 1997, MNRAS, 288, L51
7. Halpern, J. P. 1999, ApJ, 517, L105
8. Reichart, D. E., 1999, ApJ, 521, L111
9. Zuo, L., & Lu, L. 1993, ApJ, 418, 601
10. Reichart, D. E., 2000, ApJ, submitted (astro-ph/9912368)
11. Valageas, P., Schaeffer, R., & Silk, J. 1999, A&A, 345, 691
12. Fruchter, A. S. 1999, ApJ, 512, L1

The GRB/SN Connection: An Improved Spectral Flux Distribution for the Supernova Candidate Associated with GRB 970228

Daniel E. Reichart[*], Francisco J. Castander[†], and Donald Q. Lamb[*]

[*]*Department of Astronomy & Astrophysics, University of Chicago,
5640 South Ellis Avenue, Chicago IL, 60637*
[†]*Observatoire Midi-Pyrénées, 14 Av. Edouard Belin, 31400 Tolouse, France*

Abstract. We better determine the spectral flux distribution of the supernova candidate associated with GRB 970228 by modeling the spectral flux distribution of the host galaxy of this burst, fitting this model to measurements of the host galaxy, and using the fitted model to better subtract out the contribution of the host galaxy to measurements of the afterglow of this burst.

INTRODUCTION

The discovery of what appear to be SNe dominating the light curves and spectral flux distributions (SFDs) of the afterglows of GRB 980326 (Bloom et al. 1999) and GRB 970228 (Reichart 1999; Galama et al. 1999) at late times after these bursts strongly suggests that at least some, and perhaps all, of the long bursts are related to the deaths of massive stars. Here, we build upon the results of Reichart (1999) by modeling the SFD of the host galaxy of GRB 970228, fitting this model to measurements of the host galaxy, and using the fitted model to better subtract out the contribution of the host galaxy to measurements of the afterglow of this burst.

OBSERVED AND MODELED SFDS FOR THE HOST GALAXY

In Figure 1, we plot the observed SFD of the host galaxy of GRB 970228, as measured with *HST*/WFPC2, *HST*/NICMOS2, and Keck I, and converted to the standard bands (Castander & Lamb 1999a; Fruchter et al. 1999). To these measurements and a broadband measurement made with *HST*/STIS (Castander & Lamb 1999a; Fruchter et al. 1999), which is not plotted, we fit a two-parameter, spectral

CP522, *Cosmic Explosions: Tenth Astrophysical Conference,*
edited by Stephen S. Holt and William W. Zhang
© 2000 American Institute of Physics 1-56396-943-2/00/$17.00

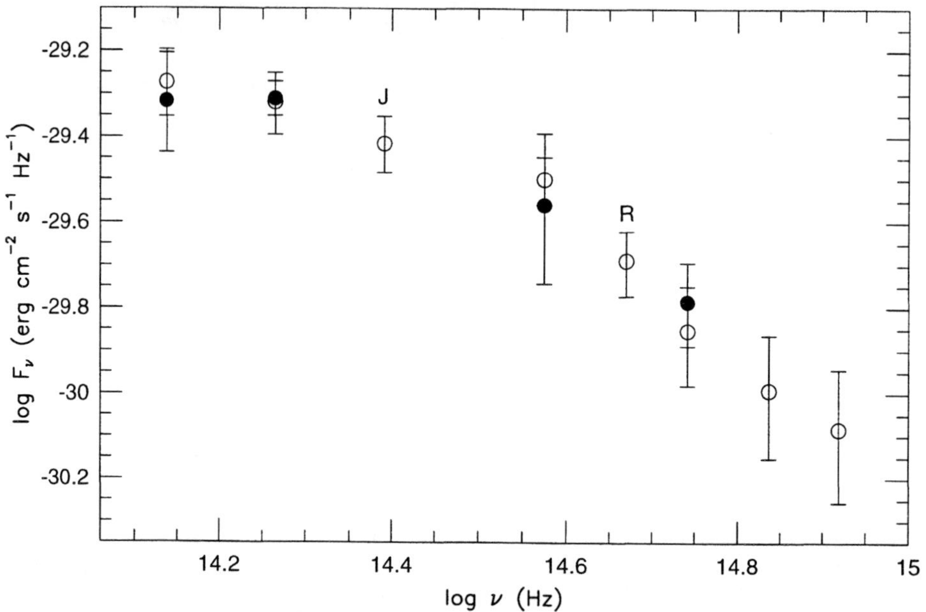

FIGURE 1. The observed (filled circles) and modeled (unfilled circles) K- through U-band SFDs of the host galaxy of GRB 970228.

synthesis model (see Castander & Lamb 1999a for details). The two parameters are the normalization of the SFD, and the age of the galaxy, defined to be the length of time that star formation has been occurring at a constant rate. Taking $A_V = 1.09$ mag for the Galactic extinction along the line of sight (Castander & Lamb 1999b), we find a fitted age of 270^{+460}_{-180} Myr; different values of A_V affect primarily the fitted age, and not the fitted SFD. Furthermore, models in which star formation slows considerably, or ceases, are generally too red to account for the measurements. Finally, we note that the fitted J- and R-band spectral fluxes are perfectly consistent with what one finds simply from linear interpolation between adjacent photometric bands.

SFD OF THE SN CANDIDATE DERIVED USING THE OBSERVED SFD OF THE HOST GALAXY

In Figure 2, we plot the SFD of the afterglow minus the *observed* SFD of the host galaxy from Figure 1. For the SFD of the afterglow, we use the revised K-, J-, and R-band measurements of Galama et al. (1999) and the I- and V-band measurements of Castander & Lamb (1999a; see also Fruchter et al. 1999); all of these measurements were taken between 30 and 38 days after the burst. We have scaled these measurements to a common time of 35 days after the burst, and have

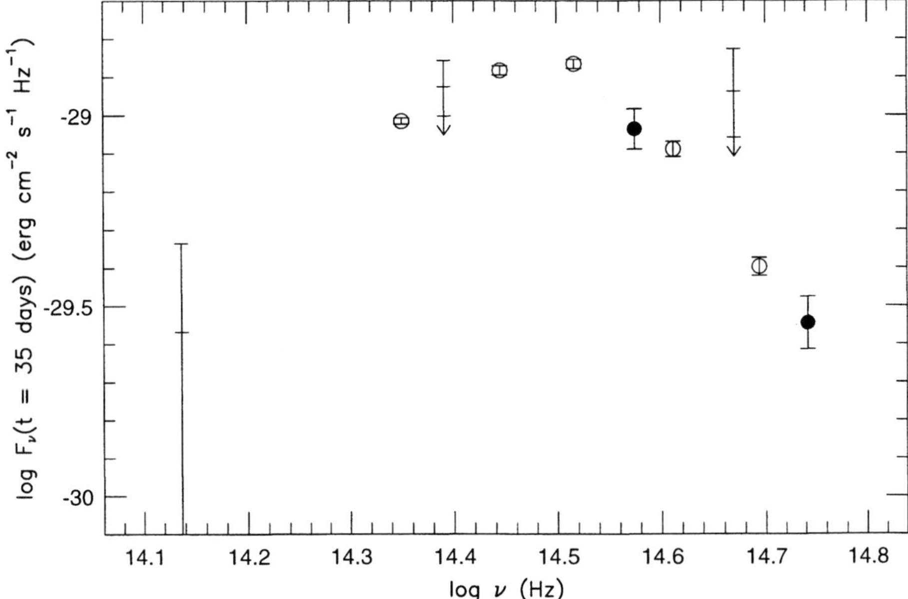

FIGURE 2. The K- through V-band SFD of the late afterglow of GRB 970228 after subtracting out the *observed* SFD of the host galaxy from Figure 1 and correcting for Galactic extinction (filled circles and upper limits), and the I- through U-band SFD of SN 1998bw after transforming to the redshift of GRB 970228, $z = 0.695$, and correcting for Galactic extinction (unfilled circles). The K-, J-, and R-band upper limits are 1, 2, and 3 σ.

corrected these measurements for Galactic extinction along the line of sight (see Reichart 1999 for details). The K-band measurement of the afterglow is consistent with that of the host galaxy (Galama et al. 1999), resulting in an upper limit in Figure 2; J- and R-band measurements of the host galaxy are not available, again resulting in upper limits in Figure 2. As originally concluded by Reichart (1999), this SFD is consistent with that of SN 1998bw, after transforming it to the redshift of the burst, $z = 0.695$ (Djorgovski et al. 1999), and correcting it for Galactic extinction along its line of sight (see Reichart 1999 for details).

SFD OF THE SN CANDIDATE DERIVED USING THE MODELED SFD OF THE HOST GALAXY

In Figure 3, we plot the same distribution, but minus the *modeled* SFD of the host galaxy from Figure 1. The SN-like component to the afterglow is detected in the J band, and possibly in the R band. The J-band measurement suggests that the SN-like component is $\approx 1/2$ mag fainter, and $\approx 1/2$ of a photometric band bluer, than SN 1998bw; however, this difference in J-band spectral fluxes is

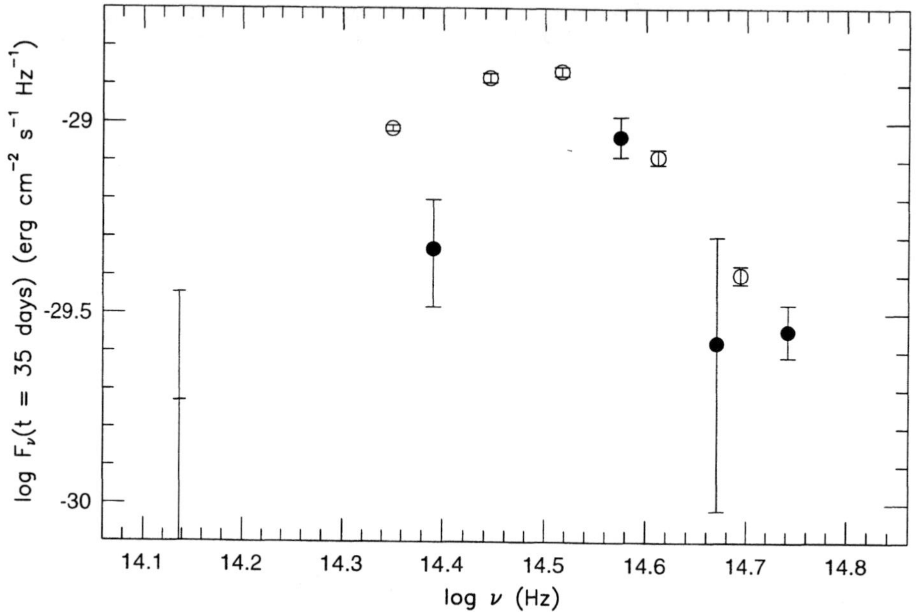

FIGURE 3. The K- through V-band SFD of the late afterglow of GRB 970228 after subtracting out the *modeled* SFD of the host galaxy from Figure 1 and correcting for Galactic extinction (filled circles and upper limits), and the I- through U-band SFD of SN 1998bw after transforming to the redshift of GRB 970228, $z = 0.695$, and correcting for Galactic extinction (unfilled circles). The K-band upper limits are 1, 2, and 3 σ.

significant only at the $\approx 2.5\ \sigma$ level. When possible photometric zero point errors and uncertainties in our spectral synthesis model of the SFD of the host galaxy are included, this difference is significant only at the $\approx 2\ \sigma$ level. However, it is suggestive of what is generally expected: the Type Ic SNe that are theorized to be associated with bursts (e.g., Woosley 1993) are not expected to be standard candles.

REFERENCES

1. Bloom, J. S., et al. 1999, Nature, 401, 453
2. Castander, F. J., & Lamb, D. Q. 1999a, ApJ, 523, 593
3. Castander, F. J., & Lamb, D. Q. 1999b, ApJ, 523, 602
4. Djorgovski, S. G., et al. 1999, GCN Report 289
5. Fruchter, A. S., et al. 1999, ApJ, 516, 683
6. Galama, T. J., et al. 1999, ApJ, submitted
7. Reichart, D. E., 1999, ApJ, 521, L111
8. Woosley, S. E. 1993, ApJ, 405, 273

A Jet-Disk Symbiosis Model for Gamma Ray Bursts: Fluence Distribution, Cosmic Rays and Neutrinos

G. Pugliese[1], H. Falcke[1], Y. Wang[2], and P. L. Biermann[1]

[1] *Max-Planck-Institut für Radioastronomie, Auf dem Hügel 69, 53121 Bonn, Germany*
[2] *Purple Mountain Observatory, Academica Sinica, Nanjing 210008, China*

Abstract.
We consider a jet-disk symbiosis model to explain Gamma Ray Bursts and their afterglows. It is proposed that GRBs are created inside a pre-existing jet from a neutron star in a binary system which collapses to a black hole due to accretion. In our model we assume that a fraction of the initial energy due to this transition is deposited in the jet by magnetic fields. The observed emission is then due to an ultrarelativistic shock wave propagating along the jet. Good agreement with observational data can be obtained for systems such as the Galactic jet source SS433. Specifically, we are able to reproduce the typical observed afterglow emission flux, its spectrum as a function of time, and the fluence distribution of the corrected data for the 4B BATSE catalogue. We also studied the relation between the cosmological evolution of our model and the cosmic ray energy distribution. We used the Star Formation Rate (SFR) as a function of redshift to obtain the distribution in fluences of GRBs in our model. The fluence in the gamma ray band has been used to calculate the energy in cosmic rays both in our Galaxy and at extragalactic distances. This energy input has been compared with the Galactic and extragalactic spectrum of cosmic rays and neutrinos. We found that in the context of our model it is not possible to have any contribution from GRBs to either the extragalactic or the Galactic cosmic ray spectra.

INTRODUCTION

Gamma-Ray Bursts are short bursts that peak in the soft γ-ray band, between 100 KeV and a few MeV. The duration of their emission goes from 10×10^{-3} s to 10^3 s, and they show variability of the order of ms. They also show persistent emissions in the X, optical, infrared and radio bands (afterglow), a spatially isotropic distribution, and a nonthermal spectrum. It is believed that GRBs are associated with relativistic shocks caused by a relativistic fireball in a pre-existing gas, such as the interstellar medium or a stellar wind/jet, producing and accelerating electrons/positrons to very high energies, which produce the gamma-emission and the various afterglows observed [1,2]. More than 30 years after their discovery, thanks to the

CP522, *Cosmic Explosions: Tenth Astrophysical Conference,*
edited by Stephen S. Holt and William W. Zhang
© 2000 American Institute of Physics 1-56396-943-2/00/$17.00

Burst and Transient Source Experiment (BATSE) and the Italian-Dutch satellite BeppoSax, the scientific community knows that Gamma Ray Bursts (GRBs) are isotropically distributed in the sky and that at least some of them are at cosmological distances. But the present data available for redshift position and host galaxy localization are still too few to give us good statistics to study the evolution of GRBs and their redshift distribution. Because of this lack of information, it is still necessary to assume that GRBs follow the statistical distribution of some other well known objets to obtain the GRBs fluence or flux distribution itself [3,4].

GRB JET MODEL: KEY POINTS

In our model [5], GRBs develop in a pre-existing jet. We consider a binary system formed by a neutron star and an O/B/WR companion in which the energy of the GRB is due to the accretion-induced collapse of the neutron star to a black hole. To fix the jet parameters we use the basic ideas of the jet-disk symbiosis model by Falcke & Biermann [6]. In this model, accretion disk, jet, and compact object are considered as an entire system. Mass and energy conservation are applied and the total jet power $Q_{\rm jet}$ is found to be a substantial fraction of disk luminosity $L_{\rm disk}$. We assume that the collapse of a neutron star to a black hole in a binary system induces a highly anisotropic energy release along the existing jet: a violent twist and jerk of the magnetic field. It initiates a relativistic shock wave, with an initial bulk Lorentz factor of about 10^4. Baryonic mass is known to be low in jets. The bulk Lorentz factor evolution derives from the sweep up of the jet material. Magnetic field and particle number density evolution are obtained from the jump conditions in the ultrarelativistic shock. We consider a power law electron energy distribution with a low energy cut-off. Pre-existing energetic electrons/positrons are further accelerated in the shock. The afterglow emission is due to synchrotron and Inverse Compton processes from the shock region. The fluence of the initial burst is determined by shock, dissipation, and γ-γ optical depth effects. The emission region is optically thin very early on and always in the fast cooling regime. There are only two parameters for the explosion: the energy in bulk flow along the jet, $E_{51} \cdot 10^{51}$erg, and the fraction δ of shock energy in relativistic particles. The parameters from the binary system jet are: the mass flow $\dot{M} \cdot 10^{-5} M_\odot$/yr, the speed of the unperturbed jet $0.3 \cdot v_{0.3}$, as well as the minimum electron Lorentz factor $100 \cdot \gamma_{\rm m,2}$. With these parameters and a distance $D_{28.5} \cdot 10^{28.5}$cm, a time $t_5 \cdot 10^5$s, and a frequency $\nu_{14} \cdot 10^{14}$Hz, we obtain the correct flux level of the afterglow:

$$F_\nu^{\rm (ob)}(t) \simeq 7.45 \times 10^{-28} \delta (E_{51}^{5/4} \dot{M}_{-5j}^{-1/4} v_{0.3}^{1/4}) \gamma_{\rm m,2} D_{28.5}^{-2} t_5^{-5/4} \nu_{14}^{-1} \ \ \mathrm{erg \ cm^{-2} s^{-1} Hz^{-1}} \quad (1)$$

CONTRIBUTION TO COSMIC RAY AND NEUTRINO FLUX

We calculated the GRB rate and compared the corresponding cumulative distribution in fluence with the data. We used the SFR as a function of the redshift presented by Madau [7] with a flatter SFR at high redshift to obtain the corresponding fluence distribution of GRBs with the redshift and to use their rate to study the eventual contribution of GRBs to the cosmic ray distribution, both in our Galaxy and in the extragalactic region. We checked if in our jet model GRBs were standard candles. The corrected data for the 4B BATSE catalogue fluence distribution [8] require the adoption of a luminosity function with a power $\phi(f) \propto f^{-1.55}$. The result of our calculations is shown in Fig. 1(left), in which the theoretical fluence distribution curve is compared with the 4B corrected data. Considering the total number of GRBs in BATSE catalogue, an observing time of 8 years, a volume scale of $h^{-3}10^{10.8}\mathrm{Mpc}^3$, with $H_0 = h\,(100 \text{ km s}^{-1} \text{ Mpc}^{-1})$ the Hubble constant, a beaming factor $\frac{4\pi}{2\pi\theta^2} = 200\,\theta_{-1j}^{-2}$, with θ the jet opening angle, the rate of GRBs is:

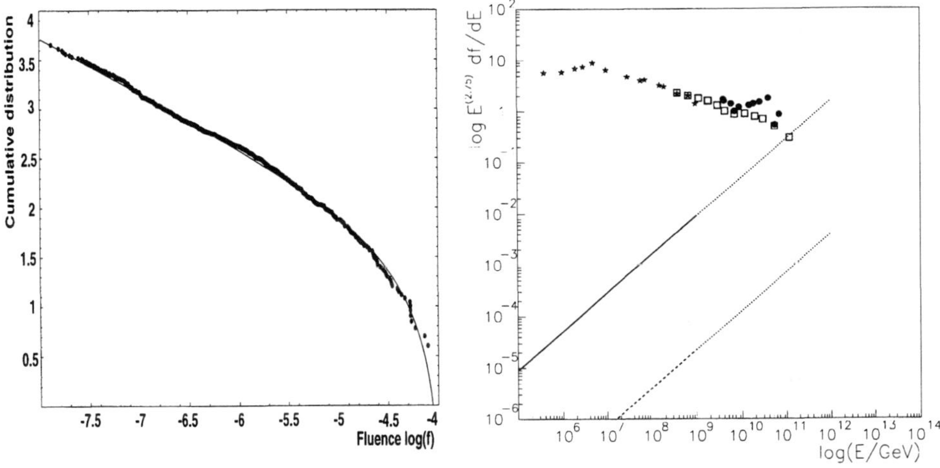

FIGURE 1. Left: comparison between the cumulative fluence distribution obtained with our model (solid line) and the corrected data of catalogue BATSE 4B (full circles), kindly provided by V. Petrosian. The fit we obtained requires a power law luminosity function distribution for GRBs $\phi(f)\,df = f^{-1.55}\,df$. Right: comparison between the extragalactic GRB contribution and the all particle cosmic ray spectrum, expressed in $[\mathrm{GeV}^{-1} \text{ cm}^{-2} \text{ s}^{-1} \text{ sr}^{-1}]$. The solid line corresponds to the assumption that each GRB gives the same contribution to the CR spectrum; this model is excluded by observations. The dashed line corresponds to the model in which each GRB gives a contribution proportional to its own fluence. Any contribution beyond 10^{18} eV is ruled out (dotted lines) considering the attenuation due to the interaction with the microwave background. The stars represent the cosmic ray data from the Akeno experiment, the open squares are the Fly's Eye data and the full circles are the AGASA data.

$$10^{-5.4}(h^3\theta_{-1j}^{-2}) \text{ GRBs per year per } 100 \text{ Mpc}^3 \qquad (2)$$

We used the GRB rate obtained with the SFR from Madau and two different approaches to calculate the contribution from GRBs to the cosmic rays and the neutrino spectra. First we considered that each GRB gives the same contribution equal to 10% of the initial energy, here 10^{51}ergs. Secondly we assume that each GRB contributes proportionally to its own fluence; the fluence distribution adopted has a power law. In Fig. 1(right) we compared the all particle energy spectrum as measured by different ground-based experiment with the spectrum from GRBs in the case that each of them gives the same contribution (dashed line) and with the one in which the contribution is proportional to the fluence (solid line) for the extragalactic case. In the jet-disk symbiosis model for GRBs any extragalactic origin at high energies for cosmic rays is ruled out considering that for energies greater than 10^{18} eV (dotted line), the interactions with the microwave background are relevant and decrease the curve substantially. A corresponding analysis for the cosmic ray contribution from GRBs inside our Galaxy leads to the same result: Near 10^{18} eV the arrival directions of CRs are observed to be ispotropic to an excellent approximation, and yet their diffusion time out of the Galaxy is much shorter than the time scale between GRBs in our Galaxy. Therefore the time for isotropization is not available, ruling out any contribution from GRBs.

CONCLUSIONS

To summarize, our model can explain the initial gamma ray burst, the spectrum and temporal behaviour of the afterglows, the low baryon load, an optical rise, and do all this with a modest energy budget. Moreover, this GRB model is developed within an existing framework for galactic jet sources, using a set of observationally well determined parameters. Using a relatively small set of parameters, the jet-disk symbiosis model applied to GRBs, a tested SFR, and the fundamental physics of the photohadronic interactions we arrive at the conclusion that GRBs are unlikely to give any contribution to the high energy cosmic ray spectrum both inside and outside our Galaxy and to the neutrino spectrum as well.

REFERENCES

1. Paczyński B., *ApJ* **308**, L43 (1986).
2. Mészáros P., and Rees M.J., *ApJ* **405**, 278 (1993).
3. Fenimore E. E., and Bloom J.S., *ApJ* **453**, 25 (1995).
4. Cohen E., and Piran T., *ApJ* **444**, L25 (1995).
5. Pugliese G., Falcke H., and Biermann P.L., *A&A* **344**, L37 (1999).
6. Falcke H., and Biermann P.L., *A&A* **293**, 665 (1995).
7. Madau P. et al., *MNRAS* **283**, 1388 (1996).
8. V. Petrosian, private communication.

The Proton-Photon Instability Model for GRBs

Demosthenes Kazanas* and Apostolos Mastichiadis[†]

*NASA/GSFC, Code 661, Greenbelt, MD 20771
† Department of Astronomy, University of Athens, GR 15784, Athens, Greece

Abstract. We propose that an instability based on the upstream reflection and reinterception of synchrotron radiation, produced in a relativistically moving blast wave, can convert the kinetic energy of its constituent protons to an $e^+ - e^-$ plasma, on time scales comparable to the light crossing time of the blast wave thickness, much in the way proposed to occur in the relativistic jets of AGN. For sufficiently large values of the blast wave Lorentz factor $\Gamma > \Gamma_c$, this process can proceed without the existence of an accelerated population of protons on the shock frame. The value of Γ_c depends on the value of magnetic field on the blast wave frame B_0 (Gauss) and it is given by $\Gamma_c \sim 600 B_0^{-1/5}$, in reasonable agreement with current estimates of the values of these parameters. The resulting $e^+ - e^-$ pairs have a Lorentz factor $\sim \Gamma_c$ and therefore produce synchrotron and IC radiation with peak emission at frequencies $\propto B_0 \Gamma_c^2$, $B_0 \Gamma_c^4$ respectively. In in the lab frame, the IC peak emission is preceived at a frequency $B_0 \Gamma_c^5 \simeq 1$ providing a possible account for the observed energy distribution of GRBs.

INTRODUCTION

The observations of GRB afterglows and the ensuing determination of their reshifts has provided an unequivocal answer to the question of their distance and, to a large degree, to their energetics. Both these quantities seem to corroborate the suggestion of Rees & Mészáros [1] that the GRB emission is due to relativistic outflowing plasma of Lorentz factors $\Gamma \sim 100 - 1000$. It was thus proposed that the GRB emission is due to relativistically expanding blast waves with Lorentz factors in the range above. Furthermore, the subsequent power-law in time evolution of the afterglow flux at various frequencies seems to also be in general agreement with the expanding relativistic blast wave model, since there is no apparent scale in the expansion of a relativistic gas.

The detailed physics of the blast wave generation is still much debated and rather unclear as yet, however, the values of the associated Lorentz factors indicate that, though the energy density of its constituents is dominated initially by energy rather than rest mass, it must also contain a small but finite amount of baryons. In the absence of this component (i.e. if it were a pure $e^+ - e^-$ plasma) radiative losses

CP522, *Cosmic Explosions: Tenth Astrophysical Conference,*
edited by Stephen S. Holt and William W. Zhang
© 2000 American Institute of Physics 1-56396-943-2/00/$17.00

would prevent the expanding fireball to grow to the sizes implied directly from radio observations or inferred indirectly on the basis of physical arguments. The baryons are therefore instrumental for the transport of energy to large distances and also for the efficient acceleration of the blast wave to relativistic Lorentz factors through conversion of the fluid's internal energy to directed motion without radiative losses (a pure $e^+ - e^-$ plasma at the inferred energy densities would be totally thermalized to a temperature $T \simeq 1$ MeV [2] and it would anninilate when the temperature at the fluid rest frame fell below $\sim 50 - 100$ keV, thus viciating its further acceleration to the requisite values of Γ.

The GRB spectra are presumably due to synchrotron radiation by relativistic electrons acclelerated in the flow. The acceleration of particles and the isotropization of the blast kinetic energy are at this point sources of controversy. While most authors seem to favor, on the basis of the observed millisecond variability internal shocks in the flow [3], the dissipation in external shocks seems to also be able to account for the observed variability [4]. However, no matter whether the variability is due to internal or external shocks, the energy is assumed to be contained in the intertia of baryons and that has to be converted with very high efficiency to electrons and eventually into γ-rays. Shocks are generally thought to be inefficient in accelerating electrons, with most of their available energy channeled presumably into the proton component of the plasma. Thus, while the relativistic electrons required to produce the observed radiation may in fact be present, the efficiency of this process adds further constraints to the already tight energy budget of GRBs.

An additional conundrum associated with the synchrotron interpertation of the observed GRB emission (in fact with any mechanism for the photon emission in GRB) is the fact that their νF_ν distribution peaks at energies $E_{pk} \simeq 0.2$ MeV with a very small variance around this value [5]. Considering that this emission originates in a relativistically moving plasma with a range of about 10 or more in the Lorentz factor, the apparently small dispersion in E_{pk} is hard to understand, even in case that, for some unknown reason, all GRBs do manage to produce the identical electron distribution functions on the fluid rest frame. If, on the other hand, the Lorentz factor of the electrons responsible for the emission is also associated with the blast wave Lorentz factor Γ, then $E_{pk} \propto \Gamma^3$ and the observed restricted range in the value of E_{pk} becomes even harder to understand.

THE PROTON-PHOTON INSTABILITY

It is apparent from the above discussion that the baryonic component in the relativistically expanding blast waves associated with GRB plays a crucial role in the conversion of the initial internal energy into directed motion and its transport away from the energy release region; if not for this process the spectra of GRB would have an almost black body form, as indicated in [2]. Instrumental in this process is very long energy loss time scales associated with the baryonic component. At the same time, however, this very property, so fundamental in producing the observed relativistic outflows, is in direct conflict with the requirement of very efficient conversion of the proton kinetic energy into γ-rays.

This situation is not very different from that encountered in the study of Active Galactic Nuclei (AGN), in which acceleration of outflows to relativistic energies ($\Gamma \simeq 10$) and emission of high energy radiation at distances several orders of magnitude larger than the ultimate energy source (the black hole) is demanded, in conjunction with the presence of rapid variability. In this latter case it has been generally assumed that the acceleration of plasma to the requisite Lorentz factors is achived through Poynting flux conversion to particle kinetic energy and the acceleration of particles (thought to be electrons and positrons in some models) takes place *in situ* through shocks. The make up of the plasma in these cases (i.e. whether it consists of electrons and protons or electrons and positrons) is not known; however, if the bulk acceleration of the plasma is through conversion of Poynting flux either composition is allowed.

There are, however, models which invoke the acceleration of the AGN jets to relativistic Γ's by the same method invoked for GRB's, namely the conversion of fluid internal energy into directed one by expansion [6–8]. These models suffer from the same problem of efficient conversion of the proton kinetic or internal energy (if one assumes the presence on relativistic protons in the outflow) to the observed γ-ray radiation. However, in a novel development [8], it was indicated that it is indeed possible to convert the energy storred in a relativistic proton population into electrons (and then to γ-rays) on time scales of order of the light-crossing of the time scale of the system. This is achieved through a combination of an instability associated with a plasma "blob" containing relativistic protons to its own synchrotron radiation, put forward in [9], and the boosting of the "blob's" synchrotron radiation energy density (if the "blob" is moving relativistically) upon an upstream reflection by a "mirror", as disucssed in [10].

The instability discussed in [9] involves the interplay of a loop of interactions: (a) Synchrotron radiation produced by a population of relativistic electrons with Lorentz factor γ. (b) pair production of the synchrotron photons by interaction with the relativistic protons of Lorentz factor γ to replace the "cooling" relativistic electron responsible for the synchrotron radiation. This loop involoves two threshold conditions: (a) A kinematic one, ensuring that the synchrotron photons are above the pair production threshold with the relativistic protons. (b) A dynamic one, ensuring that at least one of the synchrotron photons will pair produce upon its traversal of the system to replace the radiating electron which produced it.

Assuming the presence of a magnetic field of magnitute b)in units of the critical field $B_c \simeq 4.4 \cdot 10^{13}$ G), the kinematic condition is given by the requirement that a photon of energy x (normalized to the electron rest mass) be able to pair-produce on interacting with a relativistic proton of Lorentz factor γ, to produce at least another electron of the same Lorentz factor. This limits x, γ to those obeying $x\gamma \geq 2$. However, since the photons are considered to be synchrotron radiation, $x \simeq b\gamma^2$, leading to $\gamma \geq (2/b)^{1/3}$ for the minimum proton Lorentz factor required for the instability [9].

The upstream reflection and re-interception of the synchrotron radiation emitted by a relativistically moving plasma changes both thresholds: The re-intercepted photons have now energy $b\gamma^2\Gamma^2$ and energy density higher by Γ^2. This then modifies the kinematic threshold to $\gamma \geq (2/b\Gamma^2)^{1/3}$ and reduces substantially the dynamic constraint to values consistent with blazar observations [8].

GRBs, thus, appear to present an ideal setting at which these considerations should apply: The plasma is indeed moving relativistically ($\Gamma \sim 10^2 - 10^3$) and the presence of a blast wave is considered as ideal for the acceleration of particles, with protons the most efficiently accelerated component. However, in order to further restrict our considerations and remove as much freedom as possible, we consider the case of NO further acceleration of protons at the shock front other than that associated with the isotropisation of their distribution behind the shock. We therefore assume that (almost) all protons behind the shock have an mean energy Γm_p^2, i.e. a Lorentz factor similar to that of the blast wave. Thus, one can replace the Lorentz factor γ of individual protons from the threshold condition given above with the Lorentz factor of the outflow itself Γ. This then leads to the condition $\Gamma \geq (2/b\Gamma^2)^{1/3}$ or to $\Gamma \geq (2/b)^{1/5}$ or $\Gamma \gtrsim 600\,B_0^{-1/5}$ where B_0 is the magnetic field in Gauss. The values of the magnetic field and critical Lorentz factor are not far removed from those considered pertinent to GRBs.

According to this mechanism, therefore, the energy stored in isotropised protons behind the blast wave will convert to electrons of the same Lorentz factor, Γ. The resulting synchrotron and Inverse Compton radiation will peak correspondingly at energies $b\Gamma^2$ and $b\Gamma^4$, on the blast wave frame. Viewed by the observer, these components will appear blue-shifted by another factor of Γ to $b\Gamma^3$ and $b\Gamma^5$ respectively. Taking into account the expression for the threshold Lorentz factor, one obtains peaks at energies $\sim b(2/b)^{3/5}$ and ~ 2. It is worth noting that the second energy is of order of the electron rest mass and it is independent of the value of the magnetic field or that of the Lorentz factor! We believe that a feedback similar (if not identical) to that presented above lies at the heart of the radiative processes associated with the prompt GRB emission. We are currently exploring these possibilities.

REFERENCES

1. Rees, M. J. & Mśzáros, P. 1992, MNRAS, 258, P41
2. Paczyński, B. 1986, ApJ, 308, L43
3. Sari, R & Piran, T. 1997, ApJ, 485, 270
4. Dermer, C. D. & Mitman, K. E. 1999, ApJ, L513
5. Preece, R.D., et al. 1999, ApJS, in press
6. Contopoulos,J. & Kazanas, D., 1995, ApJ, 441, 521
7. Subramanian, P., Becker, P. A. & Kazanas, D., 1999, ApJ, 523, 203
8. Kazanas, D. & Mastichiadis, A. 1999, ApJ, 518, L17
9. Kirk, J. G. & Mastichiadis, A., 1992, Nature, 360, 135
10. Ghisellini, G. & Madau, P. 1996, MNRAS, 280, 67

High-Energy Transient Explorer-2

Donald Q. Lamb[a], George R. Ricker[b], Geoffrey Crew[b], John P. Doty[b], Al Levine[b], Roland Vanderspek[b], Joel Villasenor[b], Edward E. Fenimore[c], Mark Galassi[cc], Masaru Matsuoka[d], Nobuyuki Kawai[d], Atsumasa Yoshida[d], Jean-Luc Atteia[e], Gilbert Vedrenne[e], Jean-Francois Olive[e], Michel Boer[e], Jean-Luc Issler[f], Graziella Pizzichini[g], Kevin Hurley[h], J. Garrett Jernigan[h], Carlo Graziani[a], and Stanford E. Woosley[i]

[a]*Department of Astronomy & Astrophysics, University of Chicago, 5640 South Ellis Avenue, Chicago, IL 60637*
[b]*Center for Space Research, MIT, Cambridge, MA 02139*
[c]*Los Alamos National Laboratory, Los Alamos, NM 87545*
[d]*Institute of Physical and Chemical Research, 2-1 Hirosawa, Wako, Saitama, Japan 351-01*
[e]*CESR (CNRS/UPS), BP 4346, 31029 Toulouse Cedex, France*
[f]*CNES, Direction de la Recherche, 18, ave. Edouard Belin, Toulouse Cedex F-31055, France*
[g]*Te.S.R.E.-CNR, Via Gobetti 101, Bologna, Italy,*
[h]*Space Sciences Laboratory, University of California, Berkeley, CA 94720*
[i]*Astronomy and Astropysics Board, University of California, Santa Cruz, CA 95064*

Abstract. We describe the scientific goals, the spacecraft and instrumentation, and the operation of the High-Energy Transient Explorer-2 (HETE-2), which is currently scheduled to be launched in early 2000.

INTRODUCTION

The past two years have seen a dramatic breakthrough in our understanding of gamma-ray bursts (GRBs). This has come about as a result of radio and optical follow-up observations of GRBs, made possible by the rapid determination and dissemination (within half a day to a day or so) of accurate (1 - 3 arcminute) positions for the bursts by the BeppoSAX satellite team (Costa et al. 1999). Among the discoveries to come from these follow-up observations is confirmation that most GRB sources lie at cosmological distances, that GRB sources are compact and expand at relativistic speeds, that GRBs appear to be the result of the collapse of massive stars, and that GRBs can produce astonishingly bright optical emission coincident with the burst itself (see, e.g., the reviews by Lamb 1999, Wheeler 1999).

CP522, *Cosmic Explosions: Tenth Astrophysical Conference,*
edited by Stephen S. Holt and William W. Zhang
© 2000 American Institute of Physics 1-56396-943-2/00/$17.00

Here we describe the scientific goals, the spacecraft and instrumentation, and the ground operation of the High-Energy Transient Explorer-2 (HETE-2) mission, which is designed explicitly to help unravel the mystery of gamma-ray bursts (GRBs). A key feature of the mission is the determination of accurate GRB positions and their dissemination to the astronomical community in near real time.

SCIENTIFIC GOALS

The primary scientific objective of the HETE-2 mission will be to carry out multiwavelength studies of GRBs using X-ray and gamma-ray instruments mounted on a single, compact spacecraft. These studies will include the following.

1) Localizing many GRBs with < 20 arcminute accuracy in the X-ray band using the WXM and a subset of these bursts with < 20 arcsecond accuracy in the X-ray band using the SXC, in near real time aboard the spacecraft and transmitting these positions directly to a network of primary and secondary ground stations for transmission via the Gamma-Ray Burst Coordinate Network (GCN). This will allow simultaneous and rapid follow-up studies in the radio, infrared, and optical bands.

2) Localizing some GRBs with somewhat higher accuracy in the X-ray band as quickly as possible on the ground. This will allow rapid follow-up studies using the most sensitive radio, infrared, and optical telescopes.

3) Studies of the spectra of many GRBs in the energy range of 0.5 to 500 keV, allowing searches for X-ray precursors and tails, and narrow lines in the burst spectra.

4) Studies of the X-ray afterglow of GRBs.

5) Studies of the intensities, time histories, and spectra of soft gamma-ray repeater (SGR) bursts, should any SGRs be active during the HETE-2 mission.

SPACECRAFT AND INSTRUMENTATION

The HETE-2 spacecraft is a small satellite, measuring roughly a meter high by half a meter in diameter. HETE-2 retains most of the original HETE-1 design, with significant modifications made primarily to the power system. It was developed and constructed by essentially the same MIT team that completed HETE-1, relying in part on key spacecraft design and devlopment consultants.

The spacecraft consists of a spacecraft bus, in which the satellite control hardware and the spacecraft computers reside, and the science payload, which points out one end of the spacecraft. In Figure 1, the four solar panels (which are stowed, parallel to the spacecraft, for launch) are connected at the bottom of the spacecraft bus, and the science instruments point out the top. On orbit, the science instruments will always be pointing away from the sun, and the deployed solar panels will be directed toward the sun.

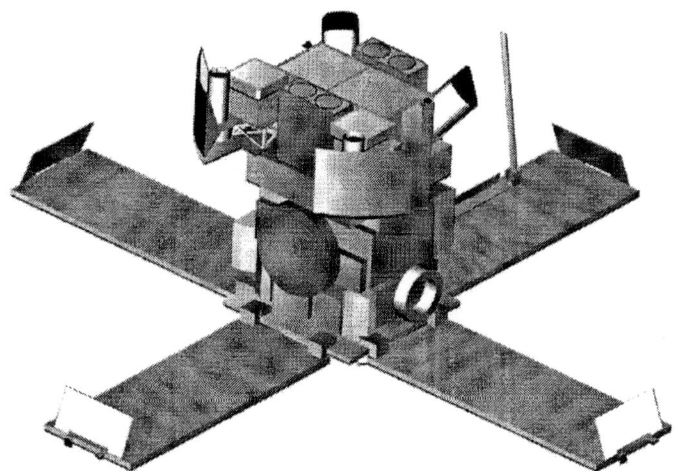

FIGURE 1. The HETE-2 satellite. In this view, the two pairs of units comprising FREGATE are located on the lower left and the upper right of the top face of the satellite; the two 1-D masks and detectors comprising the WXM are located at the upper left and lower right of the top face of the satellite; and the two cube-shaped units comprising the SXC are located at the right and bottom corners of the top face of the satellite.

The design of the French Gamma Telescope (FREGATE) is based on that of the successful GRB experiment Low-Energy Gamma-Ray Burst Detector (LILAS) on board the Russian *Phobos* mission. Its prime objectives are the detection and spectroscopy of GRBs and the monitoring of variable X-ray sources.

TABLE 1. Omnidirectional Gamma-ray Spectrometer

Built by	Centre d'Etudes de Spatiales Rayonnements (France)
Instrument type	cleaved NaI(TI)
Energy Range	6 keV to > 400 keV
Timing Resolution	10 μs
Spectral Resolution	$\sim$25% @ 620keV, $\sim$ 9% @ 662 keV
Effective Area	120 cm^2
Sensitivity (10 sigma)	$\sim 3 \times 10^{-8}$ erg cm^{-2} s^{-1} over 8 keV - 1 MeV
Field of View	$\sim$3 steradians (total for 4 units)

The Wide-Field X-ray monitor(WXM) consists of two, crossed 1-D position sensitive detectors with a coded aperture. Since it is unlikely that two transients will occur at the same time, two 1-D coded aperture imaging systems (in the x and y directions) can be used, simplifying the construction and software.

The addition of the SXCs (replacing the UV cameras on HETE-1) originated from the recent discovery of copious low-energy X-ray emission during and after GRBs. This allows finer localizaiton with a charge-coupled device (CCD)/coded

TABLE 2. Wide Field X-ray monitor

Built by	RIKEN(Japan) and Los Alamos National Laboratory
Instrument type	Coded Mask with Position Sensitive Proportional Counter
Energy Range	2 to 25 keV
Timing Resolution	1 ms
Spectral Resolution	$\sim 25\%$ @ 20 keV
Detector Quantum Efficiency	90% @5 keV
Effective Area	$\sim$175 cm^2 for each of two units
Sensitivity (10 sigma)	$\sim 8 \times 10^{-9}$ erg cm $^{-2}$s^{-1} over the 2-10 keV range
Field of View	1.6 steradians (FWZM)
Angular resolution	$\pm$11 arcmin (normal incidence 8 keV

mask combination resulting in a compact module(the volume of the SXC is 10 times smaller than that of the WXM).

The HETE SXC's are two 1-D coded mask systems using MIT-LL CCID-20's as the detecting elements. The CCID-20's used were fabricated by MIT Lincoln Lab, each consisting of 2048×4096 pixels measuring 15 microns square. The CCID-22s used for the boresight camera are 512×512 pixel devices likewise built by MIT-LL.

TABLE 3. Soft X-ray Camera

Built by	MIT CSR
Instrument type	4CCID-20 Detectors
Camera dimensions	10cmx10cmx17.5cm
Energy Range	500 eV to 14 keV
Timing Resolution	1.2s
Spectral Resolution	46 eV@ 525 eV,129 eV @ 5.9 keV
Dectector Quantum Efficiency	93% @ 5 keV,> 20% (0.514 keV)
Effective Area	6.1x6.1cm^2(each of 2 units)
Burst Sensitivity (4 sigma)	0.47 cts cm$^{-2}s^{-1}$
Steady source Sensitivity (4 sigma)	$\sim 700\times$ t$-1/2$ mCrab
Field of View	0.91 sr
Focal Plane scale	33" per CCD pixel
Localization Precision	Faint Burst, 5 σ - 15",
	Bright burst (1 Crab; 10 s), 22 σ, 3"

Additional information about the HETE-2 satellite, instrumentation, and burst-alerts can be found at the HETE-2 website: http://space.mit.edu/hete/.

REFERENCES

1. Costa, E. 1999, A&A Supp, 138, 425
2. Lamb, D. Q. 1999, A&A Supp, 138, 607
3. Wheeler, J. C. 1999, in *Cosmic Explosions*, Proceedings of the STScI Symposium, April 1999, in press

Blueshift Without Blueshift:
Red Hole Gamma-Ray Burst Models
Explain the Peak Energy Distribution

James S. Graber*

*407 Seward Square SE
Washington, DC 20003

Abstract. Gamma-ray bursts are still a puzzle. In particular, the central engine, the total energy and the very narrow distribution of peak energies challenge model builders. We consider here an extreme model of gamma-ray bursts based on highly red- and blue-shifted positron annihilation radiation. The burst emerges from inside the red hole created by the complete gravitational collapse of the GRB progenitor.

GRB MODEL BUILDING CHALLENGES

Because gamma-ray bursts vary so rapidly, they must be compact. These compact gamma-ray bursts release enormous energy, and therefore they must form an intense fireball that is optically thick, pair-producing, and thermalized. But the spectrum is not thermal, and there is no sign of pair-production attenuation at the high end of the observed spectrum [1]. This seeming self-contradiction (the opacity problem) can be solved by having the fireball power a relativistic shell or jet that collides with something (perhaps itself) to produce the observed gamma rays [2]. This fireball-driven relativistic shock model is currently the leading candidate to explain GRBs [3]. It solves the opacity problem. But like almost all other published models, it fails to explain the observed spectroscopy of GRBs, particularly the narrowness of the observed peak energy distribution [4]. Furthermore, this model does not explain the high ratio of the energy of the GRB burst itself (caused by internal shocks) to the energy in the afterglow (caused by external shocks in the fireball/shock model) [5]. Nevertheless the predictions of this model for the afterglows themselves are consistent with current observations [3].

Finally, there is the problem of the overall energetics of the GRB. The two leading candidates to produce the initial fireball or fireballs –the so-called central engine– are merging neutron stars and core-collapse supernovae [6]. Both these sources have over 10^{54} ergs of total energy available. This is more than enough energy for even the most energetic GRB, but it is not at all clear how to prevent most of it from falling into the newly created black hole that forms in the standard general relativity versions of these models.

There seems to be an inherent conflict between solving the opacity problem and solving the peak energy distribution problem. The only successful technique available to solve the opacity problem is to invoke highly relativistic bulk motion. In the relativistic frame, the gamma rays are below pair-production threshold and so do not suffer pair-production attenuation. This definitively solves the opacity problem. But unless the Lorentz gamma factor of the bulk motion can be fine-tuned to a very narrow range for all GRBs, the resulting blueshift will not only relocate the peak of the photon energy distribution; it will also substantially widen it, inconsistent with the observed narrow E-peak distribution. Thus one needs to find a way to fine-tune the Lorentz gamma factor or find some other way around this conflict. In the fireball/shock model, the gamma factor depends sensitively on the baryon loading, and hence will vary widely. Furthermore, the internal shocks model is dependent on shocks with varying Lorentz gamma factors colliding with each other. So narrowly limiting the gamma factor is not a reasonable option for this model.

A generic solution to this problem is provided if the relativistic bulk motion results not from an initial explosion, but rather from the gravitational acceleration of matter falling into a deep potential well. An arbitrarily high Lorentz gamma factor can be attained, but the accompanying blueshift will be exactly cancelled when the matter and radiation are redshifted as they emerge from the potential well. (By that time, the matter and radiation will have separated, so the opacity problem has already been solved).

A black hole can provide the necessary deep potential well. Once matter or radiation is deep in the potential well of a black hole, however, it is almost impossible for it to escape. Therefore, we will consider an alternative gravitational collapse paradigm in which it is possible to escape from deep within the potential well of a gravitationally collapsed object.

WHY CONSIDER RED-HOLE MODELS?

The problems with constructing a GRB model might be sufficient motivation to consider alternate theories of gravity. However, a stronger motivation comes from the theory of gravitation. Recent theoretical developments in string theory, quantum gravity and critical collapse strongly suggest the possibilities of both gravitational collapse without singularities (and without loss of information) and also gravitational collapse without event horizons [7–9]. If these possibilities are correct, we are forced to consider the phenomenological consequences (such as different models for GRBs and core-collapse supernovae) of alternate paradigms for gravitational collapse in which black holes do not form [10].

RED HOLES– A NEW PARADIGM

Many authors have considered the alternative in which a hard core collapsed object similar to a smaller harder denser neutron star forms in place of a black hole [11]. We here consider the alternative in which no such hard surface forms. Instead the spacetime stretching that forms a black hole in the standard model occurs, but it does not continue to the extent necessary to form an event horizon or a singularity. Instead, spacetime stretches enormously, but not infinitely, and forms a wide, deep potential well with a narrow throat. We call this a red hole.

This type of spacetime configuration was considered by Harrison, Thorne, Wakano and Wheeler (HTWW) in 1965, but only as a way station in the final collapse to a black hole (not yet then called by that name) [12]. In their version, part of the configuration is inside the event horizon, the collapse continues, and a singularity soon forms.

In the new alternate paradigm we call a red hole, no event horizon forms and no singularity forms. The gravitational collapse does not continue forever, but eventually stops. (Why? Perhaps due to quantum effects or string-theory dualities, but we cannot discuss this adequately here.) As the collapse proceeds, the collapsing matter becomes denser and denser until it reaches a critical point, after which, the distortion of spacetime is so great that the density decreases. This happens because the spacetime is stretching outward faster than the collapsing material can fall inward. (This decreasing density effect was already noticed by HTWW in their analysis of gravitational collapse in the context of standard general relativity [12]. In general relativity, this expansion of spacetime is mostly hidden behind the event horizon and does not prevent the formation of a singularity in a finite time. This is not the case in several observationally viable alternate theories of gravity [13,14].) This is why we are confident that the center of a red hole resembles a low-density vacuum more than it resembles a high-density neutron star. The decrease in density due to this enormous stretching may also be a factor in halting the gravitational collapse of the red hole before the stretching becomes infinite.

As a result, even though the stretching of spacetime is enormous, it never becomes fast enough to exceed the speed of light and cause an event horizon to form. And it stops before it reaches an infinite size or any other form of singularity. (Infinite density and infinite curvature also do not occur.) Nevertheless, it is very hard to escape from a red hole. First, there are trapped orbits inside the red hole for photons as well as massive particles, which allows permanent or nearly permanent trapping of mass and energy. Second, the Shapiro delay in crossing a red hole is very substantial, (in some cases, enormous). Hence particles which are only crossing the red hole or passing through are in effect temporarily trapped.

In fact most of the matter falling into a red hole will be trapped. However, radiation, and highly relativistic matter that falls directly into the center of the red hole and does not rescatter while inside the red hole, can travel straight through and emerge on the other side. This possibility is essential for our proposed new GRB models.

RED-HOLE BURST MODEL

Elsewhere, we have considered models based on relocating part or all of the standard fireball/shock model inside or near a red hole. Here, we want to consider an even more radical model. In this model, the central engine is the direct source of the gamma-ray burst. There is no intervening finely tuned jet of baryons. There is no sensitive dependence on the baryon loading factor, and no dependence on a later shock to retransform the energy into gamma rays. Instead the original pair-rich fireball (created by matter collapsing into a red hole) becomes rapidly thin as it falls into the interior of the red hole and expands. Because everything (photons, baryons, electrons and positrons) is falling into the red hole at almost the same highly relativistic speed, the photons are below pair-production threshold in the infalling frame. Therefore, the fireball is optically thin and the annhilation radiation escapes. The plasma is falling with highly relativistic Lorentz gamma factors up to 1000 or more. The pair-annihilation photons are emitted in opposing pairs. One is highly redshifted, while its twin is equally and oppositely highly blueshifted. The spectrum is highly broadened, but the central peak does not move significantly, since the net blueshift of the infalling electron-positron pair is balanced by the net (or average) redshift of the escaping photon pair. Thus this model can solve the narrow peak energy distribution with ease. The more critical question is whether the combined annihilation line and thermal spectrum of the pair-rich fireball can be stretched enough to create the Band spectrum, or whether more conventional reliance on synchroton shock emission and/or inverse Compton scattering is necessary.

REFERENCES

1. Band, D., et al., *ApJ* **413**, 281 (1993).
2. Rees, M. J., and Meszaros, P., *MNRAS* **258**, 41P (1992).
3. Piran, T., *Phys. Rep.* **314**, 575 (1999).
4. Preece, R. D. et al., to appear in *ApJS*(1999); astro-ph 9908119.
5. Paczynski, B., and Rhoads, J.,*ApJL* **418**, L5 (1993).
6. Woosley, S. E., *ApJ* **405**, 273 (1993).
7. Maldacena, J. and Strominger, A., *Phys. Rev.* **D55**, 861 (1997).
8. Shapiro, S. and Teukolsky, S. A., *Phys. Rev. Lett.* **66**, 994 (1991).
9. Christodoulou, D., *Ann. Math.* **140**, 607 (1994).
10. Graber, J. S., to appear in *Largest Explosions Since the Big Bang* Ed. Livio, M.; astro-ph 9908113 (1999).
11. Robertson, S. L., *ApJL* **517**, 117 (1999).
12. Harrison, B. K., Thorne, K. S., Wakano, M., and Wheeler, J. A., *Gravitation Theory and Gravitational Collapse*, Chicago: University of Chicago Press, 1965, ch. 8, pp. 69-75.
13. Yilmaz, H., *Ann. Phys.* **101**, 413 (1976).
14. Itin, Y., *Gen. Rel. Grav.* **31**, 187 (1999).

Jets

Astrophysical Jets

Mario Livio

Space Telescope Science Institute, 3700 San Martin Drive, Baltimore, MD 21218

Abstract. I show that the assumption that the acceleration and collimation mechanisms of jets are the same in all the classes of astrophysical objects which are observed to produce jets, can lead to interesting conclusions. Jets have now been observed in: active galactic nuclei, young stellar objects, massive x-ray binaries, low mass x-ray binaries, black hole x-ray transients, symbiotic systems, planetary nebulae, and supersoft x-ray sources, and possibly recurrent novae and pulsars.

An attempt is made to identify the necessary ingredients for the acceleration and collimation mechanisms. I show that most likely: (i) jets are produced at the center of accretion disks, and are accelerated and collimated hydromagnetically, (ii) the production of *powerful* jets may require a hot corona or access to an additional heat/wind source associated with the central object. Tentative explanations for the presence of jets in some classes of objects and absence in others are given. Some critical observation that can test the ideas presented in this paper are suggested.

I INTRODUCTION

Highly collimated jets are observed in many classes of astrophysical objects, ranging from active galactic nuclei (AGN) to young stellar objects (YSOs). In the present paper, like in a previous review [1], I will make the assumption that the jet formation mechanism, namely, the mechanism for acceleration and collimation, is the same in most if not all of the different classes of objects which exhibit jets. Adopting a mostly phenomenological approach, I will then attempt to determine to which constraints such an assumption can lead. Previous attempts of a similar nature were made, for example, by [2–7]. However, with the discovery of new classes of objects which produce jets (see Section II below) and with recent developments in theoretical work, the constraints become more meaningful. It should be noted right away that the emission mechanisms which render jets observable in the different classes of objects, are very different in objects like, for example, YSOs and AGN. Here, I therefore concentrate only on acceleration and collimation.

I should also note that bipolar, only weakly collimated outflows, are observed in many objects, such as: luminous blue variables, planetary nebulae, novae in outburst and post asymptotic giant branch stars. Models for the formation of

CP522, *Cosmic Explosions: Tenth Astrophysical Conference,*
edited by Stephen S. Holt and William W. Zhang
© 2000 American Institute of Physics 1-56396-943-2/00/$17.00

TABLE 1. Systems Which Exhibit Collimated Jets

Stellar	
Object	Physical System
Young Stellar Objects	*Accreting* young star
Massive X-Ray Binaries	*Accreting* neutron star or black hole
Black Hole X-Ray Transients	*Accreting* black hole
Low Mass X-Ray Binaries	*Accreting* neutron star
Symbiotic Stars	*Accreting* white dwarf
Planetary Nebulae Nuclei	*Accreting* nucleus (or "interacting winds")
Supersoft X-Ray Sources	*Accreting* white dwarf
Recurrent Novae(?)	*Accreting* white dwarf
Pulsars(?)	Spinning neutron star
Extragalactic	
Active Galactic Nuclei	*Accreting* supermassive black hole

these bipolar outflows exist (e.g. [8–10]; and see [11] for a review), but they will not be discussed in the present paper.

II   THE DISK-JET CONNECTION

In this section, I present all the classes of objects which exhibit jets, and discuss some aspects of the observational evidence for a connection between accretion disks and jets.

A   Systems Producing Collimated Jets

In Table 1, I give a list of all the types of objects in which collimated jets have been (at least tentatively) observed, and the nature of the physical system involved. A few of these objects (symbiotic stars; low mass x-ray binaries with a neutron star accretor) require a little explanation, one class (planetary nebulae), has not yet routinely made it into the jet literature, another class (supersoft x-ray sources) is relatively new, and two classes (recurrent novae and pulsars) are still only tentative.

Systems which have been traditionally associated with jets are: many AGN (e.g. [12–14]) and YSOs (e.g. [15–17]) and some massive x-ray binaries (HMXBs), such as SS 433 (e.g. [18]), Cyg X-3 (e.g. [19]) and the Galactic center source 1E140.7-2942 (e.g. [20]). More recently, black hole x-ray transients have been added as a class (e.g. GRS 1915 + 105, [21]; GRO 1655−40, [22]).

So far, the only low mass x-ray binary (LMXB) with a neutron star accretor in which a jet has been observed is Cir X-1 [23], and even in that case it is not clear how collimated the flow really is.

The only symbiotic system in which a jet has been unambiguously observed, both in the optical and in the radio, is R Aqr [24,25]. Spectroscopic evidence suggests the possible presence of a jet also in MWC 560 [26], and optical images suggest the existence of a jet in He 2−104 [27].

I now turn to the new classes of objects which should, in my opinion, be from now on routinely included in any discussion of jets. In planetary nebulae (PNe), jets have now been directly observed (in the optical) in NGC 6543 [28]. Other systems in which the data are less conclusive include K1−2 (e.g. [29,30]), M1−92 [31], and NGC 7009 (e.g. [32,33]). In addition, several "point-symmetric" PNe have been interpreted as resulting from precessing jets ([34], and references therein and see VI).

A new exciting addition to the classes of objects which produce jets are the supersoft x-ray sources (SSS). These are luminous ($L_{\rm bol} \sim 10^{37}$–10^{38} erg s^{-1}) objects, with a characteristic radiation temperature of $(1 - 10) \times 10^5$ K (e.g. [35–37]), in which probably a white dwarf accretes mass from a subgiant companion at such a high rate that it burns hydrogen steadily (e.g. [38]). Recent spectroscopic observations of the LMC source RX J0513.9−6951 reveal what is probably a bipolar collimated outflow with a velocity of ~ 3800 km s^{-1} [39–41], through the presence of blue- and red-shifted satellite emission features to the optical Hydrogen and Helium recombination lines. Similar features corresponding to a projected velocity of ~ 850 km s^{-1} have now been observed also in the SSS RX J0019.8+2156 [42,43]. In the latter case satellite lines to Pγ, Pβ, and Br γ have also been observed [44]. Such "jet lines" may also be present in CAL 83 [45]. The similarity of the spectral features corresponding to the outflow to those observed in SS 433 (e.g. [46]) is striking.

The latest class of systems which observations indicate that they may produce jets is that of recurrent novae (RNe). The hydrogen emission lines in the recurrent nova U Sco show a triple structure, with red and blue shifted satellite peaks corresponding to line-of-sight velocities of ± 1800 km s^{-1} [47,48]. These satellite peaks could correspond to an outflow with an opening angle of $\sim 6°$ [48]. Similar "jet" satellite lines were seen in the (probably recurrent) nova Nova Oph 1998 [49].

Finally, an intriguing x-ray image of the Crab pulsar shows features that may be interpreted as jets (1999; Chandra X-Ray Observatory Center press release, NASA PR 99-109).

An examination of Table 1 reveals that all the objects which exhibit jets (with the possible exception of the Crab pulsar) contain *accreting* central objects (some models for jets in PNe and YSOs do not involve accretion, see II B, but others do); this leads us naturally to the question in the next section.

B   Does the Formation of Jets Require an Accretion Disk?

Clearly a complete answer to this question is difficult, since it requires both a demonstration that disks can produce jets in all the different classes of objects and that other mechanisms cannot produce them. Since I have adopted a phenomenological approach, I will rather attempt to answer the simpler question: has

·an accretion disk been observed in all of the classes of objects which produce jets?

In the case of YSOs, the answer is clearly: *yes,* (e.g. [50–53]) with the most dramatic manifestation being the disks and jets recently observed in the Herbig-Haro object HH 30 [54], in DG Tau B [55], and in Haro 6–5B [56]. Similarly, disks have unambiguously been observed in all the classes of x-ray binaries (HMXBs, LMXBs, SSS, and black hole x-ray transients; e.g. [57,41,58,59]). Furthermore, in the case of the black hole x-ray transients it has been shown that most likely, the IR and radio emitting plasma is ejected from the inner disk [60–62]; and see also III A). The situation with AGN is somewhat more frustrating. Although almost all of the researchers in this field agree that there are accretion disks in AGN, the evidence is somewhat circumstantial (e.g. [63,64]), and every now and then there are even attempts to cast doubt on their existence (e.g. [65]). Here I would merely like to mention a few recent pieces of evidence for the presence of disks in AGN, which are fairly convincing. (i) The iron $K\alpha$ line in MCG–6–30–15, which is consistent with emission from a disk, [66–68]; but see also [69,70]) and a similar line from NCG–5–23–16 [71] and other AGNs. (ii) The fact that the fit to the double peaked Balmer lines in 3C 390.3 with an accretion disk, and the superluminal motion observed in the same source, give an inclination angle for the disk and the jet which shows that the jet is exactly perpendicular to the disk [72]. (iii) The dust torus observed in NGC 4261 [73], which is remarkably consistent with AGN unification schemes containing an accretion disk (e.g. [74]). (iv) The warped subparsec-scale molecular disk observed in the maser emitting LINER NGC 4258 [75–77]. (v) The fact that velocity-delay maps of optical and ultraviolet emission lines in objects like NGC 5458 [78,79] and NGC 4151 [80] appear much more consistent with disk kinematics than with spherical freefall.

Incidentally, for some time there has been a question whether the double-peaked Balmer emission lines observed in some (mostly radio-loud) AGN [81,82] originate in an accretion disk, or in two line emitting cones (formed by two-sided jets; [83]). However, [84]) have shown that at least in the case of 3C 390.3 the double-peaked lines cannot be produced in a two-sided jet, because the emitting region on the receding jet is expected to be obscured from view by the accretion disk, which is optically thick up to radii of $R \sim 10^{18}$ cm ($M_{\rm BH}/10^8$ M$_\odot$) (e.g. [85]; see also [34]). The fact that the red wing of a line produced in a two-sided jet may be obscured from view by the accretion disk is well known from YSOs (see e.g. [OI] λ 6300 profiles for T Tauri stars; [86]).

In the case of PNe, until recently, only theoretical arguments for the presence of disks in these systems existed [87,34]. These relied on one hand on the fact that following a common envelope phase (which is required, to form the observed close binary nuclei; see e.g. [34] for a review), the somewhat bloated secondary companion is likely to fill its Roche lobe. On the other hand, in binary systems in which the secondary star accretes from the wind of an AGB star primary, a disk can form around the secondary (e.g. [88]). Large dust disks have been observed in the optical and infrared (e.g. CRL 2688; [89,90]), and in the optical in the "Red Rectangle" nebula [91].

A word of caution is needed in relation to YSOs (and PNe). While the presence of accretion disks in the former systems is unquestionable, some models for the collimation of jets in these systems (and indeed in PNe), suggest that refraction through oblique shocks is sufficient to produce highly collimated jets, without an active role for the accretion process (e.g. [92–95]). In these models, a fast and dilute wind interacts with a slowly moving or stationary torus in the equatorial plane, and collimation is achieved via refraction through the oblique shocks in the interaction region. Further work on these models will be required, to establish whether they can indeed produce long-lived, highly collimated jets. Here, however, I will not discuss such models further, since, as explained in the introduction, I am interested in a universal model for all the classes of objects, while this mechanism ("shock focused inertial confinement") requires the presence of a torus which is not expected to exist at least in some of the systems.

To conclude this section therefore, my answer to the question: do jets *require* an accretion disk? is: probably yes, although inertial collimation and the processes operating in pulsars deserve more attention.

C   Do Accretion Disks Require Jets or Outflows?

What I mean by this question is: are outflows/jets the *main* mechanism for transport/removal of angular momentum? The suggestion that this may be the case has been made by many authors (e.g. [96–98]). The idea here is very simple, the angular momentum carried away by a disk wind is

$$\dot{J}_W = \dot{M}_W \, \Omega \, r_A^2 \ , \tag{1}$$

where $\dot{M}_W$ is the mass loss rate in the wind, Ω is the local angular velocity and r_A is the local Alfven radius (see III). At the same time, the rate at which angular momentum needs to be removed from the disk for accretion to occur is

$$\dot{J}_{\mathrm{acc}} = \frac{1}{2} \, \Omega \, r^2 \, \dot{M}_{\mathrm{acc}} \ , \tag{2}$$

where $\dot{M}_{\mathrm{acc}}$ is the accretion rate through the disk. If we require that all the angular momentum is removed by the wind, we obtain

$$\frac{\dot{M}_W}{\dot{M}_{\mathrm{acc}}} = \frac{1}{2} \left(\frac{r}{r_A} \right)^2 \ , \tag{3}$$

from which it is clear that if $r_A \sim 10r$, then only less than 1% of the accreted mass needs to be lost in the wind. Such mass loss rates are indeed observed in CVs and YSOs (e.g. [99–101]), so from this point of view, winds could in principle provide the main mechanism for removal of angular momentum. In particular, external torques (associated with the jet) could extract more angular momentum than internal torques (due e.g. to internal magnetic fields) in systems like SS 433,

·in which the jets are inferred to be more powerful than the underlying accretion disks.

One signature of potential removal of angular momentum by the outflow would be the detection of rotation in the wind or jet. Most models of hydromagnetic acceleration (III C) predict that the ratio $V_\varphi/\Omega r_o$ (where V_φ is the angular velocity in the wind and (Ωr_o) is the rotational velocity at the base of the outflow) should increase approximately linearly with distance till the Alfven radius (because the magnetic field enforces rigid corotation), and then decline, while more or less conserving specific angular momentum. Indications for rotation in the disk winds in CVs have been observed in OY Car [102], and more recently in V347 Pup [103]. In V347 Pup in particular, it was found that during eclipse, the fastest rotating wind is eclipsed. Evidence for rotation in jets has also been found in AGN (e.g. in NGC 4258; [104]; or M87, [105]). However, it is not easy to distinguish observationally between rotation in the jet material and precession of the jet, which may occur, for example, as a result of a radiation-induced warping of the accretion disk ([106,107]; and see VI).

It should be noted, however, that [103] and [100], were able to fit successfully the observed wind lines in CVs by simply assuming that the wind rotates with the rotational velocity at the base of the flow ($\Omega\ r_o$). Thus, at present, there is no clear observational evidence for the type of extraction of angular momentum that is predicted for hydromagnetically driven winds.

Furthermore, at least in the case of CVs, there exists clear observational evidence which suggests that winds *are not* the main mechanism of removal of angular momentum. This is related to the behavior of the disk radius, during dwarf nova outbursts (which are caused by a disk instability which results in the sudden accretion of a significant fraction of the disk material). The point is the following, if angular momentum is transported outwards in the disk through viscous processes (rather than being removed by the wind), then at outburst, since matter diffuses inwards, the angular momentum of that matter has to be transferred to the outer parts of the disk, and the radius is expected to expand [108,109], and then decrease slowly to its initial size. Observations of the dwarf novae U Gem and Z Cha in outburst [110–113], show that the radius behaves exactly as model calculations predict, on the basis of this scenario. Furthermore, [114] have shown that the disks in OY Car, HT Car, and Z Cha are all larger in outburst than in quiescence, which is again consistent with viscous transport of angular momentum (rather than removal by the wind). Coming back, therefore, to the question posed at the beginning of this section: Do accretion disks require jets or outflows, as their *main* mechanism for angular momentum transport/removal? In the case of CVs at least the answer appears to be: no. More observations of rotation in jets and bipolar outflows (including velocity gradients across the outflow) are needed, in order to settle this question definitively for other classes of objects.

III   CLUES ON THE JET FORMATION MECHANISM

Since we have determined that the formation of jets most probably requires the presence of an accretion disk, we can now examine some of the properties of jets, in an attempt to determine which basic ingredients must be associated with the accretion disk, for the acceleration and collimation mechanisms to operate.

A   The Jet Origin

An important conclusion can be drawn from the observed jet velocities. In Table 2, I give examples for the ratio V_{jet}/V_{escape} (where V_{escape} is the escape velocity from the central object) for the different classes of objects. It is immediately clear that in *all* cases the jet velocity is of the order of the escape velocity from the central object (or the Keplerian speed near its surface). This immediately indicates that most of the outflow originates from the *center of the accretion disk*, from the vicinity of the central object (see also [3]). This general inferrence has received impressive observational confirmation by the HST images of HH 30 [54] and the DG Tau B [55], which show clearly the jet emanating from the center of the accretion disk.

Evidence for the fact that jets originate in the inner disk is provided also by multiwavelength observations of the black hole x-ray transient ("microquasar") system GRS 1915+105 [60,115,62,61,116]. These observations have demonstrated convincingly that there exists a one-to-one correspondence between x-ray and IR (and probably radio) flares, and the constant time delay between the x-ray/IR peaks indicates that these are triggered by the same event. This, in turn, implies that initially, the emitting regions of the x-ray and IR are in close proximity to each other. The fact that subsequently the IR and x-ray emission appear to decouple suggests that the emitting regions separate significantly at later times. A picture in which the inner disk ejects a relativistic plasma which produces the IR and radio flares by synchrotron emission is consistent with the existing data (especially since GRS 1915+105 has actually been observed to eject relativistic blobs which produce synchrotron emission; [21]), although not all the details have been clarified (e.g. [117]).

B   Ingredients Which May Not Be Universally Essential for the Acceleration and Collimation of Jets

In Table 3, I list a few properties which at one time or another have been suggested as being associated with the formation of jets. It should be remembered that we are considering mechanisms which can operate in all the classes of objects, and therefore, any ingredient which has "NO" for any of the classes is considered to be not absolutely essential, unless it can be shown that an equivalent ingredient

TABLE 2. The Ratio of Jet Velocity to the Escape Velocity from the Central Object

Object	$V_{\mathrm{jet}}/V_{\mathrm{escape}}$	Example
Young Stellar Objects	~ 1	HH30, HH34 $V_{\mathrm{jet}} \sim 100\text{–}350$ km s^{-1}
Active Galactic Nuclei	~ 1	Radio sources, $\gamma \lesssim 10$ M87, $\gamma \gtrsim 3$
X-Ray Binaries	~ 1	SS 433, Cyg X–3 $V_{\mathrm{jet}} \sim 0.26c$
Black Hole X-Ray Transients	~ 1	GRO 1655-40, GRS 1915+105 $V_{\mathrm{jet}} \gtrsim 0.9c$
Planetary Nebulae	~ 1	FLIERS, Ansae, hot winds $V \sim 200\text{–}1000$ km s^{-1}
Supersoft X-Ray Sources	~ 1	RX J0513.9-6951, RX J019.8+2156 $V_{\mathrm{jet}}(\text{projected}) \sim 3800$ km s^{-1}, $V_{\mathrm{jet}}(\text{projected}) \sim 850$ km s^{-1}
Recurrent Novae	~ 1	U Sco $V_{\mathrm{jet}}(\text{projected}) \sim 1800$ km s^{-1}

is present for that class. A few words of explanation are in order. There is no question that the central object does not (generally) need to be near break-up rotation, although models relying on this property have been suggested in the past for YSOs (e.g. [118,119]; see however [120]). Similarly, it is quite clear that the central object does not need to be relativistic (the compactness of the central object merely determines the escape speed and thereby the jet speed). The question of funnels is somewhat more ambiguous, since one may argue that an ion torus or some form of an advection dominated flow in AGN (e.g. [121,122]), or a torus formed by a slowly moving wind in PNe and YSOs (e.g. [94,95]) can provide for a form of inertial collimation (see also model for SS 433 by [123] and see discussion II B). At present, however, there is no reason to suspect that a funnel is present, for example, in the SSS. Furthermore, some of the types of structures proposed in the past are believed to be either globally unstable to non-axisymmetric modes (e.g. [124]) or to generate too much radiation drag to be able to produce the observed (in black hole sources) superluminal motions (e.g. [4]). I therefore at present do not regard funnels as a necessary ingredient, but more work on this mechanism is definitely needed. It is very clear that the source luminosity does not exceed the Eddington luminosity in a number of classes of objects, and therefore it is perhaps unlikely that jets are driven by radiation pressure alone, as a universal mechanism. Nevertheless, I should note that it has been shown that radiation pressure on resonance lines can accelerate disk winds (e.g. [125–127]).

The surrounding gas pressure of the extensive hot atmosphere in elliptical galaxies has been suggested to be an essential ingredient in the production of jets in

TABLE 3. Ingredients Which May Not be Absolutely Necessary for the Formation of Jets

	YSOs	AGN	XRBs	SSS	PNe	RNe
Central object near break-up rotation	NO	NO	NO	?	NO	NO
Relativistic central object	NO	YES	YES	NO	NO	NO
"Funnel"	?	?	NO(?)	NO	YES(?)	NO
$L \gtrsim L_{edd}$ (Radiation pressure driven)	NO	NO	NO	YES	YES	NO
Extensive hot atmosphere (Gas pressure)	YES(?)	YES	NO	NO	YES(?)	NO
Boundary Layer	NO(?)	NO	NO/YES	YES(?)	YES(?)	YES(?)

radio loud AGN [128]. However, such an environment certainly does not exist in some of the classes of objects in Table 3.

The situation with the boundary layer is somewhat more problematic. A boundary layer between the accretion disk and the central object may (in principle at least) exist in all of the objects, with the exception of those containing a black hole (AGN and black hole x-ray transients). [129] suggested in fact that the origin of the energetic winds in YSOs is the boundary layer (the driving being due to shear-generated toroidal magnetic fields). In view of the fact, however, that the black hole sources (and indeed many YSOs, in which the flow is channelled onto magnetic field lines) do not contain a boundary layer, we have to conclude at this point that a boundary layer is not an essential ingredient in the formation of jets.

As a consequence of all of the above (basically, if we are looking for a *universal* mechanism then all the mechanisms in Table 3 are excluded), we are now led to examine the formation of jets in the context of what is regarded as the most promising model for jet acceleration and collimation: *an accretion disk threaded by a reasonably ordered, poloidal, large scale magnetic field.*

C   Hydromagnetic Jet Acceleration and Collimation

The suggestion that accretion disks could generate a magnetically driven outflow was first made about 20 years ago (e.g. [130–132]; see also [133]. The fact that in an expanding wind the magnetic pressure declines more slowly than the gas pressure also means that magnetic collimation is likely to dominate eventually. Much of the work done presently (e.g. [6,134–137]) relies on the seminal model of [96]. Significant progress has also been achieved in numerical simulations (e.g. [138–141]). The basic idea is that at least some fraction of the magnetic flux is in open field lines, which form some angle with the disk surface. The magnetic energy density is larger (above the disk) than the thermal and kinetic energy densities, and hence the outflowing (ionized) material is forced to follow field lines. Since these lines are corotating with their foot points in the disk, material is accelerated by the centrifugal force like a bead on a wire. It turns out that the acceleration in this model can occur only for an inclination (of the field lines to the vertical at the disk surface) larger than $30°$ [96]. Blandford and Payne have shown that for angles smaller than $30°$ there is an effective potential barrier, while for angles larger than $30°$ an outflow can be driven spontaneously (although they recognized the fact that in such a case their model of a 'cold' wind does not describe the flow in the neighborhood of the disk adequately). More recently, [137] have demonstrated (by analyzing the dynamics of a transonic flow in the disk's corona) that *a certain potential difference must be overcome even for angles larger than 30°* (see V for further discussion).

In the magneto-centrifugal model, the acceleration stops at the Alfven surface, where the kinetic energy density becomes comparable to the magnetic energy density. The collimation however, occurs in this picture, outside the Alfven surface. Since the gas is no longer attached to the field lines, the field gets wound up by the rotation, generating loops which are carried by the outflow to form a spiral field. The curvature force exerted by the toroidal magnetic field on the outflowing material acts in the direction of the rotation axis, and thus collimation of the flow by these "hoop stresses" can in principle be obtained (e.g. [142–144]). However, magnetic fields which are wound up to the point at which the toroidal component dominates, are known to be unstable to kink instabilities (e.g. [145]), similar in nature to those of a twisted rubber band. Once the instability sets in, collimation by hoop stresses is strongly reduced. Since the effects of kink instabilities have not yet been properly incorporated in numerical simulations, and therefore, it is not clear how effective collimation by hoop stresses really is, it is important to examine other possible collimation mechanisms.

A poloidal magnetic field can also act as a collimator (by exerting a magnetic pressure gradient), if the radius of the disk is large compared to the radius of the central object, and if the magnetic flux is largest at the outer disk [4,146,135]. For example, for a vertical field of the form $B_Z \sim (r^2/R_{\rm in}^2 + 1)^{-1/2}$ (where $R_{\rm in}$ is the radius at the inner edge of the disk), good collimation is obtained for $R_{\rm Alfven} \sim R_{\rm disk}$ (e.g. [6,147,135]). In this case collimation is achieved simply because the field lines have a naturally collimating shape (e.g. parabolas).

D   Additional Observational Consequences of the Model for Acceleration and Collimation

Analytic and numerical models of magnetized disks are now capable of producing collimated outflows from the disk center (e.g. [138,135,141,148,149,140]). The terminal jet velocities that are obtained are generally of the order of the Keplerian velocity at the footpoint of the jet [150,143,138,139], with a relatively weak dependence on the strength of the magnetic field (although for Alfven speeds in the disk corona larger than some critical value the jet speed could increase considerably, e.g.

[140]). As I noted in III A, the observed jet speeds are always of the order of the escape speed.

A second important consequence of poloidal collimation is that the minimum opening angle of the jet can be estimated. For a vertical field with a radial dependence of the form above, and assuming that all the collimation occurs before the Alfven surface (due, for example, to kink instabilities after the Alfven surface), the minimum jet opening angle is given by (e.g. [6])

$$\Theta_{\min} \simeq (R_{\mathrm{in}}/R_{\mathrm{out}})^{\frac{1}{2}} \ . \tag{4}$$

Here, R_{out} is the outer disk radius. In Table 4, I give the expected values of $\Theta_{\min}$ for all the classes of objects which produce jets. An examination of Table 4 reveals something extremely interesting. For YSOs, HMXBs, the black hole x-ray transients, AGN, and R Aqr, the opening angle is indeed very small, which is consistent with the observation of highly collimated, powerful jets in these systems. In the recurrent nova U Sco (which has an orbital period of 1.23 days) the opening angle is again quite small, which is consistent with the fact that a moderately collimated outflow may have been observed from this system. However, in the SSS (and possibly in the PNe) the expected opening angle is relatively large (similar to that in ordinary CVs), and yet these classes were found to produce jets (while no jets have been observed in dwarf nova systems). This suggests perhaps that while poloidal collimation is important, it may not (in its simplest form) represent the entire picture.

Very recent 43 GHz VLBI observations of the jet in the active galaxy M87 provide additional support for the accretion disk-jet connection. These observations show that the jet is broad at its base (opening angle of $\sim 60^{\circ}$), with strong collimation occurring on scales of 30–100 Schwarzschild radii from the central black hole ([151]; see also [105]). This is fully consistent with the jet being collimated by the accretion disk, and tentatively suggests that poloidal collimation may be operative.

IV   THE ORIGIN OF THE LARGE SCALE MAGNETIC FIELD

One of the important questions in relation to the model for acceleration and collimation described in the previous sections is what is the origin of the large-scale magnetic field that is assumed to thread the disk (e.g. [152]). Two main possibilities exist: (i) the field is advected inwards by the accreting matter in the disk (e.g. [96,98,153,154]), or (ii) the field is generated locally by the same disk dynamo [155,156] which is responsible for the disk viscosity (e.g. [157–160]).

In the first case one might expect a field distribution in which the vertical field is proportional to the surface density, $B_Z \sim \Sigma$, (e.g. [146,152]). If the disk is standard, then $\Sigma \sim \alpha^{-1} R^{-1/2} (H/R)^{-2}$, where α is the [161] viscosity parameter and H is the disk half-thickness. If in addition we assume that $\alpha \sim (H/R)^{1.5}$, as

TABLE 4. Minimum opening Angle of the Jet in Poloidal Collimation (see text)

Object	R_{in} (cm)	R_{out} (cm)	Θ_{min}
YSOs	10^{11}	10^{15}	0.01
XRBs, XRTs	10^{6}	10^{11}	< 0.01
AGN	10^{14}	$> 10^{17}$	~ 0.01
R Aqr	10^{9}	10^{13}	0.01
RNe	10^{3}	$\sim 10^{11}$	0.02
PNe	$\lesssim 10^{10}$	$\gtrsim 10^{11}$	$\lesssim 0.2$
SSS	10^{9}	$\lesssim 10^{11}$	$\lesssim 0.1$

may be suggested by the observed exponential decays in the outbursts of black hole transients [162,163], then we obtain (for a standard disk) $B_Z \sim R^{-81/92}$, which is very close to the $B_Z \sim R^{-1}$ used in several models (e.g. [135]). Under different assumptions (e.g. that motions in the $r - \varphi$ plane generate a magnetic diffusivity η_{mag}), one can obtain $B_Z \sim R^{-\gamma}$ with $\gamma = 3/2 \, \nu/\eta_{\mathrm{mag}}$ (where ν is the kinematic viscosity; e.g. [146]). The situation may be more complicated if the accretion process is non-stationary. This could be the case, for example, in the black hole systems (e.g. [5]). If the black hole has a high spin, then it can be a powerful source of angular momentum (see V) which can stop the accretion and expel the material away. In this way a cyclic process may arise, in which the magnetic field builds up close to the hole until the inner disk is driven away, from where the process repeats itself.

If the field is generated locally, then one still needs to explain the origin of the *large scale* (length scale of order R) field (which is required for the collimation of the hydromagnetic wind). This is required, since the dynamo generated fields have a length scale only of order H (e.g. [164,165]). It has been suggested by [156], that through reconnection of magnetic loops, an inverse cascade process results, which leads to the generation of large scale fields. Here I will adopt a strictly phenomenological approach, in an attempt to place constraints on the required model. First, it is easy to show (e.g. [3]) that if: viscosity is generated by a dynamo (and hence $\alpha \sim B_D^2/4\pi\rho_D C_S^2$, where B_D is the magnetic field in the disk, ρ_D is the density and C_S the speed of sound), $B_Z \sim B_\varphi$, and the jet velocity is of the order of the Keplerian velocity then

$$\frac{B_Z}{B_D} \sim \left[\frac{\dot{M}_j}{\dot{M}_{\mathrm{acc}}} \frac{H}{R}\right]^{\frac{1}{2}} . \tag{5}$$

Here $\dot{M}_j$ is the mass loss rate in the jet. If we now assume that the large scale field is generated by reconnection of magnetic loops with a length distribution $n(l) \sim l^{-\delta}$, then (since the length scale of B_D is H and of B_Z is R; [156])

$$\frac{B_Z}{B_D} \sim \left(\frac{H}{R}\right)^{\delta-1} . \tag{6}$$

Combining eqs. (5) and (6) we obtain

$$\frac{\dot{M}_j}{\dot{M}_{\mathrm{acc}}} \sim \left(\frac{H}{R}\right)^{2\delta-3} \ .$$ (7)

Now, from observations we know that H/R is in the range 0.03–0.3, while $\dot{M}_j/\dot{M}_{\mathrm{acc}}$ is (in many, although not all systems) in the range 0.01–0.3 (e.g. [3] and references therein). Therefore, if reconnection of magnetic loops is indeed the process through which the large scale field is generated, then irrespective of how the process works, the length distribution of loops must satisfy

$$1.7 \lesssim \delta \lesssim 3.4 \ .$$ (8)

In the particular model of [156], they obtained $\delta = 2$, which satisfies condition (8).

It is also interesting to note that in a simulation of the global development of MHD turbulence in a disk in which vertical stratification has been omitted, the magnetic field was found to contain power (in power spectra) on the largest scales (although it was patchy, with typical scale $\sim H$; [166]).

It is of course possible that the large scale field is obtained through a combination of local generation and advection. (I should note that under some conditions the field of the central star could stabilize at least some fraction of the disk against the magnetorotational (Balbus-Hawley) instability; [137].

V IS THE PICTURE COMPLETE?

At this point it is appropriate to ask whether there are additional ingredients (other than the accretion disk, threaded by a magnetic field) which are necessary, in order to produce *powerful* jets. This question is, in my opinion, inevitable, at least because of the following four puzzles (the first two of which have been known for quite some time): (i) why are there radio-loud and radio-quiet AGN? (ii) What is the difference between objects like SS 433 (which possibly contains a neutron star) and other neutron star LMXBs (of which only Cir X-1 has perhaps a jet)? (iii) Why dwarf novae appear not to produce jets, while recurrent novae, SSS and PN nuclei which are very similar systems do produce jets? (iv) Why do some of the black hole x-ray transients produce powerful jets while the others appear not to?

In an attempt to answer all of these questions the following conjecture has been proposed [7]: *Powerful jets are produced by systems in which on top of an accretion disk threaded by a vertical field, there exists an additional source of heat/wind (possibly associated with the central object).*

While the above conjecture has been originally based only on phenomenology, more recent theoretical work has given it a more rigorous basis. [137] solved for the local vertical structure of an accretion disk threaded by a poloidal magnetic field. They showed that for sufficiently strong fields, stable equilibria (with respect to the

TABLE 5. Potential Energy/Wind Sources that are Associated with the Central Objects in Jet Producing Galactic Sources

Object	Heat/Wind Source
Young Stellar Objects	Energy released at magnetosphere/disk interface or at stellar surface
SS 433	Supercritical accretion onto central object
Supersoft X-Ray Sources	Hot central star (due to steady nuclear burning)
Planetary Nebula Nuclei	Hot central star (due to nuclear burning)

Balbus-Hawley instability) could be found for any value (in the range $0 \lesssim i < 90°$) of the inclination angle between the magnetic field and the vertical to the disk surface. Most importantly however, by analyzing the dynamics of the transonic outflow in the disk corona, [137] showed that a certain potential difference must be overcome even when $i > 30°$. Thus, the launching of an outflow from an accretion disk requires a hot corona or access to an additional source of heat, in accordance with the above conjecture. One may attempt to identify the additional heat/wind source in the Galactic objects which produce jets; this is done in Table 5.

I now turn to the extragalactic sources and the long standing question of what are the differences between radio-loud and radio-quiet AGN. This problem has been recently reviewed by [167] and discussed by [168] and [128]. Generally, there are two classes of explanations for the fact that powerful jets are found in radio-loud AGN and not in radio-quiet ones: (i) the central engines in radio-louds and radio-quiets are the same, but either the formation and/or the propagation of the jets is prohibited by some external circumstances in the radio-quiets. (ii) Only the central engines in the radio-louds are capable of producing powerful jets.

Examples of possibility (i) include the suggestion by [168], that mass losing stars in spiral galaxies prevent the hydromagnetic wind from self collimating (by never becoming super-Alfvenic), and the suggestion by [128] that the gas pressure of the hot atmosphere in ellipticals is essential for the production of jets (in some sense this is similar to the suggestion of inertial collimation in PNe and YSOs, see II B). It is impossible at present to rule out such possibilities. I should note however, that while the central environments of S0 galaxies are generally similar to those in ellipticals, few S0 galaxies are powerful radio sources. At any rate, since my basic assumption has been that the formation and collimation of jets *is the same* in all classes of objects, while these types of scenarios are specific to AGN only, I will proceed under the assumption that only the central engines in radio louds are capable of producing powerful jets (see also [167], for a discussion).

It has been shown that if one plots the radio 5 GHz luminosity (which is related to $\dot{M}_j$) as a function of the [O III] $\lambda5007$ luminosity (which is related to $\dot{M}_{acc}$; e.g. [169,170]), there appears to be a separation between radio-louds and radio-quiets, with a correlation between the radio and [O III] $\lambda5007$ luminosities existing in each of the groups [171,172]. It is important to note that the *range* in the [O III] luminosity (and hence, presumably the range in $\dot{M}_{acc}$) in the two groups is similar for sources at low redshifts ($z \lesssim 0.2$; [172]). However, radio-loud sources at higher redshifts reach higher values of [O III] luminosity (higher $\dot{M}_{\mathrm{acc}}$). We may now attempt to interpret these facts in the context of the ideas presented in the present paper. First, the existence of a correlation between the power in the jet and the accretion luminosity is a direct consequence of hydromagnetic jets driven by accretion (see eq. (7)). Second, the mass of the central black hole determines the Eddington luminosity, and therefore, the *maximum* $\dot{M}_{\mathrm{acc}}$ which an object can have (how far to the right, on the [O III] axis, the object can be found). Hence, only the most massive black holes can occupy the upper right portion of the correlation for each group. Given the fact that the radio-loud sources at high redshifts extend to higher values of $\dot{M}_{\mathrm{acc}}$, this may indicate that the most massive black holes in these systems have higher masses than those of the most massive black holes in the radio-quiets. We now come to the most difficult question, namely, which parameter distinguishes the radio-louds from the radio-quiets. In the context of the new conjecture [7], *an additional heat/wind source must be identified.* A natural such source can be provided by the black hole *spin*, since the rotational energy can (in principle) be extracted from the spinning black hole by the [173] mechanism (essentially equivalent to a resistor rotating in a magnetic field). In fact, the suggestion that the spin of the black hole is what distinguishes radio-louds from radio-quiets has already been made in the past (e.g. [4,167]). However, some doubts have been raised about this suggestion by recent work which showed that the power from the accretion disk may be expected to dominate over that which can be extracted from the black hole [174,175]. The black hole spin as an energy source may still be important if the plunging orbits (inner to the marginally stable orbit) are considered (e.g. [176,5,177]). Further insights into the radio-loud radio-quiet dichotomy may be provided by recent observations suggesting that essentially all radio-quiet quasars brighter than $M_R = -24$ reside in massive ellipticals [178].

In discussing relativistic jets from AGNs it is very instructive to make comparisons with Galactic black hole X-ray transient systems (e.g. [179]). In an intriguing recent work, [180] used the observed color temperature and flux from a few black hole transients in an attempt to determine the black hole spin (by determining the radius of the innermost stable orbit.) While this determination is far from secure, the result (if taken at face value) is very intriguing. Zhang et al. found that the spin of the sources 1124+68, 2000+25, and LMC X–3 (all of which are not known to have jets) is consistent with zero, while the black holes in 1655–40 and 1915+105 (both of which have powerful jets) are maximally rotating ($a_* = 0.93$ and 0.998 respectively, where $a_* = Jc/GM^2$ is the dimensionless specific angular momentum).

VI WARPS AND PRECESSION

High resolution imaging of planetary nebulae has revealed the fact that at least 5% of them exhibit "point-symmetric" features, namely, their morphology exhibits a point reflection symmetry about their center (e.g. [181,182]). Examples of this morphology include NGC 6543 [28] and NGC 5307 [91]. I refer here to "blobs" that are point-symmetric, rather than to the global shape of the nebula. It has been suggested that the point-symmetic morphology is produced by episodic ejection of matter into a two-sided wobbling jet [183,184,87,34].

Similarly, there exists observational evidence that the jets in AGNs are frequently not perpendicular to the plane of the outer disks, or that the disk itself is warped (e.g. [185–188,77,189]; and see discussion in [190]).

A mechanism that can cause the jet to wobble (or precess) was suggested by [106,191], who showed that an accretion disk which is irradiated by a central source can become unstable to warping (see also [192,193,107]). Once the disk becomes warped, it starts to wobble and precess, and in the non-linear regime it can even undergo complete inversion. [191] and [194] have shown numerically that radiation-induced warping occurs at all radii satisfying

$$R/R_s \gtrsim 20\eta^2/\epsilon^2 \ , \tag{9}$$

where $R_s = 2GM/c^2$ is the gravitational radius of the central object, $\eta = \nu_2/\nu_1$ is the ratio of the (R, z) and (R, ϕ) shear viscosities, and $\epsilon = L_*/\dot{M}_{acc}c^2$ (L_* is the luminosity of the central source). [191,?,7,11,195], and [196] have further shown that jet wobbling and precession can occur in AGNs, in planetary nebulae, in some SSSs and some X-ray binaries.

The following point should be noted. In many systems torques other than radiative may be even more important in producing warps. For example, if a fraction ξ of the incident radiative power produces an instantaneous wind of speed v_w, then the ablative torque will exceed the radiative torque as long as $\xi \gtrsim v_w/c$. Furthermore, precessing jets may be obtained also as a result of instabilities in axisymmetric magnetic fields.

VII   CRITICAL OBSERVATIONS

Some critical observations can help to test, and further advance the ideas presented in this paper. In particular, it should be noted that to my knowledge, there is no direct observation which confirms that jets are hydromagnetically driven. In this respect Zeeman observations of the maser line in AGNs can prove very useful in placing meaningful constraints (the field strength can be determined by polarization measurements).

Similarly, reliable determinations of the collimation length scale in sources other than M87 and of the ratio $\dot{M}_{\rm jet}/\dot{M}_{\rm acc}$ can help in constraining theoretical models.

Multiwavelength observations of the Galactic black hole sources (e.g. [61]) also hold the promise of providing information about the mechanism of ejection of relativistic plasma. Searches for jets in other black hole systems are also strongly encouraged. Determinations of more black hole masses and (possibly) spins in

AGNs will certainly provide useful information toward an understanding of the differences between radio-louds and radio-quiets. In this respect, the mere fact that the question of the spin can even be addressed (e.g. via the observations of the iron Kα line; [197]) is quite remarkable.

As was noted in II A and III D a firm determination concerning the presence or absence of jets in dwarf novae can have important implications for the jet production mechanism. In a recent work, [198] derived a relation between the accretion rate through the disk and the expected equivalent width (EW) of the jet lines. The relation is approximately of the form EW (line) $\propto \dot{M}_{\mathrm{acc}}^{4/3}$. Knigge and Livio showed that for typical CVs, jet lines (if they exist) are expected to have EWs of a few hundredths to a few tenths of Angstroms. These are difficult, but not impossible to detect.

ACKNOWLEDGEMENTS

This work has been supported in part by NASA Grant NAG5-6857. I would like to thank Jim Pringle and Roger Blandford for useful discussions.

REFERENCES

1. Livio, M. 1999, Phys. Rep., 311, 225
2. Königl, A. 1986, in *Twelfth Texas Symposium on Realtivistic Astrophysics, Annals of New York Academy of Sciences, Vol. 470*, eds. M. Livio & G. Shavio (New York: New York Academy of Sciences), p. 88
3. Pringle, J. E. 1993, in *Astrophysical Jets*, eds. D. Burgarella, M. Livio, & C. P. O'Dea (Cambridge: Cambridge University Press), p. 1
4. Blandford, R. 1993, in *Astrophysical Jets*, eds. D. Burgarella, M. Livio, & C. P. O'Dea (Cambridge: Cambridge University Press), p. 15
5. Blandford, R. D. 1998, in *Accretion Processes in Physical Systems: Some Like it Hot*, eds. S. Holt & T. Kallman, (Woodbury: AIP), p. 43
6. Spruit, H. C. 1996, in *Evolutionary Processes in Binary Stars*, eds. R. A. M. J. Wijers, M. B. Davies, & C. A. Tout (Cambridge: Cambridge University Press), p. 249
7. Livio, M. 1997a, in *Accretion Phenomena and Related Outflows*, eds. D. T. Wichramasinghe et al. (San Francisco: ASP Conf. Ser.), p. 845
8. Balick, B. 1987, AJ, 94, 671
9. Nota, A., Livio, M., Clampin, M., & Schulte-Ladbeck, R. 1995, ApJ, 448, 788
10. Paresce, F., Livio, M., Hack, W., & Korista, K. 1995, A&A, 299, 823
11. Livio, M. 1997b, Space Sci. Rev., 82, 389
12. Marscher, A. P. 1993, in *Astrophysical Jets*, eds. D. Burgarella, M. Livio, & C. P. O'Dea (Cambridge: CUP), p. 73
13. Laing, R. A. 1993, in *Astrophysical Jets*, eds. D. Burgarello, M. Livio, & C. P. O'Dea (Cambridge: CUP), p. 95

14. Wilson, A. 1993, in *Astrophysical Jets*, eds. D. Burgarella, M. Livio, & C. P. O'Dea (Cambridge: CUP), p. 121

15. Ray, T. P. & Mundt, R. 1993, in *Astrophysical Jets*, eds. D. Burgarella, M. Livio & C. P. O'Dea (Cambridge: CUP), p. 145

16. Reipurth, B. & Heathcote, S. 1993, in *Astrophysical Jets*, eds. D. Burgarella, M. Livio, & C. P. O'Dea (Cambridge: CUP), p. 35

17. Bally, J. 1997, in *Accretion Phenomena and Related Outflows*, IAU Colloq. 163, eds. D. T. Wickramajinghe, L. Ferrario & G. V. Bicknell (San Francisco: ASP), p. 3

18. Margon, B. 1984, ARA&A, 22, 507

19. Strom, R. G., van Paradijs, J., & van der Klis, M. 1989, Nature, 337, 234

20. Mirabel, I. F., Cordier, B., Paul, J., & Lebrun, F. 1992, Nature, 358, 215

21. Mirabel, I. F. & Rodriguez, L. F. 1994, Nature, 371, 46

22. Hjellming, R. M. & Rupen, M. P. 1995, Nature, 375, 464

23. Stewart, R. T., Caswell, J. L., Haynes, R. F., & Nelson, G. J. 1993, MNRAS, 261, 593

24. Burgarella, D. & Paresce, F. 1992, ApJ, 389, L29

25. Dougherty, S. M., Bode, M. F., Lloyd, H. M., Davis, R. J., & Eyres, S. P. 1995, MNRAS, 272, 843

26. Hillwig, T., Livio, M., & Honeycutt, R. K. 1998, unpublished

27. Corradi, R. L. M., Livio, M., Schwarz, H. E., & Munari, U. 2000, in preparation

28. Harrington, J. P. & Borkowski, K. J. 1994, BAAS, 26, 1469

29. Bond, H. E. & Livio, M. 1990, ApJ, 355, 568

30. Pollacco, D. L. & Bell, S. A. 1996, private communication

31. Trammell, S. R. & Goodrich, R. W. 1996, ApJ, in press????

32. Schwarz, H. E., Corradi, R. L. M., & Melnick, J. 1992, A&AS, 96, 23

33. Balick, B. 1997, ST ScI press release

34. Livio, M. & Pringle, J. E. 1996, ApJ, 465, L55

35. Hasinger, G. 1994, Rev. Mod. Astr., 7, 129

36. Kahabka, P. & Trümper, J. 1996, in *IAU Symp. 165, Compact Stars in Binaries*, eds. E. P. J. van den Heuval & J. van Paradijs (Dordrecht: Kluwer), p. 425

37. Kahabka, P. & van den Heuvel, E. P. J. 1997, ARA&A, 35, 69

38. van den Heuvel, E. P. J., Bhattacharya, D., Nomoto, K., & Rappaport, S. A. 1992, A&A, 262, 97

39. Pakull, M. W. 1994, private communication

40. Crampton, D., Hutchings, J. B., Cowley, A. P., Schmidtke, P. C., McGrath, T. K., O'Donoghue, D., & Harrop-Allin, M. K. 1996, ApJ, 456, 320

41. Southwell, K. A., Livio, M., Charles, P. A., O'Donoghue, D., & Sutherland, W. J. 1996, ApJ, 470, 1065

42. Becker, C. M., Remillard, R. A., Rappaport, S. A., & McClintock, J. E. 1998, ApJ, 506, 880

43. Tomov, T., Munari, U., Kolev, D., Tomasella, L., & Rejkuba, M. 1998, A&A, 333, L67

44. Quaintrell, H. & Fender, R. P. 1998, A&A, 335, L17

45. Crampton, D., Cowley, A. P., Hutchings, J. B., Schmidtke, P. C., Thompson, I. B., & Liebert, J. 1987, ApJ, 321, 745

46. Vermeulen, R. C., et al. 1993, A&A, 270, 204

47. Bonifacio, P., Molaro, P., & Selvelli, P. 1999, IAU Colloq. 7129

48. Lépine, S., Shara, M. M., Livio, M., & Zurek, D. 1999, ApJ, 522, L121

49. Shemer, O. & Leibowitz, E. 1999, in preparation

50. Beckwith, S. & Sargent, A. I. 1993, in *Protostars and Planets III*, eds. E. H. Levy, & J. I. Lunine (Tucson: University of Arizona Press), p. 521

51. Beckwith, S. V. W. 1994, in *Theory of Accretion Disks 2*, eds. W. J. Duschl, J. Frank, F. Meyer, E. Meyer-Hofmeister & W. M. Tscharnuter (Dordrecht: Kluwer), p. 1

52. O'Dell, C. R. & Wen, Z. 1994, ApJ, 436, 194

53. Najita, J., Carr, J. S., Glassgold, A. E., Shu, F. H., & Takanaga, A. T. 1995, ApJ, 462, 919

54. Burrows, C. J., et al. 1996, ApJ, in press

55. Burrows, C. J., et al. 1999, ST ScI press release

56. Krist, J., et al. 1999, ST ScI press release

57. van Paradijs, J. & McClintock, J. E. 1995, in *X-Ray Binaries*, eds. W. H. G. Lewin, J. van Paradijs & E. P. J. van den Heuvel (Cambridge: Cambridge University Press), p. 58

58. Schandl, S., Meyer-Hofmeister, E., & Meyer, F. 1996, in *Supersoft X-Ray Sources*, ed. J. Greiner (Berlin: Springer), p. 53

59. Gänsicke, B. T., Beuermann, K., & de Martino, D. 1996, in *Supersoft X-Ray Sources*, ed. J. Greiner (Berlin: Springer), p. 107

60. Fender, R., et al. 1997, MNRAS, 290, L65

61. Mirabel, I. F., Dhawan, V., Chaty, S., Rodriguez, L. F., Marti, J., Robinson, C. R., Swank, J., & Geballe, T. R. 1998, A&A, 330, L9

62. Eikenbery, S. S., Matthews, K., Morgan, E. H., Remillard, R. A. & Nelson, R. W. 1998, ApJ, 494, L61

63. Netzer, H. 1992, in *Testing the AGN Paradigm*, eds. S. Holt, S. Neff, & M. Urry (New York: American Institute of Physics), p. 146

64. Kinney, A. L., 1994, in *The Physics of Active Galaxies*, eds. G. V. Bicknell, M. A. Dopita, & P. J. Quinn (San Francisco: ASP Conference Series), p. 61

65. Barvainis, R. 1992 in *Testing the AGN Paradigm*, eds. S. Holt, S. Neff, & M. Urry (New York: American Institute of Physics), p. 129

66. Tanaka, Y., et al. 1995, Nature, 375, 659

67. Fabian, A. C., Nandra, K., Reynolds, C. S., Brandt, W. N., Otani, C., Tanaka, Y., Inoue, H., & Iwasawa, K. 1995, MNRAS, 277, L11

68. Iwasawa, K., et al. 1996, MNRAS, 282, 1038

69. Reynolds, C. S. & Begelman, M. C. 1997, ApJ, 488, 109

70. Young, Ross, & Fabian 1998, MNRAS, 300, L1 **need initials**

71. Weaver, K. A., Yagoob, T., Mushotzky, R. F., Nousek, J., Hayashi, I., & Koyama, K. 1997, AAS, 188, 1601

72. Eracleous, M., Halpern, J. P., & Livio, M. 1996, ApJ, 459, 89

73. Ferrarese, L., Ford, H. C., & Jaffe, W. 1996, ApJ, 470, 444

74. Urry, C. M. & Padovani, P. 1995, PASP, 107, 803

75. Miyoshi, M., et al. 1995, Nature, 373, 127

76. Neufeld, D. A. & Maloney, P. R. 1995, ApJ, 447, L17

77. Herrnstein, J. R., Greenhill, L. J., & Moran, J. M. 1996, ApJ, 468, L17

78. Wanders, I., et al. 1995, ApJ, 453, L87

79. Done, C. & Krolik, J. H. 1996, ApJ, 463, 144

80. Ulrich, M.-H. & Horne, K. 1996, MNRAS, 283, 748

81. Eracleous, M. & Halpern, J. P. 1994, ApJS, 90, 1

82. Eracleous, M., Halpern, J. P., Gilbert, A. M., Newman, J. A., & Filippenko, A. V. 1997, ApJ, 490, 216

83. Veilleux, S. & Zheng, W. 1991, ApJ, 377, 89

84. Livio, M. & Xu, C. 1997, ApJ, 478, L63

85. Collin-Souffrin, S. & Dumont, A. M. 1990, AJAA, 229, 292

86. Edwards, S., Cabrit, S., Strom, S. E., Heyer, I., Strom, K. M., & Anderson, E. 1987, ApJ, 321, 473

87. Soker, N. & Livio, M. 1994, ApJ, 421, 219

88. Morris, M. 1987, PASP, 99, 1115

89. Sahai, R. & Trauger, J. T. 1996, HST press release

90. Thompson, R., Ricke, M., Schneider, G., Hines, D., Sahai, R. et al. 1997, ST ScI, press release

91. Bond, H. E., Ciardullo, R., Fullton, L. K. & Schaefer, K. G. 1998, in preparation

92. Canto, J., Tenorio-Tagle, G., & Rozyczka, M. 1988, A&A, 192, 287

93. Frank, A., Balick, B., & Livio, M. 1996, ApJ, 471, L53

94. Frank, A. & Mellema, G. 1996, ApJ, 472, 684

95. Mellema, G. & Frank, A. 1997, MNRAS, 292, 795

96. Blandford, R. D. & Payne, D. G. 1982, MNRAS, 199, 883

97. Pudritz, R. E. & Norman, C. A. 1986, ApJ, 301, 571

98. Königl, A. 1989, ApJ, 342, 208

99. Drew, J. E. 1995, in *The Analysis of Emission Lines*, eds. R. Williams & M. Livio (Cambridge: Cambridge University Press), p. 49

100. Knigge, C., Woods, J. A., & Drew, J. E. 1995, MNRAS, 273, 225

101. Lizano, S., Heiles, C., Rodriguez, L. F., Koo, B.-C., Shu, F. H., Hasegawa, T., Hayashi, S., & Mirabel, I. F. 1988, ApJ, 328, 763

102. Naylor, T., Bath, G. T., Charles, P. A., Hassall, B. J. M., & Sonneborn, G. 1988, MNRAS, 231, 237

103. Shlosman, I., Vitello, P., & Mauche, C. W. 1996, ApJ, 461, 377

104. Cecil, G., Wilson, A. S., & Tully, R. B. 1992, ApJ, 390, 365

105. Biretta, J. A. 1993, in *Astrophysical Jets*, eds. D. Burgarella, M. Livio, & C. P. O'Dea (Cambridge: Cambridge University Press), p. 263

106. Pringle, J. E. 1996, MNRAS, 281, 357

107. Maloney, P. R., Begelman, M. C., & Pringle, J. E. 1996, ApJ, 472, 582

108. Livio, M. & Verbunt, F. 1988, MNRAS, 232, 1P

109. Ichikawa, S. & Osaki, Y. 1992, PASP, 44, 151

110. Smak, J. 1984, PASP, 96, 5

111. Smak, J. 1984, AcA, 34, 317

112. Smak, J. 1984, AcA, 34, 161

113. O'Donoghue, D. 1986, MNRAS, 220, 23P

114. Harrop-Allin, M. K. & Warner, B. 1996, MNRAS, 279, 219

115. Pooley, G. G. & Fender, R. P. 1997, MNRAS, 292, 925

116. Greiner, J., Morgan, E. H., & Remillard, R. A. 1996, ApJ, 473, L107

117. Mirabel, I. F. & Rodriguez, L. F. 1998, Nature, 392, 673

118. Shu, F. H., Lizano, S., Ruden, S. P., & Najita, J. 1988, ApJ, 328, L19

119. Shu, F. H. 1991, in *Frontiers of Stellar Evolution*, ASP, p. 23

120. Shu, F., Najita, J., Ostriker, E., Wilkin, F., Ruden, S., & Lizano, S. 1994, ApJ, 429, 781

121. Rees, M. J., Phinney, E. S., Begelman, M. C., & Blandford, R. D., 1982, Nature, 295, 17

122. Narayan, R., Yi, I., & Mahadevan, R. 1995, Nature, 374, 623

123. Begelman, M. C. & Rees, M. J. 1984, MNRAS, 206, 209

124. Papaloizou, J. C. B. & Pringle, J. E. 1984, MNRAS, 208, 721

125. Murray, N., Chiang, J., Grossman, S. A. & Voit, G. M. 1995, ApJ, 451, 498

126. Proga, D. & Drew, J. E. 1997, in *Accretion Phenomena and Related Outflows*, IAU Collq. 163, eds. D. T. Wickramasinghe, G. V. Bicknell & L. Ferrario (San Francisco: ASP), p. 782

127. Proga, D., Stone, J. M., & Drew, J. E. 1998, MNRAS, 295, 595

128. Fabian, A. C. & Rees, M. J. 1995, MNRAS, 277, L55

129. Pringle, J. E. 1989, MNRAS, 236, 107

130. Blandford, R. 1976, MNRAS, 176, 465

131. Lovelace, R. V. E. 1976, Nature, 262, 649

132. Bisnovatyi-Kogan, G. & Razmaikin, A. A. 1976, Astrophys. Sp. Sci., 42, 401

133. Michel, F. C. 1973, ApJ, 180, L133

134. Königl, A. 1997, in *Accretion Phenomena and Related Outflows*, IAU Colloq. 163, eds. D. T. Wickramasinghe et al. (San Francisco: ASP), p. 551

135. Ostriker, E. C. 1997, ApJ, 486, 291

136. Matsumoto, R. & Shibata, K. 1997, in *IAU Symp. 184, The Central Region of the Galaxy and Galaxies*, 189

137. Ogilvie, G. I. & Livio, M. 1998, ApJ, 499, 329

138. Kudoh, T. & Shibata, K. 1997, ApJ, 474, 362

139. Matsumoto, R. 1998, in *13th North American Workshop on CVs*, ed. S. Howell, (San Francisco, ASP), 286

140. Meier, D., Edgington, S., Godon, P., Payne, D. G., & Lind, K. R. 1997, Nature, 388, 350

141. Ouyed, R., Pudritz, R. E., & Stone, J. M. 1997, Nature, 385, 409

142. Sakurai, T. 1985, A&A, 152, 121

143. Begelman, M. C. & Li, Z. Y. 1994, ApJ, 426, 269

144. Heyvaerts, J. & Norman, C. A. 1996, preprint

145. Parker, E. N. 1979, *Cosmical Magnetic Fields: Their Origin and Activity*, (Oxford: Clarendon Press), Ch. 9

146. Spruit, H. C. 1994, in *Cosmical Magnetism*, ed. D. Lynden-Bell (Dordrecht: Kluwer), p. 33

147. Königl, A. & Kartje, J. F. 1994, ApJ, 434, 446

148. Ouyed, R. & Pudritz, R. E. 1997a, ApJ, 482, 712

149. Ouyed, R. & Pudritz, R. E. 1997b, ApJ, 484, 794

150. Michel, F. C. 1969, ApJ, 158, 727

151. Junor, W., Biretta, J. A., & Livio, M. 1999, Nature, 401, 891

152. Begelman, M. C. 1993, in *Astrophysical Jets*, eds. D. Burgarella, M. Livio, & C. P. O'Dea (Cambridge: Cambridge University Press), p. 305

153. Lovelace, R. V. E., Romanova, M. M., & Newman, W. I. 1994, ApJ, 437, 136

154. Pelletier, G. & Pudritz, R. E. 1992, ApJ, 394, 117

155. Balbus, S. A. & Hawley, J. F. 1998, Rev. Mod. Phys., 70, 1

156. Tout, C. A. & Pringle, J. E. 1996, MNRAS, 281, 219

157. Balbus, S. A. & Hawley, J. F. 1991, ApJ, 376, 214

158. Hawley, J. F. & Balbus, S. A. 1991, ApJ, 376, 223

159. Stone, J. M., Hawley, J. F., Gammie, C. F., & Balbus, S. A. 1996, ApJ, 463, 656

160. Brandenburg, A., Nordlend, A., Stein, R. F., & Torkelson, V. 1995, ApJ, 446, 741

161. Shakuro, N. I. & Sunyaev, R. A. 1973, A&A, 24, 337

162. Cannizzo, J. K., Chen, W., & Livio, M. 1995, ApJ, 454, 880

163. Cannizzo, J. K. 1998, ApJ, 494, 366

164. Hawley, J. F., Gammie, C. F., & Balbus, S. A. 1995, ApJ, 440, 742

165. Matsumoto, R. & Tajina, T. 1995, ApJ, 445, 767

166. Armitage, P. J. 1998, ApJ, 501, L189

167. Wilson, A. S. 1996, in *Energy Transport in Radio Galaxies and Quasars Symposium*, eds. P. Hardee, A. Bridle, & A. Zensus (San Francisco: ASP Conf. Ser.), 9

168. Blandford, R. D. & Levinson, A. 1995, ApJ, 441, 79

169. Falcke, H., Malkan, M. A., & Biermann, P. L. 1995, A&A, 298, 375

170. Miller, P., Rawlings, S., Saunders, R., & Eales, S. 1992, MNRAS, 254, 93

171. Rawlings, S. 1994, in *ASP Conf. Ser. 54, The First Stromlo Symposium* eds. G. V. Bicknell, M. A. Dopita, & P. J. Quinn (San Francisco: ASP), 253

172. Xu, C., Livio, M. & Baum, S. A. 1999, AJ, 118, 1169

173. Blandford, R. D. & Znajek, R. L. 1977, MNRAS, 179, 433

174. Ghosh, P. & Abramowicz, M. A. 1997, MNRAS, 292, 887

175. Livio, M., Ogilvie, G. I., & Pringle, J. E. 1999, ApJ, 512, 100

176. Krolik, J. 1999, ApJ, 515, L73

177. Gammie, C. F. 1999, ApJ, 522, L57

178. Mclure, R. J., Kukula, M. J., Dunlop, J. S., Baum, S. A., O'Dea, C. P., & Hughes, D. H. 1999, MNRAS, 308, 377

179. Mirabel, I. F. & Rodriguez, L. F. 1997, Vistas in Astron., 41, 15

180. Zhang, S. N., Cui, W., & Chen, W. 1997, ApJ, 482, L155

181. Corradi, R. L. M. & Schwarz, H. E. 1995, A&A, 293, 871

182. Manchado, A., Guerrero, M. A., Stanghellini, L., & Serra-Picart, M. 1996, *The IAC Morphological Catalog of Northern Galactic Planetary Nebulae*, (La Laguna: IAC)

183. Raga, A. C., Canto, J., & Biro, S. 1993, MNRAS, 260, 163

184. Lopez, J. A., Meaburn, J., & Palmer, J. W. 1993, ApJ, 415, L135

185. Ulvestad, J. S. & Wilson, A. S. 1984, ApJ, 285, 439

186. Heckman, T. M., O'Dea, C. P., Baum, S. A., & Laurikanes, E. 1994, ApJ, 428, 65

187. Colbert, E. J. M., Baum, S. A., Gallimore, J. F., O'Dea, C. P. & Christensen, J. A. 1996, ApJ, 467, 511

188. Schmitt, H. R., Kinney, A. L., Storchi-Bergmann, T., & Antonucci, R. 1997, ApJ, 477, 623

189. Schreier, E. J., Marconi, A., Axon, D. J., Caon, N., Macchetto, D., Capetti, A., Hough, J. H., Young, S., & Packham, C. 1998, ApJ, 499, L143

190. Pringle, J. E., Antonucci, R. R. J., Clarke, C. J., Kinney, A. L., Schmitt, H. R., & Ulvestad, J. S. 1999, preprint

191. Pringle, J. E. 1997, MNRAS, 292, 136

192. Petterson, J. A. 1977, ApJ, 216, 827

193. Iping, R. C. & Petterson, J. A. 1990, A&A, 239, 221

194. Livio, M. & Pringle, J. E. 1997, ApJ, 486, 835

195. Southwell, K. A., Livio, M., & Pringle, J. E. 1997, ApJ, 478, L29

196. Wijers, R. A. M. J. & Pringle, J. E. 1999, MNRAS, 308, 207

197. Nandra, K., George, I. M., Mushotzky, R. F., Turner, T. J., & Yagoob, T. 1997, ApJ, 477, 602

198. Knigge, C. & Livio, M. 1998, MNRAS, 297, 1079

Blazars, Jets, and the Unification of AGN

C. Megan Urry

Space Telescope Science Institute,
3700 San Martin Drive, Baltimore Maryland 21218

Abstract. The rapid variability and high luminosity of blazars are signatures of beamed emission from powerful relativistic jets closely aligned with the line of sight. The jet emission is magnified a thousand-fold or more by relativistic beaming, providing a unique opportunity to probe how jets are formed and collimated near supermassive black holes. That such jets are common to all radio-loud active galaxies—the so-called "unification" hypothesis—is supported by observations of properties unaffected by relativistic beaming, such as host galaxies and local environments. Our recent HST studies of blazar host galaxies confirm they are luminous but otherwise normal ellipticals, well matched to the hosts of their parent radio galaxies. There is no sign that the active nucleus strongly affects the host galaxy or vice versa, and no correlation between nuclear (jet) power and host galaxy luminosity (mass), as might have been expected given an underlying relation between black hole mass and bulge mass. These results are consistent with the idea that all massive ellipticals harbor supermassive black holes, with (low) duty cycles of activity, and perhaps point to scenarios where the cosmic formation of black holes is closely linked to that of the most massive galaxies.

BLAZARS, AND RELATION TO OTHER AGN

According to the prevailing paradigm [1], all AGN share certain elements: a central supermassive black hole, an accretion disk, a broad-line region surrounded by a dusty torus or warped disk, and an extended narrow-line region; radio-loud AGN also have jets—relativistic outflows produced near the black hole, which eventually decelerate on large scales. We now understand blazars are radio-loud AGN whose inner jets are pointing toward us.

Operationally, blazars defined by their extreme activity. They are rapidly variable with large amplitudes; examples include a factor-of-10 gamma-ray flare in one day in the luminous quasar 3C 279 [2] and doubling times of an hour or less in the lower-luminosity BL Lac objects Mrk 421 [3] and PKS2155–304 [4–6]. Their apparent luminosities are high (10^{45}–10^{49} ergs s^{-1}), and their brightness temperatures and/or variability ($\Delta L/\Delta t$) greatly exceed typical AGN values. In addition,

CP522, *Cosmic Explosions: Tenth Astrophysical Conference,*
edited by Stephen S. Holt and William W. Zhang

blazar emission is often highly polarized ($\gtrsim 3\%$) and most blazars are superluminal radio sources.

Superluminal motion—the apparent faster-than-light separation of compact components in high-resolution radio maps—along with many of the other characteristics of blazars, can be explained by relativistic beaming [7]. The apparent luminosity is enhanced by the factor δ^{2-3}, where $\delta = [\gamma(1 - \beta\cos\theta)]^{-1}$ is the Doppler beaming factor, because the radiation appears beamed in the forward direction and the apparent time scales are compressed. For small angles to the line of sight ($\theta \lesssim 1/\gamma$), $\delta \sim \gamma$, where γ is the Lorentz factor of the bulk jet motion (β is the corresponding velocity), and the apparent velocity of clumps within the jet is $v_{app} \sim \gamma c$. (See [8] for a nice discussion of the various relativistic transformations involved.) Thus the defining characteristics of blazars follow naturally from their having relativistic jets oriented nearly along the line of sight, because the Doppler-boosted jet emission dominates the other AGN components (e.g., emission lines or thermal accretion disk emission).

That this picture is correct was confirmed dramatically by CGRO EGRET detection of more than 60 blazars. Blazars are the strongest extragalactic sources at GeV energies (an unexpected but in retrospect predictable fact given the beaming hypothesis). The apparent gamma-ray luminosity, together with very rapid gamma-ray variability, requires relativistic beaming; otherwise, the optical depth to pairs would exceed the limit on compactness, $l \ll 40$, above which pairs are produced copiously and gamma-rays would never escape. The compactness implied from the observed luminosity and variability time scale, $\Delta L/\Delta t$, is proportional to the fifth power of the Doppler beaming factor times the true (intrinsic) compactness, implying $\delta > 1$ in at least some cases. Typical values computed from this compactness limit are $\delta > 3\text{–}20$ [9].

Thus blazars are the rare manifestations of radio galaxies whose jets are aligned with the line of sight. Historically, the unification of radio galaxies with blazars has been expressed in two separate schemes: (1) the unification of FR I (low-luminosity) radio galaxies with BL Lac objects (weak-lined blazars), and (2) the unification of FR II (high-luminosity) radio galaxies with stronger-lined, more luminous, variable and/or polarized quasars. There is almost certainly overlap at intermediate luminosities, and the boundaries are quite fuzzy in any case. Having understood this orientation effect, we now see blazars as extremely useful laboratories for studying jets, since the jet emission is so greatly magnified.

BLAZAR EMISSION: CURRENT PARADIGM

The spectral energy distributions of blazars are morphologically simple, consisting of two broad spectral components. The synchrotron component, at low frequencies, begins in the radio (i.e., all blazars are radio-loud) and peaks (in νF_ν) at infrared to soft X-ray wavelengths. The gamma-ray component, whose origin is not unequivocally determined, rises from X-ray energies to peak in the GeV–TeV

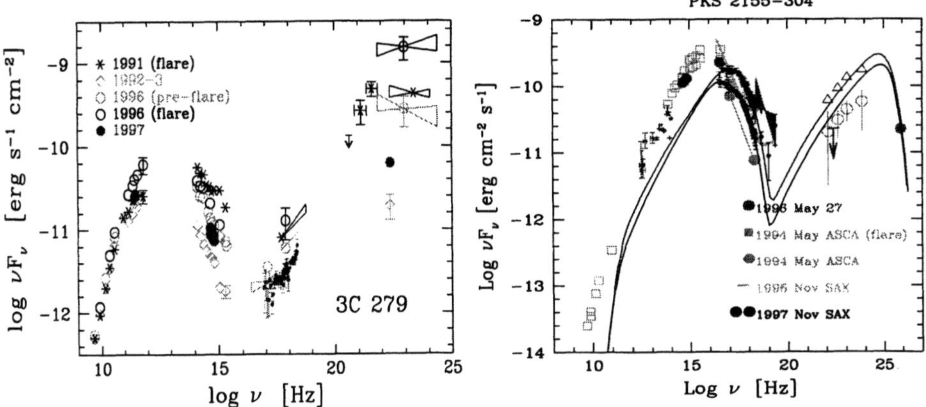

FIGURE 1. Spectral energy distributions for two typical blazars, which share a characteristic two-component morphology but differ in the details. *Left:* SED of the "red" superluminal quasar 3C 279, with peak power output at infrared and GeV energies. *Right:* SED of the "blue" BL Lac object PKS 2155–304, with peak power output at X-ray and TeV energies.

range. Long-term monitoring of blazars has shown that variability in each component is more pronounced above the peak frequency; that the two components often vary in a correlated way, perhaps with short lags; and that the gamma rays appear to vary with larger amplitude than the synchrotron emission.

In detail, there is quite a range of properties across blazars. Two examples are shown in Figure 1. With increasing luminosity, the peak frequency of the synchrotron emission and the peak energy of gamma-ray emission both decrease, and the ratio of gamma ray to synchrotron luminosity increases [10,11]. In fact, many of the blazars detected at GeV energies with CGRO EGRET were emitting most of their energy in gamma rays, at least during flare states.

A promising scenario for production of the observed emission is synchrotron radiation by energetic electrons in the jet, which also Compton scatter ambient photons to produce the gamma rays. In low-luminosity blazars, the synchrotron photons dominate the local energy density, so the synchrotron self-Compton process likely produces the gamma rays. In more luminous blazars, photons from the accretion disk (UV), disk corona (X-ray), or broad-emission-line region (mostly Lyα) are relatively more important and so are likely to dictate the gamma-ray properties.

For synchrotron radiation the frequency of the peak emission depends on the characteristic electron energy and the magnetic field: $\nu_{peak} \propto \gamma_e^2 B$. The electron energy in turn depends on the balance between injection/acceleration of energetic electrons and their cooling and/or escape. This leads to a consistent picture for blazars in which the cooling of relativistic jet electrons on ambient thermal pho-

301

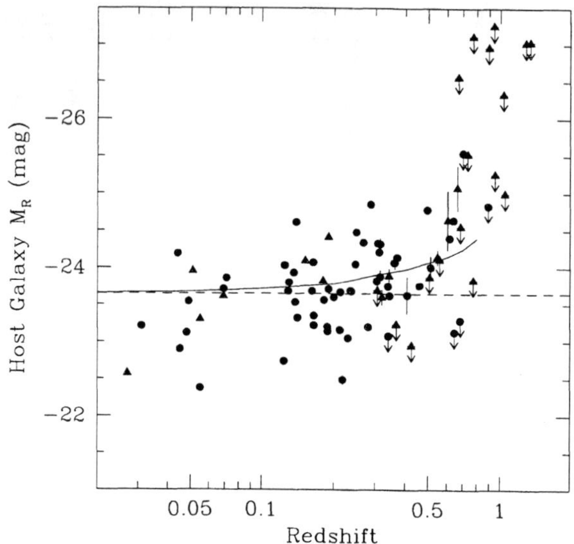

FIGURE 2. Absolute K-corrected R-band magnitudes of BL Lac host galaxies as a function of redshift. The observed values scatter near $\langle M_R \rangle = -23.7$ mag (*dashed line*). Passive evolution (*solid line* predicts an increase above $z \sim 0.5$, where there are few detections. Host galaxies of "red" (*triangles*) and "blue" (*circles*) BL Lac objects (defined by synchrotron peak emission in the infrared-optical or UV-X-ray, respectively) are similar.

tons increases with jet luminosity (since disk and BLR emission correlates with jet power), consistent with the observed trends in spectral shape with luminosity [12].

The wide range of SED shapes observed corresponds to a large range in intrinsic jet power. Present samples of blazars are too small, and are selected at flux levels too high, to determine accurately the total blazar density in the universe (which depends on the counts at faint fluxes). In particular, there is some dispute about the relative numbers of high-luminosity (GeV-bright) and low-luminosity (TeV-bright) blazars. This means we have no idea at present which kind of jet nature preferentially makes. Deeper surveys now underway should address this question.

HST SURVEY OF BLAZAR HOST GALAXIES

The connection of blazars to other AGN and to normal galaxies can be probed through direction observations of their host galaxies and environments. We used the Hubble Space Telescope (HST) to carry out a large imaging survey of low-luminosity blazars [13,14] (i.e., BL Lac objects), which despite their intrinsic faintness are visible to high redshift because of beaming. We wanted to address several issues. First, are BL Lac host galaxies (which are unaffected by beaming) the same as those of FR I radio galaxies, as required by unification? Previous studies indicated

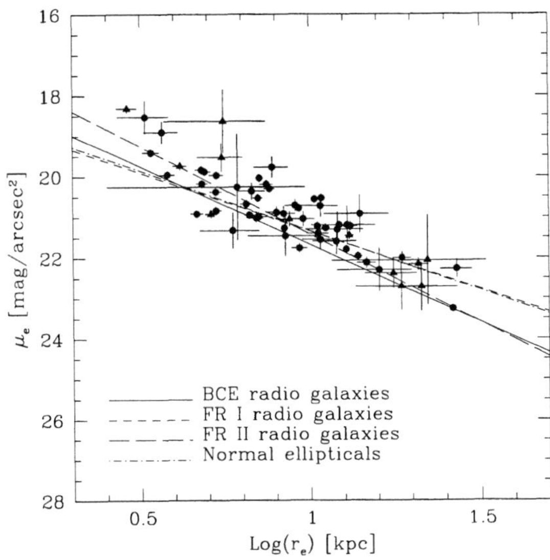

FIGURE 3. Surface brightness (μ_e) at the half-light radius versus half-light radius (r_e) for BL Lac host galaxies (*triangles*: red/luminous BL Lacs; *circles*: blue/less luminous BL Lacs). The BL Lac host galaxies follow closely the Kormendy relations of radio galaxies and normal ellipticals. (references in [13]).

the answer was largely yes [15], but with HST spatial resolution we were able to detect more host galaxies and to determine more precisely their morphologies and luminosities. Second, what is the connection of these AGN to "normal" (non-active) galaxies? Ultimately we would like to probe whether their stellar populations and evolution are similar; with the present survey, we were able to compare them to present-day ellipticals.

Our HST survey consisted of WFPC2 images of 110 BL Lac objects from 6 complete samples (mostly X-ray- and radio-selected) spanning the redshift range $0 < z < 1.4$ ($\sim 1/4$ of the redshifts are unknown). We obtained the following key results:

1. Host galaxies were detected in roughly 2/3 of the sample, including essentially all with $z < 0.5$ and about 1/4 of those with $z > 0.5$.

2. The host galaxies are luminous ellipticals, about a magnitude brighter than the characteristic luminosity of normal ellipticals (m_*). Figure 2 shows the absolute host galaxy magnitudes (K-corrected) as a function of redshift. The values scatter closely around the mean value $\langle M_R \rangle = -23.7$ mag (*dashed line*). Above $z \sim 0.5$ there are mostly upper limits so we cannot comment on the expected evolution (*solid line* shows passive evolution model for star formation at redshifts of a few). There is no difference in the host galaxies of "red" (*triangles*) and "blue" (*circles*) BL Lac objects (defined by whether

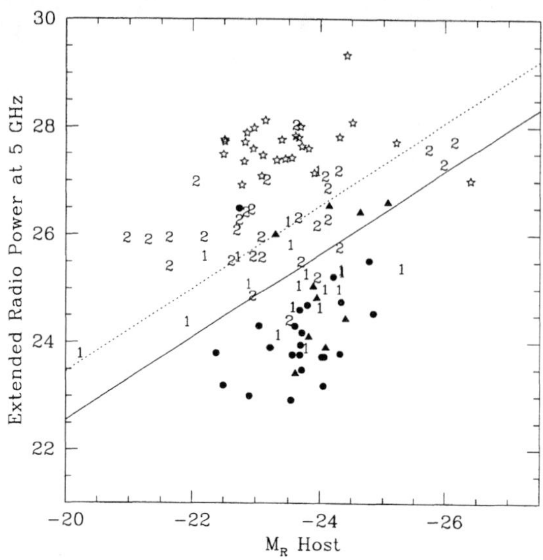

FIGURE 4. Extended radio power at 5 GHz versus R-band host galaxy magnitudes for BL Lac objects (*triangles*: red/luminous BL Lacs; *circles*: blue/less luminous BL Lacs), FR I and II radio galaxies (*1, 2 symbols*), and quasars (*stars*; references in [13]). The radio output reflects the power of the central engine and the horizontal axis correlates with the mass of the host galaxy; neither quantity is affected by beaming. Theoretical models can explain the approximately diagonal separation of FR Is and FR IIs if more powerful jets are needed to break through an increasingly dense ISM.[16] As expected from the unification scenario, quasars and BL Lacs follow the same division.

their peak synchrotron emission occurs in the infrared-optical or UV-X-ray, respectively; see Fig. 1), apart from the selection effect that red BL Lacs occur at systematically higher redshift because they are more luminous.

3. The R-band Kormendy relation (μ_e vs. r_e) for BL Lac host galaxies is very similar to those for radio galaxies and normal ellipticals. Figure 3 shows how the results from our survey (data points) follow the same Kormendy relation as radio galaxies and normal elliptical galaxies. To first order, the galaxy does not appear to know whether it has an active nucleus or not. (This is an extension of there being no difference between red and blue BL Lacs.)

4. In radio power and host galaxy magnitude, BL Lacs are an excellent match to FR I radio galaxies, providing further confirmation of the unification hypothesis (Fig. 4). This is a robust result since neither the extended radio power nor the host galaxy magnitude is affected by relativistic beaming. Figure 4 also shows that quasars overlap well with FR II radio galaxies, as predicted by unification.

5. We found no correlation between nuclear brightness and host galaxy luminosity
   in the BL Lac sample, as had been suggested in some quasar samples [17,18].
   Although the BL Lac nuclei are clearly affected by beaming, we also found
   no correlation in radio power versus host galaxy mass (Fig. 4), which relates
   nuclear and galaxy properties in a way unaffected by beaming, suggesting such
   trends or correlations are not present in low-luminosity AGN. Because there
   appears to be a correlation between the mass of the bulge in nearby galaxies
   and the central mass (thought to be a black hole), this implies Eddington
   ratios for BL Lac nuclei span two or more decades ($L/L_{Edd} \sim 0.001$–0.1).

6. The host galaxies of BL Lac objects, quasars, and radio galaxies are very similar, independent of their large disparity in (intrinsic) nuclear power. There
   is simply no sign that the luminosity of the central engine, which spans 5 or
   6 decades in power, has any effect on the global characteristics of the host
   galaxy. Brightest Cluster Galaxies are somewhat brighter, by about a magnitude, than the other AGN host galaxies, sugggesting the influence of additional
   processes more recent than the original galaxy formation (such as accretion of
   satellite galaxies in the central cluster potential).

SUMMARY

Blazars are relativistically beamed jet sources, fortuitously aligned with the line
of sight. They are just like all other radio-loud AGN, except for orientation, and
because of relativistic beaming are excellent laboratories for studying jet processes.

The observed spectral energy distributions are dominated by two components,
synchroton radiation plus (quite possibly) Compton-scattered gamma rays. Observed trends of the SED shape with luminosity can be explained in terms of a
paradigm where electron acceleration (presumably similar in all jets) is balanced
by cooling on synchrotron and external photons, thus producing the gamma-rays.
In low-luminosity sources, the synchrotron photons are the dominant source of cooling, so these are the bluest BL Lacs. At higher luminosity, there are also disk and
line photons, in addition to the synchrotron photons, so the electrons cool more,
the peak frequencies are lower, and the gamma-ray flux is higher (these are the red
BL Lacs). This blazar paradigm remains to be tested, however, since gamma-ray
observations could be biased by selection effects stemming from the high threshold
of EGRET.

Multiwavelength flaring in blazars is complex, so interpreting the observed light
curves in terms of underlying physics requires time variable codes that can probe
the complex interplay of time scales for electron injection, acceleration, cooling,
and diffusion.

Host galaxies of radio-loud AGN are luminous ellipticals, independent of nuclear
luminosity, at least at low luminosities. They are about a magnitude brighter
than typical field ellipticals. The similarity of BL Lac host galaxies and FR I
radio galaxies confirms unification, as does the correspondence between quasar

host galaxies and FR IIs. The properties of blazar host galaxies are "normal" in morphology, luminosity, and Kormendy relation, suggesting that nuclear processes may be largely de-coupled from galaxy formation and evolution.

Acknowledgements—I thank my blazar colleagues for their ideas, data, and helpful discussions. Those whose multiwavelength work greatly helped my own understanding include Laura Maraschi, Rita Sambruna, Elena Pian, Paola Grandi, Joe Pesce, Annalisa Celotti, Lucio Chiappetti, Giovanni Fossati, Gabriele Ghisellini, Jun Kataoka, Tad Takahashi, Fabrizio Tavecchio, Aldo Treves, and Ann Wehrle. I thank my collaborators on the blazar host galaxies project, particularly Riccardo Scarpa, Matt O'Dowd, Joe Pesce, Renato Falomo, Aldo Treves, Daniela Calzetti, and Mauro Giavalisco. Support for this work was provided by NASA through grant number GO6363.01-95A from the Space Telescope Science Institute, which is operated by AURA, Inc., under NASA contract NAS 5-26555, and through grant number NAG5-3313.

REFERENCES

1. Holt, S. S., Neff, S. G., & Urry, C. M. (eds.), 1992, *Testing the AGN Paradigm*, New York: AIP, 718 pp.
2. Wehrle, A. E., et al. 1998, ApJ, 497, 178
3. Gaidos, J. A., et al. 1996, Nature, 383, 319
4. Morini, M., Chiappetti, L., Maccagni, D., Maraschi, L., Molteni, D., Tanzi, E. G., Treves, A., & Wolter, A. 1986 ApJ, 306, L71
5. Pian, E., et al. 1997, ApJ, 486, 784
6. Zhang, Y. H., et al. 1999, ApJ, 527, 719
7. Blandford, R. D. & Rees, M. J. 1978, in *Pittsburgh Conference on BL Lac Objects*, ed. A. M. Wolfe, (Pittsburgh: U. Pittsburgh Press), p. 328
8. Ghisellini, G. 1999, in *Proc. XIII National Meeting on General Relativity (SIGRAV)*, (astroph 9905181)
9. Dondi, L. & Ghisellini, G. 1995, MNRAS, 273, 583
10. Sambruna, R. M., Maraschi, L., & Urry, C. M. 1996, ApJ, 463, 444
11. Fossati, G., Celotti, A., Ghisellini, G., & Maraschi, L. 1987, MNRAS, 289, 136
12. Ghisellini, Celotti, A., Fossati, G., Maraschi, L., & Comastri, A. 1998, MNRAS, 301, 451
13. Urry, C. M., Scarpa, R., O'Dowd, M., Falomo, R., Pesce, J. E., & Treves, A. 1999, ApJ, 532, in press
14. Scarpa, R., Urry, C. M., Falomo, R., Pesce, J. E., & Treves, A. 1999, ApJS, in press
15. Wurtz, R., Stocke, J. T., & Yee, H. K. C. 1996, ApJS, 103, 109
16. Bicknell, G. V. 1995, ApJS, 101, 29
17. McLeod, K. K. & Rieke, G. H. 1994, ApJ, 431, 137
18. McLeod, K. K. & Rieke, G. H. 1995, ApJ, 454, L77

Microquasars

Jochen Greiner

Astrophysical Institute Potsdam, An der Sternwarte 16, 14482 Potsdam, Germany

Abstract. Microquasars are binary systems in our Galaxy which sporadically eject matter at relativistic speeds along bipolar jets. While phenomenologically similar to quasars, their vicinity allows much more in-depth studies. I review the observational aspects of microquasars, in particular the new developments on the interplay between accretion disk instabilities and jet ejection.

1. INTRODUCTION

Accreting X-ray binaries can be classified [1] in various ways, two important being those according to the nature of the compact object (neutron star or black hole) or the mass of the donor (low-mass vs. high-mass). Most of the high-mass X-ray binaries (HMXB) show X-ray pulsations thus suggesting a magnetized neutron star as the accretor. The donors usually are giants of spectral typ O or B. Low-mass X-ray binaries (LMXB) comprise X-ray bursters, globular cluster X-ray sources, soft X-ray transients and many galactic bulge sources. The donors typically are main-sequence stars of spectral types K-M. Both, neutron stars as well as black hole candidates have been found as accreting objects in LMXBs.

Radio observations with high spatial resolution during the past decade have shown that a small number of X-ray binaries display blobs which move with an apparent velocity larger than the velocity of light (superluminal motion) away from the core of the X-ray source (Fig. 2). Though the lifetime of these blobs as radio emitters is short (few days to weeks) compared to the repetition timescale of ejections, the generally accepted notion is that two-sided jets are emitted by these X-ray sources. The phenomenological similarity to radio-loud quasars led to the naming of microquasars [2].

There is no strict limit above what jet speed a system is considered to be a microquasar. Tab. 1 lists all galactic binaries with jets faster than 0.1c (though often the speed is not exactly known). As can be seen, microquasars do not belong to one of the above X-ray source categories, since they comprise both, either neutron star or black hole accretors as well as low-mass or high-mass donors.

Previous reviews with different emphasis on individual sources, jet physics and observational constraints can be found in [3, 4].

CP522, *Cosmic Explosions: Tenth Astrophysical Conference,*
edited by Stephen S. Holt and William W. Zhang

TABLE 1. Galactic binary sources showing relativistic jets ($v > 0.1c$)

Source	X-ray[1]	Radio[1]	Accretor	Donor	D (kpc)	V_{app}[2]	V_{int}[3]	Θ[4]	Refs.
GRS 1915+105	t	t	black hole?		10–12	1.2c-1.7c	0.92c-0.98c	66°-70°	[5, 6]
GRO J1655-40 (V1033 Sco)	t	t	black hole	F3-6 IV	3.2	1.1c	0.92c	72°-85°	[7, 8, 9]
XTE J1748-288	t	t	black hole?		8–10	0.9c-1.5c	>0.93c		[10]
XTE J1819–254 (V4641 Sgr)[5]	t	t			0.5?	~0.8c			[11]
SS 433 (V1343 Aql)	t	t	neutron star	OB?	5.5	0.26c	0.26c	79°	[12, 13]
Cygnus X-3 (V1521 Cyg)	t	t	neutron star?	WR?	8–10	~0.3c	~0.3c	>70°	[14, 15]
CI Cam (XTE J0421+560)	t	t	neutron star?	Be?	~1	~0.15c	~0.15c	>70°	[16, 17]
Circinus X-1 (BR Cir)	p	p	neutron star	MS	8.5	$\geq$0.1c	$\geq$0.1c	>70°	[15, 18]
1E1740.7-2942	p	p	black hole?		8–10				[2, 19]
GRS 1758-258	p	p	black hole?		8–10				[20, 21]

[1] t $\equiv$ transient, p $\equiv$ persistent

[2] V_{app} is the apparent speed of the highest velocity component of the ejecta.

[3] V_{int} is the intrinsic velocity of the ejecta.

[4] Θ is the angle between the direction of motion of the ejecta and the line of sight.

[5] This X-ray transient has initially been related to the optical variable GM Sgr. However, GM Sgr is a different variable, and the optical counterpart of XTE J1819–254 was named V 4641 Sgr [22].

2. THE TWO MOST IMPORTANT MICROQUASARS

Our present understanding mostly rests on the results obtained over the last years for two microquasars, namely GRS 1915+105 and GRO J1655-40. Therefore, some basics concerning these two systems will be described below.

2.1. GRS 1915+105

Already shortly after the discovery of GRS 1915+105 as a strongly variable new X-ray transient in 1992 [23] it became clear that it was an unusual source. It did not show the canonical fast-rise, exponential decay light curve, and did not show a related bright optical outburst. Later observations showed it to be X-ray active even during times of BATSE-non-detections, suggesting a strongly variable X-ray spectrum [24]. With the RXTE observations since early 1996 these spectral and temporal variations were discovered [25] to extend down to timescales of seconds to minutes (Fig. 1), leading to a variety of new interpretations (see par. 3–5), including the emptying of the inner part of the accretion disk during the stalls and its subsequent replenishment [25, 26].

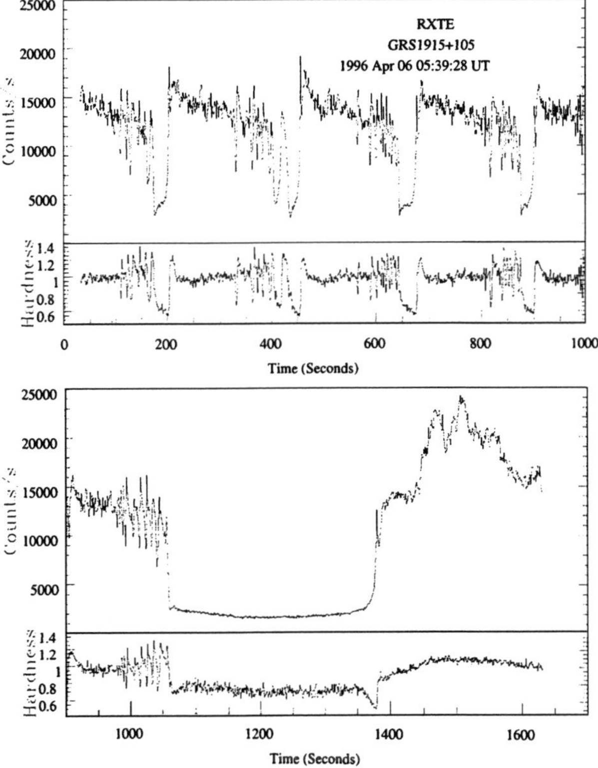

FIGURE 1. X-ray light curve of GRS 1915+105 on April 6, 1996 showing the quasi-periodically repeating pattern of 30 sec duration brightness sputters (top) and a major lull (bottom). The top panel of each plot shows the count rate while the lower panel shows the hardness ratio (count ratio in the 4.4-25 keV versus 2-4.4 keV band) at the same time resolution of 1 sec (time along the abscissa refers to the time labeled in the top panel). (Taken from [25].)

High-resolution radio observations in 1994 revealed for the first time superluminal motion of plasmons in our Galaxy [5]. Recent observations [6] revealed plasmon velocities substantially larger than those observed in 1994, and constrain the distance to less than about 11.5 kpc. Since the distance of GRS 1915+105 is not accurately known, the constraint on the bulk Lorentz factor is very loose (1.4–30). Due to its location in the galactic plane and a consequently huge visual extinction ($\sim$25–27 mag) no decent optical studies are possible. Thus, nothing is known yet about the orbital period and the binary components.

2.2. GRO J1655-40

GRO J1655-40 was discovered in July 1994 as a new X-ray transient [27], and superluminal motion of the radio blobs (Fig. 2) was discovered within a few weeks [7, 8]. However, the interpretation of the radio data is much more complicated as compared to GRS 1915+105, since the blob movement shows wiggles of about 2° around the jet axis, and the relative brightness of the receding and approaching jet does not follow the simple relativistic Doppler boosting description. Instead, the observed brightness difference is often much larger, and also flips from side to side [8]. Thus, the ejecta must be intrinsically asymmetric. Furthermore, the jet axis seems not to be perpendicular to the orbital plane, but inclined by 15° [9].

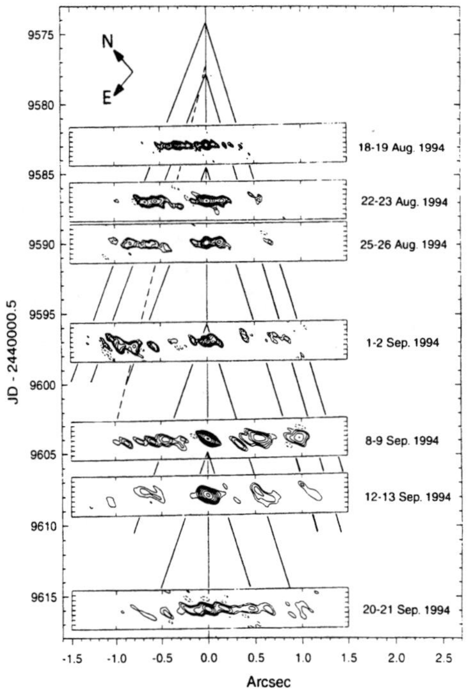

FIGURE 2. A sequence of seven VLBA images of GRO J1655-40 at 1.6 GHz, each having an agular size of $3''.0 \times 0''.4$. The vertical separation corresponds to the time elapsed between the images. The solid lines between the images identify motions of 54 mas/d (left) and 45.5 mas/d (right). The vertical line marks the central source from which the offset (in arcsec) is shown on the horizontal axis (from [8]).

The brightness ($\sim$17 mag) of the optical counterpart has allowed to determine with high precision [9] the orbital period ($2\overset{\mathrm{d}}{.}621$), the inclination ($69\overset{\circ}{.}5$), the mass (2.34 $M_\odot$) and radial velocity amplitude (228.2 km/s) of the donor, and finally the mass of the accreting object of 7.02$\pm$0.22 $M_\odot$. With this mass estimate, the accretor in GRO J1655-40 is one of the best black hole candidates in the Galaxy.

3. QUASI-PERIODIC OSCILLATIONS

RXTE observations have revealed a perplexing variety of quasi-periodic oscillations (QPO) in the X-ray power density spectra [28, 29, 30]. Some sources show up to 7 different QPOs simultaneously spanning three decades in frequency; QPOs sometimes occur with up to 3 harmonics, and their power has a diverse energy dependence. While QPOs are believed to provide a valuable means of probing the X-ray emission region, the understanding of basic properties is still rather poor.

Often, QPOs are attributed to processes in the accretion disk, but evidence for such an origin is scarce. In several cases the properties of QPOs seem to correlate much better with those of hard X-rays, commonly believed to arise in a Comptonizing corona, rather than those of soft X-rays (accretion disk). An intriguing correlation has recently been found in GRS 1915+105 [31]: during source states with QPOs the X-ray intensity variations are mostly in the hard, power law component, while during states without QPOs intensity variations are dominated by the soft, accretion disk component. In addition, there is a strong correlation between QPO frequency (2–10 Hz) and disk temperature. This suggests a delicate interplay between accretion disk and Comptonizing corona in a way where the QPO is produced in the corona, but its frequency is determined by the state of the disk.

One of the simple predictions based on the existence of a Comptonizing corona is the time lag of hard X-ray photons with respect to soft ones. This is indeed observed [32]. Moreover, the observed time lag scales roughly logarithmically with photon energy, as would be expected from a Comptonization process. However, a detailed QPO waveform analysis for 4 QPOs in GRS 1915+105 has shown that the mean waveform does not exhibit the profile smearing that would be caused at the delayed higher energies [28].

A "stable" QPO has been seen in both, GRS 1915+105 (67 Hz) and GRO J1655-40 (300 Hz) with varying strength. If associated with the Keplerian motion at the last stable orbit around a (non-rotating) black hole according to f (kHz)$= 2.2/M_{\mathrm{BH}}$ ($M_\odot$) gives $M_{\mathrm{BH}} \sim 7$ $M_\odot$ for GRO J1655-40, in surprising agreement with the optically determined mass! However, the strong increase of the fractional rms towards larger energies (up to 25 keV) is incompatible with an origin in the disk which has an effective temperature of 2 keV. Alternatively, three other models have been proposed, all resorting on relativistic effects (e.g. [32] for an overview): (i) diskoseismic oscillations [33, 34], (ii) frame dragging [35] and (iii) oscillations related to a centrifugal barrier [36]. At present, it is not clear which of these models, if any, provides a correct description.

4. ROTATING BLACK HOLES?

The X-ray spectra of microquasars consist of (at least) four different components [37]: (1) a thermal component with effective temperature of 2–3 keV which usually is attributed to the emission of the accretion disk, (2) a hard, power law component extending up to 600 keV [38] without any obvious cut-off which is generally interpreted as comptonization of the accretion disk spectrum by hot electrons in a corona above the disk, (3) iron features which have been interpreted as absorption lines of He- and H-like iron [39, 40], and (4) an additional component comprising of excess emission in the 10–20 keV range which has been interpreted as Compton reflection hump.

The effective temperature of the thermal component is very high when compared to the neutron star binaries ($\approx$1.2 keV) or canonical black hole soft X-ray transients ($\approx$0.7–1.1 keV). If interpreted according to the standard accretion disk prescription of Shakura & Sunyaev [41], where T (keV) = 1.2 $(\dot{M}/M)^{1/4}$, the observed temperatures of 2–3 keV would correspond to a mass of the central object of much less than 1 $M_\odot$. Such a low mass, however, seems unacceptable given the high luminosities and non-thermal spectra up to 600 keV. Thus, either a different cooling process is active in microquasars, or the implicit assumption of the last stable orbit at 3 Schwarzschild radii for the above $T(M)$ relation is not valid. In fact, for prograde rotating black holes the innermost stable orbit is closer to the hole, thus allowing higher temperatures of the disk. Application to GRO J1655-40 with its known parameters yields a nearly maximally rotating black hole (a=0.93) [42]. For GRS 1915+105, assuming a mass of 30 $M_\odot$ (based on the lower QPO frequency and luminosity arguments), a similar high rotation rate is deduced (a=0.998). Note that this black hole spin is not inconsistent with the Thirring-Lense interpretation of the QPOs.

5. DISK INSTABILITIES AND JET EJECTION

The variability of the X-ray intensity and spectrum can be rather complex. In GRS 1915+105 (Fig. 1) periodically repeating stalls, short outbursts, rapid oscillations with large amplitude have been observed occasionally while at other times the emission is practically constant [25]. In most cases, intensity variations are coupled to spectral variability. While not all of the variability is understood yet, some patters are most probably due to instabilities in the accretion disk. During the repeating stalls (Fig. 1) the spectrum softens dramatically (factor of $\sim$2 in effective temperature) which has been interpreted as the vanishing of the inner part of the accretion disk [25, 26] (note, however, that this interpretation has been called into question because temperature $\equiv$ inner-disk-radius variations are imitated by a change of the hardening factor T_{col}/T_{eff} when the accretion rate changes [43]). After a few minutes the disk hole is refilled, and the disk temperature and X-ray intensity goes back to its normal, high value. The whole cycle of emptying and

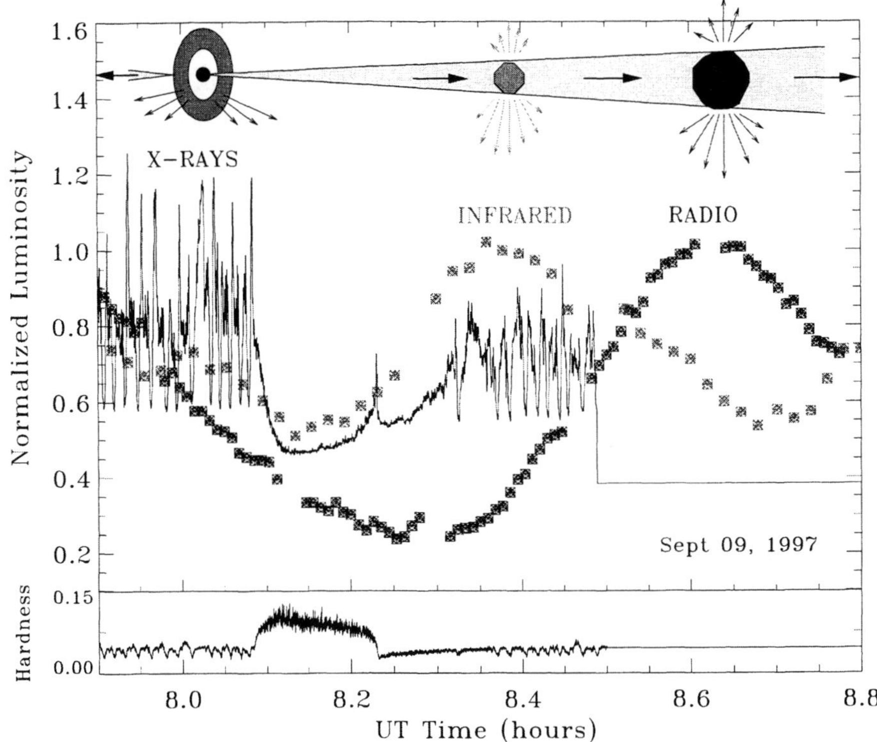

FIGURE 3. Contemporaneous X-ray (solid line) radio (dark squares), infrared (grey squares) light curves (top panel) and X-ray hardness ratio (bottom panel) for GRS 1915+105 on 9 Sep 1997 [45]. The sudden drop in X-ray intensity combined with the X-ray spectral change suggest an emptying of the inner accretion disk [25, 26]. The temporal displacement of the infrared and radio curves is consistent with being due to synchrotron emission from an adiabatically expanding plasmon. (from [45])

replenishment of the inner disk is governed by only one parameter, namely the radius of the disk ring which empties. The larger the radius, the deeper the stall, the cooler the measured temperature of the remaining disk and the longer the time to refill the hole [26], just as expected for a radiation-pressure dominated disk.

How, if at all, are these instabilities related to a transient radio emission and jet ejection? It has been known for several years that transient radio emission is associated to X-ray state transitions. Direct radio imaging, discrete peaks in the radio light curves as well as rapidly evolving radio spectra (to optically thin) suggest that the transient radio emission is related to discrete ejections. The most dramatic observation showing the relation of transient radio emission to the disk behaviour was made for GRS 1915+105 in September 1997 (Fig. 3) when radio and infrared oscillations have been found to follow large-amplitude X-ray variations [45] similar to those described above (Fig. 1). These radio/infrared oscillations have been interpreted as synchrotron emission from repeated small ejections suffering adiabatic expansion losses [44, 45]. The time delay between the infrared and radio maximum is consistent with what one expects from an adiabatically expanding plasma cloud. Together with the above described accretion disk instability cycle this strongly suggests that the inner part of the accretion disk disappears as X-ray radiating source, and possibly is accelerated and ejected away from the system. First-order estimates suggest that up to $\approx 10\%$ of the mass involved in the disk instability may get ejected [26, 44, 45].

Many details of the disk-jet coupling are still unknown. The expected time delay between the infrared flare and the ejection (X-ray event) is only 10^{-3} sec, and thus it is unclear which X-ray feature would relate to the ejection. Alternatively, one could interpret this observed time delay as the duration of a continuous ejection event [45]. Then, the ejection time scale would be related to the viscous time scale of the disk rather than the dynamical one. Is indeed a fraction of the inner disk ejected while falling onto the black hole, or does the innermost disk gets geometrically thick during the instability? What is the role of magnetic fields?

6. DIFFERENCES AND SIMILARITIES BETWEEN MICROQUASARS AND QUASARS

There are two distinct differences between microquasars and quasars:

- Jets of quasars are oriented within a few degrees towards us while those of microquasars are mostly perpendicular. This is a pure observational bias: statistically one expects the same number of objects with angles between $0°$–$60°$ and between $60°$–$90°$. For quasars, however, the jet emission is intrinsically too faint to be visible without the Doppler boosting at small angles (factor $>10^4$ flux enhancement). This is also the reason why quasar jets are one-sided only while in microquasars they are two-sided. It is interesting to speculate how a microquasar would look like if its beam is directed towards us!

- The masses of the accreting objects in microquasars are a factor of 10^{6-9} smaller than those of quasars. Most of the basic observable parameters scale with this mass. Thus, microquasars provide the fortunate circumstance that the timescale of accretion and jet ejection is much better adapted to the human lifetime than that of quasars.

The surprising finding that the two microquasars GRO J1655-40 and GRS 1915+105 may contain maximally rotating black holes actually fits nicely into our picture of quasars. Black hole rotation has long been considered as a way to explain the radio-loud vs. radio-quiet dichotomy in AGN [46]. While the impact of the black hole spin on jet ejection has still to be clarified, this relation provides one of the many examples that the similarities between microquasars and quasars go beyond their names.

7. OUTLOOK

For the near future one can expect further substantial progress because the good observability of the "right" timescales of the jet-disk coupling (seconds to minutes as compared to many years in quasars) combined with a series of new generation instrumentation at X-ray (Chandra, XMM, Astro-E) and Earth-bound (8–10 m class telescopes) optical wavelength opens new perspectives to attack some of the major unsolved questions:

- If it is possible to identify emission lines originating in the jets and to measure their Doppler motion, a new and independent method for distance determination would be available [5].

- Our present understanding allows a much better prediction of the disk-jet interaction which in turn will improve the ability to obtain much improved simultaneous coverage of the jet ejection events.

- X-ray observations with high spectral resolution will provide new insight into the dynamics of matter flow and emission processes in the strong gravitational field near black holes. One may expect that the spin and the mass of black holes could be determined thus providing the basis for the understanding of the energy source of jet ejection and acceleration to relativistic speeds.

- The new correlations found between QPO properties and spectral characteristics will eventually lead to a better theoretical understanding of the origin of QPOs which in turn promises to measure the spin and the mass of black holes and the intimate connection between accretion disk instabilities and jet ejection.

Acknowledgements: JG is supported by the German Bundesministerium für Bildung, Wissenschaft, Forschung und Technologie (BMBF/DLR) under contract Nos. 50 QQ 9602 3, and expresses gratitude for support from the organizers and DFG grants KON 1973/1999 and GR 1350/7-1.

REFERENCES

1. Lewin W.H.G., van Paradijs J., van den Heuvel E.P.J., 1995, *X-ray binaries*, Cambridge Astrophysics Series 26, Cambridge Univ. Press, Cambridge
2. Mirabel F., Rodriguez L.F., Cordier B., Paul J., Lebrun F., 1992, Nat 358, 215
3. Mirabel F., Rodriguez L.F., 1999, ARAA 37 (in press)
4. Fender R., 2000: *Astrophysics and Cosmology: A collection of critical thoughts,* Lecture Notes in Physics, Springer (in press)
5. Mirabel F., Rodriguez L.F., 1994, Nat 371, 46
6. Fender R., Garrington S.T., McKay D.J. et al. 1999, MN 304, 865
7. Tingay S.J., Jauncey D.L., Preston R.A. et al. 1995, Nat. 374, 141
8. Hjellming R.M., Rupen M.P., 1995, Nat 375, 464
9. Orosz J.A., Bailyn C.D., 1997, ApJ 477, 876

10. Hjellming R.M., 1998, (Paris workshop)
11. Hjellming R.M., Rupen M.P., Mioduszewski A.J., 1999, (see URL page http://www.aoc.nrao.edu/~rhjellmi/gmsgr.html)
12. Margon B.A., 1984, ARAA 22, 507
13. Spencer R.E., 1984, MN 209, 869
14. Molnar L.A., Reid M.J., Grindlay J.E., 1988, ApJ 331, 494
15. Stewart R.T., Caswell J.L., Haynes R.F., Nelson G.J., 1993, MN 261, 593
16. Mioduszewski A.J., 1998 (Paris workshop)
17. Garcia M., 1998, (Paris workshop)
18. Fender R.P., Spencer R., Tzioumis T. et al. 1998, ApJ 506, L121
19. Sakano M., Imanishi K., Tsujimoto M., Koyama K., Maeda Y., 1999, ApJ 520, 316
20. Rodriguez L.F., Mirabel I.F., Marti J., 1992 ApJ 401, L15
21. Marti J., Mereghetti S., Chaty S. et al. 1988, AA 338, L95
22. Green D.W.E. (ed.), 1999, IAU Circ. 7277
23. Castro-Tirado A.J., Brandt S., Lund N., 1992, IAU Circ. 5590
24. Greiner J., Harmon, B.A., Paciesas W.S., Morgan E.H., Remillard R.A., 1997, ASP Conf. Ser. 121, p. 709
25. Greiner J., Morgan E.H., Remillard R.A., 1996, ApJ 473, L107
26. Belloni T., Mendez M., King A.R., van der Klis M., van Paradijs J., 1997, ApJ 488, L109
27. Zhang N.S., Wilson C.A., Harmon B.A., et al. 1994, IAU Circ. 6046
28. Morgan E.H., Remillard R.A., Greiner J., 1997, ApJ 482, 993
29. Chen X., Swank J.H., Taam R.E., 1997, ApJ 477, L41
30. Remillard R.A., Morgan E.H., McClintock J.E., Bailyn C.D., Orosz J.A., 1999, ApJ 522, 397
31. Muno M.P., Morgan E.H., Remillard R.A., 1999, ApJ (in press; astro-ph/9904087)
32. Cui W., 1999, in *High-energy processes in accreting black holes*, eds. J. Poutanen & R. Svensson, ASP Conf. Ser. 161, p. 97
33. Perez C.A., Silbergleit A.S., Wagoner R.V., Lehr D.E., 1997, ApJ 476, 589
34. Nowak M.A., Wagoner R.V., Begelman M.C., Lehr D.E., 1997, ApJ 477, L91
35. Cui W., Zhang S.N, Chen W., 1998, ApJ 492, L53
36. Titarchuk L., Lapidus I., Muslimov A., 1998, ApJ 499, 315
37. Greiner J., Morgan E.H., Remillard R.A., 1998, eds. R.N. Ogley, J. Bell Burnell, New Astron. Rev. 42, p. 597
38. Tomsick J.A., Kaaret P., Kroeger R.A., Remillard R.A., 1999, ApJ 512, 892
39. Ebisawa K., 1997, in *X-ray imaging and spectroscopy of cosmic hot plasmas*, Univ. Acad. Press, Tokyo, p. 427
40. Ueda Y., Inoe H., Tanaka Y., et al. 1998, ApJ 492, 782
41. Shakura N.I., Sunyaev R.A., 1973, AA 24, 337
42. Zhang N.S., Cui W., Chen W., 1997, ApJ 482, L155
43. Merloni A., Fabian A.C., Ross R.R., 1999, (in press; astro-ph/9911457)
44. Fender R.P., Pooley G.G., 1998, MN 300, 573
45. Mirabel I.F., Dhawan V., Chaty S., et al. 1998, AA 330, L9
46. Wilson C.A., Colbert E.J.M., 1995, ApJ 438, 62

The Shadow of the Black Hole at the Galactic Center

Heino Falcke[*], Fulvio Melia[†#], and Eric Agol[+]

[*]*Max-Planck-Institut für Radioastronomie, Auf dem Hügel 69, D-53121, Bonn, Germany*
[†]*Physics Department and Steward Observatory, The University of Arizona, Tucson, AZ 85721*
[+] *Physics and Astronomy Department, Johns Hopkins University, Baltimore, MD 21218*
[#] *Presidential Young Investigator and Sir Thomas Lyle Fellow*

Abstract. We show that perhaps already with the next generation of long-baseline interferometers at submm-wavelengths we will able to image the shadow of the black hole in the Galactic Center. To a distant observer, the event horizon casts a relatively large "shadow" with an apparent diameter of ~ 10 gravitational radii due to bending of light by the black hole, nearly independent of the black hole spin or orientation. The predicted angular size for the Galactic Center black hole is $\sim 30\,\mu$arcseconds, a mere factor two smaller than the highest currently achieved resolution with VLBI techniques. Taking into account scatter-broadening of the image in the interstellar medium and the finite achievable telescope resolution, we show that the shadow of Sgr A* can be observed at suitably high frequencies. The main problems are possible optical depth effects for an ADAF model and Doppler boosting for a jet model. This has an influence on which dynamic range and which observing frequency is ultimately required to prove or disprove the existence of an event horizon.

INTRODUCTION

Many energetic events in the universe are considered to be connected with black holes, be it powerful radio jets, UV, X-ray, and γ-ray emission from quasars, stellar mass black hole candidates, or even Gamma-ray bursts (e.g., Pugliese et al. 1999; see also this volume). The best evidence for the existence of a supermassive black hole is found in our Galaxy (Eckart & Genzel 1996) where the compact radio source Sgr A* lies at the dynamical center of the central stellar cluster (Ghez, et al. 1998; Reid, et al. 1999; Backer & Sramek 1999). Sgr A* is surrounded by thermally radiating gas streamers (Sgr A West) spiraling into the nucleus (Zhao & Goss 1998) and a non-thermal radio shell, possibly a hypernova resulting from the tidal disruption of a star by the central black hole (Khokhlov & Melia 1996). The nature of the radio emission in Sgr A* is not clear; the latter may be due either to an outflow/jet (Falcke et al. 1993; Falcke & Biermann 1999) from the black hole or an inflow/accretion onto the black hole (Melia 1992,1994; Narayan et al. 1995).

CP522, *Cosmic Explosions: Tenth Astrophysical Conference,*
edited by Stephen S. Holt and William W. Zhang

Very Long Baseline Interferometry (VLBI) observations have confirmed that the radio emission of Sgr A* is very compact up to the highest frequencies (215 GHz; Krichbaum et al. 1998) and its radio spectrum at mm- and submm-wavelengths indicates the presence of an even more compact radio-emitting plasma component (Falcke et al. 1998; Serabyn et al. 1997). This is intriguing since the size of Sgr A* could be less than 17 Schwarzschild radii (0.11 mas at 215 GHz) for a black hole mass of $2.6 \times 10^6 M_\odot$ at a distance of 8 kpc. Of all the known black hole candidates, Sgr A* is the source where the angular size on the sky of its Schwarzschild radius is the largest. In fact, the angular resolution of ground-based VLBI experiments now comes interestingly close to the scale where significant general relativistic effects are important. We here report on calculations we have made with a general relativistic ray-tracing program which aim at clarifying what general relativistic effects might be realistically measurable in future VLBI experiments.

CALCULATIONS AND RESULTS

We consider an optically thin emission region with frequency-independent emissivity around a black hole with arbitrary spin. The intensity and structure of the emission region can be arbitrary as well, but we choose a number of generic scenarios where we have either a spherical distribution of the intensity scaling as a power-law with radius r or a jet-like distribution (hollow cylinder). We also can allow for various velocity fields, e.g., with rotation, inflow, or outflow. The appearance of the emission region for an observer at infinity, taking all general relatvistic effects into account, is then calculated using standard formalism (e.g., Thorne 1981; Viergutz 1993; Jaroszynski & Kurpiewski 1997).

To test whether general relatvistic effects would be visible we convolved the resulting images from the ray-tracing calculations with two Gaussian beams: one representing the scatter-broadening of the image due to the interstellar material along our line-of-sight towards the Galactic Center and one representing the finite resolution of VLBI with 8000 kilometer baselines. The width of the former has a ν^{-2} (e.g., Lo et al. 1998) and the latter a ν^{-1} dependence.

Regardless of the exact emission model we use, we find a characteristic structure

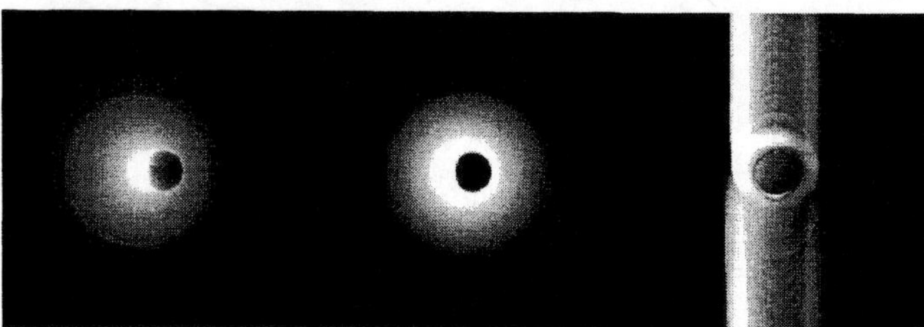

FIGURE 1. Images of the shadow of a black hole for rotating and non-rotating black holes and for spherical and jet-like emission models.

in all models: a bright ring of emission with a pronounced deficit of emission inside of that (Fig. 1). We call the deficit in the inner region the "shadow" of the black hole since it is caused by the deficit of photons emitted near the black hole that have disappeared into the event horizon or are bent away from our line of sight. The circumference of the shadow is determined by the 'photon-orbit'—a theoretical orbit where photons can circle the black hole an infinite number of times, but when perturbed may escape to infinity (Bardeen 1973). Interestingly, the size of this shadow is much larger than the event horizon—due to gravitational lensing—and is always of the order 10 R_g ($R_g = GM_\bullet/c^2$) for rotating and non-rotating black holes.

The exact intensity distribution of the bright ring depends significantly on the nature of the emission region, however. A rotating inflow would produce a slightly asymmetric ring due to Doppler-boosting of one side of the shells in Keplerian rotation. A jet would look even more asymmetric since boosting due to rotation plus fast outflow would enhance one quadrant of the ring (Fig. 1).

The relatively large size of the shadow is of particular interest for Sgr A*, since at a wavelength of around 1.3 mm the black hole shadow, the scattering disk, and the possible resolution of mm-VLBI become comparable. This is illustrated in Figure 2. It is clear that at wavelengths shortward of λ1.3 mm the shadow could actually be imaged with ground-based telescopes.

DISCUSSION

The possibility of seeing the effect of an event horizon is tantalizing. The shadow of the black hole in the Galactic Center is expected to have a diameter of $\sim 30\,\mu$as. The highest resolution so far achieved with VLBI is $\sim 50\,\mu$as. To achieve the additional improvement of a factor of two to three in resolution would require extending mm-VLBI to submm wavelengths. While this is difficult because of atmospheric effects, it is not technically impossible. Quite a number of submm-telescopes and arrays are currently under construction or consideration that could be used for such an experiment.

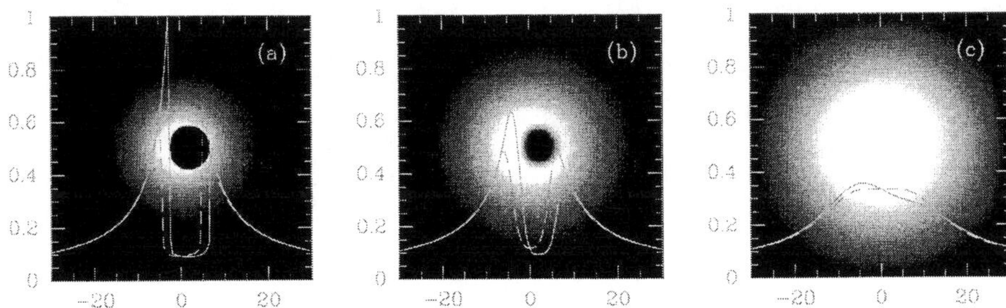

FIGURE 2. The expected shadow of Sgr A*. a) the ray-tracing simmulation; b) simmulated VLBI image at λ0.6mm; c) simmulated VLBI image at λ1.3mm (see Falcke et al. 2000 for details).

Another concern is whether the source itself could become an obstacle. Clumpiness of the emission may not be a major source of confusion because of the short rotation timescale of about 100 seconds. Optical depth effects could be a major problem. However, currently available submm spectra of Sgr A* (e.g., Serabyn et al. 1997) indicate a rather flat spectrum with a turnover towards the infrared. Hence at some wavelength between mm-radio and IR the source is bound to be optically thin. A second pitfall is anisotropic beaming. For example, in the jet model a quadrant is amplified due to relativistic beaming, making it more difficult to pick out the entire faint ring with low resolution or low dynamic range observations.

Without a better understanding of spectrum, structure, and nature of the emission it will be difficult to predict exactly at what wavelength the shadow will be unambiguously detectable and how much technical development still has to be done. In any case there is no reason to think that imaging the shadow is in principal impossible and any upcoming VLBI experiment at 1.3 mm and shorter wavelengths involving Sgr A* from now on could already show the first signs of the event horizon.

Acknowledgments This work was supported in part by a Sir Thomas Lyle Fellowship (FM), NASA grant NAG58239 (FM), DFG grants Fa 358/1-1&2 (HF), and NSF grant AST-9616922 (EA).

REFERENCES

1. Backer, D. C. & Sramek, R. A. 1999, ApJ, 524, 805
2. Bardeen, J. M. 1973, in Black Holes, ed. C. DeWitt & B. S. DeWitt, (New York: Gordon & Breach), 215
3. Eckart, A. & Genzel, R. 1996, Nature, 383, 415
4. Falcke, H. & Biermann, P. L. 1999, A&A, 342, 49
5. Falcke, H., Mannheim, K., & Biermann, P. L. 1993, A&A, 278, L1
6. Falcke, H., Goss, W. M., Matsuo, H., et al. 1998, ApJ, 499, 731
7. Falcke, H., Melia, F., Agol, E. 2000, ApJ, 528, L13
8. Ghez, A. M., Klein, B. L., Morris, M., & Becklin, E. E. 1998, ApJ, 509, 678
9. Jaroszynski, M. & Kurpiewski, A. 1997, A&A, 326, 419
10. Lo, K. Y., Shen, Z. Q., Zhao, J. H., & Ho, P. T. P. 1998, ApJ, 508, L61
11. Khokhlov, A. & Melia, F. 1996, ApJ, 457, L61
12. Krichbaum, T. P. et al. 1998, A&A, 335, L106
13. Melia, F. 1992, ApJ, 387, 25
14. Melia, F. 1994, ApJ, 426, 577
15. Narayan, R., Yi, I., & Mahadevan, R. 1995, Nature, 374, 623
16. Pugliese, G., Falcke, H. & Biermann, P. L. 1999, A&A, 344, L37
17. Reid, M. J., Readhead, A. C. S., Vermeulen, R. C. et al. 1999, ApJ, 524, 816
18. Serabyn, E., Carlstrom, J., Lay, O., et al. 1997, ApJ, 490, L77
19. Thorne, K. S. 1981, MNRAS, 194, 439
20. Viergutz, S. U. 1993, A&A, 272, 355
21. Zhao, J. -H. & Goss, W. M. 1999, in: "The Central Parsecs of the Galaxy", ASPC 186, eds. H. Falcke, A. Cotera, W.J. Duschl, F. Melia, M.J. Rieke, 224

The Eddington Luminosity Phase in Quasars: Duration and Implications

V.I. Dokuchaev[*1], Yu.N. Eroshenko[*], and L.M. Ozernoy[†‡]

*Institute for Nuclear Research of the Russian Academy of Sciences,
117312 Moscow, Russia
†George Mason University, Fairfax, VA 22030-4444, USA
‡Code 685, Laboratory for Astronomy and Solar Physics,
NASA/Goddard Space Flight Center, Greenbelt, MD 20771, USA

Abstract. Non-steady and eruptive phenomena in quasars are thought to be associated with the Eddington or super-Eddington luminous stage. Although there is no lack in hypotheses about the total duration of such a stage, the latter remains essentially unknown. We calculate the duration of quasar luminous phase in dependence upon the initial mass of a newborn massive black hole (MBH) by comparing the observed luminosity- and redshift distributions of quasars with mass distribution of the central MBHs in normal galactic nuclei. It is assumed that, at the quasar stage, each MBH goes through a single (or recurrent) phase(s) of accretion with, or close to, the Eddington luminosity. The mass distributions of quasars is found to be connected with that of MBHs residing in normal galaxies by a one-to-one corrrespondence through the entire mass range of the inferred MBHs if the accretion efficiency of mass-to-energy transformation $\eta \sim 0.1$.

INTRODUCTION

An approximate relationship $M_h \simeq (0.003 - .006)M_b$ between the central MBH mass, M_h, and that of the galactic bulge, M_b, has been established for a few dozen of galaxies, both nearby and more distant ones [1], [2]. A relationship between absolute magnitudes of quasars and their host galaxies found in [3] is reduced to the MBH to bulge mass relation in galaxies provided that [4]: (i) the central MBH shines at or near to the Eddington luminosity and (ii) the host galaxy undergoes through a starburst episode. This correlation, coupled with the known luminosity function of galaxies, can serve to obtain [5] the MBH mass distribution $\phi_1(M_h)dM_h$. The history of matter accretion onto a central MBH thought to serve as a source of quasar activity is linked to the present observable properties of each individual quasar, such as its luminosity, variability, and emission spectrum. If the bolometric luminosity of a quasar comprises a fraction λ of the Eddington luminosity, i.e.

[1)] E-mail ID: dokuchaev@inr.npd.ac.ru

CP522, *Cosmic Explosions: Tenth Astrophysical Conference*,
edited by Stephen S. Holt and William W. Zhang

$\lambda = L/L_E$, $L_E = 4\pi GM_h m_p c/\sigma_T$, the underlying accretion is accompanied by an exponential growth of the MBH mass with the characteristic time $t_E = 4.5 \cdot 10^8 \eta/\lambda$ yrs, where η is the accretion efficiency of mass-to-energy transformation. The crucial problem is the duration, t_q, of such a nearly Eddington accretion phase. Usually an effective t_q, the same for the entire black hole mass range $M_h \sim 10^6 - 10^{10} M_\odot$, is calculated by comparing the global number density of normal galaxies and quasars and is found to be $t_q = 10^6 - 5 \cdot 10^8$ years in [6]. Meanwhile the recent data on mass distribution of MBHs in galaxies provide an opportunity to solve this problem in a more detailed way, *viz.*, to calculate the dependence of t_q upon M_h, which is the major aim of this paper. We shall explore whether the distribution functions of quasars and MBHs in normal galaxies are consistent with each other, and we will do this locally in the vicinity of each mass.

THE EDDINGTON LUMINOSITY PHASE

It would be reasonable to assume that the duration of the Eddington phase t_q depends on the initial mass of a newborn MBH or, in other words, on the initial luminosity, L_i, of the quasar: $t_q = t_q(L_i)$. For simplicity, the transition to and out of the Eddington phase is supposed to occur instantaneously:

$$
L = \begin{cases}
0, & \text{if } t < t_i; \\
L_i \exp[(t - t_i)/t_E], & \text{if } t_i < t < t_i + t_q(L_i); \\
0, & \text{if } t > t_i + t_q(L_i),
\end{cases} \tag{1}
$$

where t_i is the instant of the MBH formation. Along with the distribution function of MBHs in the galactic nuclei, ϕ_1, we use the observed distribution of quasars in absolute magnitude M_B and redshift $z \leq z_e \sim 3$, $\phi_2(M_B, z)dM_B dz$, taken from [7]. The balance equation is given by

$$
\frac{2.5}{\ln 10} \int_0^\infty dz(1 + z)^{-3/2} \phi_2(M_B(L), z) \simeq \frac{t_E(L)}{t_0} \int_{M(L)}^X \phi_1(M_h)dM_h, \tag{2}
$$

where $X = M(L)\exp[t_q(L)/t_E]$ with $M(L)$ determined by equation $L = \lambda L_E$ and a relationship $t = t_0/(1 + z)^{3/2}$ for the flat cosmological model is used. While obtaining Eq. (2), we have also taken into account that t_E very slowly varies with respect to $\phi_1(M_h)$. Eq. (2) defines t_q as an implicit function of L_i, which can be translated into a relationship between t_q and the initial BH mass M_i by using equation $L = \lambda L_E$.

RESULTS

Numerical solution of Eq. (2) is found by adopting the mass distribution of MBHs $\phi_1(M_h)dM_h$ from [5], derived with the use of three relationships, *viz.*, (i)

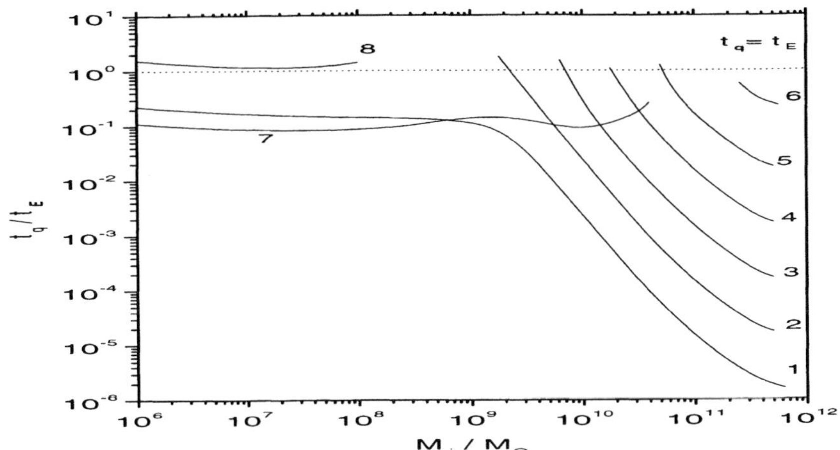

FIGURE 1. The ratio $t_q(M_i)/t_E$ as a function of the initial MBH mass M_i. Curves labeled 1 to 6 are based on distribution A for $\eta = 10^{-1}$, 10^{-2}, 10^{-3}, 10^{-4}, 10^{-5}, and 10^{-6}, respectively. Curves 7 and 8 are based on distribution B for $\eta = 10^{-1}$ and 10^{-2}, respectively.

a correlation $\log(M_h) = \log(M_b) - 2.6 \pm 0.3$ between the MBH mass M_h and the bulge mass M_b; (ii) the mass-luminosity relation for galaxies, and (iii) the Schechter luminosity function. In Fig. 1 and Fig. 2, we employ two somewhat different distributions in MBH mass from [5] and name them 'distribution A' and 'distribution B', which correspond to the power-law and log-Gaussian shape of dispersion, respectively. Fig. 1 presents the results of our numerical computation of the ratio $t_q(M_i)/t_E$ for different values of η and MBH mass distributions A and B. It should be noted that if $\eta < 0.1$, the solution exists not for all values of M_i. The domain where the solution exists is determined from the condition that the r.h.s of Eq. (2) exceeds its l.h.s. if one puts $X = +\infty$. For those M_i which lead to an opposite condition, the number of galactic nuclei with MBHs is not enough to explain, in the framework of our model, the distribution function of quasars in M_B and z, even if these MBHs stay in an active quasar state during the maximum possible time $t_q \sim 3t_E$. The solution only exists at $\eta > 7 \cdot 10^{-7}$ for BH mass distribution A and at $\eta > 6 \cdot 10^{-3}$ for distribution B. A single-valued mapping $M_i \to M_h$ breaks up on the left end of curves 2 to 6.

Fig. 1 demonstrates the main result of this work: *distributions of MBHs and quasars in mass are connected by one-to-one correspondence through the whole range of the observed masses only for $\eta \sim 0.1$*, both for the distribution A (curve 1) and B (curve 7). This concordance breaks down for a certain range of MBH masses, *viz.* the solution of Eq. (2) does not exist if $\eta \ll 0.1$. Nevertheless, the jumps of the η value are not excluded on the boundary of the domain, where the

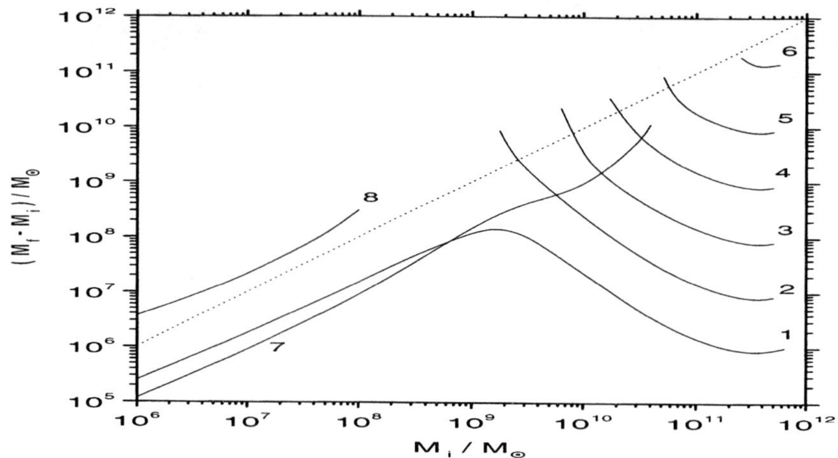

FIGURE 2. The accreted mass, $M_f - M_i$, as a function of the initial mass M_i. The curves are labeled in the same way as in Fig. 1.

solution exists. Similar jumps seem to be quite natural if MBHs in the different mass ranges are formed by different ways (e.g., by collapse of massive gas clouds, stellar clusters, etc., see review by Rees 1984) and so there are various accretion regimes with different values of η. If such jumps indeed take place, transitions between the curves of each of distributions A and B are possible. These transitions must be smoothed because the MBHs formed by different ways would coexist in some mass range(s). The *most probable value of η established above is $\eta \sim 0.1$*, which corresponds to curves 1 and 7 in Fig. 1. For both these curves, the relationship $t_q < t_E$ is carried out and therefore the BH mass growth is not substantial – it generally does not exceed a value comparable to the initial BH mass, and for $M_h > 10^7$ M$_\odot$ it is negligible.

REFERENCES

1. Kormendy J., Richstone D., 1995, ARA&A, 33, 581
2. Magorrian J. et al., 1997, astro-ph/9708072
3. Bahcall J.N., Kirhakos, S., Saxe, D.H., Schneider, D.P. 1997, ApJ 479, 642
4. Ozernoy L.M. 1998, Bull. Amer. Astr. Soc. 30, 1286
5. Salucci P. et al., 1998, astro-ph/9811102, astro-ph/9812485
6. Haehnelt M.G., Natarajan P., Rees M. J., 1997, astro-ph/9712259
7. Boyle B.J., 1991, in Proc. Texas/ESO–CERN Symp. on Relativistic Astrophysics, Cosmology, Fundamental Physics, ed. Barrow J.D., Mestel L., Thomas P.A., p. 14

The Giant, Ultra-Soft, and Luminous X-Ray Outburst From the Optically Inactive Galaxy Pair RX J1242.6-1119: Flare of a Tidally Disrupted Star ?

Stefanie Komossa[1] and Jochen Greiner[2]

[1] *Max-Planck-Institut für extraterrestrische Physik, Giessenbachstr., D-85748 Garching;*
skomossa@xray.mpe.mpg.de
[2] *Astrophysikalisches Institut Potsdam, An der Sternwarte 16, D-14482 Potsdam*

Abstract. We discuss our detection of a giant X-ray flare ($L_x \gtrsim 10^{44}$ erg/s) from the direction of the previously unknown, optically inactive galaxy pair RX J1242.6-1119 and investigate outburst scenarios. The huge peak luminosity makes RX J1242.6-1119 an excellent candidate for a tidal disruption event that occurred at the center of one of the two galaxies.

INTRODUCTION

Supermassive black holes (SMBHs) are expected to reside in the centers of many or all galaxies, and a lot of effort has been put in deriving masses of central dark objects in nearby galaxies in the last few years (see [1] for a recent review). One very efficient way to probe the *direct vicinity* of these SMBHs is to detect the flares of stars that get tidally disrupted by the SMBHs [2–4]. A few very good candidates for these kind of events in nearby, *non-active* galaxies have emerged in recent years (e.g., [5,6]; see [7] for a review). Here, we extend our discussion [8] of another such candidate: the giant X-ray flare from RXJ 1242.6-1119.

X-RAY AND OPTICAL OBSERVATIONS

The X-ray source RXJ 1242.6-1119 showed strong variability between two *ROSAT* observations taken 2 yrs apart: Whereas the source was not detected during the *ROSAT* all-sky survey (countrate $CR < 0.015$ cts/s), it is present in a pointed PSPC observation with $CR = 0.3$ cts/s, revealing variability by at least a factor of 20. The high-state spectrum is ultra-soft ($kT_{\mathrm{bb}} = 0.06$ keV when fit by a black body model). We derive an intrinsic luminosity of $L_x > 9\,10^{43}$ erg/s (corrected for cold absorption) in the (0.1–2.4) keV *ROSAT* band, using $H_0 = 50$

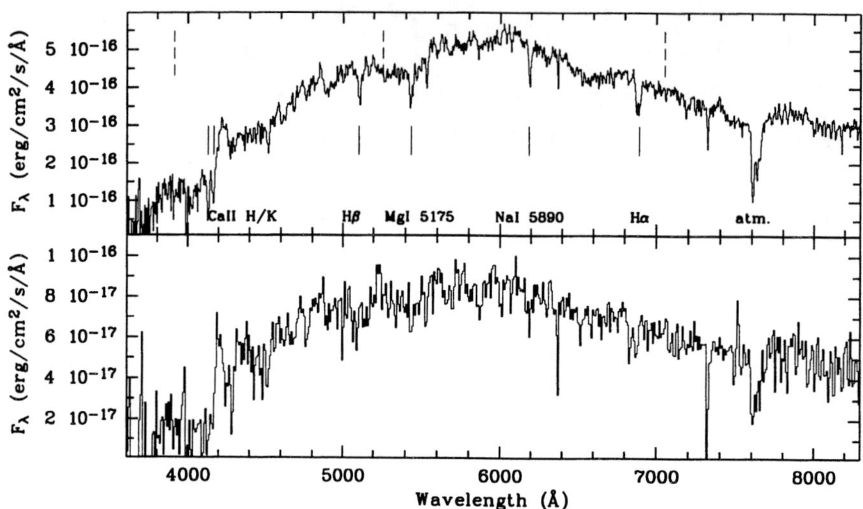

FIGURE 1. Optical spectra of the two galaxies within the X-ray error circle. Prominent absorption lines are marked, together with the expected locations of undetected emission lines of [OII]λ3727, [OIII]λ5007 and [SII]λ6717 (dashed lines, from left to right).

km/s/Mpc. Since we most likely have not caught the source exactly at maximum light, since the spectrum probably extends into the EUV, and since we have conservatively assumed no X-ray absorption intrinsic to RXJ 1242-11, the real peak luminosity is likely to be much higher. The X-ray properties of RXJ 1242-11 are summarized in Tab. 1.

Optical imaging shows two interacting galaxies in the X-ray error circle. Other objects are fainter than $\sim$22^{m}. Our optical spectroscopy performed several years after the X-ray high-state observation reveals that both galaxies are *non*-active (Fig. 1); no Seyfert-typical emission lines are detected (see [8] for details). We derive redshifts of z=0.05 for the two galaxies.

TABLE 1. Summary of the X-ray properties of RXJ 1242-11 during the giant flare. Γ_{x} is the photon index derived from a powerlaw spectral fit, T_{bb} the black body temperature for a black body fit. In both cases, we fixed the cold absorption to the Galactic value in the direction of RXJ 1242-11. L_{x} gives the intrinsic luminosity in the (0.1–2.4) keV band.

date of observation	N_{gal} [10^{20} cm^{-2}]	kT_{bb} [keV]	$L_{\mathrm{x,bb}}$ [erg/s]	Γ_{x}	$L_{\mathrm{x,pl}}$ [erg/s]
15–19/7/1992	3.7	0.06±0.01	8.8 10⁴³	−5.1±0.9	35.5 10⁴³

OUTBURST SCENARIOS

Both, the dramatic X-ray variability and the huge outburst luminosity of RXJ 1242-11 are very rare among *non-active* galaxies. The only previous cases are the UV outburst in NGC 4552 ([9]; but note that the authors find several hints for weak permanent activity in this galaxy) and the X-ray outburst of NGC 5905 [5,6]. Another candidate has been recently presented [10]. In the following, we discuss several potential outburst scenarios (see [8] for details):

- *Variable galactic foreground star:* This is unlikely, since further optical sources within the X-ray error circle are extremely weak. Given the high $L_{\mathrm{X}}/L_{\mathrm{opt}}$ value, an ISM accreting neutron star might come to mind, but the strong X-ray variability would require an extreme ISM density gradient, thus leading us to reject this possibility.

- *Stellar sources within the galaxy pair* (like X-ray binaries, or a supernova in dense medium) fall several orders of magnitude short in explaining the huge outburst luminosity. For example, the most luminous X-ray supernovae observed typically reached $L_{\mathrm{x}} \simeq 10^{39-40}$ erg/s (e.g., [11] and references therein).

- *X-ray afterglow of a GRB:* This is an unlikely explanation, since the "on" timescale of several days of RX J1242-11 is much too long as compared to all known cases of GRB afterglows which quickly faded with a t^{-1} law after detection.

- *Accretion disk instability:* If an accretion disk is present in the system, the accretion rate would have to be rather low, since there is no multi-wavelength evidence for an AGN. The disk may then settle into the advection dominated (ADAF) mode. An instability in such a disk around a SMBH at the center of one of the two galaxies may then lead to burst-like variability. However, in case of repeated such outbursts in RX J1242-11, one would expect to see permanent AGN-typical NLR emission lines (like [OI] or [OIII]) in the optical spectrum of one of the two galaxies which are not detected.

- *Accretion event onto a SMBH:* The giant peak luminosity of at least $\sim 10^{44}$ erg/s in the (0.1–2.4) keV energy band strongly suggests that we have seen an accretion event, most likely the flare of a star that was tidally disrupted by a SMBH [2–4] at the center of one of the two galaxies.

Such X-ray outbursts provide important information on the presence of SMBHs in non-active galaxies, the accretion history of the universe, and the link between active and normal galaxies. Future X-ray surveys (like the one that was planned with *ABRIXAS*, or the one that will be carried out with *MAXI*) will be valuable in finding further of these outstanding sources.

Acknowledgements: It is a pleasure to thank L.M. Ozernoi for discussions. The *ROSAT* project has been supported by the German Bundesministerium für Bildung und Wissenschaft (BMBW/DLR) and the Max-Planck-Society. JG is supported by BMBF/DLR under contract No. FKZ 50 QQ 9602 3 and expresses gratitude for support from the organizers and DFG grants KON 1973/1999 and GR 1350/7-1. This and related papers can be retrieved from http://www.xray.mpe.mpg.de/~skomossa/

REFERENCES

1. Kormendy J., Richstone D.O., 1995, ARA&A **33**, 581
2. Lidskii V.V., Ozernoi L.M., 1979, Sov. Astron. Lett. **5(1)**, 16
3. Rees M.J., 1988, Nature **333**, 523
4. Rees M.J., 1990, Science **247**, 817
5. Bade N., Komossa S., Dahlem M., 1996, A&A **309**, L35
6. Komossa S., Bade N., 1999, A&A **343**, 775
7. Komossa S., Greiner J., 2000, in *ASCA/ROSAT workshop on AGN and the X-ray background*, ISAS Report, T. Takahashi & H. Inoue (eds), in press; http://www.xray.mpe.mpg.de/~skomossa/
8. Komossa S., Greiner J., 1999, A&A **349**, L45
9. Renzini A., Greggio L., Di Serego Alighieri S., Cappelari M., Burstein D., Bertola F., 1995, Nature **378**, 39
10. Grupe D., Leighly K., Thomas H.-C., 1999, A&A **351**, L30
11. Schlegel E.M., 1999, ApJ **527**, L85

On the Possibility of an Elastic Space Model of the Metagalaxy

Valeriy P. Polulyakh

Continuum, 3150 Central Expw., Santa Clara, CA 95051-0816

Abstract. In the proposed model an inhomogeneous transition of compact ($\Lambda > 0$) space in loose 3-dimensional space (decompactification-DC) with appropriate creation of matter is a factor which determines the evolution of Met. The "elasticity" of space is a driving force of the DC. The DC start in a single causally connected domain and develops in oscillating quasi-solitons (elastons). The apparent global expansion of the Met is a summary outcome of a local expansion of space into a large number of elastons. Thus, all blazars, quasars, and AGNs have the identical central "engines" that are elastons at different stages of the evolution, and a supernova is a final burst of the elaston's energy. The radio-loud quasars and galaxies are thus in the process of elaston decay. The calculated average distance for QSO $\Delta\ln(1+z)=0.23$ is consistent with observations. A mass of a typical star is simply determined by the condition that density of energy on a surface of the "star-elaston" was nuclear $M_{star} \approx c^3 (32\pi G^3)^{-1/2} \rho^{-1/2}_{nucl}$. The particle horizon coincides with the event horizon and has temperature 3° K.

INTRODUCTION

Apparently, the structure and population of the Met are the clues for any of the current models proposed. For the sake of explanation theorists use a variety of hypotheses. The basic principles of the HBBM-birth of Met in the past along with its evolution are firmly established while the hypotheses regarding the instantaneous and homogeneous creation of matter evidently need revision [1].

THE MODEL

The wave function of the universe $\Psi[\mathrm{h}_{ij};\Phi]$ which measures the probabilistic correlation between the 3 dimensional geometry h_{ij} and the matter field Φ is a solution of the Wheeler-DeWitt equation [2]

$$(-1/2\nabla^2 + \mathrm{W})\Psi[\mathrm{h}_{ij},\Phi] = 0, \quad (1)$$

which plays a role analogous to that of Schrodinger or Klein-Gordon (KG) equations in a flat space-time model. It is known [3] that for Schrodinger equation

CP522, *Cosmic Explosions: Tenth Astrophysical Conference,*
edited by Stephen S. Holt and William W. Zhang
© 2000 American Institute of Physics 1-56396-943-2/00/$17.00

$$\Psi_{xx} + [U(t, x) + \epsilon]\Psi = 0, \quad (2)$$

where $U(t, x)$ is a finite positive potential energy which depends on time, t, as a parameter, eigenvalues ϵ do not depend of t if U satisfies the Korteweg-de Vries (KdV) equation

$$U_t + \alpha UU_x + U_{xxx} = 0. \quad (\alpha \text{ is a constant}) \quad (3)$$

Any positive initial perturbation U>0 in a finite space domain, in accordance with the KdV equation will divide into an aggregate of separate solitons. The resulting direct scattering problem requires the solution of eq.(2) to yield the potential U (0, x). As a result, one gets a spectrum of eigenvalues $\epsilon_n(t)=\epsilon_n(0)$ and hence steady energy levels and number of solitons.

Let us consider a causally connected domain in the compact space, which has positive energy and no particles. The evolution of this "bubble" results in appearance of a set of solitons whose energies are eigenvalues of eqn. (1). That is, some of solutions $W(h_{ij},\Phi)$ (from eqn. (1)) can exist in the form of solitons. The soliton possess a field configuration different from that expected for a vacuum, and which is described by a wave equation solution localized in space and having finite energy. They have energy (we will call it elastic) at the expense of compacticity of space itself, and this is not the energy of condensed matter or matter fields. Then, eqn. (1) describes nonuniform Met in superspace h_{ij} and one cannot use a one-dimensional factor, as it is the case in the Friedman-Lemaitre model.

It is known that many Lorentz invariant differential equations which have solutions that are solitons can have the form $\Box\Psi=F(\Psi)$, for example systems with either degenerate vacuum, Higgs $\Phi_-{}^4$ [$F(\Psi)= -\Psi+\Psi^3$], or Sine-Gordon (SG) [$F(\Psi)= -\sin\psi$] properties [4]. In space of more than one dimension, there is no stable steady state solution of Lorentz invariant non-linear field equations [5]. However, this does not rule out an "almost-stable" finite energy oscillating solution of "quasi-solitons" which may have some internal time dependent structure. For a real 4-dimensional world only numerical methods can be used to obtain solutions for the formation and interaction of quasi-solitons. The computer solutions [4] show features surprisingly similar to the real population of the Met. For the bubbles and kinks in two- and three-dimensional space, the SG and $\Phi_-{}^4$ models manifest behavior very similar to that of a pendulum in the presence of friction. During an oscillation some of the energy of the bubbles is lost by radiation (to infinity). Small bubbles disappear altogether, but bubbles with large initial energy can form some new, long-lived objects called pulsons. In the center of such an object one finds localized field energy having a bell-shaped radial distribution. In the first stage the formation of the bubble the pulson occurs. During the long second stage there are regular oscillations about fixed vacuum and growth of the pulson dimension. In the third stage, the amplitude of the pulson decreases rapidly as it expands. The fourth stage is characterized by a small amplitude expansion. In some models instability leads to the quasi-soliton breaking up into pieces, and its energy dispersing, an "anticollapse". Perturbation leads to decay into constituent solitons. A soliton-like solution corresponds to quasi-stationary Met, with slow evolution on the average and at the same time accompanied by the rapid alteration of the nature of some cosmic objects and in their periodicity.

It is natural to link the distortion of an empty space with the elastic energy, which manifests itself via the initial curvature that is Λ. In this way, Λ is a measure of the initial elastic energy of an empty space. Suppose, for example, that the energy of the initial compact domain is elastic one. The curvature will drop out because p<0, which leads to a DC which is a creation of quasi-flat space. The final state of the elastic space (for $t \rightarrow \infty$) is a flat space, which will have zero energy.

The hypothesis consists of the following: The creation of matter is a damper (or viscosity) of the DC and elastic energy converts into matter (the particles) with an effective rate of $k \approx (c^3/8\pi G)[(\rho_c - \rho)/\rho_c]^n$, where ρ_c and ρ are the current critical and real density of matter respectively for n>0. So, for $t \rightarrow \infty$, $\rho \rightarrow \rho_c$, and $k \rightarrow 0$ (since the tendency of space will be to become flat) one observes $\Lambda \rightarrow 0$. When all energy changes into matter we get exactly $\Omega = \rho/\rho_c = 1$, but here $\rho_c = \rho + \rho_{elast} = \rho + (\Lambda c^2/8\pi G)$. For $t \rightarrow \infty$ the Met transforms from an empty state with high elastic energy into an Eucledian geometry world in which the density of matter $\rho = \rho_c$.

Let us introduce the following definition: The elastons are the long-living concentrations of an elastic energy of space described by a quasi stationary solution of a wave equation for $\Lambda(t, x, y, z,)$: $\Box\Lambda = P(\Lambda)$ and they serve as the birthplaces for the particles as well as for free Eucledian space. The elastons have energy $(c^4/8\pi G)\int \Lambda dV$. As far as Λ, R, p, ρ, have soliton-like dependence on (t, x, y, z) then the observable values z, H also in general are not stationary and not uniform. The apparent global expansion of the Met is a summary outcome of a local expansion of space into a large number of elastons. Thus co-moving coordinate system exists with a non-uniform soliton-like expansion. The character of the expansion is determined by a rate $\partial\Lambda/\partial t$. In linear approximation, during the formation of free flat space the "resistance force" is approximately given by $c^4/8\pi G$.

To make up a 3-dimensional sphere with a radius R_0, the elasticity of space should perform work $A \approx (c^4/8\pi G)R_0$ against the prevailing viscosity. In accordance with the hypothesis, matter with energy $E_M \approx A$ is created from the expenditure of this work. For the Met with $R_0 \approx 10^{26}$m we have energy $E \approx 10^{69}$ J which coincides with the energy of the universe's visible matter. On the other hand, for $\rho << \rho_c$, $k \approx c^3/8\pi G$, and designating T_M as an age of the Met we also get $M_M \approx (c^3/8\pi G)T_M \approx (c^2/8\pi G)R_0$, and correspondingly get $T_M \approx 10^{10}$ years. The average density of matter is $\rho \approx (3c^2/64\pi^2) R_0^{-2} \approx 6.4*10^{-31}$g sm^{-3}. Why was the Met created as non-uniform and loaded heavily with elastons? To answer this, let us calculate the size domain we need to accommodate the required elastic energy for $\rho = \rho_c$ if curvature Λ is uniform ($\Lambda \sim R_0^{-2}$). From $(c^4/8\pi G) \int \Lambda dV \approx 10^{70}$ J we have $R \approx 10^{28}$ m. For a non-uniform distribution of Λ the domain may be very small and the energy density of the elastons achieves a particular value when the quantum mechanism of a particle creation begins. Evidently, the Met, which started out as uniform, could not have had as many accessories as it does now.

Within the framework of the model proposed it is natural to treat star formation as an evolution of elastons. To develop some mechanism for barion-genesis we need an elastic energy density on the order of nuclear $\approx 10^{14}$g/sm^3. It is a case for elaston with $R_{elast} \sim 10$ km. For this elaston the energy density is $\approx 5*10^{14} g/sm^3$ and the mass $M_{elast} \approx 2*10^{33}$ g, which is the mass of a typical star, such as our Sun. Since $\rho_{elast} \sim 1/R^2_{elast}$ and $M_{star} \sim R_{elast}$, the mass of a star is simply determined by the condition that density of an energy in the "star-elaston" is nuclear, and $M_{star} = c^3/12(2\pi G^3)^{1/2} \rho_{nucl}^{1/2} \approx \rho_{nucl}^{-1/2}$. For $\Lambda \approx r^{-2}$ one can limit integration over r where $\rho \approx \rho_{nucl} \approx 5*10^{14}$ g/sm^3. This distance is the radius of the star-elaston $r = R_{elast}$. The energy of this elaston is 3 times more then for $\Lambda = const = 1/R_{elast}^2$, and the mass of the newborn star is Mstar $= \int \rho dV \approx c^3(32\pi G^3)^{-1/2} \rho^{-1/2}_{nucl} = 7*10^{33}$ g. If α^{-1}, Λ_0 are the width and amplitude respectively then for $\Lambda(r) = \Lambda_0$ $ch^{-2}(\alpha r)$ we have $M_{star} \approx (c\pi)^2 \Lambda_0/24 G \alpha^3$.

In the inflation scenario [6] the universe is bounded, but this boundary is hidden beyond the event horizon. It follows from the existence of the bounded causally connected domain at the beginning of the current universe and finite T_0 that the Met is bounded. The event and particle horizons coincide: $R_{ev} = R_{part} = R_0$. It is easy to see that the horizon in our model has attributes of a black body (BB). Indeed, a photon $h\nu$, which has a velocity in the direction toward the horizon, does not make its way through it because for $r > R_0$ the space is incoherent with ours, and at the same time there is no reflector in the point R_0. Consequently, a photon (for any ν) will be "absorbed" in the vicinity of the horizon. If the horizon could absorb radiation as a BB it should radiate as BB with the temperature T_H. Thus the radiated power will be $P_H = 4\pi\sigma R_0^2 T_H^4$, where σ is the Stefan-Boltzman constant. If an equivalent radiative mass density is ρ_r we can write $P_H = (\rho_r/\rho)*kc^2$. Equating the two expressions we get $T_H^4 \approx (\rho_r/\rho) kc^2/4\pi\sigma R_0^2$. An experimental value for the background radiation equivalent mass density (BR) is $\rho_r \approx 5*10^{-34}$ g sm^{-3} and from this one can derive $T_H \approx 3$ ^0K. Thus, the apparent temperature T_H (reduced to modern epoch owing to ρ_r) represents a satisfactory means of measuring the temperature BR. The apparent dipole anisotropy of BR with $\Delta T \approx 10^{-3}$ K may be connected with our position in the Met and a motion relatively to the horizon.

The hierarchical system of clusters, superclusters and voids [7] points at the multi-stage decay of the elastons. A majority of the quasars, and about 70% of the radio galaxies are in binary systems. Seemingly, the energy of elastons decreases owing to falling apart. The elastons are situated inside the nuclei of these objects. Let us suppose that a period between decay of elastons is T and $N_0 \approx 2^{t/T}$, then after n decays, we get $N_1 \approx 2^{(t+nT)/T}$. Since the average distance between elastons is $R_i \approx (N_i)^{-1/3}$, then $R_0/R_1 \approx 2^{n/3}$, where R_0 is the initial average distance. If one also takes into account that $R_0/R_1 = 1+z$ we have ln $(1+z) = $ ln $(2^{n/3}) = (n/3)$ln 2, or Δln $(1+z) = (1/3)$ln $2 = 0.23$. Arp [8] measured average distance between quasars $0.2 < \Delta$ln $(1+z) < 0.25$, and for the Virgo Cluster Δln $(1+z) = 0.23$.

There is no direct evidence from radio and optical data of the infall into any cosmic objects; only outflows are observed. The absolutely different cosmic population shows very similar morphology. A bipolar axisymmetric structure and jets are ubiquitous in the Met: from quasars and radiogalaxies to supernovae, planetary nebulae, Herbig-Haro objects and YSO. For instance, radio image of galaxy Cygnus

A and jets from young star HH 1 / 2 are stunningly similar. It is reasonable to suppose an identical mechanism for all referred phenomena. The fragmentation of the elastic energy in this model is a continuous process. The energy of the initial domain broken into pieces. The more "fundamental" or "antique" elastons, of existing now, are those in the centers of cD galaxies. The elastons ejected from cD evolve into galaxies, which form clusters.

What repels newborn blobs in jets away from the elaston? One of the reasons may be the gradient force analogous to one that repels a diamagnetic from the non-homogeneous magnetic field. Indeed, with a sudden local creation of matter a blob of space with Λ that is less than in the surrounding elastic space is also created. Owing to the non-homogeneity of the elastic space inside of the elaston, the repulsive force which acts on the volume dV with elasticity $\Lambda(t, r,\theta,\varphi)$ is $d\mathbf{F}\approx -(c^4/8\pi G)(1- \Lambda(t, r,\theta,\varphi) / \Phi(t, r,\theta,\varphi))$ dVgrad Φ. Here $\Phi(t, r,\theta,\varphi)$ is a distribution of Λ in the elaston. A trajectory of the blob (Jet) is directed opposite to gradΦ and force depends of the amount of created matter. Each blob is ejected from elaston with its own position angle. The portions of elaston created at different time are seen as knots in the jet.

A blob that has been just ejected is a "naked" elaston. A gaseous cloud and dust do not surround it yet. We can see its "private life" with the rapid variability of radiation and γ-ray bursts. Then to the extent of growing of gaseous shell it turns into quasar. During the fragmentation of a nuclear elaston on "star" elastons we are observing the AGN. Some of AGNs (Seyfert's, for instance) still have enough energy to eject new "galactic" elastons. Astronomers now think of all these processes of birth and decay of elastons in terms of collision or the swallowing of one galaxy by another. The stage of star formation is accompanied by the gas and dust creation on all scales from new galaxies to Herbig-Haro objects. It is exactly a creation because we can see only outflows. To explain the mechanism of this creation is the goal of a future detailed physical theory. The radio-loud quasars and galaxies are thus in a process of elaston decay. During the intervening periods between decays they are radio-quiet.

Thus, all blazars, quasars, and AGNs have the identical central engines that are elastons at different stages of the evolution, and a supernova is a final burst of the elaston's energy. This is confirmed by an increasing burst frequency for a supernova in E through Ir galaxies by a factor of about one hundred.

The apparent superluminal motion [9,10] can be explained in this model by the creation of matter in a co-moving coordinate system. New matter does not move through space with super-relativistic velocity after birth. The distance between blobs can increase faster then light because we see an expansion of space itself. In case of blazars, which have very energetic elastons, this phenomenon is commonplace, but in our Galaxy only a few "microquasars" are known to have superluminal jets [10]. The superluminal jet from XTE J1748-288 was stopped suddenly because elastic energy in the blob has exhausted and turned into matter. In the center of the Galaxy probably the residual elaston settled down after it generated Galaxy itself. The process of the creation of matter and the expansion of compact space can takes place in different parts of a jet, where elastons are experiencing repulsive forces. This can explain the doubling of components, with rapid brightening and acceleration of their motion [11].

In a system that is at DC, when all spatial scales, including λ, grow, the radiation leaving the elaston will undergo a wavelength increase i.e., it will redden. The deeper in elaston the creation of the photon takes place the greater will also be z of the emitted radiation. This is a local redshift, $z_{loc} \sim f\,[\Lambda(r)]$, where r is a position of the birthplace of the photon. Since photons can be created in different depths, the spectral lines experience the broadening. If different elements were born in differing depths than their spectra have various values of z. The absorption lines were formed in the outer, more dense and cold, layers and therefore they have smaller z then emission spectra.

The creation of the free space in elastons forms the general expansion of the Met, $dV_{Met}/dt \sim k\partial(\int \Lambda dV)/\partial t$, which determines the global parameter, z_{glob}. Thus the apparent redshift is $z = z_{loc} + z_{glob}$. The expansion of the elastons may be not spherical therefore, in local scale, it may also give us quite a complicated and inhomogeneous picture of the redshift. There are many of known close associations between QSO having large redshifts and galaxies with small redshifts. Since the measurement of H uses the dependence $H \sim z$, then it follows that $H = H_{loc} + H_{glob}$. The age of the Met then is found to be $T_0' = 1/H = 1/(H_{loc} + H_{glob})$, less than $T_0 = 1/H_{glob}$.

CONCLUSIONS

In this paper I have attempted to modify the BBM in order to explain some well established facts: the similarity of cosmic jets on the different spatial scales, the evolution of galaxies and quasars, the isotropy of the measured cosmic background radiation. In order to do this I proposed a hypothesis that relates the creation of Euclidean space and matter. This hypothesis stands for, in fact, the presence of domain with positive elastic energy $c^4/8\pi G \int \Lambda(x,y,z,t)dV$ that may be transformed in particles in a process of drop of Λ. The initial domain of compact space ($\Lambda > 0$) was similar to a squeezed spring. After being released it broke up into an aggregate of separate domains (the elastons) which are the birthplaces for the particles as well for free ($\Lambda = 0$) Euclidean space. This process successively leads to appearance of QSO, galaxies, supernovas and stars.

In this way the elastons are the primary formations but the material objects are the derivatives.

The initial homogeneity of matter postulate is not needed in this new model since it includes an independent explanation for the isotropy of the CBR.

The distinguishing features of this model are:

1. The matter in the Met has never been homogeneous.

2. A factor, which determines the elastic energy of space, is a soliton-like function of space and time $\Lambda(x, y, z, t)$, here called elastons.

3. At different stages of their evolution (depending on the content of energy) the elastons give birth to a set of various cosmic objects at the expense of decay, ejection, slow expansion and anticollapse.

REFERENCES

1. Polulyakh, V.P., *astro-ph/9910305*.
2. De Witt, B. S., *Phys. Rev.* **160**, 1113 (1967).
3. Gardner, C. S., Green, J.H., Kruskal, M.D., & Miura, R.M. *Phys. Rev. Lett.* **19**, 1095 (1967).
4. Makhankov. V.G., *Soliton Phenomenology*, Boston: Kluwer Acad. Publ., 1990.
5. Derric, G. J., *Mth. Phys.* **5**, 1252 (1964).
6. Linde, A. D., *Particle Physics and Inflationary Cosmology*, NY: 1990.
7. Einasto, J., et al., *Nature* **385**, 139 (1997).
8. Arp, H. C. et al., *Astron. Astrophys.* **239**, 33 (1990).
9. Krichbaum, T.P. et al., *A & A* **230**, 271 (1990).
10. Mirabel, I. F., & Radriguez , L.F., *Nature.* **371**, 46 (1994).
11. Biretta, J. A. et al., *ApJ* **308**, 93 (1986).

Novae and X-Ray Bursts

Thermonuclear Runaways on Accreting White Dwarfs: Models of Classical Novae Explosions

Margarita Hernanz and Jordi José

Institute for Spatial Studies of Catalonia (IEEC/CSIC/UPC), Edifici Nexus-201, C/Gran Capità 2-4, 08034 Barcelona (SPAIN).

Abstract. The mechanism of classical novae explosions is explained, together with some of their observational properties. The scarce but not null impact of novae in the chemical evolution of the Milky Way is analyzed, as well as their relevance for the radioactivity in the Galaxy. A special emphasis is given to the predicted gamma-ray emission from novae and its relationship with the thermonuclear model itself and its related nucleosynthesis.

INTRODUCTION

Classical novae are a very common type of cosmic explosion. They occur very often in the Galaxy (~ 35 yr^{-1}), although only $\sim$3-5 are discovered by amateur astronomers in the Galaxy every year. The explosion occurs on the top of white dwarfs accreting mass in a cataclysmic binary system and it is related to degenerate hydrogen ignition, leading to a thermonuclear runaway. Hydrogen burning occurs mainly through the CNO cycle, which operates out of equilibrium, because radioactive nuclei of lifetimes longer than the evolutionary timescale are synthesized. Convection also plays a crucial role, since it transports some of the β^+-unstable nuclei synthesized to the outer envelope, where their subsequent decay powers the expansion of the envelope and the increase of luminosity.

Although classical novae release large amounts of energy ($\sim 10^{45}$ erg), they don't have an impact in the interstellar medium dynamics, like supernovae. Concerning the chemical evolution of the Galaxy, they account only for $\sim 1/3000$ of the Galactic disk's gas and dust content; therefore, novae scarcely contribute to Galactic abundances, except for some particular elements, like ^{13}C and ^{17}O. Also, novae contribute to the radioactivity of the Galaxy, through the emission of γ-rays related to the decay of some medium and long-lived nuclei (i.e., ^{7}Be, ^{22}Na and ^{26}Al).

Novae have been observed in all energy ranges, except in the γ-ray domain, where current instruments are not sensitive enough. From the observations in the optical,

CP522, *Cosmic Explosions: Tenth Astrophysical Conference,*
edited by Stephen S. Holt and William W. Zhang

ultraviolet and infrared, information about the physical conditions of the ejecta
and its composition has been deduced. On the other hand, soft X-ray observations,
when available, have provided some information about the turn-off of novae. A
good example of a nova observed at all wavelengths is Nova Cyg 1992.

This paper has been organized as follows: first of all, some observational properties of classical novae are presented, mainly the light curves. Then, the thermonuclear runaway model is explained and illustrated with some numerical results.
Finally, nucleosynthesis in classical novae and its relevance for the chemical evolution and the radioactivity of the Galaxy, with a spetial mention to theoretical
models of γ-ray emission from novae, is presented.

SOME OBSERVATIONAL PROPERTIES OF NOVAE

Visual light curves of classical novae have some general trends that distinguish
them from other variable stars, such as supernovae or dwarf novae, to cite two extreme examples. First of all, there is an increase in luminosity which corresponds to
a decrease of m_V (apparent visual magnitude) of more than 9 magnitudes occuring
in a few days. In some cases, a pre-maximum halt, 2 magnitudes before maximum,
has been observed (see [1] and references therein).

In order to characterize the nova light curves, their speed class is defined from
either t_2 (or t_3), which is the time needed to decay in 2 (or 3) visual magnitudes
after maximum. Novae speed classes range from very fast ($t_2 < 10$ days) and fast
($t_2 \sim 11 - 25$ days) to very slow ($t_2 \sim 151 - 250$ days) [2]. Nova Cyg 1992, for
instance, had $t_2 \sim 12$ days, which is quite fast, and Nova Her 1991 was even faster
($t_2 \sim 2$ days). An example of a slow nova is Nova Cas 1993, which had $t_2 \sim 100$
days.

There is a relationship between the absolute magnitude at maximum M_V and
the speed class of novae, in the sense that brighter novae have shorter decay times
(t_2 or t_3). The theoretical explanation of this relationship [3] is based on the widely
accepted model of nova explosions, which shows that novae reach a luminosity at
maximum which is close to the Eddington luminosity, and also that novae should
eject roughly all their envelope in a time similar to t_3. Thus, one can establish
quantitatively that L_{max} is an increasing function of M_{wd} and that t_3 is a decreasing
function of M_{wd}. From these two relationships a new one can be obtained which
relates M_V at maximum with t_3. This empirical relation (which is valid both in the
V and B photometric bands) is very often used to determine distances to novae,
once visual extinction is known. Different calibrations of the maximum magnitude-
rate of decline relationship (MMRD) exist, with that from [4] being the most usual
one (see also [5]).

It is important to mention that it is in general assumed that two different kinds
of nova populations exist [6]: the disk population (with scaleheight $z \lesssim 100$ pc)
made of bright and fast novae, and the bulge population (with scaleheight up to
$z \gtrsim 1$ kpc), made of dimmer and slower novae [7]. Also, Della Valle & Livio [8]

have established a link between these disk and thick-disk/bulge novae and the spectroscopic classification of Williams [9].

Since the launch of astronomical satellites and, specially, of the IUE (International Ultraviolet Explorer), light curves have been extended to other energetic domains away from the optical one. It was discovered from IUE observations of novae that the luminosity in the ultraviolet band increases when the optical one starts to decline: the reason is that there is a shift of the energy distribution to higher energies, because deeper and hotter regions of the expanding envelope are seen (the photosphere recedes because opacity decreases when temperature falls below 10^4K, the recombination temperature of hydrogen). Also, infrared observations when available (i.e., for novae in which dust forms) indicate an increase once the ultraviolet luminosity starts to decline, which is interpreted as the resulting reradiation by dust grains in the infrared of the ultraviolet energy they have absorbed. In summary, the bolometric luminosity of classical novae is constant for a quite long period of time, being the duration of this constant L_{bol} phase dependent on the remaining envelope mass of the nova.

The constancy of L_{bol} has been interpreted and obtained theoretically, although the concomitant mass-loss has not been well understood and modeled. The phase of constant L_{bol} corresponds to hydrostatic hydrogen burning in the remaining envelope of the nova, accompanied by a continuous mass-loss probably by an optically thick wind. Since the bolometric luminosity deduced from observations is close to or even larger than the Eddington luminosity, radiation pressure is probably the main force causing ejection of nova envelopes. An additional observational proof of this phase has come from the observations in the soft X-ray range. The EXOSAT satellite detected the nova GQ Mus (Nova Mus 1983) as a soft X-ray emitter (in the interval 0.04-2 keV), 460 days after optical maximum [10]. ROSAT detected again that source (0.1-2.4 keV), even 9 years after the explosion [11]. Nova Cyg 1992 was detected by ROSAT too as a powerful soft X-ray source, but the emission lasted in that case only for one year and a half [12]. The interpretation of the soft X-ray emission is that it is related to blackbody emission of the remaining hydrogen-burning shell, which becomes visible when the expanding envelope is transparent to it. The luminosity deduced is close to L_{Edd}, thus indicating again the constancy of L_{bol} and the hardening of the spectra. It is worth mentioning that in some novae (Nova Her 1991, Nova Pup 1991, Nova Vel 1999) hard X-ray emission has also been detected, with much smaller luminosities [13]. In these cases, the interpretation is different, since the emission mechanism is probably related to bremsstrahlung in some shocked region around the nova.

The observed turn-off times of novae, deduced from soft X-ray and ultraviolet observations, are between 1 and 5 yr [14], except for Nova Mus 1983 (9-10 yr). This is much shorter than expected from the nuclear burning timescale of the remaining envelope, thus telling that some extra mechanism besides of nuclear reactions leads to the extinction of the shell.

Spectra of novae are quite complicated and show different features related to some particular phases: four succesive systems of absorption lines, and five overlapping systems of emission lines are present in almost all novae. ¿From the emission-line spectra in the nebular phase, both in the optical and the ultraviolet, and also from infrared spectra in some cases, detailed abundance determinations of nova ejecta are available (see [15] for a recent review). A general trend is observed: in many novae there is an enhancement of metallicities above solar and, in particular, enhancements of carbon-nitrogen-oxygen (CNO) elements and/or neon. It is known since long ago (see for instance [16] and [17]) that some enrichment of the accreted matter (which is in principle assumed to be of solar composition) with the underlying white dwarf core (of the CO or ONe type) is necessary, both to power the nova explosion and to explain the observed enhancements (see the following section for details).

THERMONUCLEAR RUNAWAY MODEL OF NOVA EXPLOSIONS

The accepted scenario of classical novae explosions is the thermonuclear runaway model, in which a cold white dwarf in a cataclysmic variable accretes hydrogen-rich matter, as a result of Roche lobe overflow of the main sequence companion. If the accretion rate is low enough (e.g., $\dot{M} \sim 10^{-9} - 10^{-10}$ M$_\odot$ yr^{-1}), the accreted hydrogen is compressed up to degenerate conditions, thus leading to thermonuclear burning without control (thermonuclear runaway). The explosive burning of hydrogen produces some β^+-unstable nuclei of relatively short timescales (i.e., ^{13}N, ^{14}O, ^{15}O, ^{17}F), which are transported by convection to the outer envelope, where they are preserved from destruction until they decay. Their subsequent decay implies a huge liberation of energy in the outer shells, which originates envelope expansion, increase in luminosity and mass ejection, as required in a classical nova explosion.

In order to understand the thermonuclear runaway (TNR) mechanism for nova explosions, it is important to evaluate some relevant timescales (see [18] for a review). The accretion timescale, defined as $\tau_{\rm acc} \sim$ M$_{\rm acc}/\dot{M}$ (which is of the order of $10^4 - 10^5$ yr, depending on the accretion rate $\dot{M}$), the nuclear timescale $\tau_{\rm nuc} \sim$ C$_{\rm p}$T$/\epsilon_{\rm nuc}$ (which is as small as some seconds at peak burning), and the dynamical timescale ($\tau_{\rm dyn} \sim$ H$_{\rm p}/$c$_{\rm s} \sim (1/g)\sqrt{P/\rho}$). During the accretion phase, $\tau_{\rm acc} \leq \tau_{\rm nuc}$, accretion can proceed and increase the envelope mass. When degenerate ignition conditions are reached, degeneracy prevents envelope expansion and the TNR occurs. As temperature increases, degeneracy would be lifted (since T would become larger than T$_{\rm Fermi}$) and expansion would turn-off the explosion, but this is not so because $\tau_{\rm nuc} \ll \tau_{\rm dyn}$ (specially if the envelope is enriched in CNO elements above solar values, thus enhancing the contribution of the CNO cycle to hydrogen burning). Therefore, since the envelope can not respond by expanding, temperature and nuclear energy generation rate, $\epsilon_{\rm nuc}$, continue to increase without control. The value of the nuclear timescale is crucial for the development of the TNR and its final fate. In fact there are mainly two types of nuclear timescales: those related to β-decays, τ_{β^+}, and those related to proton capture reactions, $\tau_{(\rm p,\gamma)}$.

TABLE 1. Some properties of the explosions of CO and ONe novae

Nova type	M_{wd} ($M_\odot$)	Peak temperature (K)	Kinetic energy of the ejecta	M_{ej} ($M_\odot$)
CO	1.15	2.05×10^8	1.1×10^{45}	1.3×10^{-5}
ONe	1.15	2.31×10^8	1.5×10^{45}	2.6×10^{-5}
ONe	1.25	2.51×10^8	1.5×10^{45}	1.8×10^{-5}

In the early evolution towards the TNR, $\tau_{\beta^+} < \tau_{(p,\gamma)}$ and the CNO cycle operates in equilibrium. But as temperature increases up to $\sim 10^8$ K, the reverse situation is true ($\tau_{\beta^+} \gtrsim \tau_{(p,\gamma)}$), and thus the CNO cycle is β-limited. In addition, since the large energetic output produced by nuclear reactions can't be evacuated only by radiation, convection sets in and transports the β^+-unstable nuclei to the outer cooler regions where they are preserved from destruction and where they will decay later on ($\tau_{conv} \lesssim \tau_{\beta^+}$), leading to envelope expansion, increase in luminosity and final mass ejection if the attained velocities are large enough. Another important effect of convection is that it transports fresh unburned material to the burning shell. In summary, non-equilibrium burning occurs and the resulting nucleosynthesis will be far from that of hydrostatic hydrogen burning.

In table 1 we show some general properties of computed models, corresponding two carbon-oxygen (CO) and oxygen-neon (ONe) novae, with accretion rate 2×10^{-10} $M_\odot$ yr^{-1}. More details about these models are given in [19]. Other recent detailed theoretical models of CO and ONe nova explosions are shown in [20] and [21], respectively.

NUCLEOSYNTHESIS IN NOVA EXPLOSIONS

The main goal of studies of nucleosynthesis in novae is, of course, to reproduce the observed abundances in novae ejecta. Although both from the observational and the theoretical side some uncertainties exist (different determinations of observed abundances or uncertain initial conditions for theoretical models), a quite good fit is obtained in many cases (see table 2, and more details in [19], together with the quoted papers in the table for the analyses of the observations and derivation of the abundances).

Chemical evolution of the Galaxy

In contrast with supernovae, novae are not important contributors to the abundances observed in the interstellar medium, but it is also true that they can contribute to Galactic abundances in some particular cases, when the overproduction factors with respect to solar abundances are larger than around 10^3 (see [19] and [15]). In figure 1 we show the overproduction factors relative to solar abundances versus mass number for two typical novae: a CO and an ONe ones, with mass 1.15 $M_\odot$, 50% of mixing with core material and accretion rate 2×10^{-10} $M_\odot$ yr^{-1}.

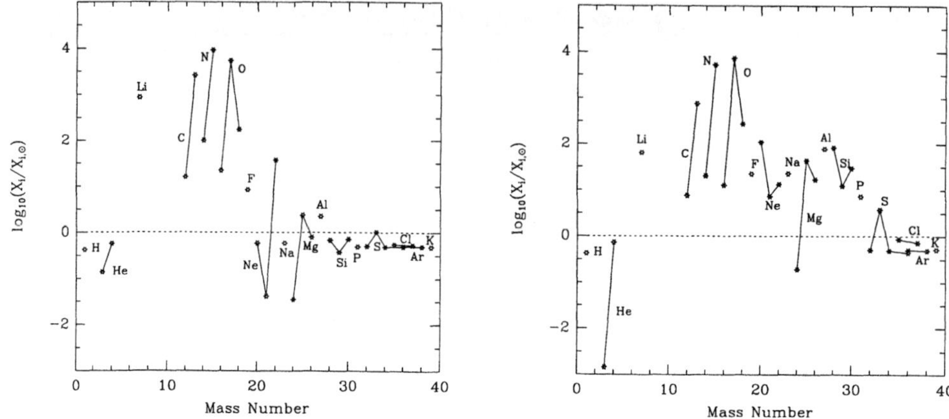

FIGURE 1. Overproduction factors, with respect to solar abundances, obtained for two nova models: a CO nova of 1.15 $M_\odot$ with 50% mixing between accreted matter and core material (left), and an ONe nova of the same mass and degree of mixing (right).

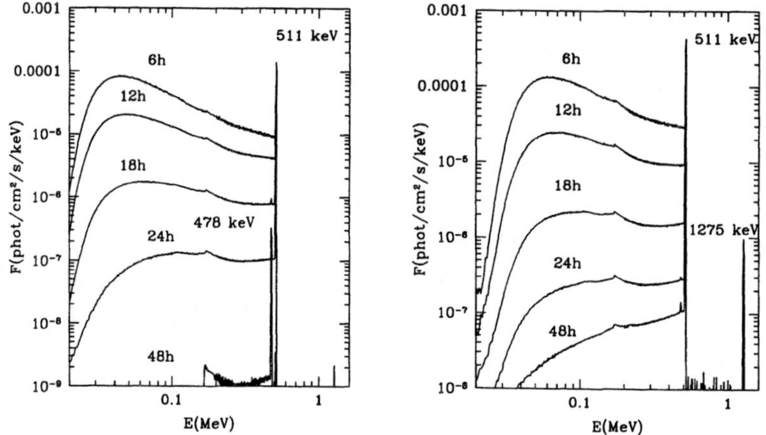

FIGURE 2. Temporal evolution of the γ-ray emission of a CO (left) and an ONe (right) nova, of 1.15$M_\odot$, at a distance of 1 kpc.

TABLE 2. Comparison between some models of CO and ONe novae and some observations

Model	H	He	C	N	O	Ne	Na-Fe	Z
				Element				
V693 CrA 1981								
Vanlandingham et al. 1997	0.25	0.43	0.025	0.055	0.068	0.17	0.058	0.32
ONe, $1.15M_\odot$, mixing 50%	0.30	0.20	0.051	0.045	0.15	0.18	0.065	0.50
Andreä et al. 1994	0.16	0.18	0.0078	0.14	0.21	0.26	0.030	0.66
ONe, $1.15M_\odot$, mixing 75%	0.12	0.13	0.049	0.051	0.28	0.26	0.10	0.75
Williams et al. 1985	0.29	0.32	0.0046	0.080	0.12	0.17	0.016	0.39
ONe, $1.25M_\odot$, mixing 50%	0.28	0.22	0.060	0.074	0.11	0.18	0.071	0.50
V1370 Aql 1982								
Andreä et al. 1994	0.044	0.10	0.050	0.19	0.037	0.56	0.017	0.86
ONe, $1.35M_\odot$, mixing 75%	0.073	0.17	0.051	0.18	0.14	0.24	0.14	0.76
Snijders et al. 1987	0.053	0.088	0.035	0.14	0.051	0.52	0.11	0.86
ONe, $1.35M_\odot$, mixing 75%	0.073	0.17	0.051	0.18	0.14	0.24	0.14	0.76
QU Vul 1984								
Austin et al. 1996	0.36	0.19		0.071	0.19	0.18	0.0014	0.44
ONe, $1.0M_\odot$, mixing 50%	0.32	0.18	0.030	0.034	0.20	0.18	0.062	0.50
Saizar et al. 1992	0.30	0.60	0.0013	0.018	0.039	0.040	0.0049	0.10
ONe, $1.15M_\odot$, mixing 25%	0.47	0.28	0.041	0.047	0.037	0.090	0.0035	0.25
PW Vul 1984								
Andreä et al. 1994	0.47	0.23	0.073	0.14	0.083	0.0040	0.0048	0.30
CO, $1.15M_\odot$, mixing 25%	0.47	0.25	0.073	0.094	0.10	0.0036	0.0017	0.28
V1688 Cyg 1978								
Andreä et al. 1994	0.45	0.22	0.070	0.14	0.12			0.33
CO, $1.15M_\odot$, mixing 25%	0.47	0.25	0.073	0.094	0.10	0.0036	0.0017	0.28
Stickland et al. 1981	0.45	0.23	0.047	0.14	0.13	0.0068		0.32
CO, $0.8M_\odot$, mixing 25%	0.51	0.21	0.048	0.096	0.13	0.0038	0.0015	0.28

In this figure some general features are distinguishable: both in CO and in ONe novae, the largest yields correspond to elements of the CNO group, whether in CO novae Li is also largely overproduced. In the case of more massive ONe novae (i.e. 1.35 $M_\odot$ [19]), intermediate-mass elements (such as Ne, Na, Mg, S, Cl) are also overproduced.

The origin of Galactic lithium (^{7}Li) is still not completely understood. Although it is widely accepted that there is some primordial lithium produced during the big bang, and, of course, that spallation reactions by cosmic rays in the interstellar medium or in flares also produce it, some extra stellar source of ^{7}Li (without generating ^{6}Li) has to be invoked. The synthesis of ^{7}Li in classical novae, by the *beryllium transport* mechanism [22], can produce large amounts of ^{7}Li, but complete hydrodynamical models are needed in order to compute correctly the yields [23,24]. ^{7}Li formation is favored in CO novae, with respect to ONe ones (see figure 1). The reason is that CO novae evolve faster (because of their larger ^{12}C content), allowing photodisintegration of ^{8}B through ^{8}B$(\gamma,\mathrm{p})^7$Be to prevent the destruction of the ^{7}Be synthesized during the first part of the TNR (by means of ^{3}He$(\alpha,\gamma)^7$Be). Large overproduction factors with respect to solar abundances are obtained (see

TABLE 3. Main radioactive isotopes ejected by novae

Isotope	Lifetime	Main disintegration process	Type of γ-ray emission	Nova type
^{13}N	862 s	β^+–decay	511 keV line & continuum	CO and ONe
^{18}F	158 min	β^+–decay	511 keV line & continuum	CO and ONe
^{7}Be	77 days	e^-–capture	478 keV line	CO
^{22}Na	3.75 years	β^+–decay	1275 keV & 511 keV lines	ONe
^{26}Al	10^6 years	β^+–decay	1809 keV & 511 keV lines	ONe

TABLE 4. Ejected masses (in $M_\odot$) of radioactive nuclei obtained from theoretical models of CO and ONe novae (^{13}N and ^{18}F 1 hr after peak temperature).

Nova type	M_{wd} ($M_\odot$)	^{13}N	^{18}F	^{7}Be	^{22}Na	^{26}Al
CO	1.15	2.3×10^{-8}	2.6×10^{-9}	1.1×10^{-10}	3.8×10^{-12}	6.2×10^{-10}
ONe	1.15	2.9×10^{-8}	5.9×10^{-9}	1.6×10^{-11}	7.0×10^{-9}	2.1×10^{-8}
ONe	1.25	3.8×10^{-8}	4.5×10^{-9}	1.2×10^{-11}	6.3×10^{-9}	1.2×10^{-8}

figure 1), but classical novae can only account for roughly $\sim 10\%$ of the global Galactic ^{7}Li [24]. It is important to stress that Romano et al. [25] have obtained (using our nova yields in a complete model of chemical Galactic evolution) that the contribution from novae is required in order to reproduce the shape of the growth of Li abundance versus metallicity.

Concerning the nuclei of the CNO group, the main isotopes produced in novae are ^{13}C, ^{15}N and ^{17}O. The Galactic ^{17}O is most probably almost entirely of novae origin [19]. Novae also contribute significantly to the Galactic ^{13}C and ^{15}N, but an extra source of ^{15}N is required.

Radioactivity in the Galaxy and γ-ray emission from novae

An important property of novae ejecta is the presence of radioactive nuclei (the role of novae as potential γ-ray emitters was mentioned long ago [26–28]).

Besides of the very short-lived isotopes responsible of the explosion itself (see above), there are other short, medium and long-lived nuclei which have some relevance for the radioactivity of the Galaxy and for the γ-ray emission of individual novae. A list of these nuclei with their main properties is displayed in table 3, whereas the ejected masses of them obtained from complete hydrodynamical models are shown in table 4. The short-lived nuclei ^{13}N and ^{18}F are produced in similar quantities in both nova types, whereas ^{7}Be is mainly produced in CO novae (see the discussion on ^{7}Li synthesis above) and ^{22}Na and ^{26}Al are produced in appreciable amounts only in ONe novae. The reason is that in nova explosions the temperatures reached (around $2 - 3 \times 10^8$ K, see table 1) are not high enough to break the CNO cycle; therefore, only if some seed nuclei (like ^{20}Ne, ^{23}Na, ^{24,25}Mg) are present in the envelope material, can the NeNa-MgAl cycles operate and synthesize those radioactive nuclei (and other intermediate-mass isotopes). As CO white dwarfs are devoid of these nuclei, it is almost impossible for them to produce large amounts of radioactive ^{22}Na and ^{26}Al.

It is worth mentioning that uncertainties still affect some nuclear reaction rates of the NeNa-MgAl cycles; this leads to uncertain theoretical determinations of the yields of ^{22}Na and ^{26}Al, but the error (from a purely nuclear point of view, and defined as the ratio between maximum and minimum productions) amounts to factors between 2 and 10 (see [29] for details). Also the final amount of ^{18}F synthesized in both CO and ONe novae is still not well known, since the nuclear reaction rates affecting ^{18}F destruction (via ^{18}F(p,α) and ^{18}F(p,γ)) are still not well determined [30].

Radioactive nuclei ejected by novae play a role in the radioactivity of the Galaxy which depends on their lifetimes. The short-lived nuclei (i.e., ^{13}N and ^{18}F) produce an intense burst of γ-ray emission, with duration of some hours, which is emitted before the nova visual maximum (see [31,30] for details). This emission is related to positron annihilation, which consists of a line at 511 keV and a continuum at energies between 20 and 511 keV, related to the positronium continuum plus the comptonization of the photons emitted in the line. In figure 2 we show an example of the spectral evolution of a CO and an ONe nova, at different epochs after peak temperature.

The emission related to medium-lived nuclei, ^{7}Be and ^{22}Na, appears later and is different in CO and ONe novae, because of their different nucleosynthesis. CO novae display a line at 478 keV, related to ^{7}Be decay, whereas ONe novae show a line at 1275 keV, related to ^{22}Na decay.

Finally, the long-lived isotope ^{26}Al is also produced by novae. The Galactic γ-ray emission observed at 1809 keV (Mahoney et al. [32] with the HEAO3 satellite; Diehl et al. [33] with the CGRO/COMPTEL) corresponds to the decay of ^{26}Al. Its distribution seems to correspond better to that of a young population and the contribution of novae is not the dominant one (see [34,35]).

In summary, classical novae explosions produce γ-rays, being the signature of CO and ONe novae different. The detectability distances for the lines at 478 and 1275 keV with the future instrument INTEGRAL/SPI will range between 0.5 and 2 kpc. To compute these distances, the width of the lines is taken into account ($\sim$ 7 keV for the 478 keV line and $\sim$ 20 keV for the 1275 keV line). The continuum and the 511 keV line are the most intense emissions, but their appeerence before visual maximum and their very short duration requires "a posteriori" analyses, with monitor-type instruments, with a large field-of-view and sensitivity up to some hundred keVs. With future instruments of these characteristics, novae would be detectable more easily in γ-rays than visually, because of the lack of extinction.

Future instrumentation in the γ and hard X-ray domain will give crucial insights on the nova theory allowing for a direct confirmation of the nucleosynthesis in these explosions, but also providing unique information about the Galactic distribution of novae and their rates.

REFERENCES

1. Warner, B. *Cataclysmic Variable Stars*, CUP, Cambridge (1995).
2. Payne-Gaposchkin, C. *Galactic Novae*, North-Holland, Amsterdam, (1957).
3. Livio, M. *ApJ* **393**, 516 (1992).
4. Della Valle, M., Livio, M. *ApJ* **452**, 704 (1995).
5. Shafter, A.W. *ApJ* **487**, 226 (1997).
6. Duerbeck, H.W. *Physics of Classical Novae*, eds. A. Cassatella & R. Viotti, Springer, p. 96 (1990).
7. Della Valle,M., Bianchini,A., Livio,M., Orio,M. *Astron.&Astrophys.* **266**,232 (1992).
8. Della Valle, M., Livio, M. *ApJ* **506**, 818 (1998)
9. Williams, R.E. *ApJ* , (1992).
10. Ögelman, H., Beuermann, K., Krautter, J. *ApJ* **287**, L31 (1984).
11. Ögelman, H., Orio, M., Krautter, J., Starrfield, S. *Nature* **361**, 331 (1993).
12. Krautter, J., Ögelman, H., Starrfield, S., Wichmann, R., Pfeffermann, E. *ApJ* **456**, 788 (1996).
13. Lloyd, H.M., O'Brien, T.J., Bode, M.F., Predehl, P., Schmitt, J.H.M.M., Trümper, J., Watson, M.G., Pounds, K. *Nature* **356**, 222 (1992).
14. González-Riestra, R., Orio, M., Gallagher, J. *Astron. & Astrophys.* **129**, 23 (1998).
15. Gehrz, R., Truran, J.W., Williams, R.E., Starrfield, S.E., *Publ. Astr. Soc. Pac.* **110**, 3 (1998)
16. Starrfield, S., Truran, J.W., Sparks, W.M., *ApJ* **226**, 186 (1978).
17. Prialnik, D., Shara, M.M., Shaviv, G., *Astron. & Astrophys.* **62**, 339 (1978).
18. Starrfield, S. *Classical Novae*, eds. M.F. Bode & A. Evans, Wiley, Chichester, p. 39 (1989).
19. José, J., Hernanz, M., *ApJ* **494**, 680 (1998).
20. Prialnik, D., Kovetz, A. *ApJ* **445**, 789 (1995).
21. Starrfield, S., Truran, J.W., Wiescher, M., Sparks, W.M., *Mon. Not. R. Astron. Soc.* **296**, 502 (1998).
22. Cameron, A.G.W. *ApJ* **121**, 144 (1955).
23. Starrfield, S., Truran, J.W., Sparks, W.M., Arnould, M. *ApJ* **222**, 600 (1978).
24. Hernanz, M., José, J., Coc, A., Isern, J., *ApJ* **465**, L27 (1996).
25. Romano, D., Matteuci, F., Molaro, P., Bonifacio, P. *Astron. & Astrophys.* **352**, 117 (1999).
26. Clayton, D.D., Hoyle, F., *ApJ* **187**, L101 (1974).
27. Clayton, D.D., *ApJ* **244**, L97 (1981).
28. Leising, M., Clayton, D.D., *ApJ* **323**, 157 (1987).
29. José, J., Coc, A., Hernanz, M., *ApJ* **520**, 347 (1999).
30. Hernanz,M., José,J., Coc,A., Gómez-Gomar,J., Isern,J., *ApJ Lett.* **526**, L97 (1999).
31. Gómez-Gomar, Hernanz, M., J., José, J., Isern, J., *Mon. Not. R. Astron. Soc.* **296**, 913 (1998).
32. Mahoney, W.A., Ling, J.C., Jacobson, A.S., Lingenfelter, R.E., *ApJ* **262**, 742 (1982).
33. Diehl, R. et al. *Astron. & Astrophys.* **298**, 445 (1995).
34. Prantzos, N., Diehl, R., *Phys. Rep.* **267**, 1 (1996).
35. José, J., Hernanz, M., Coc, A., *ApJ* **479**, L55 (1997).

Observations of Type I Bursts from Neutron Stars

Jean H. Swank

Laboratory for High Energy Astrophysics
NASA/GSFC Greenbelt, MD 20771

Abstract. Observations of Type I X-ray bursts have long been taken as evidence that the sources are neutron stars. Black body models approximate the spectral data and imply a suddenly heated neutron star cooling over characteristic times of seconds to minutes. The phenomena are convincingly explained in terms of nuclear burning of accreted gas on neutron stars with low mass companion stars. Prospects are promising that detailed theory and data from RXTE and future missions will lead to better determinations of important physical parameters (neutron star mass and radius, composition of the accreting gas, distance of the source). Among the variety of bursts observed, there are probably representatives of different kinds of explosive burning. RXTE's discovery of a 2.5 ms persistent coherent period from one Type I burster has now linked bursters indisputably to the epitome of a neutron star, a fast spinning magnetic compact object. Oscillations in some bursts had already been thought to arise from the neutron stars' rotations. Detailed observations of these oscillations are touchstones of how the explosive bursts originate and progress, as well as independent measures of the neutron star parameters.

INTRODUCTION

Bursts represent the nuclear burning energy of material accreted on neutron stars, which is only about 7% of the gravitational energy irradiated. But, concentrating many hours worth in 10 seconds, they are impressive in making the neutron star glow as brightly as the Crab Nebula for a few seconds. During those seconds it dominates the scene and we can get a good look. The scene is changing rapidly, so only observations with large area and high time resolution have a hope of detecting the kind of dynamical phenomena that could occur. Now that neutron stars are being studied with these capabilities, we are both confirming speculations and seeing new aspects we did not envision. I will briefly summarize the kind of data that have substantiated our current picture and how the new observations compare. In this, I will concentrate on the spectral measurements, since the questions of the burst energy output, the durations, and the correlations with the persistent flux are well covered in Bildsten's paper in this volume. Then I will discuss the new results about the spins of the neutron stars in low-mass X-ray binaries. Fuller details of both the earlier and the recent work may be found in recent reviews [1–4].

BRIEF UPDATE ON THE SPECTRAL OBSERVATIONS

Burst Spectra

A key indicator that bursters are neutron stars was the nature of the energy spectra. Of the three limiting types of X-ray spectra, non-thermal, optically thin high temperature gas, and high temperature optically thick surface, the spectra of bursts were a good match to the last of these and the parameters made good sense. Projected areas were within a factor of two of the canonical neutron star and peak luminosities were within factors of 2 of the Eddington limit for hydrogen or perhaps helium, for sources concentrated within the galactic bulge [5]. At first, only for long bursts and slow evolution (100–1000 s), was it possible to say that the black body model was definitely a better fit than other simple models [6]. Now it is possible for short (10 s) bursts. It still remains true that for spectra of intervals of only a few tenths of a second, the black body model gives a good fit to the data.

Theoretically, modifications of the spectrum due to scattering and to temperature distribution in the neutron star atmosphere are expected to change the spectrum, but while the color temperature is strongly affected, the shape of the spectrum is not usually very different from the black body distribution. The spectra obtained by OSO–8, SAS–3, EXOSAT, TENMA, and Ginga, time resolved to various extents, depending on the data modes and the detector area, fit black body distributions and corrected spectral forms equally well. Conceptually, the fit temperatures are higher than expected because the emissivity is less than unity.

The predicted differences are excesses at low energies, below about 2 keV, and at high energies, above about 15 keV. London, Taam, and Howard [8] first included all the most important effects, the radiative transfer, the competition of scattering and absorption, and inelastic scattering and coined the phrase "hardening factor". Comparisons of burst spectra with absorbed black body spectra have shown deviations suggestive of these effects, although usually with detectors where the response is more uncertain in those ranges, and often when fast changes in the burst made it possible that the spectrum changed during the interval.

A potentially very informative deviation from the black body form was the dip seen in TENMA selected burst spectra of 4U 1636–536 [7], 4U 1608–522, and 4U 1746–217, at 4.1 keV (in most cases). If the energy were redshifted Fe K_α, it implies a very large redshift and puts a strain on the interpretation. That the energy is close to detector features makes it more difficult to study. It has also been suggested that the cross-over between a low temperature spectral component and a higher temperature one, could cause an apparent dip. Of the current missions, those with modest spectral resolution instruments ($E/\Delta E > 10$) have low enough area to limit their constraints and the CCDs have pile-up problems with intense

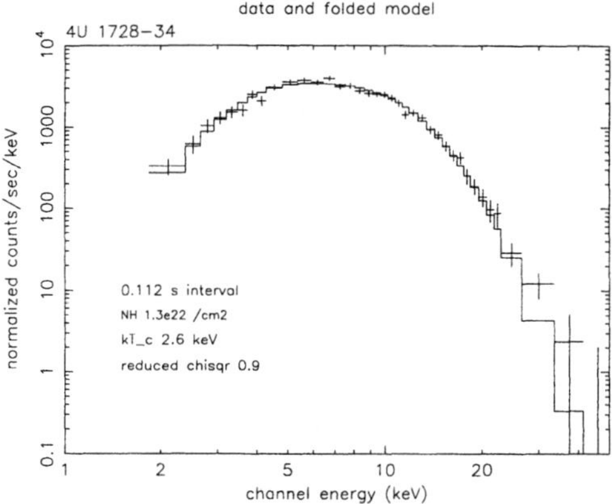

FIGURE 1. Spectrum and Fit of a Burst Interval. These spectra cannot test the low energy predictions. High energy deviations require using more data.

bursts. RXTE has the requisite area, but is a xenon detector again. It is not yet clear whether Chandra, XMM, or Astro-E can address this feature in bursts.

Typical RXTE PCA spectra for a few tenths of a second, with some exceptions, are still good fits to black body spectra, as shown in Figure 1. Additions of spectra to test for theoretical spectral differences have not been reported.

Color Temperatures and Apparent Radii

The parameters for flux, temperature, and apparent radius of these solutions are well determined (except when the color temperature puts the flux below the low energy threshold of the detectors), as shown for the example in Figure 2 for a burst with moderate radius expansion. The temperature starts down after the initial rise, the radius continues to increase, and the flux has a plateau. These are consistent with the scenario of the flux rising to the Eddington limit and lifting surface layers, such that the photosphere still emits the Eddington limit of radiation while expanding beyond the neutron star. Some bursts show evidence for both the Eddington limits of cosmic abundance gas and of hydrogen-poor/helium-rich gas. Early in the decay the photosphere contracts and the radius is approximately asymptotically consistent with that of the neutron star. The asymptotic apparent radius is not exactly constant, but slowly climbs as the flux and the temperature go down, possibly as flux dependent corrections change. The radius expands to around 16 km, less than a factor of two. This may be enough, however, for the photosphere to run into the inner part of the accretion disk, unless the inner part has been dispersed by the burst already. For other bursts the radius can be followed out to 100 km. Even greater expansions have been seen in rare long superbursts for which the average photon energy goes below the range of the detector and the rise looks like a precursor.

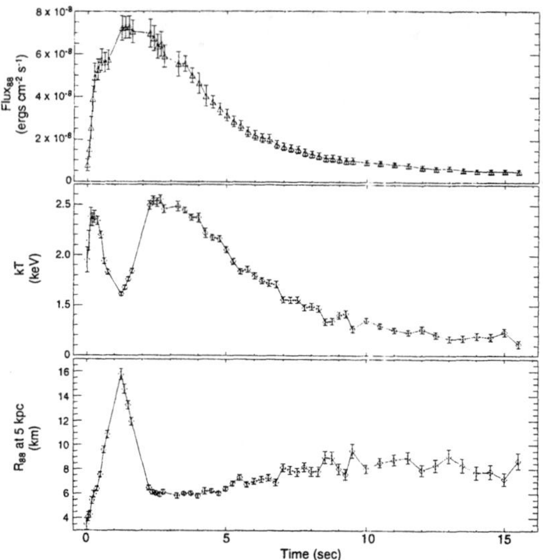

FIGURE 2. Evolution of Burst Flux, Temperature and Radius. The temperature and radius are for simple black body fits. The data mode gave rise to the gaps.

Atmospheric Models and Neutron Star Parameters

It was soon recognized that, in the rather high temperature atmosphere, scattering would dominate over free-free absorption and the emissivity would be energy and temperature dependent. The results of assuming a non-relativistic spherically symmetric neutron star led to results unlikely to be exactly correct, radii too small, luminosities exceeding the Eddington limit for He, and radius variations during bursts decays. The gravitational redshift effects are straight forward, at least if the star is not rotating near breakup. Successive calculations of the theoretical spectra increased in sophistication and applicability. [9,10]. They still depend on the assumptions that the spectrum is formed in an atmosphere in radiative and hydrostatic equilibrium.

Ebisuzaki and Nakamura [11] used the fact that the spectra calculated for a given luminosity and the spectra observed at a given luminosity fit a black body distribution for the "color" temperature), with the effective temperature a lower value. The dependence of the hardening factor (T_c/T_{eff}) on the ratio of the luminosity to the Eddington luminosity, was mapped empirically, for various atmospheric constituents and burning scenarios. In practice it was possible, in the case of radius expansion bursts, to identify the point in the decay at which for radius expansion bursts, the crust settled back onto the neutron star. That flux should correspond to the Eddington limit. Further, the observational information of the fluxes as a function of time, together with the fit color temperatures, allow determination of the temperature that the emission would have had on the surface at the touchdown moment. The latter should be a specific function of the mass and radius and provide one locus of acceptable values. For sources with known distance, the comparison of the observed flux and the theoretical Eddington limit luminosity give a second locus, to uniquely determine the mass and radius.

Efforts to fit EXOSAT data to this picture were plagued by statistical uncertainties in the determinations [12]. A long burst observed with Ginga, exhibited complications in the spectra [13]. The RXTE PCA has now obtained data which is suitable to use for such determinations. However a strategy is needed to deal with the question of the fraction of the neutron star that is emitting and the asymmetry of the emission. The ideas discussed by Bildsten, and the evidence from the bursts oscillations (discussed below) that the neutron star is not uniformly emitting (although only sometimes), show that these effects need to be considered.

Directions of Current Explorations

There are clearly several regimes of Type I bursts. Differences in characteristic recurrence times, the ratio α of the persistent to burst luminosity, and burst duration, are observed, not just between sources, but for a given source. Theoretically, as Bildsten has discussed, these imply different fuel during the bursts and for accretion of given abundances close to cosmic, depend on the mass accretion rate per unit area. While the main differences are between sources with 5–10 s bursts, separated by 3–5 hours, versus 100 s bursts separated by 12 or more hours, there are two less common types of bursts. Some anomalous faint bursts have now been confirmed [14,15]. In addition, long duration bursts are seen which last 100–1000 s and have long, but unknown, recurrence times.

The observations are not consistent if the accretion rate is the only independent parameter. Bildsten proposes that the reason for correlations opposite those expected for a spherically symmetric distribution of the burning fuel on a neutron star, may be that the fuel is not distributed uniformly and the distribution is different for different states. Taam proposes hysteresis effects due to changes in internal temperature for changes of accretion conditions.

For the spectra, persistent flux contributions during the bursts need to be resolved. Radii of expansion put the photosphere outside where the kilohertz quasi-periodic oscillations (kHz QPO) indicate the inner disk can be (for the atoll bursters at least). Contributions of reflection, which must exist as counterparts to corresponding optical bursts, need to be taken into account.

NEW INFORMATION ABOUT THE BURSTERS

A Pulsing Type I Burster

The Wide Field Camera Experiment on BeppoSAX has identified at least 16 new bursters, most in the galactic center region. Most were also transient persistent sources with peak luminosities less than 100 mCrab. SAX J1808.5-3658 was one of these [17]. The RXTE ASM could see the transient flux during the 2 week interval that it is was above 20 mCrab.

In May 1998, the RXTE PCA, in slewing between two targets, crossed the source and flagged that SAX J1808.5-365 was again bright [18]. This time the RXTE PCA observed, Wijnands [19] immediately found the signal had 401 Hz coherent pulsations and Chakrabarty and Morgan [20] found it had a 2 hour binary Doppler modulation with a tiny mass function ($3.8 \times 10^{-5} M_\odot$). Consideration of possible orbits and companions suggested the companion could be several times less than $\approx 0.15 M_\odot$ and the orbit possibly viewed nearly face-on.

Although the flux after 2 weeks dropped within 2 days by a factor of 5 the energy spectrum did not change and pulsations continued to be observed. Naively at least, accretion onto the neutron star was able to occur through the observed range of accretion rate. Quantitative limits with various assumptions about the laws describing the gas in the accretion disk (e.g. cold gas pressure or radiation pressure supported) led to estimates of the magnetic field within a factor of two of 5×10^8 Gauss [21]. The X-ray pulsar's period of 2.5 ms is in the middle of the periods of a half dozen low-mass binary radio pulsars (1.6–7.5 ms) and a magnetic field $10^8 - 10^9$ is proximate to the magnetic fields of the radio pulsars. It appears likely that such a pulsar source could evolve to a similar radio pulsar.

Flux Oscillations during Bursts

The dynamical power spectrum of a burst from 4U 1728–34 is shown in Figure 3. Strohmayer discovered both the kHz QPO and the burst oscillations in the first observations of this source. The number of sources for which oscillations have been seen during some bursts is still six, as it was by the end of the first year of RXTE operations [3]. KHz QPO in the persistent flux have been seen in a dozen atoll bursters at frequencies of 300-1200 Hz. There are twice as many bursters with weak persistent flux. A small number of these have exhibited the QPO, but for many of these the flux is very weak and the limits are not very constraining. Bursts have also been seen from them. In some cases it is known that no oscillations were seen in the bursts. But in others the results are still pending. No oscillations have been seen in bursts that have been observed from the Z-sources, GX 17+2 and Cyg X–2.

Which bursts exhibit oscillations? Some interesting correlations have been noted, but the question has not yet been addressed systematically. Definitely some bursts with strong radius expansion do not exhibit oscillations. On the other hand for bursts with mild radius expansion, there may be oscillations before the expansion starts and after it has subsided. This is in agreement with the oscillations being associated with the neutron star surface.

For the atoll bursters the position in the color-color diagram for the persistent flux has long been argued to be response to the accretion rate [22], exactly because

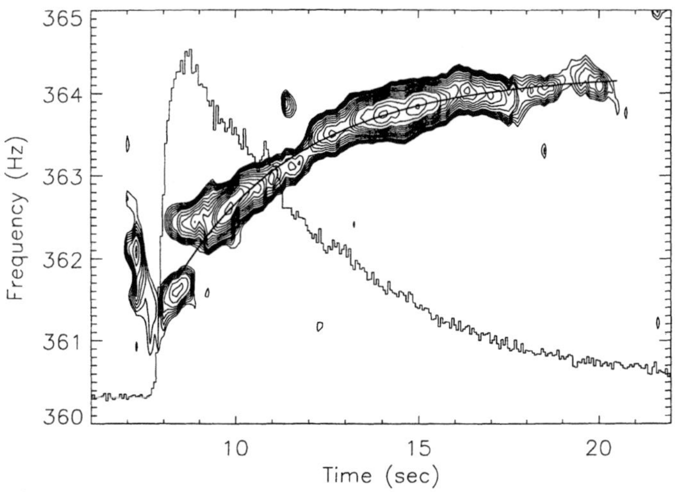

FIGURE 3. Contours of Power Spectral Density versus Time for a Burst from 4U 1728-34. Power is calculated for 2 s intervals starting every 0.25 s. for smoothed contours. An exponential recovery fit to the frequency is superposed as well as the light curve.

burst properties correlated with the position. One of the properties that could be correlated is the inhomogeneity that shows up in oscillations.

KiloHertz Quasi-periodic Oscillations in Persistent Flux

The characteristic pattern of kiloHertz oscillations is two frequencies 250-350 Hz apart, where the difference stays approximately constant, while the frequencies can change by a factor of two. Figure 4 compares the distribution of the differences with the frequencies of oscillations seen in the bursts. Of the four sources for which the burst frequency is above 500 Hz, two have the pair of kHz QPO and the burst frequency is approximately twice the difference frequency. For the two with the lowest burst frequencies, these are very close to their difference frequencies.

Models with one of these frequencies being a beat with a lower frequency would provide a natural explanation. If the burst oscillations are the neutron star's spin, it provides an obvious candidate to beat with one of the QPO. In this case, complications have to explain discrepancies of about 5%. Miller shows that the physical mechanisms in the "sonic-point model" can accommodate them [23].

If the difference has nothing to do with the spin, but instead reflects frequencies entirely in the inner accretion disk, some effect of these during the burst has to explain the burst oscillations. Otherwise the closeness of the numbers seems an unlikely coincidence. The spins could be deduced from more subtle effects on the Kepler and apsidal motion to make the data fit [24].

If the differences are the spins, the spins have a narrow distribution (257-358 Hz,

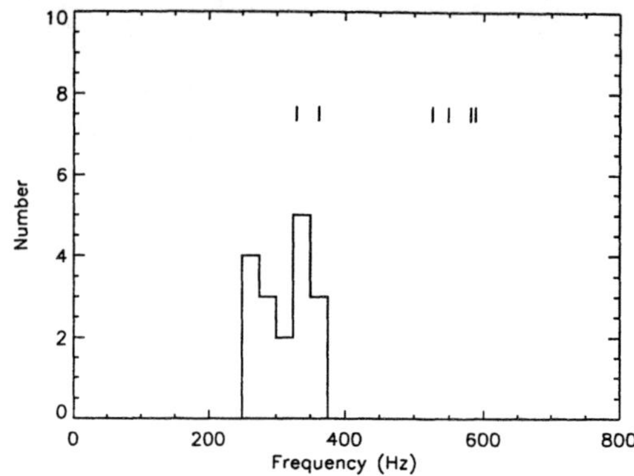

FIGURE 4. Histogram of KHz QPO Differences and Burst Oscillation Frequencies. The frequencies of greatest power during bursts are shown for the six bursters, 4U 1728–34, 4U 1636–53, 4U 1702–429, KS 1731–26, Aql X-1, and MXB 1743–29, with a histogram of the difference between the two kHz QPO in persistent fluxes. .

corresponding to 2.8-3.8 ms). The sources have widely differing accretion rates (currently). Bildsten has explained this pile-up of the spins as due to the equilibrium between spin-up torque and gravitational radiation of the neutron star angular momentum [26]. This avoids a conspiracy between the luminosity and the magnetic field that is implied for equilibrium accretion at a magnetosphere.

If the differences do represent the spins, then the bursts in some cases show the fundamental and in others the signal is dominated by the first harmonic. For at least 4U 1636-53, a weak amplitude at the fundamental appears at the beginning of some bursts, as would be the case if the burning starts at one magnetic pole and very quickly spreads to the optical pole.

Neutron Star Spin Periods

Strohmayer has led investigation of several predictions of the model in which a hot spot on the neutron star surface causes the flux oscillations as the star rotates [4,25]. They have appeared to confirm the interpretation in terms of the rotation period of the neutron star.

(1) In the bursts in which oscillations appear near the beginning of bursts and then die out, the amplitude of the oscillations can reach as high as 70 %. The amplitude dies down as the burst progresses in a way consistent with the spread of a hot spot. This increase in the burning area agrees with the spectral analysis.

(2) The asymptotic oscillations during the declining phase of the burst are usually well defined, suggesting a frequency characteristic of the neutron star (although the cause of asymmetry in this phase is not known). For some sources the asymptotic values vary little from burst to burst, even from year to year, independent of differences in the bursts. This seemed to have the promise of being a way to discover the

binary motion, because differences corresponded to Doppler motions of 20 km s^{-1} which would be reasonable for binary motion. However, as the number of bursts studied has increased, evidence has appeared for larger differences.

(3) The oscillations in the burst tails are consistent with being coherent. There is a characteristic exponential approach to the asymptote ν_a, with the frequency being a function of time: $\nu(t) = \nu_a(1 - \delta e^{-t/\tau})$ [27]. With the frequency variation taken into account, the coherence is consistent with the duration of the signal.

The change in frequency as a function of the decay cries out for a theory, if the spin of the neutron star is being deduced from it. It has been pointed out that if the heated crust is lifted enough by the radiation to be released from viscous drag, so that it can rotate freely and conserve its own angular momentum, only 10-20 M of lift is required to explain the 0.3-1 % change in frequency as the burst cools and the crust presumably subsides. Bildsten calculated such a lift in the development of bursts [28]. ¿From this point of view it appears physically reasonable. However, the implied frequency difference between the neutron star's interior and the crust, of 1-3 Hz, implies that in a few seconds the hot spot at the photosphere has rotated several times around a spot fixed on the neutron star, such as a magnetic pole. If the magnetic field and the crust material are coupled, the magnetic field would have been wound up several times, and be an important factor.

CONCLUSIONS

In Type I bursts, we can truly say that the explosion throws light on the objects and on the process causing the explosion. Limits on the neutron star masses, radii, and distances come from the spectra, and if oscillations are from a rotating hot spot, from the oscillation amplitudes. At least one burster definitely is a low magnetic field ($10^8 - 10^9$ Gauss) fast rotating pulsar and the evidence is strong for millisecond periods in 6 other bursters. The spectral and timing data tell a story of the dynamics of the crust as it is lifted, expelled sometimes, and settles back down on the neutron star.

If oscillations are found in bursts from a pulsing burster like SAX J1808.4-3658, comparison to the pulse period would be very revealing. Finding kHz QPO in the persistent flux from such a source would also be revealing, allowing comparison of the difference frequency with the known pulse frequency. In both cases, it might be that the phenomenon does not occur under the conditions of field and accretion rate pertaining for the pulsar. It is clear that the phenomena depend on the accretion rate and on the magnetic field. More work is needed on whether more sources have burst and persistent flux oscillations and possibly coherent pulsations at some level.

The newly found phenomena have the potential for clarifying the geometry and flow patterns of the accretion and independently constraining the neutron star parameters. Definitive conclusions about the burster neutron stars based on the spectra and luminosities have been hampered by statistical as well as possible systematic errors and by evidence that over determined results were not consistent.

Now that new degrees of freedom are known (fast spin, asymmetrical temperature distribution, disk very close to the neutron star), a verifiable corrected picture should be pursued.

I greatly appreciated the in-depth discussions at the 1999 Aspen Summer workshop "X-Ray Probes of Relativistic Effects near Neutron Stars and Black Holes."

REFERENCES

1. *X-Ray Binaries*, New York: Cambridge Univ. Press, 1995, Chapt. 2, pp 58-125.
2. Bildsten, L. *The Many Faces of Neutron Stars*, Dordrecht: Kluwer, 1998, Chapt. 4, pp 419-449.
3. Strohmayer, T. E., Swank, J. H., and Zhang, W., *The Active X-ray Sky: Results from BeppoSAX and RXTE*, 1998, New York:Elsevier, pp 129-134.
4. Strohmayer, T. E., *X-ray Astronomy '99: Stellar Endpoints, AGN and the Diffuse X-ray Background*, 2000, Bologna, in press.
5. van Paradijs, J.,*Nature.* **274**, 650–653 (1998).
6. Swank, J. H. Becker, R. H., Boldt, E. A., Holt, S. S., Pravdo, S. H., and Serlemitsos, P. J., *ApJ.* **212**, L73 (1977).
7. Waki et al., *Publ. Astron. Soc. Japan.* **36**, 819-830 (1984).
8. London, R. A., Taam, R. E., and Howard, W. M. , *Ap. J.* **287**, L27–30 (1984).
9. Ebisuzaki, T., *Publ. Astron. Soc. Japan.* **39**, 287–308 (1987).
10. Titarchuk, L. , *Soviet Astr. Lett.* **14**, 229 (1988).
11. Ebisuzaki, T., and Nakamura, N., *Ap.J.* **328**, 251–255 (1988).
12. Ebisuzaki, T., and Nakamura, N., *Ap. J.* **328**, 251–255 (1988).
13. van Paradijs, J., Dotani, T., Tanaka, Y., and Tsuru, T., *Publ. Astron. Soc. Japan.* **42**, 633–660 (1990).
14. Goffhelf, E., and Kulkarni, S. R., *Ap. J.* **490**, L161–164 (1997).
15. Tomsick, J., Halpern, J. P., Kemp, J. Kaaret, P., *Ap. J.* **521**, 341–350 (1988).
16. Chakrabarty, D., & Morgan, E. H., *Nature.* **394**, 346 (1998).
17. in't Zand, J. J. M., Heise, J., Muller, J. M., Bazzano, A., Cocchi, M., Natalucci, L., and Ubertini, P., *Astron. and Astrophys..* **331**, L25–2, (1998).
18. Marshall, F. E., Wijnands, R., and van der Klis, M. ., *IAUC.* **6876** (1998).
19. Wijnands, R., and van der Klis, M., *Nature.* **394**, 344 (1998.)
20. Chakrabarty, D., & Morgan, E. H., *Nature.* **394**, 346 (1998).
21. Psaltis, D., and Chakrabarty, D., *Ap. J.* **521**, 332-340 (1999).
22. van der Klis, M., Hasinger, G., Da,en, E., Penninx, W., van Paradijs, J., and Lewin, W. H. G. , *Ap. J.* **360**, L19-22 (1990).
23. Miller, C., *X-ray Astronomy '99: Stellar Endpoints, AGN and the Diffuse X-ray Background.*, 2000, Bologna, in press.
24. Stella, L., Vietri, M., and Morsink, S. M., *Ap. J.* **524**, L63–66 (1999).
25. Strohmayer, T. E., *this volume*, 2000, in press.
26. Bildsten, L., *Ap. J.* **501**, L89–93 (1998).
27. Strohmayer, T. E., and Markwardt, C. B., *Ap. J.* **516**, L81–85 (1999).
28. Bildsten, L., *Ap. J.* **438**, 852–875 (1995).

Theory and Observations of Type I X-Ray Bursts from Neutron Stars

Lars Bildsten

Institute for Theoretical Physics and Department of Physics
Kohn Hall, University of California, Santa Barbara, CA 93106

Abstract. I review our understanding of the thermonuclear instabilities on accreting neutron stars that produce Type I X-Ray bursts. I emphasize those observational and theoretical aspects that should interest the broad audience of this meeting. The easily accessible timescales of the bursts (durations of tens of seconds and recurrence times of hours to days) allow for a very stringent comparison to theory. The largest discrepancy (which was found with *EXOSAT* observations) is the accretion rate dependence of the Type I burst properties. Bursts become less frequent and energetic as the global accretion rate $(\dot{M})$ increases, just the opposite of what the spherical theory predicts. I present a resolution of this issue by taking seriously the observed dependence of the burning area on $\dot{M}$, which implies that as $\dot{M}$ *increases*, the accretion rate per unit area *decreases*. This resurrects the unsolved problem of knowing where the freshly accreted material accumulates on the star, equally relevant to the likely signs of rotation during the bursts summarized by Swank at this meeting. I close by highlighting the Type I bursts from GS 1826-238 that were found with *BeppoSAX* and *RXTE*. Their energetics, recurrence times and temporal profiles clearly indicate that hydrogen is being burned during these bursts, most likely by the rapid-proton (rp) process.

I INTRODUCTION

The gravitational energy release from matter accreted onto a neutron star (NS) of mass M and radius R is $GMm_p/R \approx$ 200 MeV per nucleon, much larger than that released from thermonuclear fusion ($E_{nuc} \approx 5$ MeV per nucleon when a solar mix goes to iron group elements). Hansen and Van Horn (1975) showed that the burning of the accumulated material in the NS atmosphere occurred in radially thin shells and so was susceptible to a thermal instability. Evidence of the instability came soon after with the discovery of recurrent Type I X-ray bursts from low accretion rate ($\dot{M} < 10^{-9}$ $M_\odot$ yr^{-1}) NSs. The successful association of the thermal instabilities found by Hansen & Van Horn (1975) with the X-ray bursts made a nice picture of a recurrent cycle that consists of fuel accumulation for several hours followed by a thermonuclear runaway that burns the fuel in $\sim 10-100$ seconds (see Lewin, van Paradijs and Taam 1995 for an overview and references).

CP522, *Cosmic Explosions: Tenth Astrophysical Conference,*
edited by Stephen S. Holt and William W. Zhang
© 2000 American Institute of Physics 1-56396-943-2/00/$17.00

The observational quantity α (defined as the ratio of the time-averaged accretion luminosity to the time-averaged burst luminosity) is close to the value expected (i.e. $\alpha = (GM/R)/E_{nuc} \approx 40$) for a thermonuclear burst origin.

Though our basic understanding from 25 years ago is unchanged, we now know much more about how the thermal instability depends on $\dot{M}$, both theoretically and observationally. It is this comparison that I emphasize, as it provides many important lessons that are likely applicable to thin shell flashes on accreting white dwarfs (classical novae); where 100-1000 yr recurrence times prohibit such detailed comparisons. I focus solely on NSs accreting at $\dot{M} > 10^{-10}$ $M_\odot$ yr^{-1}, which is appropriate for most persistently bright Low Mass X-ray Binaries (in particular the "Z" and "Atoll" sources of Hasinger & van der Klis 1989). These NSs are weakly magnetic, with $B < 10^{10}$G.

I start by reviewing the simplest aspects of the physics of the accumulation and ignition of the fresh fuel on the NS (leaning heavily on results from my Bildsten 1998 review article, to which I refer the reader for the complete set of original references). I then discuss the *EXOSAT* observations of the $\dot{M}$ dependence of the Type I X-Ray burst properties and speculate that a solution to these puzzles is possible if freshly accreted matter accumulates near the equator. This problem, as well as the observations of nearly coherent oscillations during the burst (summarized by Swank at this meeting) are the first good indicators of the breaking of spherical symmetry. I close with a detailed discussion of the Type I bursts from the binary GS 1826-234, which is a beautiful example of limit-cycle mixed hydrogen/helium burning.

II ACCUMULATION, IGNITION, EXPLOSION

Once the freshly accreted hydrogen and helium has thermalized and become part of the "star", it undergoes hydrostatic compression from the new material that is continuously piled on. The extreme gravity on the NS surface compresses the fresh fuel to ignition densities and temperatures within a few hours to days.[1] The short thermal time in the atmosphere (only ~ 10 s at the ignition location, $P \approx 10^{22} - 10^{23}$ erg cm^{-3}) compared to the time to accumulate the material (hours to days) makes the compression far from adiabatic. Indeed, the temperature contrast from the photosphere to the burning layer is a factor of ten; whereas the density contrast exceeds 10^4.

The temperature exceeds 10^7 K in most of the accumulating atmosphere, so that hydrogen burns via the CNO cycle and we can neglect the pp cycles. At high temperatures ($T > 8 \times 10^7$ K), the timescale for proton captures becomes shorter than the subsequent β decay lifetimes, even for the slowest ^{14}N(p,γ)^{15}O reaction. The hydrogen then burns in the "hot" CNO

---

[1] The physics of the compression and burning depends on the accretion rate per unit area, $\dot{m} \equiv \dot{M}/A_{acc}$, where A_{acc} is the covered area of fresh material. I sometimes quote numbers for both $\dot{m}$ and $\dot{M}$. When I give $\dot{M}$, I have assumed $A_{acc} = 4\pi R^2 \approx 1.2 \times 10^{13}$ cm^2.

$$^{12}\text{C}(p,\gamma)^{13}\text{N}(p,\gamma)^{14}\text{O}(\beta^+)^{14}\text{N}(p,\gamma)^{15}\text{O}(\beta^+)^{15}\text{N}(p,\alpha)^{12}\text{C}, \qquad (1)$$

cycle and is limited to $5.8 \times 10^{15} Z_{\text{CNO}}$ ergs g^{-1} s^{-1}, where Z_{CNO} is the mass fraction of CNO and is *independent of temperature*. The hydrogen burns this way in the accumulating phase when $\dot{m} > 900$ g cm^{-2} s$^{-1}(Z_{\text{CNO}}/0.01)^{1/2}$ and is thermally stable. The amount of time it takes to burn the hydrogen is $\approx (10^3/Z_{\text{CNO}})$ s, or about one day for solar metallicities. For lower $\dot{m}$'s, the hydrogen burning is thermally unstable and is the trigger for the Type I burst.

The slow hydrogen burning during the accumulation allows for a unique burning regime at high $\dot{m}$'s. This simultaneous H/He burning occurs when $\dot{m} > (2 - 5) \times 10^3$ g cm^{-2} s$^{-1}(Z_{CNO}/0.01)^{13/18}$, as at these high rates the fluid element is compressed to helium ignition conditions long before the hydrogen is completely burned (Lamb and Lamb 1978, Taam and Picklum 1978). The strong temperature dependence of the helium burning rate (and lack of any weak interactions) leads to a strong thin-shell instability for temperatures $T < 5 \times 10^8$ K and causes the Type I X-Ray burst for these $\dot{m}$'s. The critical condition of thin burning shells ($h \ll R$) is true before burning and remains so even during the flash (when temperatures reach 10^9 K) as the large gravitational well on the neutron star requires temperatures of order 10^{12} K for $h \sim R$. Stable burning sets in at higher $\dot{M}$'s (comparable to the Eddington limit) when the helium burning temperature sensitivity finally becomes weaker than the cooling rate's sensitivity (Ayasli & Joss 1982 and Taam, Woosley & Lamb 1996).

For solar metallicities, there is a narrow window of $\dot{m}$'s where the hydrogen is completely burned before the helium ignites. In this case, a pure helium shell accumulates underneath the hydrogen-burning shell until densities and pressures are reached for ignition of the pure helium layer. The recurrence times of these bursts must be longer than the time to burn all of the hydrogen, so pure helium flashes should have recurrence times in excess of a day and $\alpha \approx 200$. To summarize, in order of increasing $\dot{m}$, the regimes of unstable burning we expect to witness from NSs accreting at sub-Eddington rates ($\dot{m} < 10^5$ g cm^{-2} s^{-1}) are (Fujimoto, Hanawa & Miyaji 1981, Fushiki and Lamb 1987):

1. Mixed hydrogen and helium burning triggered by thermally unstable hydrogen ignition for $\dot{m} < 900$ g cm^{-2} s^{-1} ($\dot{M} < 2 \times 10^{-10} M_\odot$ yr^{-1}).

2. Pure helium shell ignition for 900 g cm^{-2} s$^{-1} < \dot{m} < (2-5) \times 10^3$ g cm^{-2} s^{-1} following completion of hydrogen burning.

3. Mixed hydrogen and helium burning triggered by thermally unstable helium ignition for $\dot{m} > (2-5) \times 10^3$ g cm^{-2} s^{-1} ($\dot{M} > 4.4 - 11.1 \times 10^{-10} M_\odot$ yr^{-1}).

The transition $\dot{m}$'s are for $Z_{CNO} \approx 0.01$. Reducing Z_{CNO} lowers the transition accretion rates and, more importantly, makes the $\dot{m}$ range for pure helium ignition quite narrow. We now discuss what happens as the thermal instability develops into a burst and what observational differences are to be expected between a pure helium ignition and a mixed hydrogen/helium ignition.

The flash occurs at fixed pressure, and the increasing temperature eventually allows the radiation pressure to dominate. For an ignition column of 10^8 g cm^{-2}, the pressure is $P = gy \approx 10^{22}$ ergs cm^{-3}, so $aT_{max}^4/3 \approx P$ gives a maximum temperature $T_{max} \approx 1.5 \times 10^9$ K. For pure helium flashes, the fuel rapidly burns (since there are no limiting weak interactions) and the local Eddington limit is often exceeded, leading to a radius expansion burst and a duration set mostly by the time it takes the heat to escape, $\sim 5 - 10$ seconds.

When hydrogen and helium are both present, the high temperatures reached during the thermal instability easily produces elements far beyond the iron group (Hanawa et al. 1983; Wallace & Woosley 1984; Hanawa and Fujimoto 1984) via the rapid-proton (rp) process of Wallace and Woosley (1981). This burning starts a few seconds after the initial helium flash (see Hanawa and Fujimoto 1984 for an illuminating example) that makes new seed nuclei and increases the temperature. The rp process burns hydrogen by successive proton captures and β decays. The seed nuclei move up the proton-rich side of the valley of stability (much like the r-process which occurs by neutron captures on the neutron rich side) more or less limited by the β-decay rates. Theoretical work shows that the end-point of this time-dependent burning is at elements far heavier than iron (Hanawa and Fujimoto 1984, Schatz et al. 1997, Koike et al. 1999).[2] The long series of β decays allows for energy release 10-100 seconds after the burst has started. We thus expect a mixed hydrogen/helium burst to last much longer than a pure helium burst.

III   OBSERVATIONS OF $\dot{M}$ DEPENDENCIES

The 3.8 day orbit of *EXOSAT* was an excellent match for the long-term monitoring of the Type I bursters needed to reveal the dependence of their nuclear burning behavior on $\dot{M}$.[3] While in a particular burning regime, we expect that the time between bursts should decrease as $\dot{M}$ increases since it takes less time to accumulate the critical amount of fuel at a higher $\dot{M}$. Exactly the *opposite* behavior was observed from many low accretion rate ($\dot{M} < 10^{-9} M_\odot$ yr^{-1}) NSs. A particularly good example is 4U 1705-44, where the recurrence time increased by a factor of ≈ 4 when $\dot{M}$ increased by a factor of ≈ 2 (Langmeier et al. 1987, Gottwald et al. 1989). If the star is accreting matter with $Z_{CNO} = 10^{-2}$ then these accretion rates are at the boundary between unstable helium ignition in a hydrogen-rich environment at high $\dot{M}$ and unstable pure helium ignition at lower $\dot{M}$. The expected change in burst behavior as $\dot{M}$ increases would then be to more energetic and more frequent bursts. This was not observed.

---

[2] In steady-state burning at $\dot{M} > 10^{-8} M_\odot$yr^{-1} Schatz et al. (1999) showed that the rp-process burns all of the hydrogen and ends at nuclei with A near 100.

[3] I hope that the equally well-matched *Chandra* and *XMM* satellites will devote as much time to Type I bursters. As I will make clear from this discussion, detailed spectroscopy during the bursts would be very informative.

Other NSs showed similar behavior. van Paradijs, Penninx & Lewin (1988) tabulated this effect for many bursters and concluded that increasing amounts of fuel are consumed in a less visible way than Type I X-ray bursts as $\dot{M}$ increases. The following trends were always found as $\dot{M}$ increases:

- The recurrence time increases from 2-4 hours to > day.

- The bursts burn less of the accumulated fuel, with α increasing from ≈ 40 to > 100 (see top panel of Figure 1).

- The duration of the bursts decrease from ≈ 30 s to ~ 5 s.

The low $\dot{M}$ bursts look like mixed hydrogen/helium burning (namely, energetic and of long duration from the rp-process) whereas the high $\dot{M}$ bursts look like pure helium burning (not so energetic, recurrence times typically long enough to have burned the hydrogen to helium before the burst and short duration due to the lack of any weak interactions). The simplest explanation would be to say that the NS has transitioned from the low $\dot{M}$ mixed burning regime (noted as 1 in §II) to the higher $\dot{M}$ pure helium burning (noted as 2 in §II). For this to be true, these NSs should be accreting at $\approx 10^{-10} M_\odot$ yr^{-1} in the lower $\dot{M}$ state and a factor of 4-5 higher in the high $\dot{M}$ state. However, these estimates are not consistent with the observations.

van Paradijs et al. (1988) estimated the $\dot{M}$ from those bursters which have shown Eddington limited radius expansion bursts. For these systems, the ratio of the persistent flux to the flux during radius expansion measures the accretion rate in units of the Eddington accretion rate $(2 \times 10^{-8} M_\odot$yr$^{-1})$. They showed that most bursters accrete at rates $\dot{M} \approx (3 - 30) \times 10^{-10}$ $M_\odot$ yr^{-1}, at least a factor of three (and typically more) higher than the calculated rate where such a transition should occur. Moreover, if the accretion rates were as low as needed, the recurrence times for the mixed hydrogen/helium burning would be about 30 hours, rather than the observed 2-4 hours. Fujimoto et al. (1987) discussed in some detail the challenges these observations present to a spherically symmetric model, while Bildsten (1995) attempted to resolve this by having much of the thermally unstable burning occur via slow deflagration fronts that do not lead to Type I bursts, but rather slow hour-long flares.

Another comparably embarrassing conundrum is the lack of regular bursting from the six "Z" sources (Sco X-1, Cyg X-2, GX 5-1, GX 17+2, GX 340+0, GX 349+2) which are accreting at $3 \times 10^{-9} - 2 \times 10^{-8}$ $M_\odot$ yr^{-1}. These NSs very rarely show Type I bursts, and when they do, they are so infrequent that the resulting α values are usually $> 10^3$ (see Kuulkers et al. 1997 and Smale 1998 for examples and discussions). In other words, these bursts are clearly not responsible for burning all of the accreted fuel, whereas theory clearly says that these objects should be burning nearly all of their fuel unstably in the mixed hydrogen/helium regime (noted as 3 above). The same mystery holds for the Atoll sources with $\dot{M} \sim 10^{-9} M_\odot$ yr^{-1} (GX 3+1, GX 13+1, GX 9+1 and GX 9+9), which at best are infrequent bursters.

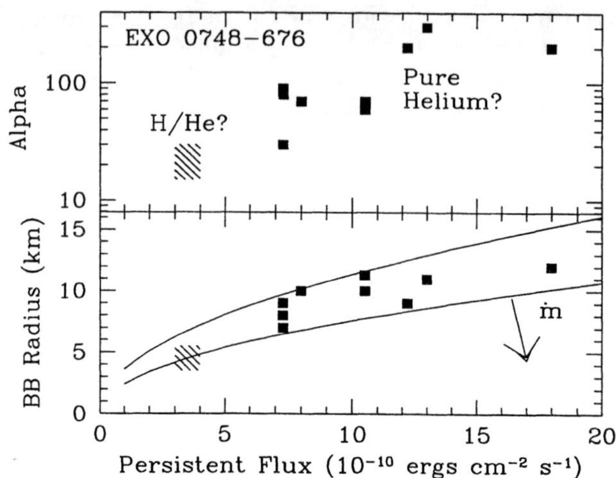

FIGURE 1. Properties of the Type I X-ray bursts from EXO 0748-676 from Gottwald et al. (1986). The value of α (top panel) and the apparent black-body radius (R_{app} for $d = 10$ kpc) in the tail of the burst, are shown as a function of the persistent X-ray flux, $F_x = L_x/4\pi d^2$. The solid squares denote those that are well measured (for some of these, the α parameters and R_{app} are lower limits; see Gottwald et al. 1986). The hatched region is where burst properties cluster at low F_x. The lower (upper) solid line on the bottom panel denotes where the accretion rate per unit area is constant at $\dot{m} = 8.3 \times 10^3$ g cm^{-2} s^{-1} ($\dot{m} = 3.7 \times 10^3$ g cm^{-2} s^{-1}). The arrow points in the direction of increasing $\dot{m}$.

IV ACCUMULATION IN THE EQUATOR?

Many of these puzzles are resolved by relaxing our spherical symmetry presumption and allowing the fresh material to only cover a fraction of the star prior to igniting. There are observational hints that this is happening, as the other clear trend (in addition to those noted in the previous section) found by *EXOSAT* was an increase in the apparent black-body radius (R_{app}) as $\dot{M}$ increased (see bottom panel in Figure 1). This parameter is found by spectral fitting in the decaying tail of the Type I bursts and is susceptible to absolute spectral corrections (see discussion in Lewin, van Paradijs and Taam 1995) that will hopefully be resolved with *XMM* observations. In a similar vein, van der Klis et al. (1990) found that the temperature of the burst at the moment when the flux was one-tenth the Eddington limit decreased as $\dot{M}$ increased (hence a larger area) for the Atoll source 4U 1636-53. In total, these observations raise the distinct possibility that the covered area increases enough with increasing $\dot{M}$ so that the accretion rate per unit area actually *decreases*.

By interpreting the measured R_{app} as an indication of the fraction of the star that is covered by freshly accreted fuel, we can measure directly $\dot{m} = \dot{M}/4\pi R_{app}^2$, which is independent of the distance to the source, as $F_x = GM\dot{M}/4\pi d^2 R$ gives $\dot{m} \approx (F_x R/GM)(d/R_{app})^2$. The bottom panel of Figure 1 shows data for the burster EXO 0748-676. The lower (upper) solid lines are curves where the local accretion rate is constant at $\dot{m} = 8.3 \times 10^3$ g cm^{-2} s^{-1} ($\dot{m} = 3.7 \times 10^3$ g cm^{-2} s^{-1}). The arrow points in the direction of increasing $\dot{m}$. The points at higher $\dot{M}$ (as inferred from F_x) tend to lie at comparable or slightly lower $\dot{m}$. The radius increase appears adequate to offset the $\dot{M}$ increase. In addition, for this source, the inferred values of $\dot{m}$ are in the range where the NS is transitioning from the mixed H/He burning at high $\dot{m}$ to the pure helium case at lower $\dot{m}$. More physically stated, the data point to the possibility that, as $\dot{M}$ increases, the area increases fast enough to allow the hydrogen to complete its burning before high enough pressures are reached for helium ignition. If such small covering areas persist to the higher $\dot{M}$'s of the Z sources, then their apparently stable nuclear burning is easily explained.

We know that these NSs accrete from a disk formed in the Roche lobe overflow of the stellar companion. However, there are still debates about the "final plunge" onto the NS surface. Some advocate that a magnetic field controls the final infall, while others prefer an accretion disk boundary layer. This is now an important issue to resolve, both for the reasons I have noted here as well as for the oscillations seen during the bursts. If material is placed in the equatorial belt, it is not clear that it will stay there very long. If angular momentum was not an issue, the lighter accreted fuel (relative to the ashes) would cover the whole star quickly. However, on these rapidly rotating neutron stars, the fresh matter added at the equator must lose angular momentum to get to the pole. This competition (namely understanding the spreading of a lighter fluid on a rotating star) has just been recently investigated by Inogamov and Sunyaev (1999), to which I refer the interested reader.

365

V   AN EXAMPLE OF MIXED H/HE BURSTS

Despite the complications I discussed in the previous sections, there are times when Type I bursters behave in a near limit cycle manner, with bursts occurring nearly periodically as $\dot{m}$ apparently stays at a fixed value for a long time. The most recent (and beautiful!) example of such a Type I burster is GS 1826-238. Ubertini et al. (1999) show that during 2.5 years of monitoring with the *BeppoSAX* Wide Field Camera, 70 bursts were detected from this object with a quasi-periodic recurrence time of 5.76 ± 0.62 hours. The persistent flux during this bursting period was $F_x \approx 2 \times 10^{-9}$ erg cm^{-2} s^{-1} (Ubertini et al. 1999, In 't Zand et al. 1999, Kong et al. 2000), which when combined with the measured ratio of R_{app}/d (about 9 km at 8 kpc; Kong et al. 2000) gives $\dot{m} \approx 8 \times 10^3$ g cm^{-2} s^{-1}, safely in the mixed H/He burning regime.

If at the upper limit distance of 8 kpc (in 't Zand et al. 1999) the global accretion rate is $\dot{M} \approx 10^{-9} M_\odot$ yr^{-1} and apparently the whole star is covered with fresh material. No coherent pulsations have been detected in these bursts, so we do not know this NS's rotation rate.

The type I bursts from this source are a "textbook" case for the mixed hydrogen/helium burning expected at these accretion rates. The estimated $\dot{m}$ gives an accumulated column on the NS prior to the burst of 1.6×10^8 g cm^{-2}, just what is expected from theory. These quasi-periodic bursts allow for a very secure measurement of $\alpha \approx 50$, which implies a nuclear energy release of 4 MeV per accreted nucleon for a $1.4 M_\odot$, 10 km NS. Energy releases this large can only come about via hydrogen burning and the long (> 100s) duration of the bursts are consistent with the expected long-time energy release from the rp-process. Figure 2 shows the time profile of such a burst seen with RXTE (Kong et al. 2000) in a few different energy bands. Though these data were taken to study the optical reprocessing (top panel is the simultaneous optical burst), they provide an important confirmation of the delayed energy release expected when hydrogen is burning via an rp-process. The resemblence of these profiles to Hanawa and Fujimoto's (1984) theoretical results are striking.

The upcoming launch of the *High Energy Transient Explorer* should provide comparable long-term coverage as the Wide Field Camera on *BeppoSAX* and gather more information on such nice bursts from many more LMXB's.

VI   CONCLUSIONS

I hope I have made the case that Type I bursts from neutron stars are still very interesting to study in their own right and provide important lessons to those studying thin shell flashes in other astrophysical contexts. The detailed comparison provided by the neutron star systems is likely telling us to seriously consider the possibility and repercussions of fuel preferentially accumulating in the equatorial region. Another place where this might prove immediately applicable are the recur-

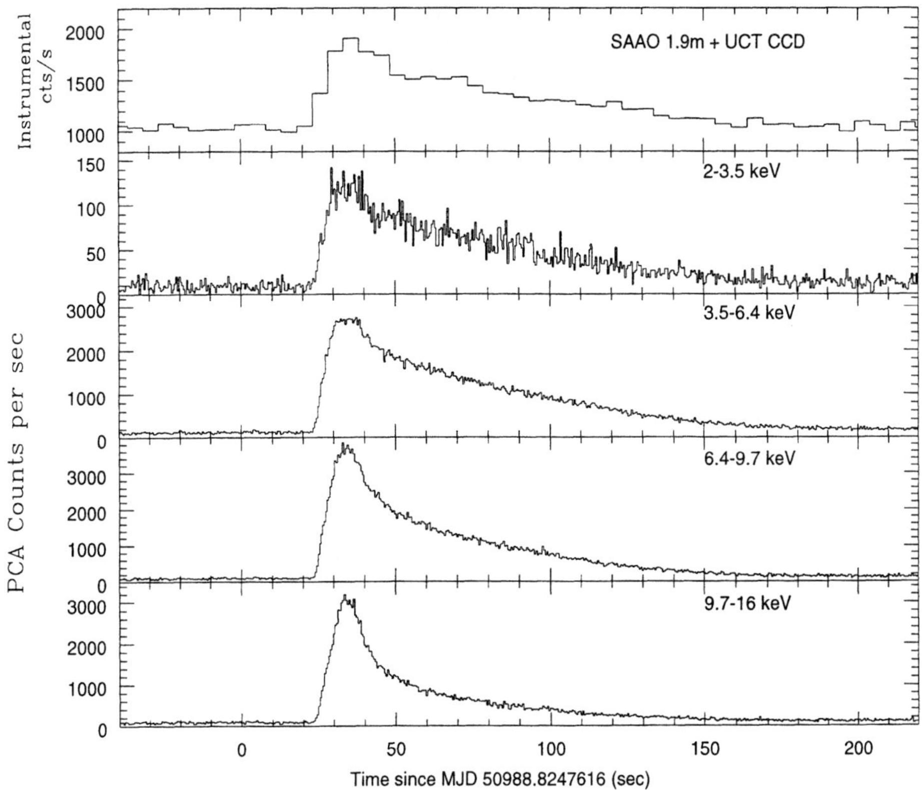

FIGURE 2. A montage showing the Type I burst profile from GS 1826-238 from RXTE as well as the reprocessed optical emission (from Kong et al. 2000). The long duration of the burst is indicative of the delayed energy release from the rapid proton (rp) process.

rent novae, where currently one infers high accretion rates and white dwarf masses in order to get the short recurrence times of 20-50 years (Livio 1994). These constraints are relaxed if we allow for a smaller covering fraction. I am not the first to say this, but hopefully the Type I burst observations make such a warning harder to ignore!

Jean Swank reviewed the observations of nearly coherent oscillations in the 300-600 Hertz range during many Type I bursts and so I will not summarize those results here. Though I am convinced that these modulations are intimately connected to rapid stellar rotation, there are still important unresolved questions. The ones that bother me the most are:

1. What causes the asymmetry at late times in the burst, long after the peak?

2. Why does the modulation appear sometimes at twice the spin frequency?

3. How does the burning front really spread on a rapidly rotating star? Ignition at one spot is plausible, but we do not understand how the ignited/hot fuel spreads around a rapidly rotating star.

It is even an open question as to why, from a particular NS, only some bursts show these oscillations. Before any meaningful theoretical work can be carried out, what is needed is the phenomenology of the oscillations in the context of the well established burst phenomenology I have discussed here. My current mental tabulations point to a complete abscence of oscillations during the long bursts (even during the long rise) indicative of mixed H/He burning. All reported detections of oscillations during bursts that I am aware of are from short duration, high α bursts.

I thank Andrew Cumming, Erik Kuulkers, and Michiel van der Klis for discussions and comments on the manuscript. My recent plunge into the *EXOSAT* Type I burst literature occurred when I was the CHEAF Visiting Professor at the Astronomical Institute "Anton Pannekoek" of the University of Amsterdam. I thank them for the hospitality and Michiel for reminding me to eventually place his 1990 article on 4U 1636-53 in a meaningful context. This research was supported by NASA via grant NAG 5-8658 and by the NSF under Grant PHY 94-07194. L. B. is a Cottrell Scholar of the Research Corporation.

REFERENCES

1. Ayasli, S., & Joss, P. C. 1982, ApJ, 256, 637
2. Bildsten, L. 1995, ApJ, 438, 852
3. Bildsten, L. 1998, in "The Many Faces of Neutron Stars" ed. R. Buccheri, J. van Paradijs, & M. A. Alpar (Dordrecht: Kluwer), p. 419
4. Fujimoto, M. Y., Hanawa, T., & Miyaji, S. 1981, ApJ, 247, 267
5. Fujimoto, M. Y., Sztajno, M., Lewin, W. H. G. & van Paradijs, J. 1987, ApJ, 319, 902

6. Fushiki, I. & Lamb, D. Q. 1987, ApJ, 323, L55

7. Gottwald, M., Haberl, F., Parmar, A. N. & White, N. E. 1986, ApJ, 308, 213

8. Gottwald, M. et al. 1989, ApJ, 339, 1044

9. Hanawa, T. & Fujimoto, M. Y. 1984, PASJ, 36, 199

10. Hanawa, T., Sugimoto, D. & Hashimoto, M. 1983, PASJ, 35, 491

11. Hansen, C. J. & Van Horn, H. M. 1975, ApJ 195, 735

12. Hasinger, G. & van der Klis, M. 1989, A&A, 225, 79

13. Inogamov, N. A. & Sunyaev, R. A. 1999, Astron. Letters., 25, 269, astro-ph/9904333

14. in 't Zand, J.J.M., Heise, J., Kuulkers, E., Bazzano, A., Cocchi, M., & Ubertini, P. 1999, A&A, 347, 891

15. Koike, O., Hashimoto, M., Arai, K., Wanajo, S. 1999, A&A, 342, 464

16. Kong, A.K.H et al. 2000, MNRAS, 311, 405

17. Kuulkers, E., van der Klis, M., Oosterbroek, T., van Paradijs, J., Lewin, W.H.G., 1997, MNRAS, 287, 495

18. Lamb, D.Q. & Lamb, F.K. 1978, ApJ, 220, 291

19. Langmeier, A., Sztajno, M., Hasinger, G., Trümper, J., & Gottwald, M. 1987, ApJ, 323, 288

20. Lewin, W. H. G., van Paradijs, J., & Taam, R. E. 1995, in "X-Ray Binaries", ed. W. H. G. Lewin, J. van Paradijs & E. P. J. van den Heuvel (London: Cambridge), p. 175

21. Livio, M. 1994, in "Interacting Binaries", ed. S. N. Shore, M. Livio, & E. P. J. van den Heuvel (Berlin: Spinger-Verlag), p. 135

22. Schatz, H., Bildsten, L., Gorres, J., Wiescher, M., & Thielemann, F.K., 1997, in "Intersections between Particle and Nuclear Physics", ed. T. W. Donnelly (New York: AIP), p. 987

23. Schatz, H., Bildsten, L., Cumming, A. and Wiescher, M. 1999, ApJ, 524, 1014

24. Smale, A. P. 1998, ApJ, 498, L141

25. Taam, R. E. & Picklum, R.E. 1978, ApJ, 224, 210

26. Taam, R. E., Woosley, S. E., & Lamb, D. Q. 1996, ApJ, 459, 271

27. Ubertini, P. et al. 1999, ApJ, 514, L27

28. van der Klis, M. et al. 1990, ApJ, 360, L19

29. van Paradijs, J., Penninx, W. & Lewin, W. H. G. 1988, MNRAS, 233, 437

30. Wallace, R. K., & Woosley, S. E. 1981, ApJS, 43, 389

31. Wallace, R. K., & Woosley, S. E. 1984, in "High Energy Transients in Astrophysics", ed. S.E. Woosley (New York: AIP), p. 273

Extracting Neutron Star Properties from X-ray Burst Oscillations

Nevin Weinberg[*], M. Coleman Miller[†], and Donald Q. Lamb[*]

[*]*University of Chicago, Department of Astronomy and Astrophysics,*
5640 South Ellis Ave., Chicago, IL 60637
[†]*University of Maryland, Department of Astronomy, College Park, MD 20742-2421*

Abstract. Many thermonuclear X-ray bursts exhibit brightness oscillations. The brightness oscillations are thought to be due to the combined effects of non-uniform nuclear burning and rotation of the neutron star. The waveforms of the oscillations contain information about the size and number of burning regions. They also contain substantial information about the mass and radius of the star, and hence about strong gravity and the equation of state of matter at supranuclear densities. We have written general relativistic ray-tracing codes that compute the waveforms and spectra of rotating hot spots as a function of photon energy. Using these codes, we survey the effect on the oscillation waveform and amplitude of parameters such as the compactness of the star, the spot size, the surface rotation velocity, and whether there are one or two spots. We also fit phase lag versus photon energy curves to data from the millisecond X-ray pulsar, SAX J1808–3658.

INTRODUCTION

Shortly after the launch of the *Rossi* X-ray Timing Explorer (RXTE) in late 1995, single kilohertz brightness oscillations were discovered in RXTE countrate time series data from thermonuclear X-ray bursts in several neutron-star low-mass X-ray binaries. These oscillations are remarkably coherent and their frequencies are very stable from burst to burst in a given source [1]. These oscillations are therefore thought to be at the stellar spin frequency or its first overtone. This suggests that the oscillations are caused by rotational modulation of a hot spot produced by non-uniform nuclear burning and propagation. Analysis of these oscillations can therefore constrain the mass and radius of the star and yield valuable information about the speed and type of thermonuclear propagation. In turn, this has implications for strong gravity and dense matter, and for astrophysical thermonuclear propagation in other contexts, such as classical novae and Type Ia supernovae.

A comparison of theoretical waveforms with the observations is required to extract this fundamental information. Here we exhibit waveform calculations that we have produced using general relativistic ray-tracing codes. We survey the effects of

CP522, *Cosmic Explosions: Tenth Astrophysical Conference,*
edited by Stephen S. Holt and William W. Zhang

parameters such as the spot size, the stellar compactness, and the stellar rotational velocity, and demonstrate that our computations fit well the phase lag data from SAX J1808–3658.

COMPUTATIONAL METHOD

To compute observed light curves, we do general relativistic ray tracing from points on the surface to the observer at infinity in a way similar to, but more general than, [2] and [3]. For simplicity, we assume that the exterior spacetime is Schwarzschild, that the surface is dark except for the hot spot or spots, and that there is no background emission. The amplitudes would be reduced by a constant factor if there were background emission. The angular dependence of the specific intensity at the surface depends on both radiation transfer effects and Doppler boosting (see [4]). For each phase of rotation we compute the projected area of many small elements of a given finite size spot. We then build up the light curve of the entire spot by superposing the light curve of all the small elements. The grid resolution of the spot is chosen so that the effect of having a finite number of small elements can alter the value of the computed oscillation amplitudes by a fraction no larger than $\sim 10^{-4}$. After computing the oscillation waveform using the above approach, we Fourier-analyze the resulting light curve to determine the oscillation amplitudes and phases as a function of photon energy at different harmonics.

RESULTS

Panel (a) of Figure 1 shows the fractional rms amplitudes at the first two harmonics as a function of spot size and stellar compactness for one emitting spot centered on the rotational equator, as seen by a distant observer in the rotational plane. These curves demonstrate that a common result of the hot-spot model is large-amplitude brightness oscillations with a high contrast in strength between the dominant harmonic and weaker harmonics, as is observed in several sources. The curves for the first harmonic illustrate the general shape of most of the first harmonic curves. Initially, the amplitude depends only weakly on spot size. However, once the spot grows to an angular radius of $\sim 40°$ there is a steep decline in the oscillation amplitude which flattens out only near the tail of the expansion. This expected behavior appears to be in conflict with the decline in amplitude observed by Strohmayer, Zhang, & Swank (1997) from 4U 1728–34, in which the initial decline is steep. The cause of this could be that the initial velocity of propagation is large, or that the observed amplitude is diminished significantly by isotropization of the beam due to scattering (Weinberg, Miller, & Lamb 1999).

Panel (b) of Figure 1 shows the fractional rms amplitude at the second harmonic under the same assumptions but for two identical, antipodal emitting spots. The range in spot size here is $0° - 90°$ since two antipodal spots of $90°$ radii cover the entire stellar surface. Note that in this situation, there is no first harmonic.

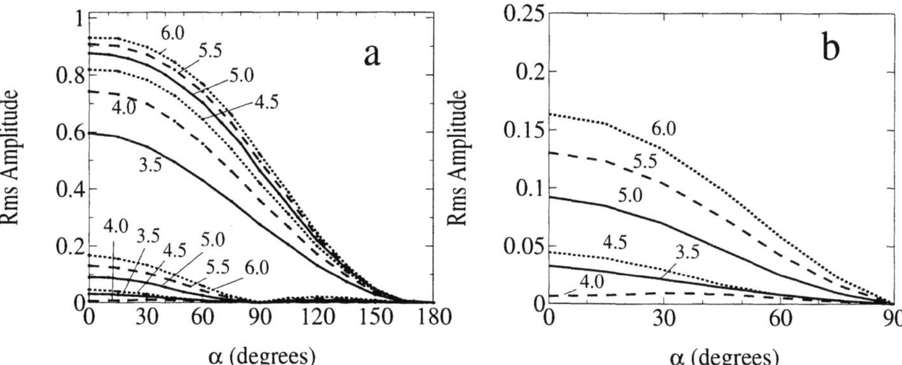

FIGURE 1. (a) Bolometric rms amplitude vs. the angular radius α of the spot at the first harmonic (upper curves) and the second harmonic (lower curves) from a single emitting spot centered on the rotational equator as seen by a distant observer in the rotational plane. Numbers denote values of R/M, where we use geometrized units in which $G = c \equiv 1$. (b) Rms amplitude vs. α at the second harmonic from two antipodal emitting spots, with both the spots and the distant observer in the rotational plane. Note the change in vertical scale in panel (b). In both cases we assume a nonrotating star and an isotropic specific intensity as measured by a local comoving observer.

These figures show that when there is only one emitting spot, the fundamental is always much stronger than higher harmonics. Thus, a source such as 4U 1636–536 with a stronger first overtone than fundamental [6] must have two nearly antipodal emitting spots. As described in detail in [4], we confirm the results of [2] and [3] that the rms amplitude decreases with increasing compactness until $R/M \approx 4$, then increases due to the formation of caustics. We also find that a finite surface rotational velocity increases the amplitude at the second harmonic substantially, while having a relatively small effect on the first harmonic (left panel of Figure 2).

As an application to data, in the right panel of Figure 2 we use our models to fit phase lag data from the millisecond accreting X-ray pulsar SAX J1808–3658. The waveforms from the accreting spot are expected to be similar to the waveforms from burst brightness oscillations, and the signal to noise for this source greatly exceeds that from burst sources such as Aql X-1 [7]. As is apparent from the figure, the fit is excellent. Further data, especially from a high-area timing mission, could be used to constrain the stellar mass or radius from phase lag data.

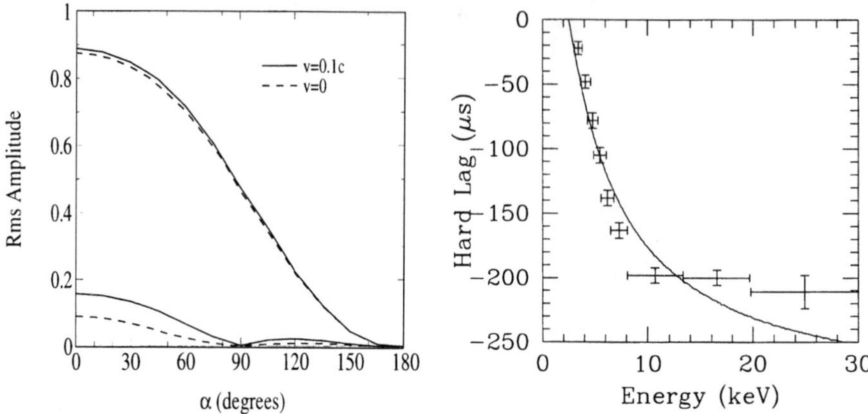

FIGURE 2. (left panel) Rms amplitude versus spot angular radius α at the first harmonic (upper curves) and the second harmonic (lower curves) from a single emitting spot with $v = 0.1c$ (*solid lines*) and $v = 0$ (*dashed lines*), $R/M = 5.0$, and a spot and a distant observer both in the rotational equator. (right panel) Phase lags versus photon energy for the millisecond X-ray pulsar SAX J1808–3658. The data (crosses) are from [8], where the reference energy is the 2–3 keV band. The solid line shows the phase lags in a Doppler shift model, assuming a gravitational mass of $1.8\,M_\odot$ and a rotational velocity of 0.1 c as measured at infinity, which would be appropriate for the observed 401 Hz spin frequency and $R = 11$ km. The angular and spectral emission at the surface are that of a grey atmosphere with an effective temperature of 0.6 keV as measured at infinity. The excellent fit apparent in this figure supports the Doppler shift explanation for the soft lags in this source.

REFERENCES

1. Strohmayer, T. E., Swank, J. H., & Zhang, W. 1998 (astro-ph/9801219)
2. Pechenick, K. R., Ftaclas, C., & Cohen, J. M. 1983, ApJ, 274, 846
3. Miller, M. C., & Lamb, F. K. 1998, ApJ, 499, L37
4. Weinberg, N., Miller, M. C., & Lamb, D. Q. 1999, in preparation
5. Strohmayer, T. E., Zhang, W., & Swank, J. H. 1997, ApJ, 487, L77
6. Miller, M. C. 1999, ApJ, 515, L77
7. Ford, E. C. 1999, ApJ, 519, L73
8. Cui, W., Morgan, E. H., & Titarchuk, L. G. 1998, ApJ, 504, L27

Phase Resolved Spectroscopy of Burst Oscillations: Searching for Rotational Doppler Shifts

Tod E. Strohmayer

Laboratory for High Energy Astrophysics
NASA's Goddard Space Flight Center
Mail Code 662, Greenbelt, MD 20771

Abstract. X-ray brightness oscillations with frequencies from 300 - 600 Hz have been observed in six low mass X-ray binary (LMXB) bursters with the Rossi X-ray Timing Explorer (RXTE). These oscillations likely result from spin modulation of a hot region on the stellar surface produced by thermonuclear burning. If this hypothesis is correct the rotational velocity of the stellar surface, $\approx 0.1\ c$, will introduce a pulse phase dependent Doppler shift such that the rising edge of a pulse should be harder (blue shifted) than the trailing edge (red shifted). Detection of this effect would both provide further compelling evidence for the spin modulation hypothesis as well as providing new observational techniques with which to constrain the masses and radii of neutron stars. In this work I present results of an attempt to search for such Doppler shifts.

INTRODUCTION

A rotating hot spot (or spots) seems to be the most simple, consistent scenario proposed to date to explain the presence of large amplitude modulations of the X-ray brightness during thermonuclear bursts. The presence of large amplitudes at burst onset combined with spectral evidence for localized X-ray emission supports this hypothesis [1]. Additional evidence for spin modulation is provided by the fact that the oscillation frequencies are stable over year timescales and within bursts the oscillations are highly coherent [2,3]

Detailed studies of the burst oscillations hold great promise for providing new insights into a variety of physics issues related to the structure and evolution of neutron stars. For example, within the context of the rotating hot spot model it is possible to determine constraints on the mass and radius of the neutron star from measurements of the maximum observed modulation amplitudes during X-ray bursts as well as the harmonic content of the pulses [4,5]. Phase resolved X-ray spectroscopy of the burst oscillations also holds great promise, and could yield methods to constrain the radii of neutron stars, a quantity which is extremely difficult to

CP522, *Cosmic Explosions: Tenth Astrophysical Conference,*
edited by Stephen S. Holt and William W. Zhang
2000 American Institute of Physics 1-56396-943-2

infer on its own. For example, a 10 km radius neutron star spinning at 400 Hz has a surface velocity of $v_{spin}/c \leq 2\pi\nu_{spin}R/c \approx 0.084$ at the rotational equator. This motion of the hot spot produces a Doppler shift of magnitude $\Delta E/E \approx v_{spin}/c$, thus the observed spectrum is a function of pulse phase [6]. Measurement of this phase dependent Doppler shift would provide further compelling evidence supporting the spin modulation model and also a means of constraining the neutron star radius, since for a known spin frequency the velocity, and thus magnitude of the Doppler shift, is proportional to the stellar radius. In addition, both the magnitude of the Doppler shift and the amplitude of spin modulation decrease as the lattitude of the hot spot increases (Both approaching zero when the spot is located at the rotational pole). Detection of this correlation in a large sample of bursts would provide definitive proof of the spin modulation hypothesis.

PHASE RESOLVED SPECTROSCOPY

Detailed searches for a Doppler shift signature are just beginning to be carried out. Studies using the oscillations in single bursts have shown that 4-5 % modulations of the fitted black body temperature are easily detected in the tails of bursts [7], and are consistent with the idea that a temperature gradient is present on the stellar surface, which when rotated produces the flux modulations. Phase lag studies in a burst from Aql X-1 indicates that softer photons lag higher energy photons in a manner which is qualitatively similar to that expected from a rotating hot spot [8].

There are several ways to address the issue of how best to search for Doppler shifts. Near burst onset the oscillation amplitude is high and the hot spot is well

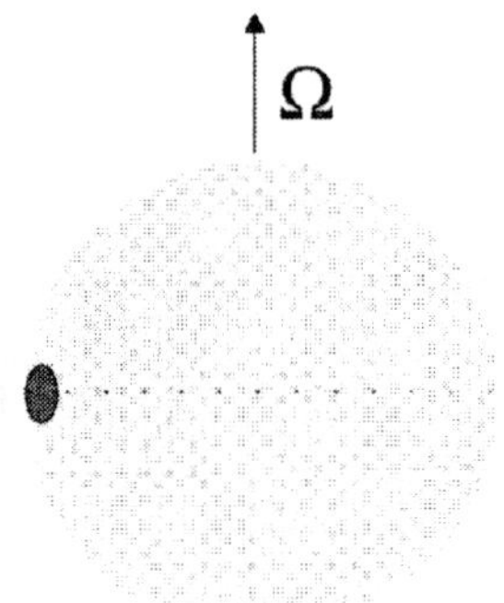

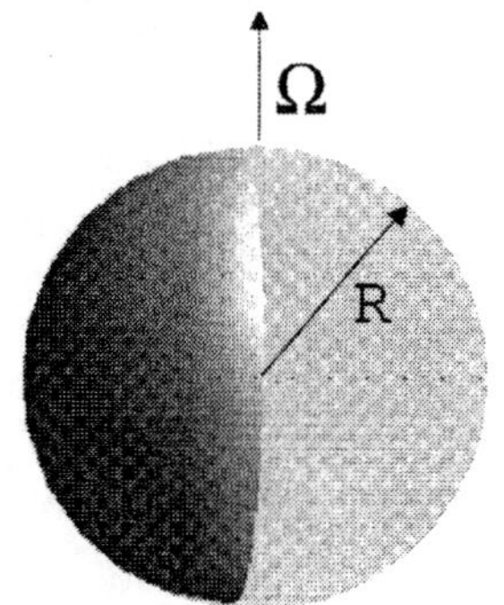

Neutron star with small hot spot near burst onset, producing large amplitudes and a maximal Doppler shift.

Neutron star with broad temperature anisotropy at late times which produces lower amplitude oscillations and a smaller Doppler shift.

FIGURE 1.

localized. If it were possible to examine the oscillations with exquisite detail nearer and nearer to the burst onset then the situation should approach that of a point-like hot spot on the star. A point spot located on the rotational equator produces the largest modulation amplitude as well as a maximal Doppler shift. Thus, in principle this would be an ideal interval to examine, however, the difficulty is that the burst flux is still weak near onset and the interval during which the spot is small (to accumulate a spectrum) is short. This means that the desired signal to examine is very weak. An alternative is to examine pulse trains in the decaying tails of bursts [3]. In this case the oscillation interval is much longer, the burst flux is still substantial and there are many more photons available. However, the oscillation amplitude is lower and at late times in bursts we expect that a well localized spot is probably not present. More likely the modulations are caused by a broad temperature anisotropy over the stellar surface (see Figure 1. This means that the observed Doppler shift represents an integral over the rotating surface of the star, much of which is moving with a lower line of sight velocity than the maximum at the rotational equator. Thus the Doppler shift will be weaker than what we might expect near burst onset. To some extent the lack of signal in a single burst can be offset by summing data from several bursts coherently [3,9].

RESULTS

To search for a Doppler shift I have examined the phase resolved spectrum of 4 bursts with 330 Hz oscillations from 4U 1702-429 [10]. Intervals in the tails of these bursts were analysed first since these have much more signal in the pulsations than intervals near onset. For each burst I fit an exponential model of the frequency drift, as in [3], and then used the phase information for each X-ray event, based on the best model, to accumulate spectra in 12 phase bins. The data were recorded in an event mode which provides the arrival time and energy channel of each X-ray. The data mode has 64 energy channels in the $\approx$ 2 - 100 keV PCA band. As a measure of spectral hardness versus pulse phase I used the mean PCA energy bin for each phase interval. In principle one could also fit the accumulated spectra with a black body function and use the color temperature as a hardness measure, however, the spectra are accumulated over an interval long compared to the burst cooling time so that the spectra are not well fit by black body functions with a constant temperature. In each burst there is a strong modulation of the mean PCA energy channel with pulse phase, with the peak of the phase folded lightcurve having the highest mean channel. This result is similar to that found for oscillations at 580 Hz in a burst from 4U 1636-53 [7], that is, the peak of the modulations are harder than the minima.

The single burst data show obvious modulations of the mean PCA channel with pulse phase, however, they do not show evidence for any asymmetry such as could be introduced by the rotational Doppler shift. This is mainly due to the lack of sufficient signal in a single burst. In an attempt to overcome these limitations I co-added the data for all 4 bursts in phase [3]. This amounts to finding a phase offset, relative to one burst, which maximizes the pulsation signal of the sum. I then computed the mean PCA channel for the summed spectra in the same way as for the individual bursts. The results are given in figure 2. There is a strong 4 % modulation of the mean PCA channel in the combined data. As a simple test for asymmetry I fit a sine function to the mean PCA channel data and find that the data are adequately described by a simple sine wave, so we do not find any strong evidence for a Doppler shift in the combined data. It still remains to conduct more sophisticated tests for asymmetry on these data, and more bursts will be added as new data becomes available.

REFERENCES

1. Strohmayer, T. E., Zhang, W. & Swank, J. H. 1997, ApJ, 487, L77
2. Strohmayer, T. E., Zhang, W., Swank, J. H. & Lapidus, I. 1998b, ApJ, 503, L147
3. Strohmayer, T. E. & Markwardt, C. B. 1999, ApJ, 516, L81
4. Miller, M. C. & Lamb, F. K. 1998, ApJ, 499, L37
5. Strohmayer, T. E., Zhang, W., Swank, J. H., White, N. E. & Lapidus, I. 1998a, ApJ, 498, L135
6. Chen, K. & Shaham, J. 1989, ApJ, 339, 279
7. Strohmayer, T. E., Swank, J. H., & Zhang, W. 1998, Nuclear Phys B (Proc. Suppl.) 69/1-3, 129-134
8. Ford, E. C. 1999, ApJ, 519, L73
9. Miller, M. C. 1999, ApJ, 515, L77
10. Markwardt, C. B., Strohmayer, T. E., & Swank, J. H. 1999, ApJ, 512, L125

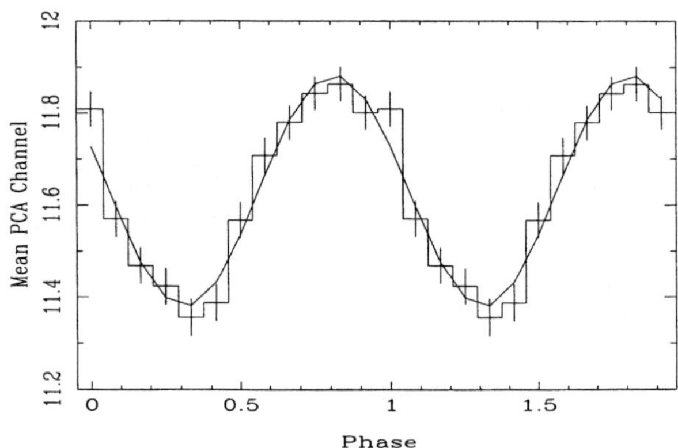

FIGURE 2. Variation of the mean PCA channel from spectra accumulated as a function of pulse phase in 4U 1702-429. Data from 4 bursts were added coherently in order to increase the signal to noise ratio. There is no strong evidence for a Doppler shift induced asymmetry.

The Effects of Metallicity on Simulations of the Nova Outburst

S. Starrfield*, G. Schwarz*, J. W. Truran[†], and W. M. Sparks[‡]

*Department of Physics and Astronomy, Arizona State University
P.O. Box 871504, Tempe, AZ 85287-1504
[†]LASR and E. Fermi Institute, University of Chicago,
933 E. 56th Street, Chicago, IL 60637
[‡]X-4, M.S. F664, Los Alamos National Laboratory, Los Alamos, NM 87545

Abstract.
We explore the possible effects of the accretion of low metallicity material, on white dwarfs in nova binary systems, on the progress of the outburst. We are motivated both by the observation that an unusually high fraction of the classical novae in the Large Magellanic Cloud (Z $\approx$ 0.005) are fast novae and a detailed study of LMC 1991 which shows it has Z=0.1Z$_\odot$ (or less). Our preliminary results indicate the importance, to both theoretical and observational studies of the nova outburst, of determining the mechanism by which the accreted hydrogen-rich envelope is enriched in white dwarf core matter.

I   INTRODUCTION

The standard model for the outbursts of classical novae attributes them to hydrogen thermonuclear runaways occurring in accreted envelopes on white dwarfs (WDs) in close binary systems. Constraints on the detailed nature of these events are provided by spectroscopic observations of individual novae. Until relatively recently, this was possible only for novae in the Galaxy. The International Ultraviolet Explorer satellite (IUE) provided as well the opportunity for detailed scrutiny of nova explosions in the Large Magellanic Cloud (LMC), with six classical novae studied in the period 1988-1995. One notable feature of these LMC events was that they were "faster" than novae studied both in our Galaxy and in M31. Specifically, of the 15 LMC novae tabulated by Della Valle and Livio (1995), 14 were fast novae (i.e. brighter than M$_V$ =7.1 at maximum).

One physical characteristic of the LMC that may be relevant to this issue is its low metallicity (Z $\approx$ 0.005) as compared to solar material. Ignoring how this might have influenced the earlier evolution of the proto-nova binary star system, clearly a low metallicity (and the implied lower opacity) will influence the accretion process. The amount of material accreted by the WD is determined by the competition

CP522, *Cosmic Explosions: Tenth Astrophysical Conference,*
edited by Stephen S. Holt and William W. Zhang
2000 American Institute of Physics 1-56396-943-2

TABLE 1. Initial Parameters and Evolutionary Results for 1.00M$_\odot$

Sequence	1	2	3	4	5	6
WD Mass	1.00M$_\odot$	1.00M$_\odot$	1.00M$_\odot$	1.00M$_\odot$	1.00M$_\odot$	1.00M$_\odot$
C+O	0.5	0.5	LMC	Solar	0.5	0.5
Z	0.001	0.02	0.001	0.02	0.001	0.02
L/L$_\odot$(10^{-3})	3.3	3.3	3.3	3.3	3.3	3.3
T$_{\rm eff}$ (10^4K)	1.58	1.59	1.58	1.59	1.58	1.59
Radius(km)	5359	5368	5360	5368	5360	5368
$\dot{M}$ (10^{-10}M$_\odot$yr^{-1})	16	16	1.6	1.6	1.6	1.6
$\tau_{\rm tnr}$ (10^3 yr)	11	8	950	488	160	154
M$_{\rm acc}$ (10^{-5}M$_\odot$)	1.7	1.2	15.1	7.7	2.6	2.4
$\epsilon_{\rm nuc-max}$(erg gm^{-1}s^{-1})	9×10^{14}	5×10^{14}	1×10^{13}	1×10^{14}	2×10^{15}	2×10^{15}
T$_{\rm peak}$ (10^6K)	152	140	226	175	165	163
L$_{\rm peak}$ (10^4L)	3.3	3.3	2.8	2.0	7.0	4.4
T$_{\rm eff-peak}$ (10^5K)	6.0	6.1	3.8	3.7	5.7	5.7
M$_{\rm ej}$ (10^{-5}M$_\odot$)	0	0.03	0	0	0.04	0.03
M$_{r>10^{11}cm}$(10^{-5}M$_\odot$)[1]	.6	.6	.2	.03	0.2	1.2
V$_{\rm max}$(km s^{-1})	16	270	0	155	1261	1220

[1]Amount of mass lying above a radius of 10^{11}cm at the end of the evolution.

between the rate at which energy is input into the accreted layers (initially by the proton-proton reaction chain) and the rate at which it is transported away from the nuclear burning region by a combination of electron conduction and radiation diffusion. The opacity is important since the larger the opacity, the greater the amount of heat trapped in the nuclear burning region and the more rapid the rise in temperature. Because of the presence of novae far from the Galactic plane and in the Magellanic Clouds, it is important to study the evolution of the outburst in material that has a smaller metallicity than Solar. (We note that the extraordinary nova LMC 1991 has a metallicity Z= 0.1Z$_\odot$: Schwarz et al. 2000, in preparation)

In this paper, we report results of a study of the effects of the accretion of low metallicity matter on simulations of the nova outburst. Hydrodynamic evolutionary calculations have been carried out for both 1.00M$_\odot$ and 1.25M$_\odot$ WDs.

II   CHARACTERISTICS OF THE MODELS

The results of our calculations are summarized in Tables 1 and 2. The fully implicit, Lagrangian, hydrodynamic code which incorporates a large nuclear reaction network that was used for this study is described in Starrfield et al (1998, and references therein). In each table, the first two columns are for $\dot{M} = 1.6 \times 10^{-9}M_\odot$yr^{-1} and the last four are for $\dot{M} = 1.6 \times 10^{-10}M_\odot$yr^{-1}. While the higher mass accretion rate is in better agreement with the observations, the lower mass accretion rate is used in order to achieve a larger accreted mass. See Starrfield et al (1998) for a discussion of the accreted versus ejected mass problem. The composition of the

TABLE 2. Initial Parameters and Evolutionary Results for $1.25M_\odot$

Sequence	1	2	3	4	5	6
WD Mass	$1.25M_\odot$	$1.25M_\odot$	$1.25M_\odot$	$1.25M_\odot$	$1.25M_\odot$	$1.25M_\odot$
C+O	0.5	0.5	LMC	SOLAR	0.5	0.5
Z	0.001	0.02	0.001	0.02	.001	0.02
$L/L_\odot(10^{-3})$	3.1	3.0	3.1	3.0	3.1	3.0
T_{eff} (10^4K)	1.93	1.92	1.93	1.92	1.93	1.92
Radius(km)	3484	3493	3484	3493	3484	3493
$\dot{M}$ $(10^{-10}M_\odot yr^{-1})$	16	16	1.6	1.6	1.6	1.6
τ_{tnr} $(10^3$ yr)	5.2	3.1	464	201	115	69
M_{acc} $(10^{-5}M_\odot)$	.83	.49	7.3	3.2	1.8	1.1
$\epsilon_{nuc-max}(erg\ gm^{-1}s^{-1})$	2×10^{16}	7×10^{15}	2×10^{13}	2×10^{14}	1×10^{17}	4×10^{16}
T_{peak} (10^6K)	215	191	312	246^2	257	229
L_{peak} (10^4L)	14.4	4.8	4.0	3.1	167	4.4
$T_{eff-peak}$ (10^5K)	8.6	8.9	6.5	6.1	8.0	8.3
M_{ej} $(10^{-5}M_\odot)$	0.03	.002	0	0.006	1.5	.3
$M_{r>10^{11}cm}(10^{-5}M_\odot)^1$	.2	.1	.01	.05	1.5	.4
$V_{max}(km\ s^{-1})$	1711	276	107	300	4995	3381

[1]Amount of mass lying above a radius of 10^{11}cm at the end of the evolution.
[2]The TNR occurred two zones above the core-envelope interface.

accreted material is either Z=0.001 or Z=0.02. However, for Sequences 1,2, 5, and 6 we have also assumed that the accreting material has mixed with the core and the resulting composition includes 25% C and 25% O. In row order, we provide the luminosity, effective temperature, radius, accretion rate, accretion time, accreted mass, peak rate of energy generation, peak nuclear burning temperature, peak luminosity, peak effective temperature, ejected mass, amount of material lying above a radius of 10^{11}cm at the end of the evolution (outside the Roche lobe of typical nova binaries), and the maximum velocities of the ejecta.

III   DISCUSSION

Two critical model characteristics that influence the violence of nova outbursts are (1) the degree of degeneracy at the base of the envelope and (2) the abundance of C and O in the burning region. High concentrations of C and O are required to fuel the runaway during the first ≈ 180 seconds of the expansion, during which time CNO cycling is constrained by the β^+-decay time scales (Truran 1982; Starrfield 1989).

Let us first assume both (1) that no significant contamination of the envelope in core matter occurs and (2) that the transferred material in the case of LMC novae will have low metallicity. For these assumptions, the masses of the accreted envelopes at the point of runaway are increased - as expected. The relevant columns in the accompanying tables are columns 4 (solar composition) and column 3 (Z = 0.001). Note, however, that although the envelope masses are increased by factors of 2-3, and the peak temperatures are significantly higher, the outbursts themselves are considerably less violent: the maximum velocities are much lower and the peak rates of nuclear energy generation are an order of magnitude lower. This would at first seem to indicate that lower metallicity populations should systematically have slower novae - and the predominance of fast novae in the LMC is certainly not explained.

This situation is complicated, however, when we attempt to take into account the influence of envelope enrichment via dredge-up of core matter. The table columns of interest here (for the same accretion rate) are columns 5 and 6, which compare the evolutions for the case where envelope enrichments in C+O are included. Here the violence of the runaway is significantly greater, both for the $Z=Z_\odot$ case and for the $Z=0.001$ case, than for the corresponding cases with no heavy element enrichment. However, the effects of the reduced metallicity of the accreted matter are less pronounced, particularly for the $1.0M_\odot$ cases. The problem here is that, for our calculations, we have effectively accreted matter of $Z \approx 0.5$, the metallicity we chose as attributable to enrichment via dredge-up. Here the characteristics of the enrichment mechanism itself become of critical importance, recognizing that such mixing will increase the opacity, enhance the heating, trigger runaway at an earlier stage, and thus act to reduce the envelope mass at ignition. If the enrichment occurs prior to runaway, the envelope (and most critically the envelope opacity) will reflect the enriched abundances, and the masses of accreted matter at the point of runaway will be reduced. In contrast, if the enrichment accompanies the early stages of runaway, larger envelopes may be accreted and the violence of the events will be increased. To date, multi-dimensional models for novae have not yielded a clear answer as to whether such late dredge-up is possible. (Kercek et al 1999, and references therein).

We must therefore conclude that it will be necessary to determine the mechanism by which accreted material mixes with core material before we can explain the observed high frequency of very fast novae in the LMC.

REFERENCES

1. Della Valle, M., Livio, M. 1995, APJ, 452, 704.
2. Kercek, A, Hillebrandt, W., Truran, J. W. 1999, A&A, 345, 831.
3. Starrfield, S., 1989, in The Classical Novae, ed. M. Bode and A. Evans, (Wiley: NY), 39.
4. Starrfield, S., Truran, J.W., Wiescher, M.C. and Sparks, W.M. 1998, MNRAS, 296, 502.
5. Truran, J. W. 1982, in Essays in Nuclear Physics, eds. C. A. Barnes, D. D. Clayton and D. N. Schramm (Cambridge: Cambridge U. Press), p. 467.

The Impact of Nuclear Reaction Rate Uncertainties on Evolutionary Studies of the Nova Outburst

W. Raphael Hix[†§], Michael S. Smith[§], Anthony Mezzacappa[§], Sumner Starrfield[*], and Donald L. Smith[‡]

[†]*Department of Physics & Astronomy, University of Tennessee, Knoxville, TN 37996-1200*
[§]*Physics Division, Oak Ridge National Laboratory[1] Oak Ridge, TN 37831-6354*
[*]*Department of Physics & Astronomy, Arizona State University, Tempe, AZ 85287-1504*
[‡]*Technology Development Division, Argonne National Laboratory, Argonne, IL 60439*

Abstract. The observable consequences of a nova outburst depend sensitively on the details of the thermonuclear runaway which initiates the outburst. One of the more important sources of uncertainty is the nuclear reaction data used as input for the evolutionary calculations. A recent paper by Starrfield, Truran, Wiescher, & Sparks [1] has demonstrated that changes in the reaction rate library used within a nova simulation have significant effects, not just on the production of individual isotopes (which can change by an order of magnitude), but on global observables such as the peak luminosity and the amount of mass ejected. We will present preliminary results of systematic analyses of the impact of reaction rate uncertainties on nova nucleosynthesis.

I NUCLEAR UNCERTAINTIES IN NOVAE

Observations of nova outbursts have revealed an elemental composition that differs markedly from solar. Theoretical studies indicate that these differences are caused by the combination of convection with explosive hydrogen burning which results in a unique nucleosynthesis that is rich in odd-numbered nuclei such as ^{15}N, ^{17}O and ^{13}C. These nuclei are difficult to form in other astrophysical events. Many of the proton-rich nuclei produced in nova outbursts are radioactive, offering the possibility of direct observation with γ-ray instruments. Potentially important γ-ray producers include ^{26}Al, ^{22}Na, ^{7}Be and ^{18}F.

The observable consequences of a nova outburst depend sensitively on the details of the thermonuclear runaway which initiates the outburst. One of the more important sources of uncertainty is the nuclear reaction data used as input for the

[1] Research at the Oak Ridge National Laboratory is supported by the U.S. Department of Energy under contract DE-AC05-96OR22464 with Lockheed Martin Energy Research Corp.

evolutionary calculations [1]. A number of features conspire to magnify the effects of nuclear uncertainties on nova nucleosynthesis. Many reactions of relevance to novae involve unstable proton-rich nuclei, making experimental rate determinations difficult. For hydrodynamic conditions typical of novae, many rates depend critically on the properties of a few individual resonances, resulting in wide variation between different rate determinations. Statistical model (Hauser-Feshbach) calculations, which are employed with great success for a large number of reactions [2], are unreliable for rates dominated by individual resonances. The similarity of the nuclear burning and convective timescales results in nuclear burning in novae which is far from the steady state which typifies quiescent burning.

II MONTE CARLO ESTIMATES OF UNCERTAINTIES

Though analysis of the impact of variations in the rates of a few individual reactions has recently been performed using one-dimension hydrodynamic models [3], analysis of the impact of the complete set of possible reaction rate variations in such hydrodynamic models remains computationally prohibitive. We therefore begin by examining in detail the nucleosynthesis of individual zones, using hydrodynamic trajectories (temperature and density as a function of time) drawn from nova outburst models. Such one zone, post-processing nucleosynthesis simulations are a common means of estimating nova nucleosynthesis (see e.g., [4,5]). For this presentation we are using a hydrodynamic trajectory for an inner zone of a $1.35M_\odot$ ONeMg WD which is similar to that described in [6]. These calculations were performed using a nuclear network with 87 species, composed of elements from n and H to S, including all isotopes between the proton drip line and the most massive stable isotope.

Figure 1 shows the abundances of each species at the end of the simulation, 5.2×10^5 sec after peak temperature. Because of the long time which has elapsed, the unstable proton-rich nuclei have decayed, reducing their abundances to less than 10^{-20}. To investigate the extent to which nuclear reaction uncertainties translate into abundance differences, we use a Monte Carlo technique which assigns to each reaction rate in the nuclear network a random enhancement factor. The error bars displayed in Fig. 1 are the 90% confidence intervals which result from 992 Monte Carlo iterations. Monte Carlo methods have been employed with great success in the analysis of Big Bang nucleosynthesis [7], but have not previously been applied to other thermonuclear burning environments.

The reaction rate enhancement factors are distributed according to the lognormal distribution, which is the correct uncertainty distribution for quantities like reaction rates which are manifestly positive [8],

$$p_{log-normal}(x) = \frac{1}{\sqrt{2}\,\pi\beta x}\exp\left(-\frac{(\ln x - \alpha)^2}{2\beta^2}\right) , \tag{1}$$

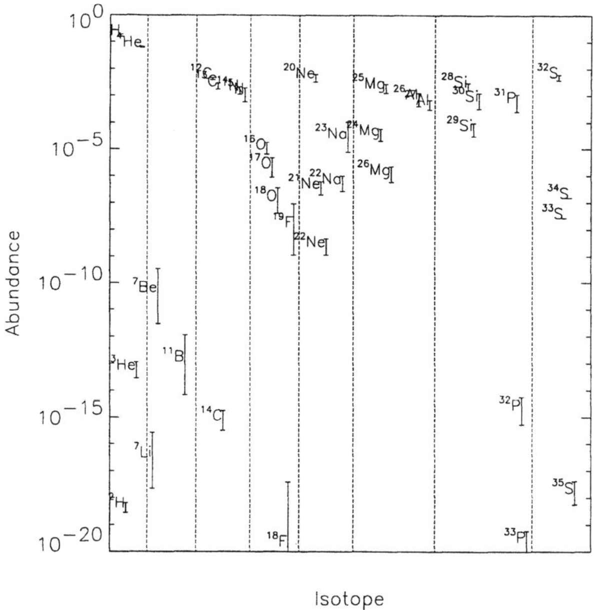

FIGURE 1. Final Abundances (elapsed time $= 5.2 \times 10^5$ sec after peak).

where α and β are the (logarithmic) mean and standard deviation. For small uncertainties ($< 20\%$), the difference between the log-normal distribution and the normal (Gaussian) distribution is small. However, for uncertainties of larger sizes such as those encountered in this problem, the difference is important. For this preliminary analysis, we have chosen to assign uncertainties of $\sim 50\%$ ($\beta = \ln(1.5)$) both to rates calculated by Hauser-Feshbach methods and also to rates whose measurement require radioactive ion beams. For all other rates we assign $\beta = \ln(1.2)$. Figure 2 plots the resulting abundance distributions for two representative nuclei. Fig. 2 also demonstrates the differences between normal and log-normal distributions for widths of these sizes. These are very conservative uncertainties; relatively few reactions, especially among unstable nuclei, have measurement uncertainties this small.

III RESULTS

As evidenced by the error bars in Fig. 1, the impact of even our conservatively chosen variations in reaction rates on the nucleosynthesis is large. While broader conclusions will require analysis of additional hydrodynamic trajectories, a number of interesting points can be made from the analysis of this single trajectory. The impact on the rate of energy production is small. At the 90% confidence level,

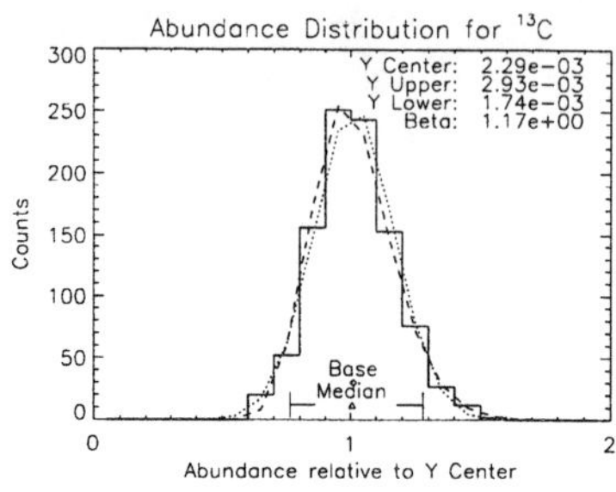

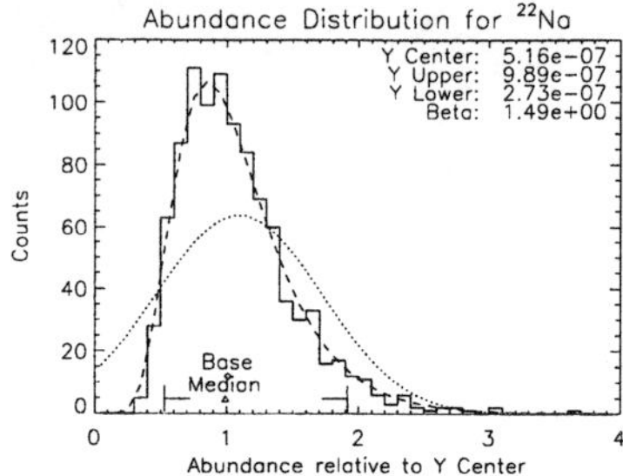

FIGURE 2. The histograms showing deviations in abundance from the mean. The dotted curves are normal distributions with the same (arithmetic) mean and standard deviation as the Monte Carlo distribution. The dashed curves are log-normal distributions with (logarithmic) mean and standard deviation from the Monte Carlo. Y Center, Y Upper, and Y Lower correspond, respectively, to the central abundance and the upper and lower error bars from Fig. 1.

variations in the amount of hydrogen consumed represent $\sim 10\%$ variation in the thermonuclear energy released. For the most abundant metals (those which represent more than 1% of the mass), 2σ variations by factors of 1.1 to 1.4 are common, with some of these nuclei showing 2σ variations as large as a factor of 2, for example, ^{15}N (2.1×) and ^{30}Si (2.3×). For the γ-ray source nuclei ^{22}Na and ^{26}Al, the 90% confidence interval includes variations of nearly a factor of 2, representing almost a factor of 4 difference in the distance from which novae may be observed by γ-ray telescopes. For ^{7}Be, the 90% confidence interval spans more than 2 orders of magnitude.

Such large uncertainties in the nucleosynthesis, resulting from poorly known nuclear reaction rates, constrain our ability to make detailed comparisons between theoretical models for the nova outburst and astrophysical observations to a degree which is often ignored. Improved knowledge of these uncertain rates, both experimental and theoretical, is necessary to provide tight constraints on the nova outburst from its nucleosynthetic products.

REFERENCES

1. S. Starrfield, J. W. Truran, M. C. Wiescher, and W. M. Sparks, MNRAS **296**, 502 (1998).
2. T. Rauscher, F.-K. Thielemann, and K.-L. Kratz, Nucl. Phys. A **621**, 331 (1997).
3. J. José, A. Coc, and M. Hernanz, ApJ **520**, 347 (1999).
4. W. Hillebrandt and F.-K. Thielemann, ApJ **255**, 617 (1982).
5. M. Wiescher, J. Gorres, F.-K. Thielemann, and H. Ritter, A&A **160**, 56 (1986).
6. M. Politano *et al.*, ApJ **448**, 807 (1995).
7. M. S. Smith, L. H. Kawano, and R. A. Malaney, ApJS **85**, 219 (1993).
8. D. L. Smith, *Probability, Statistics, and Data Uncertainties in Nuclear Science and Technology* (LaGrange Park: Am. Nuc. Soc., 1991).

Stellar Explosions

Stellar Flares: How Common? How Important?

Jeffrey L. Linsky

JILA, University of Colorado and NIST
Boulder Colorado 80309-0440

Abstract. Flares are often observed from a wide variety of stars that are called active by virtue of their rapid rotation, young age, and strong magnetic fields. Although the total energies of these flares cover many orders of magnitude in range, they typically follow a common pattern with impulsive and gradual stages and the distribution of energy among the different radiating channels is remarkably similar. The similar phenomenology points to a common scenario of magnetic reconnection, electron beam heating, and evaporation of the lower atmosphere followed by gradual cooling of the hot coronal plasma. The solar two-ribbon flare is often cited as the prototypical model. However, this conclusion must be viewed as tentative because the radiative response of an atmosphere to the impulsive input of energy could be the same for a variety of different energy input mechanisms.

In flagrante delicto.

Caught in the act.

I   INTRODUCTION

Most stars with convective zones and strong magnetic fields show transient brightenings called *flares*. These flares are most easily detected in X-rays, ultraviolet emission lines, and at radio wavelengths where the contrast between the flare emission and the quiescent (nonflare) emission from the star is large. The low photospheric temperatures of M dwarfs allows flares to be detected easily in the optical, especially in the U band. The most energetic flares on the Sun are also detected in broad-band optical observations, a phenomena called *white light flares*. Flares on the Sun and nondegenerate stars are characterized by an enormous range of energies, 10^{22} erg solar *nanoflares* to $10^{38} - 10^{40}$ erg huge flares on RS CVn and premain sequence stars. Despite their 18 orders of magnitude range in energy, flares generally show similar phenomenology, suggesting a common physical origin.

At this Workshop on Cosmic Explosions, stellar flares are small firecrackers compared to supernovae and gamma ray bursts. Nevertheless, there are elements in

CP522, *Cosmic Explosions: Tenth Astrophysical Conference,*
edited by Stephen S. Holt and William W. Zhang

common between the big and small explosions, and flares on nearby stars and especially the Sun can be studied with high temporal and spatial resolution, revealing some of the physical processes responsible for the explosions. Thus prudence requires that one should learn from the nearby relatively small flares at the same time as tackling the distant great events at the edge of the universe.

II FLARE PHENOMENOLOGY

Although flares on the Sun and different classes of stars vary greatly in energy, time scales, and volumes, they generally proceed through the same sequence of stages. The following is a very simplified overview of these stages of stellar flares. For a more comprehensive and detailed review of stellar flares, I call your attention to a number of excellent review papers [6,13,15,16,29,24].

Buildup: For stars with convective envelopes, convective motions below the photosphere, where the kinetic energy exceeds the magnetic energy ($\beta = P_{gas}/P_{mag} > 1$), displace and shuffle the footpoints of coronal magnetic fields. This footpoint motion spontaneously creates complex interlaced coronal magnetic field topologies with magnetic free energy; that is, magnetic energy over and above that of a potential field. Parker [33,35] argued that the interlaced topology of the coronal field relaxes to a static equilibrium with internal surfaces of *tangential discontinuity* (also called *current sheets*) that must eventually dissipate this free energy. The buildup phase of a flare thus describes the conversion of heat produced by nuclear reactions in the stellar core through the subsurface convective motions into the buildup of magnetic free energy in the corona awaiting a trigger to convert this energy rapidly into heat, kinetic motions, and relativistic particles.

Nanoflares: If resistive reconnection of field lines in current sheets driven by plasma inflows were inherently a continuous smooth process, then one would expect the heating of stellar coronae to be steady and uneventful. Numerical simulations and laboratory plasma experiments, however, provide a very different picture in which reconnection processes occur on very small scales (≤ 1m), are irregular in both space and time, and the dissipation is episodic [34,35]. The energies of these reconnection events vary over many orders of magnitude from nanoflares [34] ($10^{22} - 10^{24}$ erg) to the most energetic solar flares (10^{32} ergs).

Impulsive phase of large flares: The large release of energy that we call flares rather than episodic heating requires some trigger to produce an avalanche of magnetic field reconnection on a fast time scale. Theoreticians have searched for years to find the enigmatic trigger that would start the avalanche. One scenario [15,27] commonly proposed to explain the so-called *two-ribbon* flares, is the eruption of a *filament* (a magnetically confined region of cool plasma) caused by a loss of equilibrium of the mass loaded magnetic structure. These flares are long lived as they involve the disruptive opening up of large magnetic arcades over time. Whatever the cause of the eruption might be, the ejection of the filament stretches many field lines leading to enhanced field gradients at the current sheet and greatly enhanced reconnection in a large magnetic arcade. By comparison, *compact* flares involve magnetic reconnection in a few magnetic loops without disrupting a large magnetic configuration [13]. In this scenario much of the magnetic energy is converted into the acceleration of relativistic electrons and protons by the electric field. These beams then precipitate down the field lines generating gyrosynchrotron radiation as they go and hard nonthermal x-rays and ribbons of Hα emission when they impact the chromosphere. These short lived radiations characterize the *impulsive* phase of the flare.

Explosive evaporation of the lower atmosphere: A range of models has been proposed over the years to describe how energy is transmitted from the initial flare site in the corona to the dense layers of the chromosphere and upper photosphere. Possible energy transfer mechanisms include X-radiation, conduction, blast waves, and particle beams (electron or proton). At present electron beams are considered to be the most likely mechanism, so I will describe this mechanism in a little detail. Abbett and Hawley [1] modeled the response of the solar chromosphere to a high energy beam of electrons using a radiative hydrodynamic code that includes non-LTE radiative transfer in important emission lines. Initially, when the chromosphere can still radiate away the nonthermal heat input by small increases in temperature, the atmosphere remains near equilibrium with no mass motions. They refer to this stage as the *gentle phase.* When the beam heating ionizes hydrogen, the main source of radiation from the chromosphere, the atmosphere explodes with large material flows and shocks carrying away the excess energy input that radiation alone cannot accomplish. During this *explosive phase*, one observes large Doppler shifts and the *evaporation* of chromospheric gas as very hot blueshifted coronal plasma. Blue shifts with velocities up to 400 km s^{-1} are often detected in the Ca XIX lines ($T_e \geq 10^7$ K) at 3.2 Å during the first minute of solar flares, and models of beam heated evaporation predict these large upflows of the evaporated formerly chromospheric gas [2].

Gradual phase: After the initial fireworks (relativistic beams, explosive evaporation, and blast waves), the nonthermal energy thermalizes resulting in a hot corona that cools by expansion (when not confined by the magentic field), X-radiation, and thermal conduction. For solar flares the peak emission measure and temperature (often 2–3$\times 10^7$ K) of the soft X-ray emission follows the hard X-rays and 400 km s^{-1} upflows seen in the Ca XIX lines by 1–2 minutes [2], but for stellar flares the time delays are often longer [12] and the peak temperature 10^8 K or hotter [11]. During the gradual phase, the gyrosynchrotron emission from an expanding cloud of relativistic electrons can extend for hours or days in RS CVn binary systems [39,40].

III WHAT TYPES OF STARS FLARE?

A   Sun

The first star observed to flare was the Sun, and the first detected flare emission was optical (white light flares and brightenings in the Balmer emission lines). The limited red sensitivity of early emulsions meant that the first photographs of solar flares were in the Hβ line. Solar flares are now routinely studied in soft and hard X-rays, ultraviolet emission lines, Hα and other emission lines observable from the ground, radio wavelengths, and even gamma rays. Relatively short duration *impulsive* or *compact* flares and the longer duration *two-ribbon* flares are generally considered as prototypes for the more energetic short and long duration flares on stars. Transition region *explosive events* observed as broad or Doppler shifted profiles in transition region lines like C IV 1549 Å are believed to be signatures of nanoflares. These nanoflares are small contributors to the heating of these layers on the Sun, but they are important contributors to the heating in active stars [42]. The solar corona is likely heated by many small magnetic reconnection events.

B   dMe Stars

The first nonsolar flare was observed serendipitously on the star UV Ceti (dM6e)[1] [21] from an enhancement of both the optical line and continuum emission. Essentially all dMe stars are observed to flare and many of the less active dM stars also flare. Flares are easily detected at X-ray, UV, and radio wavelengths, and in the optical (especially in the U band and the Balmer lines) because of the large contrast with the dim stellar photosphere. Optical flares typically show very fast rise times (seconds to a few minutes in the U band) during the impulsive phase followed by exponential decay times of some tens of minutes to several hours during the decay phase [28]. Many dMe star flares with a wide variety of energies have now been observed and modelled. The properties of these flares show a remarkable consistency despite many orders of magnitude range in energy [17]. I list here some of their characteristic properties:

- The time evolution of the flare emission generally follows a similar pattern with the energy radiated during the impulsive phase (dominated by the optical and UV continuum) two to three times that of the energy radiated during the gradual phase (dominated by the Balmer and UV emission lines and the soft X-ray emission).

---

[1] The letter "e" is added to the spectral type of those stars that typically show the Hα line in emission. dM stars do not show the Hα line in emission.

- Multiwavelength observing campaigns show that the total energy of a flare emitted through different channels (Balmer lines, optical and UV continuum, and X-rays) are roughly proportional over a wide range of energy. In particular, the total soft X-ray emission has about the same energy as the U band continuum, corresponding to about 1/6 of the total optical and UV continuum emission.

- The main difference between giant flares (e.g., the 1985 April 12 flare on the dM3.5e star AD Leo with a U band enhancement of 4.5 magnitudes and total flare radiated energy in excess of 10^{34} ergs [17]) and much lower energy flares is probably due to the longer duration of the heating and larger area coverage of the large flares rather than to enhanced heating per unit area and time. Thus the heating rate and magnetic recombination rate per unit area on the star appear to be roughly the same [17], indicating that the physics of magnetic field reconnection limits to within a narrow range the rate at which magnetic energy can be converted to other forms of energy.

- Since the radiative output of different flares on different stars tend to follow a consistent pattern, many authors (e.g., [12,41]) have argued that flares on dMe stars are scaled up versions of solar compact or two-ribbon flares. This conclusion must be viewed as tentative because the radiative output during a flare is the response of a stellar atmosphere to the impulsive input of energy, and the character of this radiative output may only tell us that the flaring plasmas are following the rules of hydrodynamics, atomic physics, and radiative transfer. The radiative response of the atmosphere to the impulsive input of energy could be the same for a variety of different energy input mechanisms.

- Intense radio emission is commonly detected during flares. This is usually explaned as gyrosynchrotron emission from relativistic electrons, although some radio flares are naturally explained by coherent maser or plasma emission processes [4,7]. During the 1983 February flare on YZ CMi (dM4.6e) [41], the 6 cm emission was observed to peak 7 minutes after the U band peak, and the radio emission remained bright long after the end of the U band emission.

- Although beams of nonthermal electrons cannot be observed directly, their presence can be inferred by the so-called *Neupert effect*. According to presently accepted models, magnetic field reconnection accelerates beams of relativistic electrons down into the dense chromosphere and upper photosphere. There collisions with the plasma produce hard X-rays by nonthermal bremsstrahlung, the white light continuum emission (most easily seen in the U band), and evaporation of the ambient cool plasma into the corona where it later emits soft X-rays. During the impulsive phase of solar flares, the white light emission and hard X-ray emission are coincident in space and time [19], indicating their common origin from the electron beams, but hard X-rays are difficult to observe during stellar flares. This model predicts that the soft X-ray emission during the gradual phase should equal the time integral of the hard X-ray or

U-band emission (representing the heating and evaporation up to that time). The verification of the *Neupert effect* during the 1993 March 1–3 flare of AD Leo [18], using the U-band emission as a proxy for the hard X-ray emission, provides support for electron beam heating of flares on dMe stars.

- X-ray flares on dMe stars have been observed with the Einstein, EXOSAT, ROSAT, and ASCA satellites. Typically one sees X-ray fluxes increase by a factor of 10–100 over quiescent values, flare plasma peak temperatures of roughly 3×10^7 K, and the maximum in the coronal emission measure occuring about 5 minutes after the temperature maximum as is often observed in solar flares [12]. Most authors have modelled the flare X-ray plasma using solar two-ribbon models but with much larger energies. For example, the best fit model for the 1991 May 9 flare on AD Leo observed by ROSAT has a maximum temperature of 3.5×10^7 K and a magnetic loop length about 1/7 of the stellar radius [36]. Recent analyses of X-ray emission from AD Leo and EV Lac (dM4.5e) observed with ROSAT and SAX show very high temperature plasmas, perhaps above 10^8 K, with the presence of plasma at $\log T \approx 7.6$ during a flare on EV Lac. An important, but as yet unanswered, question is whether the very hot plasma is thermal or a nonthermal power law distribution that may be the low energy tail of the MeV electrons producing the gyrosynchrotron emission.

C Brown dwarfs

Stars with mass $< 0.08 M_\odot$ are called *brown dwarfs* (spectral types L and T) because their core temperatures are too low to sustain hydrogen burning and their energy source is deuterium burning for a limited time and gravitational contraction. The lowest mass M dwarfs like VB10 have been observed to flare in the optical, ultraviolet [26], and X-rays, indicating that rapid field reconnection events, presumably driven by convective motions below the photosphere, occur in the coronae of these stars. Since brown dwarfs likely have similar internal structures and magnetic dynamos as the late-M dwarfs, they should also flare by the same process. Recently an optical flare was detected in the 2MASS object J0149090+295613 classified as M9.5V [23], and ROSAT detected an X-ray flare in a young brown dwarf in the Chamaeleon I Dark Cloud [30]. The 2MASS and DENIS surveys and infrared monitoring of young clusters will likely show that brown dwarf flares are very common events.

D RS CVn systems

These are close but detached binaries containing typically a G dwarf or subgiant and an early-K subgiant that are tidally locked. Thus the stellar rotation periods equal the obital periods, which can be as short as one day or as long as about 20 days. Systems with orbital periods longer than about 20 days tend to rotate slower than synchronous and be less active. Many powerful flares have been detected in these systems in the ultraviolet [29], x-ray [24], and radio [5]. Time sequences of ultraviolet spectra of RS CVn systems, primarily in the Mg II 2796 and 2803 Å lines, observed with the IUE satellite, show large flux enhancements with decay times typically 8–12 hours. Large mass flows in the chromosphere are indicated by very broad Mg II line profiles and Doppler shifts (typically downflows) that often exceed 100 km s^{-1}.

The total radiative energy loss (for plasma in the range $4.0 \leq \log T \leq 5.3$) can be as large as 2×10^{36} ergs and the mass flows contain comparable energy [25]. In the X-ray region the total energy losses for well studied flares lie in the range 10^{34} to a few times 10^{37} ergs, and the flares have been modelled as very energetic analogs of both solar compact and two-ribbon flares [24]. For example, the 1994 August 29 flare on UX Ari observed by ASCA and EUVE [11] showed plasma temperatures near 10^8 K and metal abundances that rise from 0.17 times solar to nearly solar during the flare, suggesting evaporation of solar abundance gas from the lower atmosphere into the flaring loop. Güdel et al. [11] modelled this flare with the Kopp and Poletto [22] two-ribbon flare model with the flaring loop covering 30% of the visible surface. If this flare on UX Ari is representative of flares on RS CVn systems, the high X-ray luminosities of RS CVn flares compared to the Sun are due to much larger loop sizes and higher densities. Analysis of 12.2 Ms of EUVE photometry of 16 RS CVn systems [31] showed that these stars show obvious flares 40% of the time. Osten and Brown [31] argued that the flares are not simply scaled-up versions of solar two-ribbon flares because the rise times are far longer than are observed in solar flares.

Very luminous radio flares are detected from the relativistic plasma that expands during a flare until it becomes as large as the binary separation. The emission process is most likely gyrosynchrotron emission as the initial brightness temperature is a few times 10^9 K and the electron energies decrease with time as a result of radiative and collisional losses [39,40].

E   Algol systems

These close but detached binary systems usually contain a B- or A-type dwarf and a K giant. The Algol prototype, β Per (B8 V + K2 III), has a 2.87 day period. Using the BeppoSAX satellite, Schmitt and Favata [37] detected the giant 1997 August 31 X-ray flare on β Per with total energy 1.5×10^{37} erg in the 0.1–10 kev X-ray band. The enormous energy of the flare (about 1.2% of the K2 III star's bolometric luminosity) suggests a different flare scenario than for the relatively weak solar flares, perhaps the intertangling of the magnetic fields of both stars. However, the observation of an eclipse of the flare by the B8 V star shows that the flaring plasma was confined to a region within 0.6 radii of the surface of the K2 III star and near its rotational pole. The location of the flare close to the stellar surface, the high plasma temperature ($T \approx 1 \times 10^8$ K), and moderate strength magnetic fields (500–1000 G) suggest that even this giant flare could be a magnetic reconnection event analogous to solar flares albeit far more energetic. The 1992 August flare on Algol observed by ROSAT [32] appears to be a two-ribbon type flare with similar dimensions and temperatures to the 1997 August 31 flare, but with possible abundance variations during the flare.

F   Premain sequence stars

Intense flares from very young protostars embedded in the dust and gas of their nascent clouds have been detected by X-ray satellites from many star forming regions [8]. One well studied flare is that observed by the ROSAT HRI on YLW15, a class I protostar about 10^5 years old in the ρ Oph cloud [9]. The X-ray luminosity of the flare was in the range 10^{34} to 10^{36} erg, depending on the value of the extinction that is very large. This enormous energy corresponds to 1–100 times the bolometric luminosity of the star. The total X-ray energy of the flare, which had an e-folding decay time of five hours must have exceeded 10^{38} ergs and could have been as large as 10^{40} ergs! Grosso et al. [9] estimated that the flare dimension was 0.02–0.3 AU, much larger than the star. The very large volume and extreme energy of this flare and the somewhat less energetic flares detected on other premain sequence stars are strong arguments that these flares are not scaled up solar two-ribbon flares. They are more likely magnetic reconnection events that occur when the fields of the protostar and the accretion disk shear and interconnect as forced by the different rotation rates of the protostar and accretion disk to which the flux tubes are anchored and gas flows from the disk to the protostar along interconnecting field lines [38].

G   Single giants

Evolved single giants of spectral type G–M are not expected to flare because they are slow rotators and likely do not have strong magnetic fields. A few radio transients have been reported from single dish radio telescopes, but modern interferometers like the VLA have not detected any flares from such stars. A search of the ROSAT observations of apparently single late-type giants [14,20] revealed only a handful of flares, but these flares could be due to previously unknown binary companions. A more compelling case is that of μ Vel A, a G6 III star crossing the Hertzsprung gap for the first time. A large flare (5×10^{35} ergs) with a decay time of 1.5 days was observed by EUVE on 1998 March 10. Ayres et al. [3] argue that the flare was not due to the F dwarf companion located 2 arcseconds away, but there is a remote possibility that μ Vel A has a previously unknown late-type dwarf companion that flared. If the flare is actually from the G6 III star, then flares may play a major role in the loss of angular momentum from these rapidly evolving stars.

IV SUMMARY

Flares are often observed from a wide variety of stars that are called active by virtue of their rapid rotation, young age, and strong magnetic fields. Although the total energies of these flares cover many orders of magnitude in range (see Table 1), they typically follow a common pattern with an impulsive stage followed by a gradual stage, and the distribution of energy among the different radiating channels is remarkably similar. The similar phenomenology points to a common scenario of magnetic reconnection, electron beam heating, and evaporation of the lower atmosphere followed by gradual cooling of the hot coronal plasma. The solar two-ribbon flare is often cited as the prototypical model. However, this conclusion must be viewed as tentative because the radiative output during a flare is the response of a stellar atmosphere to the impulsive input of energy. The character of this radiative output may only tell us that the flaring plasmas are following the rules of hydrodynamics, atomic physics, and radiative transfer. The radiative response could be the same for a variety of different energy input mechanisms.

While flares on most active stars may be explicable by the two-ribbon flare model, premain sequence stars are probably the exceptions. For these stars the extreme energy and large volume point to a very different model in which magnetic reconnection events occur when the fields of the protostar and the accretion disk shear and interconnect as forced by the different rotation rates of the protostar and accretion disk to which the flux tubes are anchored and gas flows from the disk to the protostar along interconnecting field lines. New observations are need to test this model.

Flares are often thought of as isolated transient events, but there is evidence that low energy flares, often called nanoflares, occur all the time and are responsible for most of the heating in the coronae and transition regions of the most active stars. Chandra, XMM, and future very large area X-ray telescopes may be able to measure the distribution function of flares down to very low energies and thus tie together the large flares and nanoflares.

How important are stellar flares? They are essential in removing strong magnetic fields that are created by the dynamos operating in most late-type stars. In the process of reconnecting these fields, the observed energetic flares and the far more numerous nanoflares are probably responsible for most of the energy that heats the coronae of late-type stars. The high temperature and pressure of these coronae accelerate winds that replenish the interstellar medium with chemically processed matter and carry away angular momentum that decelerates the stellar rotation. In active stars like the young Sun, flares produced strong ultraviolet radiation that impacted the atmosphere of the young Earth and may have played a role in forming the organic molecules out of which life formed. Perhaps flares are important.

Finally, this Workshop on Cosmic Explosions serves to tie together transient events in astrophysics from the low energy flares on stars to the highest energy events, the γ ray bursts. The study of stellar flares shows that similar phenomenology and likely a common series of physical processes involving tangled magnetic

TABLE 1. Comparison of characteristics of Stellar Flares and Gamma Ray Bursts.

Property	Stellar Flares	Gamma Ray Bursts
Energy (ergs)	Sun: $10^{22} - 10^{32}$ dMe: $10^{31} - 10^{34}$ RS CVn: $10^{34} - 10^{38}$ PMS: $10^{36} - 10^{40}$	$10^{51} - 10^{54}$
Time scale	Sun: ms$\rightarrow$ 1 hour dMe: s$\rightarrow$ hours RS CVn: hours$\rightarrow$ days PMS: hours	ms$\rightarrow$ 1000 s
Energy peak	X-ray: $\geq$ 10 keV Radio: $\geq$ 1 MeV	MeV range
Primary energy source	convection	gravity
Jet energy source	magnetic field	gravity and rotation
Jet direction	downward to photosphere	outward from supernova
Observed "explosion"	thermal instability when jet hits chromosphere	blast wave when jet reaches "photosphere"

fields, small scale instabilities, hydrodynamics, particle acceleration, and radiation can explain most stellar flares and even coronal heating. While we understand much of the phenomenology of flares, we do not yet have good models for the physics of rapid magnetic field reconnection or a detailed understanding of the observed correlation between the thermal X-ray emission and nonthermal radio emission of stars while flaring and outside of obvious flares [10]. Studies of transients elsewhere in astrophysics will eventually reveal their physical processes. We may then find that there is much in common among these physical principles, although the observed transient phenomena occur on very different spatial and temporal scales.

This work is supported by NASA grant H-04630D to NIST and the University of Colorado.

REFERENCES

1. Abbett, W.P., and Hawley, S.L., *Astrophys. J.* **521**, 906 (1999).
2. Antonucci, E., Dodero, M.A., Martin, R., Peres, G., Reale, F., and Serio, S., *Astrophys. J.* **413**, 786 (1993).
3. Ayres, T.R., Osten, R.A., and Brown, A., *Astrophys. J.* **526**, 445 (1999).
4. Bastian, T.S., in *Radio Emission from the Stars and the Sun*, ed. A.R. Taylor and J.M. Paredes, San Francisco: Astron. Soc. Pacific, pp. 447–454 (1996).
5. Bookbinder, J., *Mem. Soc. Astron. Italiana* **62**, 307 (1991).
6. Butler, C.J., *Mem. Soc. Astron. Italiana* **62**, 243 (1991).
7. Dulk, G.A., *Ann. Rev. Astron. Astrophys.* **23**, 169 (1985).
8. Feigelson, E.D., and Montmerle, T., *Ann. Rev. Astron. Astrophys.* **37**, 363 (1999).
9. Grosso, N., Montmerle, T., Feigelson, E.D., André, P., Casanova, S., and Gregorio-Hetem, J., *Nature* **387**, 56 (1997).

10. Güdel, M., and Benz, A.O., *Astrophys. J. Let.* **405**, L63 (1993).

11. Güdel, M., Linsky, J.L., Brown, A., and Nagase, F., *Astrophys. J.* **511**, 405 (1999).

12. Haisch, B.M., Linsky, J.L., Bornmann, P.L., Stencel, R.E., Antiochos, S.K., Golub, L., and Vaiana, G.S., *Astrophys. J.* **267**, 280 (1983).

13. Haisch, B.M., and Rodonò, M., *Solar Phys.* **121**, Nos 1–2 (1989).

14. Haisch, B., and Schmitt, J.H.M.M., *Astrophys. J.* **426**, 716 (1994).

15. Haisch, B.M., Strong, K.T., and Rodonò, M., *Ann. Rev. Astron. Astrophys.* **29**, 275 (1991).

16. Hawley, S.L., *Mem. Soc. Astron. Italiana* **62**, 271 (1991).

17. Hawley, S., and Pettersen, B.R., *Astrophys. J.* **378**, 725 (1991).

18. Hawley, S., Fisher, G.H., Simon, T., Cully, S.L., Deustua, S.E., Jablonski, M., Johns-Krull, C.M., Pettersen, B.R., Smith, V., Spiesman, W.J., and Valenti, J., *Astrophys. J.* **453**, 464 (1995).

19. Hudson, H.S., Acton, L.W., Hirayama, T., and Uchida, Y., *Pub. Astron. Soc. Japan* **44**, L77 (1949).

20. Hünsch, M., and Reimers, D., *Astron. Astrophys.* **296**, 509 (1995).

21. Joy, A.H., and Humason, M.L., *Pub. Astron. Soc. Pacific* **133** (1949).

22. Kopp, R.A., and Poletto, G, *Solar Phys.* **93**, 351 (1984).

23. Liebert, J., Kirkpatrick, J.D., Reid, I.N., and Fisher, M.D., *Astrophys. J.* **519**, 345 (1999).

24. Linsky, J.L., *Mem. Soc. Astron. Italiana* **62**, 307 (1991).

25. Linsky, J.L., Neff, J.E., Brown, A., Gross, B.D., Simon, T., Andrews, A.D., Rodonò, M., and Feldman, P.A., *Astron. Astrophys.* **211**, 173 (1989).

26. Linsky, J.L., Wood, B.E., Brown, A., Giampapa, M.S., and Ambruster, C., *Astrophys. J.* **455**, 670 (1995).

27. Martens, P.C.H., and Kuin, N.P.M., *Solar Phys.* **122**, 263 (1989).

28. Moffett, T.J., and Bopp, B.W. *Astrophys. J. Suppl.* **31**, 61 (1976).

29. Neff, J.E., *Mem. Soc. Astron. Italiana* **62**, 291 (1991).

30. Neuhauser, R., and Comeron, F., *Science* **282**, 83 (1998).

31. Osten, R.A., and Brown, A., *Astrophys. J.* **515**, 746 (1999).

32. Ottmann, R., and Schmitt, J.H.M.M., *Astron. Astrophys.* **307**, 813 (1996).

33. Parker, E.N., *Astrophys. J.* **318**, 876 (1987).

34. Parker, E.N., *Astrophys. J.* **407**, 342 (1993).

35. Parker, E.N., *Solar Phys.* **169**, 327 (1996).

36. Reale, F., and Micela, G., *Astron. Astrophys.* **334**, 1028 (1998).

37. Schmitt, J.H.M.M., and Favata, F., *Nature* **401**, 44 (1999).

38. Shu, F.H., Shang, H., Glassgold, A.E., and Lee, T., *Science* **277**, 1475 (1997).

39. Trigilio, C., Leto, G., and Umana, G., *Astron. Astrophys.* **330**, 1060 (1998).

40. Trigilio, C., Umana, G., and Migenes, V., *Astron. Astrophys.* **260**, 903 (1993).

41. van den Oord, G.H.J., Doyle, J.G., Rodonò, M., Gary, D.E., Henry, G.W., Byrne, P.B., Linsky, J.L., Haisch, B.M., Pagano, I., and Leto, G., *Astron. Astrophys.* **310**, 908 (1996).

42. Wood, B.E., Linsky, J.L., and Ayres, T.R., *Astrophys. J.* **478**, 745 (1997).

High Energy Processes in Solar Flares

Reuven Ramaty* and Natalie Mandzhavidze[†]

*NASA/GSFC, Greenbelt, MD 20771
[†]NASA/GSFC and USRA, Greenbelt MD 20771

Abstract.
We review the highlights of the high energy processes in solar flares. Particle acceleration is an essential ingredient of the flare energy release process. Among all cosmic high energy sources, flares are unique in that both the accelerated particle and the radiations that they produce are observed. In this paper we emphasize the radiations, gyrosynchrotron emission, bremsstrahlung and gamma ray lines. These provide information on flare models, magnetic fields, energy content in the accelerated particles and abundances. The abundances are not only critical for constraining the acceleration mechanisms, but also provide unique information on solar atmospheric dynamics and mixing, solar wind acceleration, and Galactic chemical evolution.

INTRODUCTION AND OVERVIEW

Major cosmic explosion are essentially always accompanied by particle acceleration. In the solar system, solar flares are the most powerful particle accelerators, capable of accelerating ions to at least a GeV and electrons to at least hundreds of MeV. Evidence for particle acceleration at the Sun comes from both direct particle observations in interplanetary space (e.g. [1]), and electromagnetic radiations which can only be produced by accelerated particle interactions in the solar atmosphere, specifically radio emission (e.g. [2]), hard X-rays (e.g. [3]) and gamma rays, (e.g. [4]). Such interactions also produce neutrons which are detected in space and near Earth (e.g. [5]). The ability to observe both the accelerated particles themselves and the radiations that they produce is a unique feature of solar flare acceleration. For other cosmic sites, the presence of accelerated particles can only be inferred from the radiations they produce.

Unlike some cosmic explosions (e.g. gamma-ray bursts) whose luminosities peak in the hard X-ray and gamma-ray bands, the radiative output of solar flares is maximal in a band ranging from optical to EUV wavelengths (e.g. see figure 1 in [6]). Nevertheless, there are compelling arguments, based on the impulsiveness [7] and energetics of the hard X-ray and gamma-ray emissions, that flares are fundamentally high-energy phenomena. The energy contents in flare accelerated electrons of energies $\gtrsim 20\,\mathrm{keV}$, derived from the X-ray data [8], and ions of energies

CP522, *Cosmic Explosions: Tenth Astrophysical Conference,*
edited by Stephen S. Holt and William W. Zhang

$\gtrsim$ 1MeV/nucleon, obtained [9,4] from gamma ray data [10], can exceed about 10^{32} erg, comparable to the total radiative output of the flare. How the flare releases this energy, most likely stored in the coronal magnetic fields, and how it rapidly accelerates electrons and ions with such high efficiency, is not yet completely understood.

Studies of solar energetic particle (SEP) observed in interplanetary space have shown that these SEP events can be classified as impulsive and gradual [11–14]. Impulsive events, for which the associated soft X-ray emission is of relatively short duration, have large electron-to-proton ratios (e/p), large ^{3}He/^{4}He, large heavy ion (particularly Fe) to C or O ratios, and charged states corresponding to temperatures which significantly exceed the $(1-2)\times10^6$K coronal temperature. Gradual events, for which the soft X-rays last longer, exhibit smaller e/p, and have heavy ion abundances, ^{3}He/^{4}He, and charged states which are consistent with coronal abundances and temperatures. It is reasonable to associate the above two classes of SEP events with different acceleration mechanisms. Because of the strong association with coronal mass ejections (CMEs), particles in gradual events are probably accelerated out of coronal gas by the CME driven shock (e.g. [15]). On average, $\sim10^{15}$ g of material is ejected per CME with speeds ranging from ~100 to ~2000 km/sec (see [16]). The maximal kinetic energy of a CME is thus comparable to the energy released in large flares. Some of the CMEs travel fast enough in the corona to drive collisionless shock waves which can accelerate particles. On the other hand, particle acceleration in impulsive events is probably due to gyroresonant interactions with plasma waves (see [17,18]).

In this paper on high energy processes in solar flares, we focus on the radiations that reveal the existence of the accelerated particles at the Sun. We first discuss the bremsstrahlung and gyrosynchrotron radiation and next the gamma ray line emission.

RADIATIONS PRODUCED BY ACCELERATED PARTICLES IN SOLAR FLARES

The radiations produced by the particles accelerated in solar flares are observed at high energies, as hard X-rays and gamma rays, and at very low energies, as radio waves extending in frequency up to the millimeter region. The bulk of the hard X-rays is bremsstrahlung produced by accelerated electrons. This bremsstrahlung extends from about 20 keV, where it merges with thermal bremsstrahlung produced by hot flare plasma, up to energies of hundreds of MeV [19]. In the energy region from about 0.3 to 8 MeV there are nuclear lines superimposed on the bremsstrahlung continuum [7]. Above about 50 MeV, in some flares there is emission resulting from the decay of pions, which are produced in the interactions of the highest energy protons and α particles with the ambient solar atmosphere (e.g. [20]).

Bremsstrahlung and Gyrosynchrotron Radiation

In the hard X-ray region, the central issue is that of the energy content in the sub-relativistic electrons. Within the well accepted electron bremsstrahlung paradigm, the fact that the energy lost by the electrons to bremsstrahlung is only a small fraction of the energy lost to Coulomb collisions, leads to an energy content in the electrons which is comparable to the total radiative output of the flare. That the hard X-rays are indeed bremsstrahlung is strongly supported by the fact that other X-ray producing mechanisms, specifically inverse Compton scattering, cannot account for the observations, mainly because of the low target photon densities. On the other hand, the question of whether the bremsstrahlung is driven by accelerated electrons or ions has often been raised (e.g. [21]). Most recently, it has been shown that at least some hard X-ray production could be due to the electrons in neutralized ion beams [22]. But the near simultaneous brightening of loop footpoints in hard X rays [23] is strong evidence that for these flares relativistic electron driven bremsstrahlung is the dominant process.

At hard X-ray energies, above about 100 keV and well into the gamma ray region, the solar flare continuum is definitely electron driven bremsstrahlung. Perhaps the best evidence for this is the strong correlation between the high energy continuum and the microwave/millimeter emission. The top panel of Figure 1 shows the time profiles of the gamma ray emission above 1 MeV and the 80 GHz millimeter emission observed near the impulsive peak of the 1991 June 4 flare [24,25]. A series of 6 very large (X-class) solar flares occurred in June 1991 and gamma ray emission was observed from all of them. These very large flares also produced very intense millimeter emission. The millimeter flux observed from the 1991 June 4 flare, shown in Figure 1, is in fact the largest ever observed [26]. The gamma ray emission observed from this flare was in fact so intense that it saturated all four instruments on the Compton Gamma Ray Observatory (CGRO), except the Charged Particle Detector (CPD) which is part of the BATSE instrument on CGRO. The CPD had no energy resolution, which is the reason that the gamma ray time profile shown in Figure 1 is the integral flux above 1 MeV. The middle panel in Figure 1 shows the measured ratio between the millimeter and gamma ray fluxes. The dip in the ratio is due to the delay (a few seconds) of the gamma ray emission relative to the millimeter flux.

A relatively simple model [25] that can account for these observations is one in which the millimeter emission is thin target gyrosynchrotron radiation produced in the upper parts of magnetized coronal loops, while the gamma ray emission is thick target bremsstrahlung produced at the loop footpoints. Thin and thick target refer to the behavior of the electrons, thin target meaning that the electrons precipitate from the upper, gyrosynchrotron production region before they lose any appreciable amount of energy into the thick target footpoint regions where they slow down and stop as they produce the bremsstrahlung. Because of the high frequency considered (80 GHz), the gyrosynchrotron source is optically thin. The lower panel in Figure 1 shows the calculated ratio of the 80GHz gyrosynchrotron emission to the >1 MeV bremsstrahlung as a function of the power law index S of the accelerated electrons, for several values of the magnetic field in the upper, thin target region. Another parameter is the electron trapping time $\tau = 15\,\mathrm{sec}$ in this region, which follows from

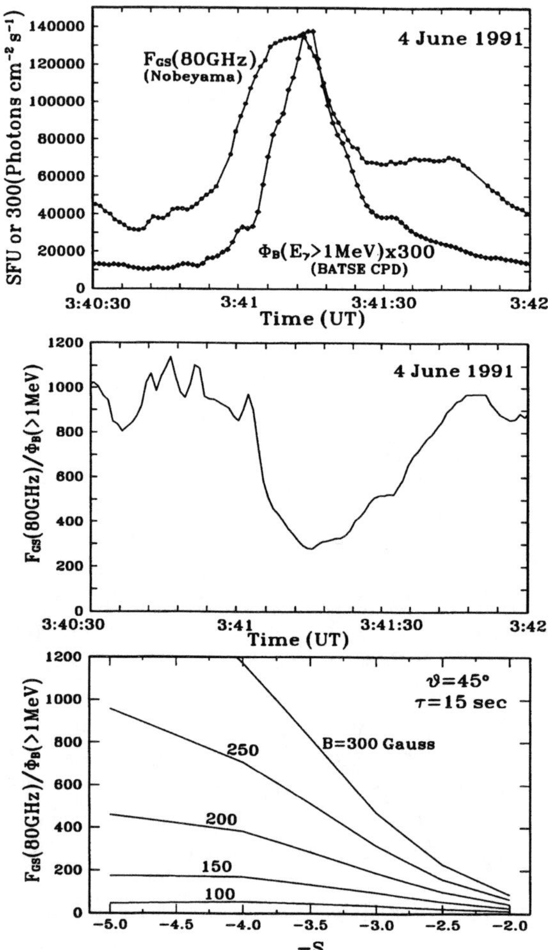

FIGURE 1. Top panel: gamma ray and millimeter wave time profile near the impulsive peak of the flare which produced the largest millimeter wave emission ever observed; middle panel: the ratio of the millimeter and gamma ray fluxes showing that the displacement seen in the top panel corresponds to a dip in the ratio with a minimum at the time of maximum of the gamma ray flux; bottom panel: calculations of the millimeter to gamma ray flux ratio, where the millimeter emission is produced by isotropic electrons in a uniform magnetic field viewed at 45° to the field direction.

the observed decay of the 80 GHz emission. The bremsstrahlung decays with about
the same time constant, because the electron energy loss time in the thick target
region is shorter than the time dependence of the electron source for this region,
which is the precipitation, so that the decay is controlled by the same precipitation
which controls the time behavior of the gyrosynchrotron radiation. As can be seen,
for a given magnetic field, $F_{GS}(80GHz)/\Phi_B(>1MeV)$ decreases as the spectrum
hardens. The observed delay of the bremsstrahlung relative to the gyrosynchrotron
emission can thus be understood in terms of a hardening and subsequent softening
of the accelerated electron spectrum, with the spectrum being hardest at the peak
of the bremsstrahlung. The observed variation of the ratio from about 1000 to
250 thus implies a magnetic field of about 250 Gauss. A higher magnetic fields
would follow from a recent analysis [27] which employed CGRO/OSSE data and
found that the bremsstrahlung flux from the 1991 June 4 flare actually was lower
by about a factor of 3 than that shown in Figure 1. The OSSE data, however, are
still consistent with the BATSE/CPD time profile shown in Figure 1.

Gamma Ray Line Emission

Gamma ray lines were first observed from solar flares in 1972 with the NaI
scintillator on OSO-7 [28]. But it was not until 1980 that routine observations
became possible with the much more sensitive spectrometer on the Solar Maximum
Mission (SMM [19,7]). Subsequently, gamma ray line observations were caried out
with GRANAT/Phebus [29], CGRO [27], and YOHKOH [30]. Starting in 2000,
observations with the imaging Ge spectrometer HESSI [31] should become available.
Gamma-ray line emission in solar flares results from nuclear deexcitations, neutron
capture and positron annihilation.

1 Deexcitation Lines:
$FIP, (He/H)_{acc}, (He/H)_{amb}, (^3He/^4He)_{acc}, (Li/H)_{amb}$

Narrow deexcitation lines are produced by accelerated protons, α particles and
^{3}He nuclei interacting with ambient He and heavier nuclei, while broad lines result
from interactions of accelerated C and heavier nuclei with ambient H and He. A
theoretical nuclear deexcitation spectrum is shown in Figure 2. Starting from high
energies, there are strong narrow lines at 6.13 MeV from ^{16}O, 4.44 MeV from ^{12}C,
1.79 MeV from ^{28}Si, 1.63 MeV from ^{20}Ne, 1.37 MeV from ^{24}Mg and 0.847 MeV from
^{56}Fe. All of these lines result from deexcitations in nuclei of the relatively abundant
constituents of the solar atmosphere. This has allowed the determination of the
abundances of these elements [32,9,33], showing that in the gamma ray production
region the low FIP (first ionization potential) elements Mg, Si and Fe are enriched
relative to C and O, a result which implies that the FIP bias, known to exist in the
corona (e.g. [1]), already sets in at lower altitutes, presumably in the chromosphere.

Other important lines are formed in reactions between abundant constituents of both the accelerated particles and the ambient medium leading to excitations in nuclei which are otherwise rare in the solar atmosphere. The feature marked $\alpha\alpha$ is a blend of two lines, at 0.429 and 0.478 MeV from ^{7}Be and ^{7}Li, respectively. The excited states of these isotopes are populated by interactions of accelerated α particles with ambient He [34]. The line marked ^{3}He at 0.937 MeV and the narrow ^{59}Ni line at 0.339 MeV result, respectively, from interactions of accelerated ^{3}He nuclei with ambient ^{16}O and accelerated α particles with ^{56}Fe, leading to excited states in ^{18}F and ^{59}Ni [35]. The line complex marked $(\alpha,^3\text{He})$ just above 1 MeV consists of 4 lines, at 1.04 and 1.08 MeV from the same ^{3}He induced reaction as that which produces the 0.937 MeV line, and at 1.00 and 1.05 MeV due to α particle interactions with ^{56}Fe leading to excited states in ^{58}Co, ^{58}Ni and ^{59}Ni. This complex was observed, but not resolved, with SMM [36,37]. We anticipate that with the Ge detectors on HESSI such resolution will be possible.

The various lines produced by accelerated α particles and ^{3}He nuclei provide information [38] on the He abundances in both the accelerated particles (the α/p ratio) and the ambient solar atmosphere, as well as the accelerated particle ^{3}He abundance (the ^{3}He/^{4}He ratio). In 5 flares out of 20 α/p was found to exceed the standard value of 0.1. The possibility of determining the accelerated α/p and its variation with time during the acceleration process from future more detailed gamma ray data will be an important ingredient for the understanding of solar flare particle acceleration [18]. In 4 flares out of 20 the ambient He abundance

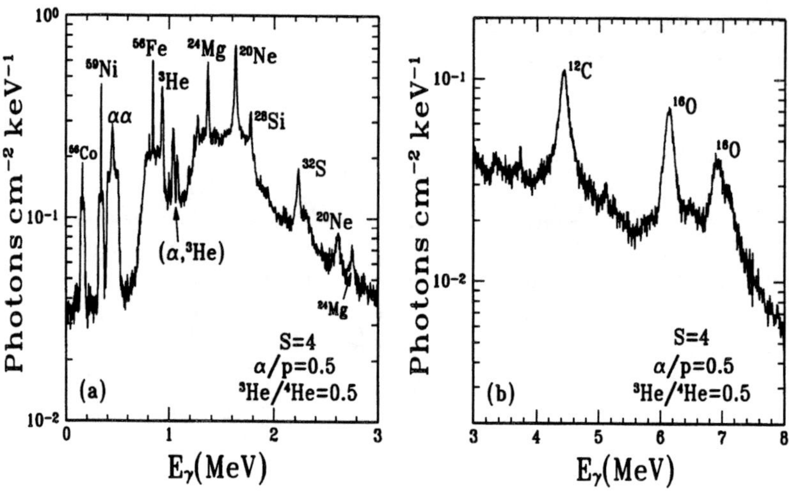

FIGURE 2. Theoretical nuclear deexcitation gamma-ray spectrum in two energy ranges. The accelerated particles energy spectrum is an unbroken power law with spectral index $S = 4$. Impulsive flare abundances are used for the accelerated particles, with $\alpha/\text{p}=0.5$ and ^{3}He/^{4}He=0.5. The ambient medium abundances are those of the average corona, except that Ne/O=0.25.

Accretion Disk Eruptions in the FU Orionis Variable Stars

Scott J. Kenyon

Smithsonian Astrophysical Observatory
60 Garden Street, Cambridge, MA 02138 USA

Abstract. The pre-main sequence FU Orionis variables undergo 3–6 mag outbursts lasting from decades to centuries. The outbursts occur in a large, self-luminous accretion disk surrounding a young star with an age of 1–10×10^5 yr. Models of thermally unstable disks generally account for the scale and duration of the outburst. The accretion rate through the disk is large enough to add $\sim 1\%$ of a solar mass to the central star in each outburst. Repetitive eruptions could add $\gtrsim 20\%$ of a solar mass to each pre-main sequence star in the solar neighborhood.

INTRODUCTION

Most pre-main sequence stars vary in brightness. Fluctuations of $\lesssim 1$ mag are a defining feature of T Tauri stars and Herbig AeBe stars [30]. These variations are often due to dark spots rotating with the stellar photosphere [8,34] or bright spots at the base of a magnetic accretion column [8]. Modest 1–3 mag eruptions have been observed in several young stars, such as EX Lup and DR Tau [33]. Spectacular 3–6 mag eruptions occur in the FU Orionis variables, also known as FUors [29,32].

G. Herbig first associated FUor eruptions with pre-main sequence stars [31,32]. FU Ori, which lies at the apex of a fan-shaped nebula within the dark cloud B35, brightened by 5 mag or more in ~ 200 days [31,32,35,79,80]. Thirty years later, Welin discovered the 5 mag eruption of V1057 Cyg within an eccentric ring of reflection nebulosity [32,37,44,83]. Herbig later noted the similarity between the optical spectra of these two stars with spectra of V1515 Cyg, a faint variable star embedded in arc-shaped nebulosity [30]. He collected archival photographic photometry and identified a slow rise from $m_{pg} \approx 15.5$ in the late 1940s to $m_{pg} \approx 13.5$ in the late 1970s [32]. This brightness increase continued until 1980, when the star experienced a dramatic decline and slow recovery [41].

Herbig's demonstration that FU Ori – and other FUors – is a pre-main sequence star is straightforward. FUors are clearly associated with the dark molecular clouds where stars form: they have radial velocities indistinguishable from the cloud velocity and they have extinctions similar to those of other young stars in the cloud.

CP522, *Cosmic Explosions: Tenth Astrophysical Conference,*
edited by Stephen S. Holt and William W. Zhang
2000 American Institute of Physics 1-56396-943-2

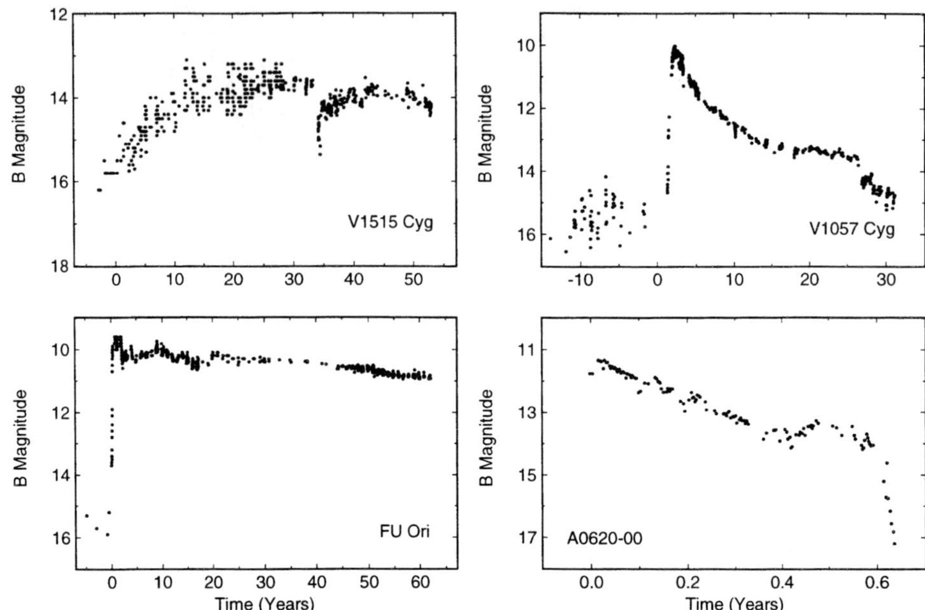

FIGURE 1. Light curves for three FU Ori variables and an X-ray transient, A0620−00.

Their optical spectra, including strong Li I absorption lines, are similar to spectra of pre-main sequence T Tauri stars having ages of $\sim 10^6$ yr. One FUor, V1057 Cyg, had a pre-outburst spectrum resembling optical spectra of T Tauri stars. The event statistics are also plausible for pre-main sequence stars, as summarized below. Finally, the rise in brightness is a real luminosity increase that is *not* a classical nova outburst, the main alternative.

A dozen pre-main sequence stars are now recognized as FUors. Most have been observed to rise 3–5 mag in brightness in less than one year. V1515 Cyg is the only known example to require a decade to rise to visual maximum, but the historical light curves for some systems are poorly documented. A few objects have been called FUors based on distinctive properties common to the class. Many recently discovered FUors are more intimately associated with the densest dark clouds than the first members of the class, suggesting that some eruptions have been missed.

BASIC PROPERTIES OF FU ORIONIS OBJECTS

FUors share distinctive morphological, photometric, and spectroscopic character-istics. Three are visual binary stars [3,38,45,71,77]. Most have delicate fan-shaped or comma-shaped reflection nebulae with sizes of $\sim$ 1000 AU [23,52,57,63]. Optical jets and bright emission knots (HH objects) are common [16,29,63]. Most appear associated with large-scale molecular outflows [19,20,51,53,86,87]. These features –

together with broad, blue-shifted Na I and H I absorption features [4,6,15,17,25,84] -- demonstrate that FUors drive powerful winds which interact with the surrounding medium [4,15,41,85].

Most FUors have F–G giant or supergiant optical spectra [28,32,64,65]. The optical photosphere has a radius $\sim$ 10–20 R$_\odot$ [32]. The optical reflection nebulae of several embedded FUors also show G-type absorption features [74,75]. All FUors but Z CMa and V346 Nor have very deep CO absorption bands on near-IR spectra [6,11,18,45,56,68,72,76]. These features resemble the CO absorption bands observed in red giants and are much stronger than those observed in other pre-main sequence stars [27]. Many FUors display strong water absorption features, strengthening the evidence for a $\sim$ 2000 K photosphere [9,72].

Several FUors show photometric variations similar to the flickering of cataclysmic variables and X-ray binaries [42]. The 0.03–0.04 mag fluctuations in FU Ori, V1057 Cyg, and V1515 Cyg occur on time scales of one day or less. This variable source has the colors of an F–G supergiant. The optical source responsible for the eruption therefore varies on a time scale comparable to its dynamical time scale, $\tau_d \sim (R_{opt}^3/GM_{opt})^{1/2} \sim$ 0.8–2.0 day for $R_{opt} \approx$ 10–20 R$_\odot$ and $M_{opt} \approx$ 0.5 M$_\odot$.

All FUors show large *excesses* of radiation over normal G supergiants at UV and IR wavelengths. The near-IR excess is clearly photospheric in origin, because the CO and H$_2$O absorption features are strong. The UV excesses in Z CMa and FU Ori result from an A- or F-type photosphere that is hotter than the G-type photosphere observed at longer wavelengths [40]. In addition to significant far-IR and submm emission [37,82], many FUors are strong radio continuum sources at cm wavelengths [14,50,69–71]. This emission is not photospheric in some FUors and may be produced in the outflow or the jet.

Herbig first noted broad absorption lines on optical spectra of several FUors [32]. This property is now characteristic of the class; all FUors have broad optical and/or IR absorption lines with $v \sin i \approx$ 15-60 km s^{-1} [26,28,38,73,74]. Assuming a random distribution of $\sin i$, the velocities are close to the breakup velocity of a 0.5–1.0 M$_\odot$object with a radius of 10–20 R$_\odot$. The near-IR CO lines in FU Ori and V1057 Cyg have significantly smaller rotational velocities than the optical lines [27]. The optical rotational velocity smoothly increases with decreasing wavelength in Z CMa and V1057 Cyg [84]. In FU Ori itself, the powerful wind masks weak absorption lines and makes it difficult to detect any variation of rotational velocity with wavelength should one exist.

Many FUors display *doubled* absorption lines on optical and near-IR spectra [26–28,73,74]. The two absorption components in V1057 Cyg are separated by 30–40 km s^{-1}; these features have much larger separations in Z CMa and FU Ori. The long-term stability of the doubled absorption lines indicates that the lines are not produced by two stellar components in a binary system [39].

Finally, FUor eruptions must be repetitive [26,32]. The event statistics of known FUors suggest that a young star must undergo 10–20 FUor eruptions before reaching the main sequence. This estimate may be a lower limit, because some outbursts have certainly been missed. Reipurth's proposal that FUor eruptions produce HH objects suggests a recurrence time scale of ~ 1000 yr, based on the identification of multiple bowshocks (ejection events) with dynamical separations of 500–2000 yr [2,24,60,61,67]. The discovery of giant HH flows with dynamical time scales of 10^4–10^5 yr [66] allows 10–100 eruptions if these eruptions power the outflow.

FU ORIONIS OBJECTS AS ACCRETION DISKS

The observations of FUors place severe constraints on possible outburst mechanisms. With a luminosity of 50–500 $L_\odot$, a typical FUor emits 10^{45} to 10^{46} erg during the course of a 10–100 yr eruption. Outbursts also recur every 10^3–10^5 yr, producing a rapidly rotating F-G supergiant in the optical and a more slowly rotating M giant in the near-IR. This process must be unique to young stars, because FUors are not observed in older systems.

Accretion is the most plausible energy source for FUor eruptions [26,47]. Accretion generates 10 times more energy, $\sim 5 \times 10^{14}$ erg g^{-1}, than deuterium burning, the primary fusion source for pre-main sequence stars. FUor outbursts, resemble those of other accreting systems such as dwarf novae and X-ray transients. In particular, the light curve of V1057 Cyg is very similar to the decline of the X-ray transient A0620-00 (Figure 1; [81]). By analogy with interacting binary systems, FUor eruptions are the high state of a thermal instability in a disk surrounding a pre-main sequence star. This instability occurs if the accretion rate from the molecular cloud core into the disk lies between the stable accretion rates in the low state and the high state. The system cycles between these two states as long as the surrounding cloud can replenish the disk between outbursts.

To make a rough estimate of the accretion rate needed to power a FUor, I assume a steady disk with $L_{disk} = 157$ $L_\odot$ $(M_\star \dot{M}/R_\star)$ and $T_{max} \approx 6500$ K $(M_\star \dot{M}/R_\star{}^3)^{1/4}$. If $T_{max} = 6500$ K to account for the F-G spectral type, $R_\star \approx (L_{disk}/157 \ L_\odot)^{1/2}$ and $M_\star \dot{M} \approx (L_{disk}/157 \ L_\odot)^{3/2}$. These relations yield $R_\star \approx 1$–2 $R_\odot$ and $M_\star \dot{M} \sim 1$–2×10^{-5} $M_\odot^2$ yr^{-1}. Using a pre-main sequence mass-radius relation for an age of 10^5 yr, $M_\star \approx 0.2$–0.3 $M_\odot$ and $\dot{M} \approx 0.5$–1×10^{-4} $M_\odot$ yr^{-1}. Thus, a young star accretes material at $\sim 10^{-4}$ $M_\odot$ yr^{-1} during a FUor eruption. For an outburst duration of 100 yr, this accretion rate results in a total accreted mass of ~ 0.01 $M_\odot$ per eruption. This mass is replenished during a low state lasting $\sim 10^3$–10^5 yr, resulting in an infall rate of $\sim 10^{-5}$–10^{-7} $M_\odot$ yr^{-1}, close to the infall rates envisioned – and in some cases observed – for typical cloud cores [1,55,58].

OBSERVATIONAL TESTS OF DISK MODELS

Figure 2 shows light curves for V1057 Cyg, one of the best-studied FUors [5,12,13,39,37,43,49,78] . The amplitude of the decline is largest in the UV and decreases monotonically from 0.3 μm to 5 μm. This behavior is characteristic of accreting systems, which evolve towards cooler temperatures as the luminosity declines [5,10,48]. The spectral type variation also agrees with disk model predictions.

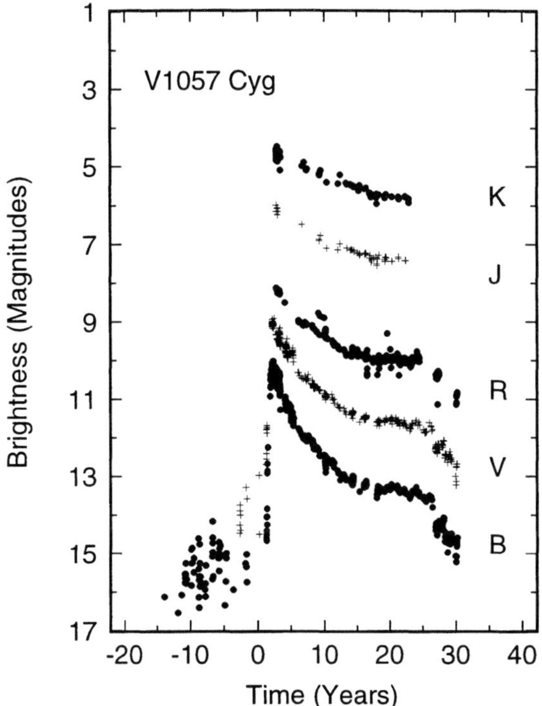

FIGURE 2. V1057 Cyg light curves. The amplitude of the decline decreases with wavelength.

If we assign stellar effective temperatures to the optical spectrum at maximum (A5) and at the current epoch (G5), the UBV decline indicates a source with a roughly constant radius. This requirement is possible with an accretion disk if the radius of the central star remains constant. Few stars evolve at constant radius when their effective temperatures change.

The light curves of other FUors also generally agree with disk model predictions. The rise times of ~ 200 days in FU Ori and V1057 Cyg are comparable to the thermal time scale, as expected for an eruption that begins at the inner edge of the disk [5,43]. The several decade rise in V1515 Cyg is too long for such an "inside-out" instability, but could be caused by an outburst that begins in the outer part of the disk. These "outside-in" eruptions can result from perturbations of the disk by a companion star or planet, in addition to the disk instability mechanism [5,7,12,13]. In all systems, the decay times are comparable to the viscous time scale, 10–1000 yr, at 0.1–1 AU.

Disk models also successfully explain FUor SEDs [26,39,37]. Figure 3 shows a dereddened SED for FU Ori. The deep UV and near-IR absorption features indicate that the UV and IR excesses over a G-type supergiant spectrum are photospheric [9]. If disks radiate like stars at the effective temperatures appropriate for the

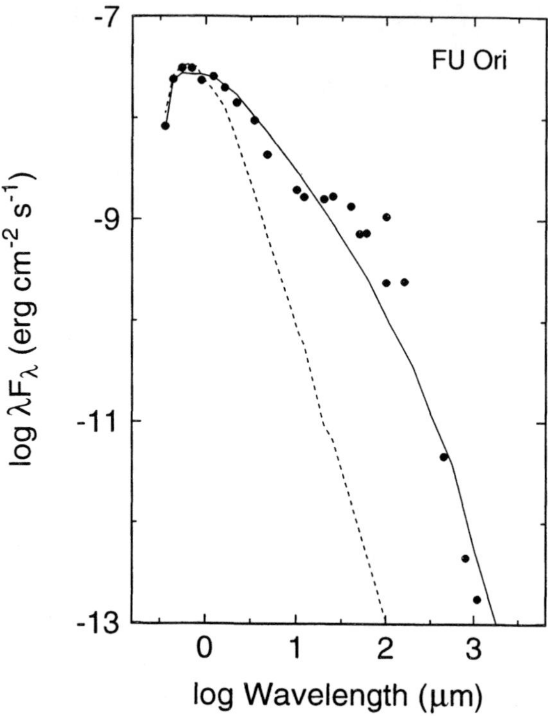

FIGURE 3. Spectral energy distribution for FU Ori. An accretion disk (solid) line fits the data better than a G star (dashed line). A 10,000 AU reflection nebula produces the modest far-IR excess over the disk spectrum at ~ 100 μm.

observed absorption features, the SEDs for FU Ori and other FUors require that the surface area of the emitting region increases with increasing wavelength. The surface area of the near-IR source must be 10–20 times larger than the surface area of the optical source; the 300 K material responsible for the 10 μm excess must have roughly 100 times the emitting area of the optical continuum region. A disk in which the temperature decreases radially outward naturally explains this observation. Disk models account for the 0.3–10 μm SEDs of V1057 Cyg, V1515 Cyg, and FU Ori quite well (Figure 3; [5,26,39,37,78]). Several other FUors with modest reddening have similar SEDs [37].

Recent interferometric observations have resolved FU Ori at near-IR wavelengths. The angular size at 2.2 μm agrees with predictions of the accretion disk model [54]. Current interferometers have the ability to make more tests by resolving the disks of V1057 Cyg and V1515 Cyg. The shape of the disk yields a stern test. The rotational velocity data predict that the inclination of the disk increases from $i \approx$ 10–20° for V1515 Cyg to $i \approx$ 30–40° for V1057 Cyg to $i \approx$ 45–60° for FU Ori. The disk in FU Ori thus should be much more elongated than the V1515 Cyg disk.

Accretion disks also naturally explain the optical and IR line profiles observed in some FUors. The gradual decrease in the rotational velocity with increasing wavelength occurs because longer wavelength emission is produced in more slowly-orbiting material at larger disk radii than the more rapidly moving inner disk material responsible for short wavelength emission. At a given disk temperature, a larger fraction of the disk surface rotates at large line-of-sight velocities and the lines appear doubled [9,39,59].

Finally, 10–100 μm observations support the notion that infall from a surrounding cloud fuels the disk in between outbursts. The 10–100 μm SEDs of several FUors and the decline of the 10–20 μm light of V1057 Cyg indicate that the mid-IR radiation is produced by optical light absorbed and reradiated by a surrounding envelope [37]. An infall rate of 1–5 $\times$ 10^{-6} $M_\odot\,\mathrm{yr}^{-1}$ produces an optically thick envelope that can reprocess optical light from the inner disk and account for the 10–100 μm SED [5,37]. This rate is sufficient to replenish the disk in 1000 yr, which allows recurrent FUor eruptions. The observational evidence for infall at comparable rates in two FUors, Z CMa [46] and L1551 IRS5 [55,58], [55], lends further support to this interpretation.

THE IMPORTANCE OF FU ORI ERUPTIONS

Young low mass stars accrete much of their final mass in FUor events. A typical eruption adds ~ 0.01 $M_\odot$ to the central star; 10–20 eruptions per pre-main sequence star implies a total accreted mass of ~ 0.1–0.2 $M_\odot$. These eruptions eject mass and momentum into the surrounding cloud. FU Ori itself has lost material with a momentum of ~ 0.3 $M_\odot$ km s^{-1} since its eruption began. With 10–20 such eruptions, the total momentum of the FU Ori outflow, 3–6 $M_\odot$ km s^{-1}, is at the lower end of the momenta, 1–100 $M_\odot$ km s^{-1} [21], observed in a typical molecular outflow.

A young star can accrete *all* of its mass in FUor events if (i) the disk undergoes regular thermal instabilities, cycling between the low and high states on a 1000–2000 yr time scale, and (ii) FUors are young enough, $\lesssim$ a few $\times$ 10^5 yr, to give the surrounding cloud time to fuel recurrent eruptions. FUors are very young objects. They have distinctive reflection nebulae and large far-IR excesses. Most are associated with jets, HH objects, and molecular outflows. These properties are more common in stars with ages of $\sim 10^5$ yr than in stars with ages of $\sim 10^6$ yr [22,60,62]. A reasonable age for the class is $\sim$ a few $\times$ 10^5 yr [37,82]. The conditions needed for the disk to cycle between the high and low states may be satisfied in many pre-main sequence stars. For an infall rate of 1–10 $\times$ 10^{-6} $M_\odot\,\mathrm{yr}^{-1}$, the disk spends most of its time in a low state where the mass accretion rate through the disk is smaller than the infall rate [5,49,43]. The disk requires $\sim 10^3$–10^4 yr to accumulate the ~ 0.01 $M_\odot$ needed to power a FUor eruption. The disk cycles between the low and high states as long as the cloud supplies mass to the disk.

If young stars accrete most of their mass in FUor events, FUors can power many molecular outflows. In most outflow models, the mass ejected by the central star is $\sim 10\%$–30% of the accreted mass. For a wind velocity of ~ 200 km s^{-1}, the momentum in the wind is $(Mv)_w \sim 20\ (M_*/M_\odot)$ M$_\odot$ km s^{-1} for a typical wind velocity of 200 km s^{-1}. The observed range of outflow momenta – $(Mv)_o = 1$–100 M$_\odot$ km s^{-1} [21] – requires stellar masses of $M_* \sim 0.1$–5 M$_\odot$ if the flows conserve momentum. The winds of FUors can power this outflow **if** a young star accretes most of its mass in FUor events.

These simple estimates establish FUors as important events if they recur on time scales of $\sim 10^3$ yr during the earliest phases of pre-main sequence stellar evolution. The fraction of time in the FUor state is the ratio of the infall rate to the FUor accretion rate: $f \sim \dot{M_i}/\dot{M}_{FU}$. FUor disk models require $\dot{M}_{FU} \sim 10^{-4}$ M$_\odot$ yr^{-1}, so $f \sim 0.05$ for a typical infall rate of $\dot{M_i} \sim 5 \times 10^{-6}$ M$_\odot$ yr^{-1}. This prediction currently agrees with the statistics of FUors among protostars: 1 out of ~ 20 protostars in the Taurus dark cloud contains a FUor (L1551 IRS5) and 5 out of 50–60 protostars with HH objects contains a central star that resembles known FUors spectroscopically [65]. Sensitive photometric and spectroscopic surveys of nearby molecular clouds can improve these statistics. If the FUor frequency among protostars turns out to be more than a few per cent, then FUors may represent the main accretion phase of early stellar evolution.

I thank J. McClintock and P. Zhao for help with light curves of X-ray transients.

REFERENCES

1. Adams, F. C., Lada, C. J., & Shu, F. H. 1987, ApJ, 308, 788
2. Bachiller, R., Tafalla, M., & Cernicharo, J. 1994, ApJo, 425, L93
3. Barth, W., Weigelt, H., & Zinnecker, H. 1994, A&A, 291, 500
4. Bastian, U., & Mundt, R. 1985, A&A, 144, 57
5. Bell, K. R., Lin, D. N. C., Hartmann, L., & Kenyon, S. J. 1995, ApJ, 444, 376
6. Biscaya, A. M., Rieke, G. H., Narayanan, G., Luhman, K. L., & Young, E. T. 1997, AJ, 491, 359
7. Bonnell, I., & Bastian, P. 1992, ApJ, 401, L31
8. Bouvier, J., & Bertout, 1989, A&A, 211, 99
9. Calvet, N., Hartmann, L., & Kenyon, S. J. 1993, ApJ, 402, 623
10. Cannizzo, J. K., & Mattei, J. A. 1998, ApJ, 505, 344
11. Carr, J. S., Harvey, P. M., & Lester, D. F. 1987, ApJL, 321, L71
12. Clarke, C. J., Lin, D. N. C., & Pringle, J. E. 1990, MNRAS, 242, 439
13. Clarke, C. J., & Syer, D. 1996, MNRAS, 278, L23
14. Cohen, M., Beiging, J. H., & Schwartz, P. R. 1982, ApJL, 289, L5
15. Croswell, K., Hartmann, L., & Avrett, E. 1987, ApJ, 312, 227
16. Davis, C. J., Mundt, R., Eislöffel, J., & Ray, T. P. 1994, AJ, 110, 766
17. Eislöffel, J., Hessman, F. V., & Mundt, R. 1990, A&A, 232, 70
18. Elias, J. H. 1978, ApJ, 223, 859
19. Evans, II, N. J., Balkum, S., Levreault, R. M., Hartmann, L., & Kenyon, S. J. 1994, ApJ, 424, 793

20. Fridlund, C. V. M., & Liseau 1998, ApJ, 499, L75

21. Fukui, Y. 1989, in *ESO Workshop on Low Mass Star Formation and Pre-Main Sequence Objects,* edited by B. Reipurth, Garching, ESO, p. 95

22. Gómez, M., Whitney, B. A., & Kenyon, S. J. 1997, AJ, 114, 1138

23. Goodrich, R. 1987, PASP, 99, 116

24. Hartigan, P., Raymond, J., & Meaburn, J. 1990, ApJ, 362, 624

25. Hartmann, L., & Calvet, N. 1995, AJ, 109, 1846

26. Hartmann, L., & Kenyon, S.J. 1985, ApJ, 299, 462

27. Hartmann, L., & Kenyon, S.J. 1987, ApJ, 322, 393

28. Hartmann, L., Kenyon, S. J., Hewett, R., Edwards, S., Strom, K. M., Strom, S. E., & Stauffer, J. R. 1988, ApJ, 338, 1001

29. Hartmann, L., & Kenyon, S. J. 1996, ARA&A, 34, 205

30. Herbig, G. H. 1960, ApJS, 4, 33

31. Herbig, G. H. 1966, Vistas in Astr, 8, 109

32. Herbig, G. H. 1977, ApJ, 217, 693

33. Herbig, G. H. 1989, in *ESO Workshop on Low-Mass Star Formation and Pre-Main Sequence Objects* ed. B. Reipurth, Garching, ESO, p. 233

34. Herbst, W., Herbst, D. K., & Grossman, E. J. 1994, AJ, 108, 1906

35. Ibragimov, M. A. 1997, Pis'ma Astr. Zh., 1, 125

36. Kawazoe, E., & Mineshige, S. 1993, PASJ, 45, 715

37. Kenyon, S. J., & Hartmann, L. 1991, ApJ, 383, 664

38. Kenyon, S. J., Hartmann, L., Gómez, M., Carr, J., & Tokunaga, A. 1993b, AJ, 105, 1505

39. Kenyon, S. J., Hartmann, L., & Hewett, R. 1988, ApJ, 325, 231

40. Kenyon, S.J., Hartmann, L.W., Imhoff, C.L., & Cassatella, A. 1989, ApJ, 344, 925.

41. Kenyon, S. J., Hartmann, L., & Kolotilov, E. A. 1991, PASP, 103, 1069

42. Kenyon, S. J., Kolotilov, E. A., Ibragimov, M. A., & Mattei, J. A. 2000, ApJ, in press

43. Kley, W., & Lin, D. N. C. 1999, ApJ, 518, 833

44. Kolotilov, E. A., & Kenyon, S. J. 1998, IBVS, No. 4494

45. Koresko, C. D., Beckwith, S. V. W., Ghez, A. M., Matthews, K., Neugebauer, G. 1991, AJ, 102, 2073

46. Liljeström, T., & Olofsson, G. 1997, ApJ, 478, 381

47. Lin, D. N. C., & Papaloizou, J. 1985, in *Protostars and Planets II,* ed. D. C. Black and M. S. Matthews, Tucson, University of Arizona Press, p. 981

48. Lin, D. N. C., & Papaloizou, J. C. B., 1995, ARA&A, 33, 505

49. Lin, D. N. C., & Papaloizou, J. C. B., 1996, ARA&A, 34, 703

50. Looney, L. W., Mundy, L. G., & Welch, W. J. 1997, ApJ, 484, L157

51. López, R., *et al.* 1998, AJ, 116, 845

52. Lucas, P. W., & Roche, P. F. 1996, MNRAS, 280, 1219

53. McMuldroch, S., Blake, G. A., & Sargent, A. I. 1995, AJ, 110, 354

54. Malbet, F., *et al.* 1998, ApJL, 507, 149

55. Mardones, D., Myers, P. C., Tafalla, M., Wilner, D. J., Bachiller, R., & Garay, G. 1997, ApJ, 489, 719

56. Mould, J. R., Hall, D. N. B., Ridgway, S. T., Hintzen, P., & Aaronson, M. 1978,

ApJL, 222, L123

57. Nakajima, T., & Golimowski, D. A. 1995, AJ, 109, 1181

58. Ohashi, N., Hayashi, M., Ho, P.T.P., Momose, M., & Hirano, N. 1996, ApJ, 466, 957

59. Popham, R., Narayan, R., Kenyon, S. J., & Hartmann, L. 1996, ApJ, 473, 422

60. Reipurth, B. 1985, A&A, 143, 435

61. Reipurth, B. 1989, Nature, 340, 42

62. Reipurth, B. 1990, in *Flare Stars in Star Clusters, Associations, and the Solar Vicinity*, IAU Symposium No. 137, Dordrecht, Kluwer, p. 229

63. Reipurth, B. 1991, in *Physics of Star Formation and Early Stellar Evolution*, NATO Adv. Study Inst., edited by C. J. Lada and N. D. Kylafis, p. 497

64. Reipurth, B. 1997, in *Low Mass Star Formation - from Infall to Outflow*, Poster Proceedings of IAU Symposium No. 182 on Herbig-Haro Objects and the Birth of Low Mass Stars, edited by F. Malbet & A. Castets, p. 309

65. Reipurth, B., & Aspin, C. 1997, AJ, 114, 2700

66. Reipurth, B., Bally, J., & Devine, D. 1997, AJ, 114, 2708

67. Reipurth, B., & Heathcote, S. 1992, A&A, 257, 693

68. Reipurth, B., Olberg, M., Gredel, R., & Booth, R. S. 1997, A&A, 327, 1164

69. Rodriguez, L., Hartmann, L. W., & Chavira, E. 1990, PASP, 102, 1413

70. Rodriguez, L., & Hartmann, L. 1992, Rev. Mex. A&A, 24, 135

71. Rodriguez, L., D'Alessio, P., Wilner, D.J., Ho, P.T.P., Torrelles, J.M., Curiel, S., Gómez, Y., Lizano, S., Pedlar, A., Canto, J., & Raga, A.C. 1998, Nat, 395, 355

72. Sato, S., Okita, K., Yayamshita, T., Mizutani, K., Shiba, H., Kobayashi, Y., & Takami, H. 1992, ApJ, 398, 273

73. Staude, H. J., & Neckel, Th. 1991, A&A, 244, L13

74. Staude, H. J., & Neckel, Th. 1992, ApJ, 400, 556

75. Stocke, J. T., Hartigan, P. M., Strom, S. E., Strom, K. M., Anderson, E. R., Hartmann, L. W., & Kenyon, S. J. 1988, ApJS, 68, 229

76. Teodorani, M., Errico, L., Vittone, A. A., Giovanelli, F., & Rossi, C. 1997, A&AS, 126, 91

77. Thiebaut, E., Bouvier, J., Blazit, A., Bonneau, D., Foy, F.-C. & Foy, R. 1995, A&A, 303, 795

78. Turner, N. J. J., Bodenheimer, P., & Bell, K. R. 1997, ApJ, 480, 754

79. Wachman, A. A. 1939, Beob. Zirk., 21, 60

80. Wachman, A. A. 1954, Zs. f. Ap., 35, 74

81. Webbink, R. F. 1978, *A Provisional Optical Light Curve of the X-Ray Recurrent Nova V616 Mon = A0620−00*, Univ. of Illinois preprint

82. Weintraub, D. A., Sandell, G., & Duncan, W. D. 1991, ApJ, 382, 270

83. Welin, G. 1971, A&A, 12, 312

84. Welty, A. D., Strom, S. E., Edwards, S., Kenyon, S. J., & Hartmann, L. W. 1992, ApJ, 397, 260

85. Whitney, B. A., Clayton, G. C., Schulte-Ladbeck, R. E., Calvet, N., Hartmann, L. & Kenyon, S. J. 1993, ApJ, 417, 687

86. Yamashita, T., & Tamura, M. 1992, ApJL, 387, L93

87. Yang, J., Ohashi, N., & Fukui, Y. 1995, ApJ, 455, 175

Eta Carinae

Kris Davidson

Astronomy Department, University of Minnesota, 116 Church St. SE, Minneapolis, MN 55455

Abstract. The very massive star η Car presents a number of significant, well-defined, disconcertingly basic, unsolved problems in several branches of astrophysics. This and related objects now constitute a rather large topic with diverse implications and many recent surprises. Being relatively unexplored ground, the subject offers extraordinary opportunities for theoretical work.

I   INTRODUCTION

At this meeting I've been impressed by how far most of the topics have been developed: GRB's, for instance, have received a truly large amount of observational and theoretical effort, until they no longer seem quite so miraculous. The subject we come to now, however — eruptive behavior *à la* Eta Carinae — is relatively unexplored territory, with entertaining surprises and wider implications than one might have expected. Despite a wealth of good data (most of it acquired rather cheaply, compared to extragalactic projects), a number of puzzles are Not Understood at an almost alarmingly basic level. They concern diverse branches of astrophysics; until η Car has been explained better, astrophysicists cannot prudently claim to understand the most massive stars, extreme stellar winds, gas dynamics in slow and fast ejecta, stellar X-ray production, or certain types of spectroscopic excitation processes. And there's a certain romanticism in Eta's role as the only naked-eye star whose basic nature remains unsettled, and the only one except the Sun which, according to a fevered extrapolation of some talks at this meeting, might do us harm! (Technological and economic harm above Earth's atmosphere, that is, not biological below it; a workable premise for a mass-market paperback thriller?) In this review I hope, frankly, to encourage fresh theoretical work.

For our purposes here, "Eta Carinae" is a wonderfully broad *topic,* not just one peculiar star. The problems are diverse, unfamiliar, and unsolved as I just said; a few special extragalactic objects appear to belong to the same category; and modern instruments revolutionized our knowledge in the 1990's even more than enthusiasts expected. A combination of technical factors has made η Car one of the most productive of all targets for HST and some X-ray telescopes, producing several discoveries which are among the most notable developments in recent stellar

CP522, *Cosmic Explosions: Tenth Astrophysical Conference,*
edited by Stephen S. Holt and William W. Zhang
© 2000 American Institute of Physics 1-56396-943-2/00/$17.00

astrophysics. The topic is big enough to fill a whole meeting or a sizable book, and here I can attempt only a sketchy view of the main problems and ideas.

First, though, three references together give a decent introduction to the subject, pre-1999. (I had something to do with all of them, but honestly they are the best reviews available.) For general information on η Car, citing most essential papers before 1997, see our *Annual Reviews* article [1]. Many authors recount the developments of 1996–1998 in [2]. This object *may* be the most extreme known example of a class of very massive stars somewhat inadequately called Luminous Blue Variables, and the best review of LBVs is still a 1994 article [3]. The last sections of that review mention several theoretical points that were later noted by other authors. Many problems of five or more years ago remain unsolved while new ones have appeared; but this is partly because only a corporal's guard of theorists have even discussed them yet.

II   BASIC FACTS AND PARAMETERS

Let's begin with a few essential, long-known, uncontroversial facts. With apologies to many authors, the primary references are too numerous to cite here but are listed in [1]. First, η Car is located about 2300 pc from us, near the big "Carina Nebula" NGC 3372. Several other stars with $M > 50\ M_\odot$ are there too, exemplifying an old puzzle: why do freakishly massive stars tend to occur in groups?

We can estimate Eta's luminosity by measuring thermal IR from circumstellar dust that absorbs most of the UV and visual-wavelength light. $L \approx 5\times10^6\ L_\odot$ is often quoted, but $L > 3\times10^6\ L_\odot$ is a more robust statement. The luminosity implies a zero-age mass of $\sim150\ M_\odot$ or more and an Eddington-limit minimum $\sim100\ M_\odot$. Since we know the star is evolved and has lost mass, often we adopt a compromise guesstimate of, say, 130 $M_\odot$ for the present-day value. The L/M ratio is probably within 30% of the Eddington limit, maybe within 15%. If this object is a binary system, then the values just quoted probably remain more or less valid for the primary component, because the hypothetical secondary star is probably much less massive. (In this review, the name "η Car" usually refers to just the primary component if a companion star exists.) Surface temperature and radius are poorly known because the stellar wind is opaque at most wavelengths, but most likely the underlying star is 20000 K or hotter with $R \sim 150\ R_\odot \approx 0.7$ AU.

We know that η Car is moderately evolved because its ejecta are observed to contain more nitrogen than oxygen and carbon; some stuff has been in the CNO cycle, transported to the surface, and ejected. Helium seems moderately, but not extremely, overabundant. The star's total lifetime should be roughly 3 million years, and since it resembles an LBV with its dense wind, evolved composition, and eruptive behavior, we usually assume that it's near the end of that lifetime. The stellar wind speed is around 500 km s^{-1} with considerable variation, and the current mass-loss rate is something like 10^{-3} to $3\times10^{-3}\ M_\odot$ yr^{-1}. This rate,

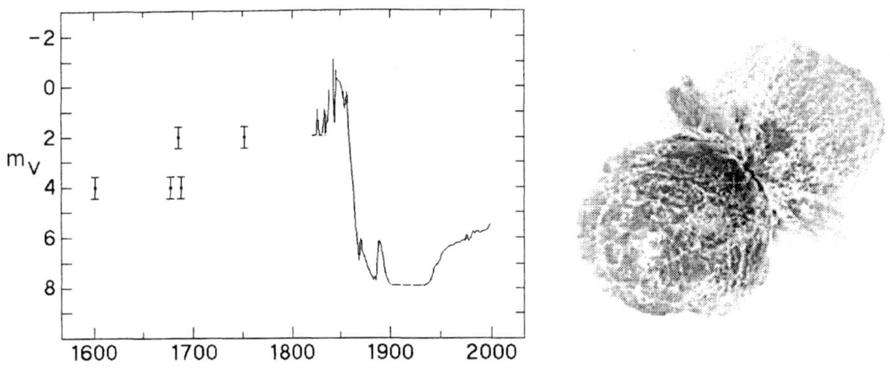

FIGURE 1. Left: Historical light curve of η Car. Right: The Homunculus.

though prodigious compared to any normal stars, seems to represent a relatively "quiescent" state of the object. Imagine what the *active* state must be like!

— Which brings us to the role of η Car in this meeting on Cosmic Explosions: Its famous "Great Eruption" observed 160 years ago was the biggest non-terminal stellar outburst that we know much about. That 1830–1860 event is conspicuous in the historical light curve, Fig. 1. To fully appreciate it, note that astronomers before 1830 sometimes reported the brightness as fourth magnitude but sometimes as second, a big difference. Today this object would appear as a fourth-magnitude blue star if its dusty circumstellar ejecta were not present. The second-magnitude episodes are thought to have been LBV-style eruptions when the wind became denser and more opaque, causing the photosphere to move outward until its characteristic temperature fell below 9000 K. Then most of the light emerged at visual rather than UV wavelengths, so the star appeared brighter to the eye with little or no increase of total luminosity. It was flip-flopping between two stellar-wind modes. Here's the important point for the Great Eruption: The preceding second-magnitude outbursts, with their photospheric temperatures around 7500 K, represented nearly the highest visual brightness allowed by the star's normal luminosity. Between about 1837 and 1858, however, η Car was appreciably brighter, first or zeroth magnitude. Its luminosity must then have increased by a factor of two or more, so *L/M was substantially above the classical Eddington limit for about twenty years.* The phenomenon resembled a slow-motion supernova, with a photosphere as big as the orbit of Saturn and a mass-loss rate of the order of 0.1 $M_\odot$ yr^{-1} at an ejection speed around 500 km s^{-1}. Unlike a supernova, though, it fluctuated rapidly and irregularly by a magnitude or more.

The star survived the eruption, temporarily hidden by dust that formed in the ejecta. After a somewhat puzzling aftershock thirty years later [4], it appears to have been more stable in the 20th century than it was in the 17th and 18th. A gradual brightening can be attributed to decreasing circumstellar extinction as the

423

ejecta expand, and 100 years in the future the star should appear much as it did 400 years ago — unless, of course, a new unexpected event occurs.

The bipolar ejecta-nebula formed by the Great Eruption (Fig. 1) was made famous by the oft-reprinted color HST/WFPC2 images, especially those prepared by Jon Morse. Called the "Homunculus" for reasons that needn't concern us here, at visual wavelengths it shines mainly by reflecting light from the star rather than by intrinsic emission, though some emission is present. Today the polar diameter is 6.5×10^{17} cm ≈ 0.7 ly ≈ 0.2 pc, corresponding to a speed of about 660 km s^{-1} at each pole. An ejected mass of 2 or 3 $M_\odot$ is often quoted, based on IR emission from the dust; but this, like the luminosity, is really a lower limit and no one will be very surprised if additional material with poorly heated dust is present, especially near the equatorial plane [5]. If earlier giant eruptions occurred many hundreds of years ago, as suggested by the new AXAF/Chandra X-ray image and other data, then the dust in their ejecta is now too cool and thin to be obvious. The Homunculus is even more complex than it looks in pictures, and I'll say more about it later.

Concerning the energy budget of the Great Eruption: The kinetic energy of ejecta was comparable to the extra luminous energy radiated, each of the order of 10^{49} ergs. A quick crude estimate shows that 2×10^{49} ergs of thermal energy should be stored in, very roughly, the outer 10 $M_\odot$ of the star. The thermal timescale for those layers is normally of the order of 30 years.

III   THE CENTRAL PROBLEM

Why did the Great Eruption occur? We don't know its cause or mechanism. In fact we don't understand the more moderate LBV-type events either, which occur also in stars like S Doradus. The LBV phenomenon is probably crucial for the exterior evolution of stars with $M > 50M_\odot$ and modern "theoretical" evolutionary tracks need empirical allowances for the resulting mass loss, though most authors don't often say so very clearly. We usually suppose that the instability occurs in the outer layers of the star and moves downward during a major eruption ("as in a geyser"). Two plausible locations have been suggested for the instability, and both may play a role. Again I omit many older references listed in [1,3].

First, the "modified Eddington limit" is a hypothetical instability in or near the photosphere. "Modified" signifies a temperature- and density-dependent opacity; more generally the Eddington limit is less straightforward than we used to imagine, and I'm ignoring some other related effects [6,7]. In the classical Eddington limit for L/M, opacity is due entirely to scattering by free electrons. Absorption opacity, however, becomes significant as a photosphere with a somewhat smaller L/M ratio evolves below 30000 K. Eta Car, LBVs, and the empirical upper boundary in the H-R diagram might all be explained if an instability arises for L/M perhaps 5% to 20% below the classical limit, at temperatures around 20000 K. This conjecture is decidedly non-trivial, for reasons noted in section 5 of [3]. (Ref. [8] invents a new name for the idea but repeats the same points made earlier in [3,11].) Alas, by now the modified Eddington limit has grown old without ever getting a proper analysis. I mentioned it in print as long ago as 1971, then repeatedly after 1979 (e.g. [9–11]), and Appenzeller, Lamers, Fitzpatrick, et al. discussed it in the late 1980's. But *no one has ever seriously attempted to demonstrate the hypothetical instability,* so far as I know, either by quasi-analytic analysis or by gasdynamic simulation. Specific appeals to theorists long ago [11,12] failed to inspire suitable efforts. This remains an opportunity for theoretical exercise; an instability may prove easy to demonstrate, or may not exist. Surely the opacity behavior causes some interesting effects; "bi-stable stellar winds" [13,14] are closely related to the modified Eddington limit.

Other instabilities which arise below the photosphere seem even more promising, but they're bewilderingly complex. Very high opacity around $T\sim2\times10^5$ K, due to iron, increases the local radiation pressure and allows vigorous local convection. The outer layers are therefore dynamically undamped by interior material, and can be unstable in various ways proposed by Stothers, Glatzel, Kiriakidis, Cox, Guzik, et al.; see [15–18] and other refs. cited there. Unfortunately this subtopic is so intricate that a non-specialist can't easily identify the points of agreement and disagreement. My own impression is that one or more of these theories may offer the current Best Bet to explain Eta's Great Eruption and/or ordinary LBV eruptions. Such instabilities may operate along with other effects.

Rotation has long been recognized as a potentially critical factor for η Car [3]. Langer's "Omega limit" [19,20] is really the Eddington limit affected by rotation, and the bi-stable wind effect can obviously lead to latitude-dependent effects in the presence of rotation [21–23]. Rotation has a particularly synergistic relationship with surface instabilities that grow with decreasing temperature, so an eruption may begin near the equator with low ejection speeds [24]. A surprisingly slow rotation rate is adequate.

Guzik et al. [16,17] have noted a reason to suspect that Eta's instability may not arise near the surface as assumed above. Observed timescales range from several years to a century or more (Fig. 1), far too long to be dynamical times. If these are thermal timescales, then a substantial fraction of the star's mass is represented, perhaps suggesting an instability deep within the star. I bet, on the other hand, that the instability is indeed near the surface, that each eruption produces temporary stability, and that thermal and/or angular momentum flow from a considerable fraction of the star eventually returns the surface to an unstable state. Eta Car must have been far from thermal equilibrium after its Great Eruption, and may still be recovering. Incidentally, the timescale for diffusion of angular momentum, required to spin up the outer layers, is most likely comparable to the thermal timescale for much of the star. As for magnetic effects, please let's not acknowledge them yet! — If we're lucky they may be insignificant. Finally, ideas for deep-seated instabilities remain welcome despite what I've just said.

IV   A 5.5-YEAR CYCLE

Long ago astronomers noticed occasional "spectroscopic events" in η Car, when, among other things, the highest-ionization emission lines in nearby ejecta disappeared for a few weeks. These were thought to be irregular shell-ejection events that reduced the UV radiation [25], resembling brief LBV eruptions but not exactly the same. Whitelock et al. then found a pronounced 5- or 6-year cycle in near-IR photometry [26]. In 1996 Damineli, aided by new data, recognized that the spectroscopic events recur faithfully with a period of 5.5 years [27]. His specific and well-founded prediction that such an event would occur near the end of 1997, though disbelieved by at least two telescope allocation committees, was dramatically vindicated when the spectrum changed and the X-ray flux abruptly crashed in late November 1997. Observations of the event produced too much new information to recount here [2]. The 5.5-year cycle promises to be one of our most important clues to the structure of η Car, but we haven't deciphered it yet! This period surprised us because it's far too long for either the star's dynamical timescale or a close interacting binary with a circular orbit, but shorter than thermal and diffusion timescales that had been discussed before 1996.

An obvious possible interpretation, of course, is that η Car has a companion star with a highly eccentric 5.5-year, $a \approx 16$ AU orbit, and that each spectroscopic event occurs near periastron. Colliding stellar winds may then account for the unusual hot thermal X-rays [28–30]. The binary hypothesis, however, is far less straightforward than some of its advocates usually acknowledge, and produced a set of <u>un</u>successful predictions for the 1997–1998 spectroscopic event and aftermath [31]. Our HST/STIS data [35] show that either η Car had an outburst in 1997–1998 as expected in the old single-star discussions [25], or else the ionization structure in its wind changed in a way that mimicked such an outburst. The same observations also show that "orbital velocity" variations measured in ground-based data [32,33] and used to derive orbit solutions [31,34] are probably illusory. Extravagant claims have been made for binary interpretations of the 1997–1998 X-ray and spectroscopic behavior, but in fact the models have been vague, qualitative, and variable as new data have appeared. In summary, and contrary to widespread assertions, the binary idea has *not* been proven. My own opinion is that a binary scenario is fairly likely but needs theoretical development, and its parameters will differ from specific models proposed so far. If a companion object exists, it's most likely an unevolved O-type star much less massive than the primary. A compact object (black hole?) is conceivable, but in that case we'd expect accretion to produce stronger X-rays. A companion star at periastron may trigger Eta's basic instability (though it's not easy to make tidal forces adequate), but probably doesn't account for the instability itself. Most of Eta's other puzzles are primarily opportunities for theoretical work, but for the binary question we equally need more observations.

The single-star possibility seems, from a theoretical point of view, more novel and therefore more fun even though most astronomers feel that it's less likely. If there is no relevant companion star, then the 5.5-year period presumably represents either a thermal timescale for some outer fraction of the star, or an angular-momentum-diffusion timescale, or, heaven forfend, a magnetic timescale. Stothers and Chin have found cycles with similar periods in stellar models adapted to η Car [15]. Complex surface activity and/or fast wind streams would probably be necessary to produce the X-rays. No detailed or quantitative effort has yet been made to develop a model of this type [31].

V   THE RECENT BRIGHTENING

Studies of η Car in recent years have moved in a frenetic, variegated, surprise-after-surprise manner that a Hollywood director would appreciate. Whenever our attention has focussed on a discovery like the 5.5-year period (several preceding ones, mostly involving the ejecta, haven't been described here), we've been hit from the side by a new, utterly unexpected development. This happened again when the star brightened by a factor of two in 1998 [36].

Nothing similar has been seen in any very massive star during the past few decades. This was not a normal LBV-like eruption; the spectrum didn't behave that way, and the apparent brightening was a comprehensive UV-to-near-IR affair. Close to the Eddington limit, the luminosity obviously cannot double without inducing mass loss on a Great Eruption scale. We think, instead, that the circumstellar extinction decreased as dust grains were rapidly destroyed in the innermost ejecta. (Motion of the ejecta can't do the trick.) But what destroyed them? — The most obvious possibility is that the luminosity increased by some fraction of the order of 10 percent. Such a phenomenon would seem to violate traditional lore concerning very massive stars. We truly don't know yet what has happened or whether it will continue.

VI   THE HOMUNCULUS EJECTA-NEBULA

Familiar HST/WFPC2 images printed in books and magazines don't show the full complexity of the Homunculus; one really needs to view these data in computer displays, noting many fine-scale details and changes that have occurred [37]. We expected the bipolar lobes, but their granulated structure and especially the skirt-like equatorial debris surprised most people and are not yet understood [1,2,37]. The gasdynamics of narrow radial structures and of small, slow-moving condensations there merit further study, while other, even narrower radial structures exist far outside the Homunculus.

Emission-line spectra at some locations near the star may be unique among known astronomical objects (Fig. 3 of [24]); I have an unverified hunch that some of these may be related to the UV spectra of AGNs. The strangest well-defined puzzle, though, concerns a set of narrow Fe II emission features near 2507 Å, seen in dense slow blobs a few hundred AU from the star. The brightest two of these lines are amazingly bright, and the intensity ratios defy the known fundamental branching ratios. Johansson has proposed stimulated emission — a natural UV laser! — to explain them [38]. By normal astrophysical standards this idea seems preposterous, because one would expect the relevant photon occupancy numbers to be insufficient by factors of the order of 10^7; but we have been unable to devise any other remotely plausible explanation. This and a host of other problems in and around the Homunculus deserve theoretical studies quite different from those involving the star [2].

Considering the high level of weirdness acknowledged throughout this brief review, some astronomers may feel that "Homunculus" would be a good name for Eta Carinae in general [39].

VII   OTHER CONSIDERATIONS

Many additional, possibly essential facts and puzzles can only be acknowledged in a few words here. For instance, one reason why we don't regard η Car as merely an isolated freak is that giant eruptions have also been seen in a few other objects [3,4]. The most grandiose was "SN 1961v" in NGC 1058, which was even brighter than Eta's Great Eruption but was also briefer so its energy was probably comparable [40,41]. Long ago Zwicky called η Car and SN 1961v "type V supernovae" [42].

In this review I have alluded to, but haven't described, the immense volume of STIS data obtained in only a few HST orbits [35,43]. Initial examinations of these data have already produced several discoveries concerning the stellar wind and the central parts of the Homunculus. Similarly the X-ray observations deserve far more attention [28,29]. Aside from the 1997–1998 event, they appear to show a periodicity of roughly 90 days which may, if real, be almost as significant as the 5.5-year cycle [44].

I've also had to neglect the radio and mm-wavelength story, as well as the 1890 secondary eruption and a current dispute about whether some equatorial material was ejected at that time [2]. One of the first images obtained with AXAF/Chandra shows the soft X-rays which extend out to about 3 times the radius of the Homunculus; these remind one of an old question, did η Car have earlier giant eruptions hundreds or thousands of years ago? There is a shadowy, faint but tantalizing possibility that ancient observations may have survived but we don't know where to look for them [3].

In summary: I used to think of η Car as "the Crab Nebula of the southern hemisphere," i.e., as the one particular, famous object that continues to reward new observational and theoretical efforts more and longer than we had any right to expect. After many years, Eta now appears to surpass even the Crab in that sense, and so far has been inexhaustible — in some respects it's still a fresh topic. Observational astronomers have been doing pretty well with many projects, but *theoretical* work has been scarce. I earnestly recommend the mysteries outlined above to theorists of all persuasions.

VIII   ACKNOWLEDGEMENTS

My own view of this subject has been colored by innumerable discussions with, especially (in alphabetical order) B. Balick, J. Cassinelli, M.F. Corcoran, D. Ebbets, A. Frank, R.D. Gehrz, T.R. Gull, J. Guzik, F. Hamann, D.J. Hillier, R.M. Humphreys, K. Ishibashi, S. Johansson, J.A. Morse, S.N. Shore, N. Smith, N.R. Walborn, and P. Whitelock. Many of the comments here have been based on HST projects, most recently GO-7302, supported by grants from the Space Telescope Science Institute, which is managed by the Association of Universities for Research in Astronomy, Inc., under NASA contract NAS5-26555.

REFERENCES

1. Davidson, K., and Humphreys, R.M., *Ann. Revs. Astr. Astrophys.* **35**, 1 (1997).
2. Morse, J.A., Humphreys, R.M., and Damineli, A. (eds.), *Eta Carinae at the Millennium*, ASP Conf. Ser. 179 (1999).
3. Humphreys, R.M., and Davidson, K., *Publ. Astr. Soc. Pacific* **106**, 1025 (1994).
4. Humphreys, R.M., Davidson, K., and Smith, N., *Publ. Astr. Soc. Pac.* **111**, 1124 (1999).
5. Smith, N., Gehrz, R.D., and Krautter, J., *Astron. J.* **116**, 1332 (1998).
6. Shaviv, N.J., *Astrophys. J.* **494**, L193 (1998).
7. Shaviv, N.J., in *Variable and Non-spherical Stellar Winds in Luminous Hot Stars*, IAU Colloq. 169, ed. B. Wolf et al., Springer, Heidelberg, p. 155 (1999).
8. Lamers, H.J.G.L.M., in *Luminous Blue Variables: Massive Stars in Transition*, ASP Conf. Ser. 120, ed. A. Nota and H. Lamers, p. 76 (1997).
9. Humphreys, R.M., and Davidson, K., *Science* **223**, 243 (1984).
10. Davidson, K., in *Instabilities in Luminous Early Type Stars*, ed. by H. Lamers and C. de Loore, Reidel, Dordrecht, p. 127 (1987).
11. Davidson, K., in *Physics of Luminous Blue Variables*, IAU Colloq. 113, ed. by K. Davidson et al., Kluwer, Dordrecht, p. 204 (1989).
12. Appenzeller, I., in *Physics of Luminous Blue Variables*, IAU Colloq. 113, ed. K. Davidson et al., Kluwer, Dordrecht, p. 195 (1989).
13. Pauldrach, A.W.A., and Puls, J., *Astron. Astrophys.* **237**, 409 (1990).
14. Lamers, H.J.G.L.M., Snow, T.P., and Lindholm, D.M., *Astrophys. J.* **255**, 269 (1995).
15. Stothers, R.B., and Chin, C.-W., *Astrophys. J.* **489**, 319 (1997).
16. Guzik, J.A., Cox, A.N., and Despain, K.M., in *Eta Carinae at the Millenium*, ASP Conf. Ser. 179, ed. J.A. Morse et al., p. 347 (1999).
17. Guzik, J.A., Cox, A.N., Despain, K.M., and Soukup, M.S., in *Variable and Non-spherical Stellar Winds in Luminous Hot Stars*, IAU Colloq. 169, ed. B. Wolf et al., Springer, Heidelberg, p. 336 (1999).

18. Glatzel, W., in *Variable and Non-spherical Stellar Winds in Luminous Hot Stars,* IAU Colloq. 169, ed. B. Wolf et al., Springer, Heidelberg, p. 345 (1999).

19. Langer, N., *Astron. Astrophys.* **329**, 551 (1998).

20. Langer, N., in *Variable and Non-spherical Stellar Winds in Luminous Hot Stars,* IAU Colloq. 169, ed. B. Wolf et al., Springer, Heidelberg, p. 359 (1999).

21. Lamers, H.J.G.L.M., and Pauldrach, A.W.A., *Astron. Astrophys.* **244**, L5 (1991).

22. Lamers, H.J.G.L.M., Vink, J.S., de Koter, A., and Cassinelli, J.P., in *Variable and Non-spherical Stellar Winds in Luminous Hot Stars,* IAU Colloq. 169, ed. B. Wolf et al., Springer, Heidelberg, p. 159 (1999).

23. Bjorkman, J.E., in *Variable and Non-spherical Stellar Winds in Luminous Hot Stars,* IAU Colloq. 169, ed. B. Wolf et al., Springer, Heidelberg, p. 121 (1999).

24. Zethson, T., Johansson, S., Davidson, K., Humphreys, R.M., Ishibashi, K., and Ebbets, D., *Astron. Astrophys.* **344**, 211 (1999).

25. Zanella, R., Wolf, B., and Stahl, O., *Astron. Astrophys.* **137**, 79 (1984).

26. Whitelock, P.A., Feast, M.W., Koen, C., Roberts, G., and Carter, B.S., *Monthly Not. Roy. Astron. Soc.* **270**, 364 (1994).

27. Damineli, A., *Astrophys. J. Letters* **460**, L49 (1996).

28. Corcoran, M.F., Petre, R., Swank, J.H., et al., *Astrophys. J.* **494**, 381 (1998).

29. Ishibashi, K., Corcoran, M.F., Davidson, K., et al., *Astrophys. J.* **524**, 983 (1999).

30. Pittard, J.M., Stevens, I.R., Corcoran, M.F., and Ishibashi, K., *Monthly Not. Roy. Astron. Soc.* **299**, L5 (1998).

31. Davidson, K., in *Eta Carinae at the Millenium,* ASP Conf. Ser. 179, ed. J.A. Morse et al., p. 304 (1999).

32. Damineli, A., Conti, P., and Lopes, D.F., *New Astron.* **2**, 107 (1997).

33. Damineli, A., Conti, P., and Lopes, D.F., in *Eta Carinae at the Millenium,* ASP Conf. Ser. 179, ed. J.A. Morse et al., p. 288 (1999).

34. Davidson, K., *New Astron.* **2**, 387 (1997).

35. Davidson, K., Ishibashi, K., Gull, T.R., and Humphreys, R.M., in *Eta Carinae at the Millenium,* ASP Conf. Ser. 179, ed. J.A. Morse et al., p. 227 (1999).

36. Davidson, K., Gull, T.R., Humphreys, R.M., et al., *Astron. J.* **118**, 1777 (1999).

37. Morse, J.A., Davidson, K., Bally, J., et al., *Astron. J.* **116**, 2443 (1998).

38. Johansson, S., Leckrone, D.S., and Davidson, K., in *The Scientific Impact of the GHRS,* ASP Conf. Ser. 143, ed. J.C. Brandt et al., p. 155 (1998).

39. Tirpitz, A. von, *My Memoirs,* Dodd Mead, New York, **1**, p. 204 (1919).

40. Goodrich, R.W., Stringfellow, G.S., Penrod, G.D., and Filippenko, A.V., *Astrophys. J.* **342**, 908 (1989).

41. Filippenko, A.V., Barth, A.J., Bower, G.C., et al., *Astron. J.* **110**, 2261 (1995).

42. Zwicky, F., in *Stellar Structure,* ed. L. Aller and D.B. McLaughlin, University of Chicago Press, Chicago, p. 140 (1965).

43. Gull, T.R., Ishibashi, K., and Davidson, K., in *Eta Carinae at the Millenium,* ASP Conf. Ser. 179, ed. J.A. Morse et al., p. 144 (1999).

44. Corcoran, M.F., Ishibashi, K., Swank, J.H., et al., *Nature* **390**, 587 (1997).

RXTE X-ray Monitoring of the Supermassive Star Eta Carinae: Colliding Wind Emission in a Pre-Hypernova Candidate Binary?

M. F. Corcoran*[,$], A. C. Fredericks[†], R. Petre[$], J. H. Swank[$], S. A. Drake*[,$], K. Davidson[#], K. Ishibashi[#], S. White[†], and A. Damineli[†]

*Universities Space Research Association, 7501 Forbes Blvd, Ste 206, Seabrook, MD 20706
[$]Laboratory for High Energy Astrophysics, Goddard Space Flight Center, Greenbelt MD 20771
[†]Dept. of Astronomy, University of Maryland, College Park, MD 20742
[#]Dept. of Astronomy, University of Minnesota, 116 Church St., SE, Minneapolis, MN 55455
[†]Instituto Astronomico e Geofisico da USP, CP 9638, 01065-970- Sao Paulo-Brazil

Abstract. *ROSAT, ASCA, RXTE* and now *Chandra* X-ray observations of the supermassive star η Carinae obtained over the past 7 years chronicle the inordinate variability of the high-energy emission. The most important characteristics are these: 1) the hard X-ray "low state" evidently recurs with a period consistent with the "Damineli Cycle" of 5.5 years; 2) the X-ray emission exhibits "flares" which occur quasi-periodically; 3) the overall X-ray brightness in the current "cycle" is brighter than at the same phase in the last cycle; 4) the hard, variable "core" source is apparently resolvable at lower energies but unresolvable at energies above $\sim$ 4 keV. The X-ray data appear consistent with a model of X-ray generation via colliding winds in a massive binary system coupled with dust scattering, though the "flaring" activity probably requires that the wind from at least one of the stars is azimuthally structured. It is difficult to account for all the X-ray properties by activity in a single star.

INTRODUCTION

Massive stars are key astronomical objects due to their important role in cosmic chemical enrichment and galactic evolution. They mark the end of their stellar lives as supernovae whose peak luminosity can equal the entire radiant output of a galaxy of a trillion stars. The extreme members of this class might produce "hypernovae" [1], cataclysms hundreds of times more energetic still, and a postulated source of gamma ray bursts. Such extraordinary explosions require stellar precursors of unusually large mass, and so should be rare. The Milky Way contains at least one possible member of this putative class of hypernova progenitors, the massive, luminous, and relatively nearby star, η Carinae.

CP522, *Cosmic Explosions: Tenth Astrophysical Conference,*
edited by Stephen S. Holt and William W. Zhang
© 2000 American Institute of Physics 1-56396-943-2/00/$17.00

Recent observations of periodic variations in some emission lines of the η Carinae spectrum suggest [2] that the star has a companion in a long-period ($P = 5.52$ years) eccentric ($e > 0.6$) orbit. Such a companion potentially provides a crucial key to understanding η Carinae. At the very least, the presence of a companion offers the most direct measure of the current mass of η Carinae (and so would provide an important empirical point for the mass-luminosity relation in the upper HR diagram). In addition, the companion can possibly interact with η Carinae and so change the course of the evolution of the star. In principle the companion's presence could help resolve some outstanding mysteries (such as the nature of the large eruptions which took place during the 1840's and 1890's), especially if tidally-induced mass transfer or other interaction effects are important.

Unfortunately, the spectrum of the star and its close neighborhood is contaminated by emission lines which are formed in thick clouds of circumstellar ejecta, so the interpretation of the spectral variations as a signature of binarity is by no means clear. Periodic mass ejections from a single star have also been suggested as a possible explanation of the observed emission-line changes [3]. Clearly, to understand the present and future state of the star, we need to determine whether η Carinae has a companion.

PERIODIC X-RAY EMISSION

A direct consequence of the presence of a companion is that the wind from η Carinae should collide with the wind or surface of this companion. This collision would produce substantial amounts of hot, shocked gas [4,5] and observable X-ray emission. These "colliding wind" X-rays would be characteristically hard ($kT \sim$ a few keV), absorbed ($N_H >$ few $\times 10^{22}$), and, most importantly, the colliding wind X-rays would vary with the orbital cycle. Thus the X-ray emission from η Carinae provides important evidence to prove or disprove the binary hypothesis. Since the optical spectrum of the star is so contaminated with circumstellar emission, X-ray variability could be the *best* way to determine the orbital properties if the star indeed has a companion.

To characterize the variability of the X-ray spectrum from η Carinae in detail, we have used the Proportional Counter Array (PCA) on the Rossi X-ray Timing Explorer (*RXTE*) and the *ASCA* X-ray observatory to monitor the 2-10 keV X-ray emission from the source.

RESULTS OF THE X-RAY MONITORING

A preliminary analysis of the *RXTE* data through mid-1997 [6] already noted an overall increase in the mean *RXTE* X-ray flux which accelerated after January 1997, with unexpected periodic "flares" occurring every 84.8 days. Our more recent data show that the X-ray variability through January 1998 proved to be more extreme than previously suspected.

The most fundamental result of this monitoring campaign is that the X-ray "low state" first observed in 1992.5 [7] did indeed recur near 1998.0, i.e. after the "Damineli cycle" of 5.5 years. Figure 1 shows the X-ray image of η Car in the 1992–1999 interval. The point-like core at the center of the image is invisible during the "low states" in 1992.5 and 1998.0, but clearly seen (by both *ROSAT* and *Chandra*) outside the "low state".

The observed X-ray flux in the 2–10 keV energy band reached its maximum ($F_{x,obs} \approx 2 \times 10^{-10}$ ergs s^{-1} cm^{-2}) on 9 November 1997, and plummeted thereafter. In only a month's time the observed X-ray flux dropped by 2 orders of magnitude ($F_{x,obs} \approx 2 \times 10^{-12}$ ergs s^{-1} cm^{-2}) and the source became nearly undetectable to the PCA. The star remained faint for more than 2 months before the X-ray emission began to recover substantially.

As discussed in [8], the X-ray variability mimics the expected variations which would be seen if the X-rays are produced by wind-wind collisions in a binary system, thus supporting the idea that η Car is indeed a binary system. However there is a significant discrepancy in the observed low-state spectrum, in which the observed

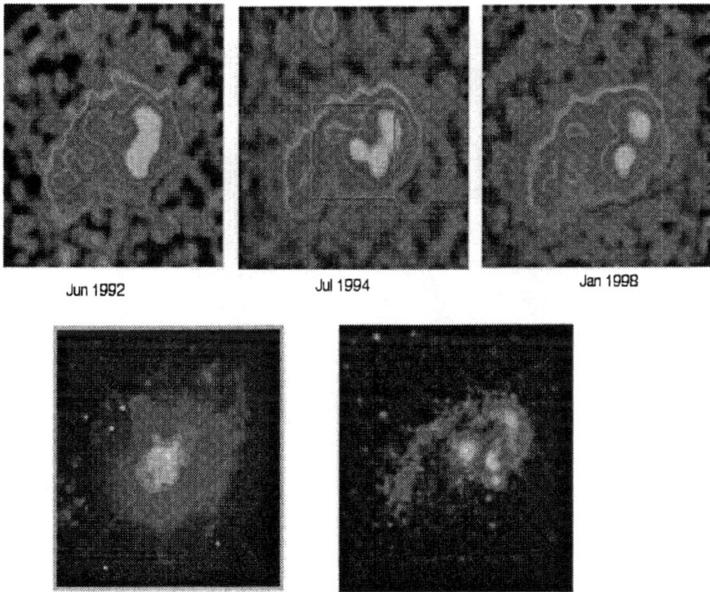

FIGURE 1. High-Resolution images showing the variability of the central source at the center of the η Car image at 4 epochs: top left, *ROSAT* HRI Jun 1992 (during the last "low state"); top middle, *ROSAT* HRI Jul 1994 (the "normal" state); top right, *ROSAT* HRI Jan 1998 (the most recent "low state"). The image on the lower right is the *Chandra*/ACIS observation from Sep 1999 (courtesy *Chandra* X-ray Observatory Center). The HST WFPC2 image is lower left.

N_H is much lower ($\approx 5 \times 10^{22}$ cm^{-2}) than model predictions ($\geq 10^{24}$ cm^{-2} at the time of minimum flux). The recent *Chandra* observation may solve this discrepancy, since a preliminary analysis of the ACIS data suggests that significant amounts of flux near 2 keV are produced in an extended region around the core (and possibly in the "shell" as well). If so then this non-varying emission would contaminate the *RXTE* and *ASCA* spectra during the low-flux state and cause the spectral analysis to severely underestimate the actual N_H to the core.

REFERENCES

1. Paczynski, B. 1998, *Ap. J.*, **494**, L45.
2. Damineli, A., Conti, P. S., Lopes, D. F., 1997, *New Astr.*, **2**, 107.
3. Davidson, K., 1997, *New Astr.*, **2**, 387.
4. Usov, V. V., 1992, *Ap. J.*, **389**, 635.
5. Stevens, I. R., Blondin, J. M., and Pollock, A. M. T., 1992, *Ap. J.*, **386**, 265
6. Corcoran, M. F., et al., 1997, *Nature*, **390**, 587.
7. Corcoran, M. F., et al., 1995, *Ap. J.*, **445**, L121.
8. Ishibashi, K., et al., 1999, *Ap. J.*, **524**, 983

Variability in η Car's Circumstellar Nebula

Nathan Smith[1], Jon A. Morse[2], Kris Davidson[1],
Roberta M. Humphreys[1], and Robert D. Gehrz[1]

1. *Astronomy Dept., University of Minnesota, 116 Church St. SE, Minneapolis, MN 55455*
2. *CASA, University of Colorado, Campus Box 389, Boulder, CO 80309*

Abstract. We discuss recent HST imaging of η Carinae obtained in near-UV and optical light with WFPC2 and in the near-infrared with NICMOS. Comparison with images obtained at earlier epochs reveals non-uniform photometric variations in the light from the many gas and dust condensations, as well as morphological changes in the inner regions of the nebula. During this same time period, the central object has brightened considerably at all wavelengths, and we discuss the relationship between the brightening of the central star and the variations in the surrounding nebula.

INTRODUCTION

η Carinae is probably the most massive and luminous evolved star in our Galaxy. During its famous Great Eruption in the 19th century [1,2] it ejected more than 2 $M_\odot$ of material [3], which later formed dust that obscured the star. These expanding dusty ejecta form a bipolar reflection nebula around the star, absorbing most of the star's optical/UV flux and re-emitting this energy in the infrared (IR).

In the context of this meeting on cosmic explosions, η Car is particularly interesting because it will (soon?) explode as a supernova, or perhaps something even more exotic. Thus, the nebula surrounding η Car gives us an example of what the circumstellar environments of supernova or hypernova progenitors might look like. Because of its proximity ($\sim$2.3 kpc), we are able to study η Car's circumstellar nebula in great detail using HST, and we are provided with a glimpse of the complex structure that may surround similar but much more distant objects when they explode. For distant objects, the circumstellar nebulosity is not spatially resolved from the central object, so the effect that this circumstellar material may have on the light curve from an unresolved object is uncertain.

η Car is currently undergoing an outburst; in the last few years the optical flux from the central star has increased by a factor of $\sim$2 [4]. Below we discuss photometric and morphological variability in η Car's circumstellar reflection nebula that has been observed in optical and IR light with HST during this same time

period, and the relationship between the variability of the star and the nebula
that it illuminates. This discussion may be relevant to certain other objects that
increase their brightness considerably, and are surrounded by stellar ejecta.

OBSERVATIONS AND ANALYSIS

HST observations of η Car have been obtained on several occasions over the past
5 years in a variety of wavelengths and with a variety of instruments; details of the
observations and data reduction can be found elsewhere [5,6].

Figure 1(a) shows the HST F631N filter image obtained in June 1999, and Figure
1(b) shows the flux ratio between this image and one obtained with the same filter in
1995. The 1995 image has been expanded in scale to correct for proper motion due
to the expansion of the nebula. The flux ratio image indicates how the brightness
has varied as a function of position in the nebula. While the bright cells that are
prominent in optical images have only increased by a factor of $\sim$1.3 in the ratio
image in Fig. 1(b), the dark lanes between these bright cells have increased by a
factor of $\sim$2. Some of the anomolous brightening and fading in the core region and
equatorial ejecta may be due to changes in intrinsic emission lines, even though
the circumstellar nebula is seen primarily in scattered light at optical wavelengths.
The changes in emission lines ([S III]λ6312 in the F631N filter; [Fe II] and [N II] in
the F336W and F658N filters, respectively) are most prominent in the core region,
within $1''$ to $2''$ of the central star, and seem to be emitted from a limb-brightened

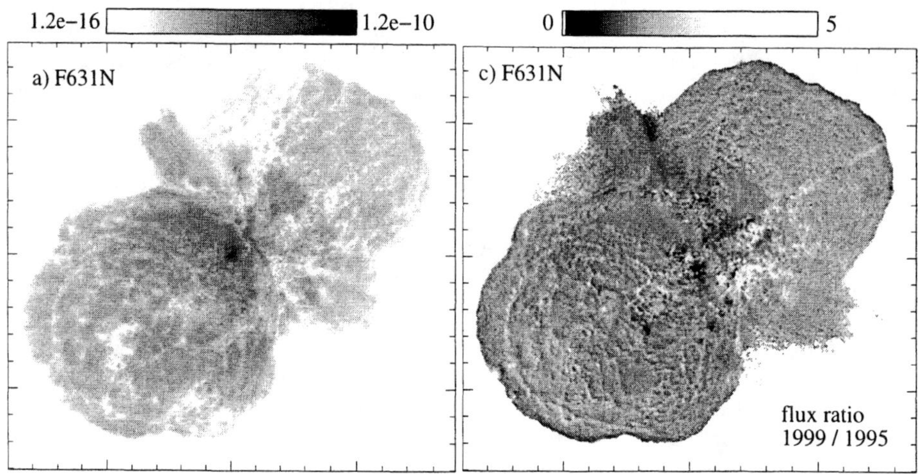

FIGURE 1. (a) WFPC2 red continuum (F631N filter) image of η Car, and (b) 1999/1995 flux
ratio image using the same filter. Negative greyscale in (a) has a logarithmic stretch, and the max
and min values in erg s^{-1} cm^{-2} Å^{-1} arcsec^{-2} are given next to the color bar above the figure.
The greyscale values for the flux ratio in (b) are also shown at the top of the figure.

disk or torus in the equatorial plane. Changes in the intensity of this line emission may be partially responsible for some of η Car's peculiar spectroscopic changes noted in large aperture ground-based observations.

Figure 2 shows near-IR continuum images of the central few arcseconds around η Car in (a) a 1995 ground-based image obtained at ESO [3], and (b) a 1998 NICMOS image [6]. It is clear that the central object has changed its morphology between the two epochs, appearing more compact (but brighter) in 1998. These changes resemble those observed in the radio continuum [7]. If the changes follow η Car's 5.5 year spectroscopic cycle, the implausibly high velocities implied by the expansion of these features indicate that the observed changes probably result from illumination effects [6]. Various lines of evidence suggest that the extended structure around the central object in Fig. 2(b) consists of a circumstellar equatorial torus with a radius of a few thousand AU [6]. This extended structure shows strong intrinsic line emission at optical and IR wavelengths which varies conspicuously with η Car's 5 year period [6]. However, this line emission would contribute less than 10% to the flux in a normal broad-band near-IR filter [6].

Figure 3(a) shows 1998 (NICMOS) and 1999 (WFPC2) imaging photometry for a range of aperture sizes. The encircled flux is similar for all continuum filters through apertures $>1''$ (the 2.15 μm data are multiplied by 2); the central $1''$ contributes roughly 1/3 to 1/2 of the total flux from η Car at all wavelengths. When resolved, the central star contributes significantly less; a few to 10% of the total flux. The central star contributes less to the total flux in the near-UV than in red continuum light, probably due to reddening from circumstellar dust.

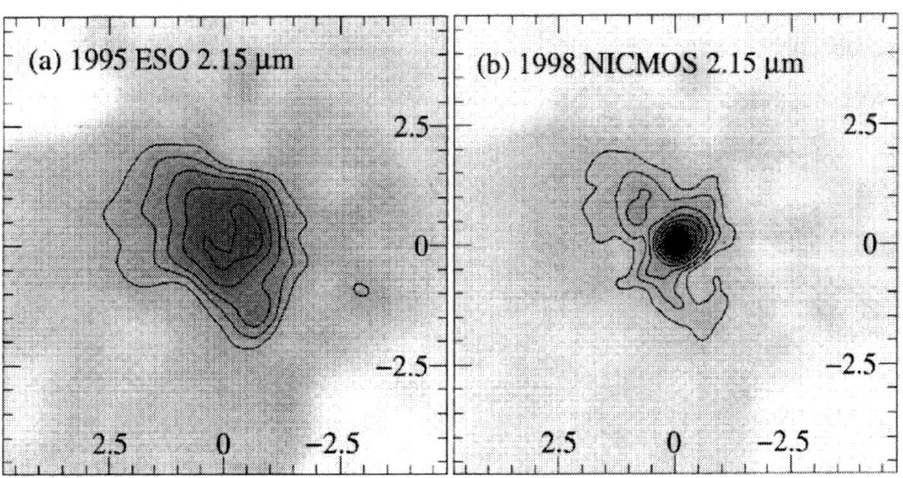

FIGURE 2. (a) 1995 ESO and (b) 1998 NICMOS images of η Car in the 2.15 μm continuum. Axes are RA and DEC offsets in arcsec. Contours are drawn at 1, 2, 4, 10, 20, 40, and 100×10^{-15} erg s^{-1} cm^{-2} Å^{-1}.

The high spatial resolution of the WFPC2 images allows us to resolve the central star from the surrounding nebulosity, and to compare the photometric variability of these two components in our multi-epoch data. Figure 3(b) compares the flux from the central star (0.14″ aperture) with the flux from the entire nebula measured in a 20″ aperture. While the central star has brightened by a factor of ∼2 in the last few years at near-UV/optical wavelengths, consistent with previous measurements [4], Fig. 3(b) indicates that *the variability of the nebula measured in a large aperture is much less dramatic.* This phenomenon may have important consequences for interpreting the light curves of extragalactic variable stars and supernovae, if they are surrounded by circumstellar material. We should expect this effect to be particularly important for variable sources that are thought to be similar to η Car, such as SN1961v and some other objects that Filippenko [8] identifies as Type IIn supernovae.

REFERENCES

1. Davidson, K., this volume.
2. Humphreys, R.M., Davidson, K., & Smith, N., *PASP*, 111, 1124 (1999).
3. Smith, N., Gehrz, R.D., & Krauter, J., *AJ*, 116, 1332 (1998).
4. Davidson, K., et al., *AJ*, 118, 1777 (1999).
5. Morse, J.A., et al., *AJ*, 116, 2443 (1998).
6. Smith, N., & Gehrz, R.D., *ApJ Letters*, in press (2000).
7. Duncan, R.A., White, S.M., & Lim, J., *MNRAS*, 290, 680 (1997).
8. Filippenko, A., this volume.

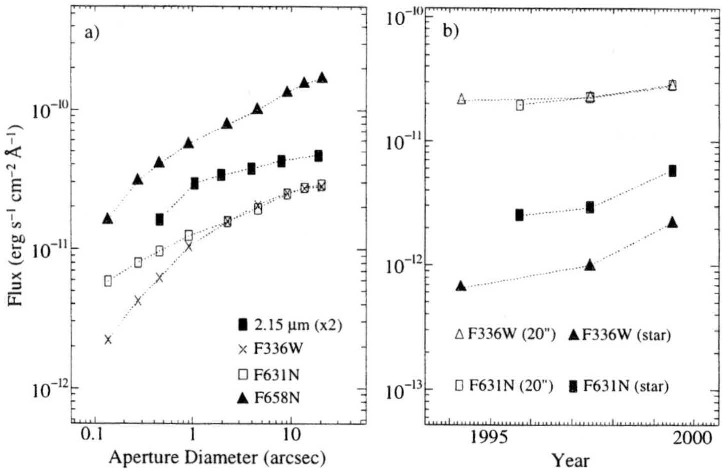

FIGURE 3. (a) aperture photometry for the recent HST data, and (b) multi-epoch continuum photometry of both the central star and the entire nebula (20″aperture).

Eta Carinae: The Iron Curtain Lifts!

T.R. Gull[*], K.Davidson[†], and K. Ishibashi[†]

[*] *Goddard Space Flight Center*
[†] *University of Minnesota*

Abstract. Observations of Eta Carinae and the surrounding ejecta nebulae were done with the Space Telescope Imaging Spectrograph (STIS) on the Hubble Space Telescope (HST) during the past two years using a $52''$x $0.1''$aperture and gratings with $R \approx 4000 - 8000$. In March 1998 and February 1999 complete coverage of the spectrum from $1640\mathring{A}$ to $10400\mathring{A}$ was accomplished at the same position angle. Not only did Eta Carinae brighten in that interval by 60 percent, but also the nebulosities within the central two arcseconds changed considerably. Most notable is the overall decrease in [Fe II] emissions and Fe II absorptions with increases in doubly-ionized ionic emissions including [Fe III]. At this time, we do not know if the changes are due to a 5.5 year cyclic variation or an overall brightening of Eta Carinae. Continued monitoring with STIS is planned for the near future and proposals are submitted for the full 5.5 year spectroscopic cycle.

INTRODUCTION

For an overview of Eta Carinae, we refer the reader to the paper in this volume by K. Davidson. Eta Carinae is a Luminous Blue Variable (LBV) that is near enough that we can resolve the star from the nebular ejecta surrounding it. Based upon its luminosity, Eta Carinae is a very massive star, possibly being 100 solar masses. There is disagreement as to whether there is a massive companion. Eta Carinae is sufficiently unstable that, over the past several centuries, there has been at least one major ejection (in 1840-1860) and other minor ejections. A 5.52 year spectroscopic cycle was noted as evidence of a very massive binary system [1]. Early observations with Chandra detected a hard x-ray central source with a faint extended structure close to the star [2].

The surrounding nebulosity, called the Homunculus, is a double-lobed structure with a large disk located between the lobes. Spectroscopy of the Homunculus demonstrates that outer structure is primarily a reflection nebulosity. The mass of the Homunculus is estimated to be 3 solar masses [3]. Mass loss rates are estimated to be about 10^{-3} solar masses per year [4]. Buried within the Homunculus are several substructures [7] including the Integral Nebula, the Bubble Nebula and nebular clumps with very unusual spectra.

CP522, *Cosmic Explosions: Tenth Astrophysical Conference,*
edited by Stephen S. Holt and William W. Zhang

Eta Carinae may be an extreme LBV based upon its mass, luminosity and current state of activity. An outer shell of emission gas extending nearly 20 arcseconds surrounds the Homunculus [5]. Most curious of all are the high velocity strings detected in the outskirts of the Homunculus [6]. These filaments show a steady increase in velocity leading from the center of the Homunculus to the ends with gradients of 30 $kms^{-1}/''$, or approaching 3000 kms^{-1} at the ends. These strings are consistent with ejecta occurring during the 1843 event.

Obviously the star and nebular ejecta are of great interest both for the evolution of massive stars leading to their death throes and for the possibility that supernovae of very massive stars may be the gamma ray bursters. The more that we understand about the very massive LBV's can lead to improved insight on whether or not GRB's are related to SN's.

OBSERVATIONS

A Guest Observer program (GO 7302, Davidson, PI) and a STIS Guaranteed Observer program (GTO 8036, Gull, PI) were combined to observe Eta Carinae and its nebulosities from December 1997 to February 1999. Partial spectral coverage occurred in December 1997 and November 1998. Full coverage from $1640\mathring{A}$ to $10400\mathring{A}$ was accomplished in March 1998 and February 1999. Based upon unfiltered CCD STIS acquisition images, the star brightened twofold and some nebulosities tripled between December 1997 and February 1999. More recent examination of the WF/PC2 images [8] indicates that other nebulosities did not brighten significantly. All observations were used the $52''x0.1''$ aperture. The angular resolution along the slit is limited primarily by the CCD pixel size, being $0.0509''$/pixel with two pixel sampling from $1640\mathring{A}- \approx 7000\mathring{A}$. Beyond $47000\mathring{A}4$, angular resolution is diffraction limited, scaling with wavelength to $10400\mathring{A}$. The position angle was 260.3 degrees north through east for both observations. Exposure times were selected throughout the spectrum to obtain reasonable signal to noise ratios, but not be limited by saturation effects. Bracketed exposures were necessary for a few bright lines on the star, most notably for $H\alpha$ and $H\beta$. Data reduction and analysis were done using IDL routines developed by the STIS Instrument Development Team (IDT).

RESULTS

The Homunculus, being primarily a reflection nebulosity, is seen by scattered starlight. $H\alpha$ is the brightest feature of the stellar spectrum and is easily detected from the tip of the Southeastern lobe to the tip of the Northwestern lobe. The scattered stellar emission line shifts in velocity to the red with increasing distances due the recessional velocity of the dust, which is in Hubble expansion. Measured proper motion and velocities are consistent with the ejecta being thrown out in the 1840's. A very strong nebular absorption, which varies slowly in velocity with position, is superimposed upon the stellar $H\alpha$ emission line reflected by the Northwest lobe. The absorber is neutral hydrogen located in a central disk and excited to the n=2 energy level (The 2S state is metastable with a lifetime of order $10^7 sec$). With electron densities estimated to exceed $10^9 cm^{-2}$, collisional excitation and Lyman α absorption are populating the metastable level. The absorption is detected in Balmer lines up to the seventh member (H(n=9 to n=2)). This is likely the first detection of Balmer absorption in a nebula. (Simple estimates suggest a total column density N(n=2) that exceeds 10^{13} cm^{-2} [9].)

Deep within the Homunculus and close to the central star, we find a very bright emission nebulosity seen in hundreds of emission lines throughout the blue and near ultraviolet spectral region. The velocity structure is fairly constant at -40 kms^{-1} from 1.7″ southeast of the star to 1.7″ northwest. However at the extreme ends of the emission structure, the velocity shifts $300 kms^{-1}$ to the red in the Northwest and $300 kms^{-1}$ to the blue in the Southwest. Because this shape is remiscent of the integral sign [$\oint$], complete with a circular stucture at the center (see next paragraph), we call this the Integral Nebula. Most of the emission lines are [Fe II] and are brightest at this slit orientation on the Weigelt blobs B and D. Between March 1998 and February 1999 the emission structure brightened in the outer regions and the extreme ends indicate some proper motion. Many [Fe II] lines in B and D dropped in intensity, and even disappeared, especially in the near ultraviolet. Some [Fe III] lines appear in the February 1999 spectra, especially in Weigelt blobs B and D.

Closer to the star, an even smaller nebular structure is detected. In the March 1998 spectrum, the spatial extent of many [Fe II] lines shows a shell structure that extends about 0.4″ in radius and with a $400 kms^{-1}$ expansion velocity. At the STIS spatial scale and medium spectral dispersion, the shape is bubble-like in appearance. Therefore we call it the Bubble Nebula. The Balmer lines show a filled sphere bounded by this shell structure. By February 1999, the [Fe II] lines no longer frame this shell, but the Balmer lines continue to fill the interior.

A series of very strong absorption line complexes are superimposed upon the star and the emission structures. Three velocity components are detected most notably in the Na D, Ca H, Ca K, Mg I and Mg II lines: a constant velocity across the dust-scattered stellar continuum corresponds to the the foreground ISM; a velocity at $\approx -300 kms^{-1}$ curves gently in velocity with the contour of the Southeast lobe; and a velocity component at -40 kms^{-1} is relatively constant and corresponds to the disk structure. Between March 1998 and February 1999 many absorption lines at -40 kms^{-1} decreased in strength. Those identified thus far are due to Fe II and indicate that the iron in the central region has converted from singly-ionized iron to doubly-ionized iron. Likewise many [Fe II] emission lines in the integral nebula and especially the Weigelt blobs B and D decreased in intensity or disappeared altogether while emission lines of [Fe III] and other doubly-ionized species appeared.

That the change in near ultraviolet absorption is due to Fe II abundance changes is demonstrated by dividing the March 1998 stellar spectrum by the February 1999 stellar spectrum. A very rich absorption spectrum of Fe II appears that is very strong in the near ultraviolet. Only a few spectral intervals are still totally saturated in the February 1999 stellar spectrum while very significant spectral intervals were totally absorbed especially around $2100\overset{\circ}{A} - 2300\overset{\circ}{A}$.

DISCUSSION

We attribute the changes in nebular absorption and nebular emission lines to increased stellar ultraviolet fluxes which leads to doubly-ionized species. What is the source of the shell structure that we call the Bubble? We note that a very simple calculation of size of the shell with the expansion velocity indicates that the shell was ejected around 1992.5 when the previous spectroscopic minimum occurred. There may be additional evidence of other, outer and much more diffuse shells which may correspond to earlier spectroscopic minima. If so, then there is evidence that during each spectroscopic minimum, Eta Carinae experiences a major lifting of material into the stellar wind. What causes this periodic ejection is not known. We must look to future observations to give us further clues.

We are not yet certain that the iron curtain will continue to lift. We suspect that the iron curtain goes up and down modulated by the 5.52 year periodicity. However the ejecta continuously flows outward with the brightness of Eta Carinae continuously increasing. The recent doubling of the stellar flux is anomalously large compared to the gradual brightening trend followed for the last fifty years. Likely, the iron curtain is caused by the periodic mass ejections every 5.52 years which causes the innermost curtain, or veil, to drop. As that ejecta gets pushed out by the stellar wind, the column density of Fe II drops and more Fe III is in line of sight. The inner veil then effectively rises part way but then drops again. Is this a constant modulation? We are unsure but suspect that the degree of iron absorption of that inner veil is dependent upon the total ejecta at each spectroscopic minimum. The southeast Homunculus lobe contributes dust absorption which is decreasing with the lobe expansion. This dust column density is likely the source of the long term unveiling of Eta Carinae, who dances with an additional inner veil.

REFERENCES

1. Damineli, A., *ApJ* **460**, L49(1996).
2. Seward, F. Private Communication, 1999.
3. Davidson, K., *IAU Colloq* **113**, Physics of Luminous Blue Variables, ed. K. Davidson,A.F.J.Moffat, & H.J.G.L.M.Lamers (Dordrecht: Kluwer),101 (1989).
4. Davidson, K., *PASP Conference Series* **179**, Eta Carinae at the Millennium, ed. J. Morse, R. Humphreys, & A. Damineli (ASP),(hereafter PASP CS **179**), 6(1999).
5. Morse, J., *PASP CS* **179**, 13 (1999).
6. Weis, K. & Duschl, W., *PASP CS* **179**, 155 (1999).
7. Gull, T., Ishibashi, K., Davidson, K. et al., *PASP CS* **179**, 155 (1999).
8. Smith, N. et al. this volume.
9. Hamann, F., etal. *PASP CS* **179**, 116 (1999).

Rapporteur Summary

Cosmic Explosions:
Rapporteur Summary of the 10th
Maryland Astrophysics Conference

J. Craig Wheeler

Department of Astronomy
The University of Texas
Austin, Texas 78751
wheel@astro.as.utexas.edu

Abstract.

This meeting covered the range of cosmic explosions from solar flares to γ-ray bursts. A common theme is the role of rotation and magnetic fields. New information from the Sun shows that a "magnetic carpet" contains most of the surface field that feeds the corona. Disk instabilities in protostellar disks may provide most of the growth of a protostar. Type Ia supernovae continue to give evidence for an accelerating Universe, and a rigorous examination is underway to characterize systematic effects that might alter the results. The binary evolution origin of Type Ia that explode as carbon/oxygen white dwarfs at nearly the Chandrasekhar limit remains a thorny problem. The discovery of the central point of X-ray emission in Cas A by CXO should give new insight into the core collapse problem in general and the nature of the still undetected compact remnant in SN 1987A in particular. Jets were described from protostars to microquasars to blazars to γ-ray bursts. Polarization studies of core-collapse supernovae lead to the conclusion that core collapse is not merely asymmetric, but strongly bi-polar. To account for normal core-collapse supernovae, the explosion must be jet-like in routine circumstances, that is, in the formation of neutron stars, not only for black holes. Given the observed asymmetries, estimates of explosion energies based on spherically-symmetric models must be regarded with caution. The strong possibility that at least some γ-ray bursts arise from massive stars means that it is no longer possible to decouple models of the γ-ray burst and afterglow from considerations of the "machine." Although it began the supernova/γ-ray burst connection, it is difficult to fit SN 1998bw/GRB 980425 into any statistical picture that incorporates the high redshift events. The implied correlation of γ-ray bursts with star formation and massive stars does not distinguish a black hole collapsar model from models based on the birth of a magnetar. Current jet models do not discriminate the origin of the jet. Calorimetry of at least one afterglow suggests that γ-ray bursts cannot involve highly inefficient internal shock models. Essentially all γ-ray burst models involve the "Blandford Anxiety," the origin of nearly equipartition magnetic fields in the associated relativistic shocks. It is important to discriminate the origin of the γ-ray burst energy as thermal energy, kinetic energy or perhaps Poynting flux.

CP522, *Cosmic Explosions: Tenth Astrophysical Conference,*
edited by Stephen S. Holt and William W. Zhang
© 2000 American Institute of Physics 1-56396-943-2/00/$17.00

I INTRODUCTION

The Tenth October Maryland conference celebrated the rapid release of energy from objects ranging from protostars to active galactic nuclei. We were treated to an inimitable history as only Virginia Trimble can do it from her personal knowledge of the players and places and her voracious reading of the literature. Roger Blandford gave an overview with a focus on unsolved problems.

In this summary, I will present a brief summary of the invited review talks and selected poster presentations. I regret, especially, that I cannot do justice to all the latter. To a certain extent, I will present the great span of exploding objects we covered as reflected in the mirror of supernovae. There are three reasons for this: (i) this is the subject I know best; (ii) various aspects of supernova research were presented at length; and (iii) supernovae, I believe, are related to everything—and everything to supernovae. The degrees of separation are, in any case, rarely so great as six. A more specific theme I will underline is that to an ever-growing extent we are forced to consider rotation, magnetic fields, and asymmetries. In Section II, I will present some work from my collaborators in Texas and elsewhere that relate to this general theme. Section III gives some perspective and unresolved issues.

II SUMMARY OF PRESENTATIONS

A Stellar and Solar Flares

Blandford emphasized that there has been great progress recently in the study of the Sun that has led to a new and different understanding of the magnetic field of the Sun and its role in coronal heating. In particular, solar flares come from regions of relatively weak magnetic field where plasma can intrude. The magnetic field is higher in a "magnetic carpet" across the solar surfaces and it is from this layer of stronger field that "nanoflares" deliver energy to power the solar corona.

Linsky reviewed our understanding of flares from a variety of stellar sources. These flares are of great interest, but if other stars operate in analogy to the Sun, then flares may be only the tip of otherwise unseen magnetic activity in other stars. Ramaty emphasized the manner in which flares can be used to study the acceleration of particles. Kenyon outlined the great progress that has been made in terms of understanding the FU Ori phenomenon as a disk instability in a protostellar disk, pointing out that all the growth of a protostar may arise during the phases of high luminosity and high mass accretion rate. Clearly, study of these phenomena, especially in the detail afforded by the Sun, can give us valuable insight into the processes of reconnection and coronal formation that are, in turn, crucial in order to understand accretion disks, winds, jets, and particle acceleration.

B Type Ia Supernovae

The use of Type Ia supernovae as "calibrated candles" by means of empirical brightness-decline relations has been startlingly successful. The tentative conclusion that there is a low matter density and a finite, positive cosmological constant has sent reverberations throughout astronomy and physics (Riess et al. 1998; Perlmutter, et al. 1999). Type Ia are more complex than can be described by the first versions of one-parameter brightness decline relations: $\Delta M(15)$ (Hamuy, et al. 1996), Multicolor Light Curve Shape (Riess, Press, & Krishner 1996), or stretch (Perlmutter et al. 1999). Theorists can, and have, invented many reasons why Type Ia might vary with look back times. On the other hand, the first order corrections have been remarkably successful in reducing the scatter in the data and providing constraints on cosmological parameters. In this conference, Kirshner and Aldering presented data showing that no significant evolutionary effects have yet appeared in the data. In particular, Aldering argued that refined analysis of error bars yields no statistically significant evidence that the early rise times are different in nearby events and those at redshift z~0.5 (e.g. Reiss et al. 2000).

Nomoto outlined models for Type Ia evolution and possible systematic effects. Of particular interest is the recent calculation of Höflich et al. (1999) showing that, at lower metallicity, a star of a given mass will produce a smaller C/O ratio. All else being the same (ignition density, density of transition to detonation), a lower C/O ratio will lead to a somewhat brighter event with a faster decline. The one parameter brightness-decline relations would then interpret such an event as somewhat dim, just the effect interpreted as evidence for a cosmological constant. More modelling of this sort and comparison with data is necessary to elucidate these sorts of physical systematic effects.

There were also discussions of possible sources of observational systematic bias. Suntzeff gave an excellent review of the observations and a cautionary note about comparing photometric results from different observatories that use slightly different filters sets and thus get slightly different results for standard stars in standard filters. There is a tendency for the deep searches to not follow up events that are at larger distance from potential host galaxies because of the ambiguity of association of the supernovae candidate with the host. Filippenko pointed out that there is some complementary bias against candidates that are close to the centers of host galaxies because those are sometimes passed over when classification spectra are obtained due to concerns about galactic contamination. These selection biases in terms of the radial position on the galaxy may in turn be important at some level because the distribution of intrinsic luminosity and decline rate is known to vary with galactocentric radius (Wang, Höflich & Wheeler 1997; Riess et al. 1999; Howell, Wang & Wheeler 1999). A bias against events at low galactocentric radius might give a bias toward events that are more homogeneous in their properties, but less able to discriminate subtle differences in progenitor dependence. The success of the light curve brightness/decline relations to remove potential evolutionary effects by including sample events that span the full range of progenitor ages and metallicity in nearby events in spirals and ellipticals might thus be subtly affected.

My personal answer to the question of whether the Universe is acclerating is "probably yes." My answer to the query of do we know for sure is "not yet."

The physics of Type Ia supernovae was discussed, without which there will remain some doubt concerning the purely empirical treatment of Type Ia as cosmological probes. Pinto gave a summary of the basic understanding of why there is a brightness/decline relation. He noted that the observed relations can be produced in white dwarfs of constant mass, near the Chandrasekhar mass in the most successful models. In this context, if the nickel mass is increased the peak luminosity naturally goes up. Less obvious is that the temperature goes up resulting in a increase in the line opacity. This traps the energy from γ-ray decay, resulting in a slower decline, the process amply illustrated in previously published models (Höflich Khokhlov & Wheeler 1995; Höflich & Khokhlov 1996; Höflich et al. 1996).

The physical question that emerges from such an analysis is why the nickel mass should vary when the white dwarf progenitor mass is essentially fixed. This is presumably a product of initial conditions and the thermonuclear combustion of the white dwarf. The physics of the combustion was described by Hillebrandt and by Niemeyer. There has been great progress in recent years in applying concepts from terrestrial combustion physics to the Type Ia problem. Observations and theory suggest that the combustion begins with a relatively slow, subsonic, turbulent deflagration. To account for the distribution of intermediate mass elements in velocity in the outer layers of Type Ia explosions, there must be a transition to a much faster burning. The issue of whether there is a natural transition from subsonic deflagration to supersonic detonation has been discussed (Khokhlov, Oran & Wheeler 1997a,b,c; Niemeyer & Woosely 1997; Khokhlov, Oran, Wheeler & Chtchelkanova 1999; Montgomery, Khokhlov, & Oran 1998). Hillebrandt and Niemeyer raised the question of the difficulty of making a direct transition from deflagration to detonation and discussed the possibility that a speed up to a very rapid, but still subsonic deflagration was possible and adequate to account for the observations. It is not clear that a very rapid deflagration would not be unstable to evolution to a detonation. One thing that is clear is that the turbulent deflagration must be studied in three dimensions to get the sign of the turbulent cascade (from large scales to small scales) correct and to understand that process in the context of spherical dilution and related effects (Khokhlov 1995).

Another major issue that arises in the context of the standard, Chandrasekhar mass model for Type Ia supernovae is the question of their prior evolution. How do white dwarfs grow to the Chandrasekhar mass sufficiently often to account for the observed rates of Type Ia? Nomoto described the models that currently seem to come closest to solving this long-standing issue. These models invoke a wind from the white dwarf so that a relatively rapid mass transfer rate from the companion does not glut the white dwarf with mass to yield a surrounding hydrogen-rich envelope, in violation of the observations. Rather, the excess mass can be blown off in the wind. Models show that the net accretion onto the white dwarf can be rapid enough to avoid degenerate hydrogen or helium ignition. Such ignition is inimical to the Type Ia process since a nova explosion will reduce the mass of the white dwarf, as discussed by Hernanz. Degenerate helium ignition produces a supernova explosion of the wrong properties, as outlined by Nomoto. In the wind models, the amount of hydrogen on the surface of the white dwarf when it explodes is sufficiently small to escape detection. For a given wind model, there are solutions that will give the desired properties of Type Ia progenitors with unstable mass loss from main sequence companions and stable mass loss on the thermal time scale from red giant companions with masses in the range 1 - 3 $M_\odot$. At lower metallicities, the wind could be less efficient and it may not be possible to produce Type Ia. This would give an epoch of turn-on of Type Ia with look back time.

One of the issues raised by these binary wind models is the reservoir problem. The amount of mass that is lost from the secondary is related to the mass that accretes onto the white dwarf, promoting its growth. This can be expressed by:

$$\Delta M_2 = \Delta M_{wd}\left(1 + \frac{\dot{M}_{wind}}{\dot{M}_{wd}}\right).$$

If the secondary only has a mass of 1 - 3 $M_\odot$, the wind model may work if the rate of loss to the wind is comparable to the rate of accretion onto the white dwarf. The companion loses only two grams for every gram that lands on the white dwarf. On the other hand, wind mass loss rates are not known very well in most circumstances, certainly not in this rather exotic one, and factors of a few may be important. If, for instance, the rate of loss to the wind is several times the growth rate of the white dwarf, then if the white dwarf must accrete several tenths of a solar mass to reach the Chandrasekhar limit, the small mass companion might not be able to provide enough. This, of course, depends on the initial mass of the white dwarf. If the supernova progenitor must grow from the mass of a field white dwarf, 0.6 $M_\odot$, the task is nearly insurmountable. If the initial mass of the white dwarf is substantially higher, the task is easier. An important issue then becomes the initial mass distribution of the white dwarfs, a function that is not well known at higher white dwarf masses and which is undoubtedly affected by the very condition of being in a binary system. Hernanz and Starrfield pointed out that there are white dwarfs in binary systems with rather large masses, for instance that in the recurrent nova system U Sco at about 1.3 $M_\odot$, so nature can do this, at least occasionally.

It is still a high priority to obtain any information that will give us hints of the nature of the binary system underlying Type Ia supernovae. If they are hydrogen accretors, as the binary wind models suppose, the explosion should take place next to a hydrogen-rich companion and within the wind. The search for the stripped companion is important, but handicapped because the hydrogen tends to lag the ejecta and is expected to show up, if at all, in the nebular phase when the spectrum is complex and weak Hα might be difficult to detect. Suntzeff pointed out that the use of sensitive eschelle detectors on the new generation of large aperature telecopes might give a new way to search for the hydrogen swept up in the wind.

C Collapse

The process of core collapse is of great current interest both for its intrinsic importance as a supernova triggering mechanism and for its potential connection to γ-ray bursts. Fryer gave an update on work to understand collapse, especially the difficult problem of neutrino transport and the critical issue of fall back which may determine whether a neutron star or black hole is left behind in a successful supernova explosion. Fallback considerations suggest that black hole formation could be common in stars with mass as low as 20 $M_\odot$, making SN 1987A right on the ragged edge of going either way, neutron star or black hole. Fryer predicted that after a decade of concentration on multidimensional hydrodynamics and specifically protoneutron star convection, the next decade would be one devoted to the effects of rotation and neutrino transport. I think the former is unambiguously true and would add magnetic fields to the mix. As for the latter, it is clear the neutrinos will continue to play a large role since they must carry off the bulk of the binding energy of the neutron star. It is not so clear that they will emerge as the final arbitrating physics of the success or failure of the explosion itself as the role of jets becomes more clear, as discussed in §D.

A great new step toward understanding the outcome of core collapse came with the launch of the new Chandra Observatory. The first obtained and released image of Cas A represented an incredible debut. Although there were hints of a central object in ROSAT data, the Chandra image showed with unambiguous clarity the dim point of X-ray emission in the center of the remnant. The issue of whether this is a cooling neutron star or an accreting neutron star or black hole is now under debate. One thing is clear. This image gives us a new slant on similar issues in SN 1987A where the expected neutron star has still not revealed itself. If the object left behind in SN 1987A is similar to that in Cas A, then it is no wonder we have not seen it. The object in Cas A is very faint, less than a few $L_\odot$, perhaps as little as 10^{32} erg s^{-1} (Tanenbaum, et al. 1999). One possibility to detect the central object in SN 1987A is to register the bolometric emission from the absorption and re-emission of any source within the ejecta. Heroic efforts to measure the bolometric luminosity still place it at more than 10^3 $L_\odot$, more than 1000 times brighter than the object in Cas A. It may be that by searching diligently in the continuum between emission lines tighter limits can be obtained in SN 1987A, but detecting an object as faint as that in Cas A will be a challenge. Figuring out the nature of the object in Cas A will immediately give us new perspectives on SN 1987A. One of the issues will be to understand why this object is 10^4 to 10^5 times dimmer than the 1000 year old pulsar in the Crab nebula. If neutron stars with rapid rotation and strong magnetic fields are necessary to make jets in supernovae, Cas A and SN 1987A do not seem to qualify. There is, however, obvious evidence for some jet-like activity in Cas A and SN 1987A has its rings and asymmetric ejecta, so this story has a long way to run.

To take a step back, Davidson regaled us with his recriminations that we have worked so little on, and understood so little of, η Carinae. He is exactly right. We have made a lot of progress modeling massive stars as spherically symmetric, but somewhere we may have taken a drastically wrong turn since we are very far from predicting the properties of η Carinae from first principles. The lack of progess in understanding η Carinae is not due to lack of interst or curiousity, but simply a lack of knowledge of where to start in this complex beast. Surely we must try. The mass of the star is likely to exceed 100 $M_\odot$, and so it must surely collapse or explode. It might be a γ-ray burst waiting to happen. The lobes and skirt are compelling in their beauty and simplicity and strongly reminiscent of the rings of SN 1987A. The details of ropes, strings, and jets are bewildering, as is the great outburst of the last century and the recent brightening. Once again the Chandra image of Cas A brings a new vision of this key object.

There is also much to entertain after the explosion. McCray outlined the three-ring circus that is beginning to ensue with the first lighting of the first blob in the inner ring of SN 1987A as the fastest ejecta collide with the most prominent protrusions. McCray revealed that the "knot" that has lit up may not have the enhanced N abundance reflecting CNO processing that is ascribed to the rings, but rather may be more representative of the ISM of the LMC. This would be an intriguing result. In any case, the next decade will be a three-ring circus replete with fireworks as the ejecta continue to interact with the ring. There is much to be learned about interstellar medium shock physics as well as the nature of the progenitor star and the ejecta.

D   Jets

The topic of jets has long been of central interest to astrophysics. There is a new concentration on relativistic jets because of their possible role in γ-ray bursts.

Livio gave an overview, pointing out that jets are ubiquitous, from protostars to γ-ray bursts. He suggested that they have a common mechanism powered by accretion, with the ejection velocity being comparable to the escape velocity from the surface of the central star or the inner edge of a surrounding accretion disk and the collimation by a surrounding magnetic field. This may even apply to jets from the nuclei of planetary nebulae, although it is not clear what the source of accreted matter is in that context. Possible jets from pulsars may represent an exception to this general mechanism. Again the images of the Crab pulsar from CXO show a jet-like protrusion that may help our understanding of these issues.

Another lesson is that the jets, especially those associated with black holes, are frequently relativistic. This was emphasized in the reviews of jets from blazars by Urry and of those from the binary black holes in "microquasars" by Greiner. Studies of these relativistic jets on both galactic and stellar scales with time-dependent, multiwavelength campaigns has great potential to teach us about how the jets form, are collimated, and propagate. The stellar cases are especially important because the activity plays out over a shorter timescale and is especially amenable to practical, detailed study.

Some of the greatest interest in jets is the possibility that they play a role in the origin of γ-ray bursts. Jets are discussed as a way of moderating the great energy requirements of the brightest bursts and there is some circumstantial evidence for collimation in the change of slope of some afterglow light curves (Rhoads, 1999; Sari, Piran, & Halpern 1999; Stanek, et al. 1999; Harrison et al. 1999). On the theoretical side, the models of Woosely, MacFadyen, and their collaborators have drawn great interest. These models are computed in the context of a "collapsar" model where a black hole forms by gravitational collapse in the center of a massive star. Rotation could lead to the formation of a disk surrounding the black hole. The disk, in turn, could generate strong neutrino fluxes that might provide an energy source by neutrino/antineutrino annihilation, or it could be the source of an MHD flow. Yet another alternative is that the spin of the black hole threaded by magnetic fields could generate energy by the Blandford-Znajek effect (Blandford & Znajek 1977). In the models discussed by MacFadyen & Woosley (1999), energy from neutrino annihilation is presumed to be deposited as thermal energy in a small region along the rotation axis. In their two-dimensional simulation, the expansion from this point is channeled by the density gradient of the surrounding Keplarian disk and is forced to proceed up the rotation axis. As the density at the jet base declines, thermal energy input at a given rate tends to provide an ever larger specific energy to the matter. The result is to promote the formation of a relativistic jet that is confined and collimated by the structure of the star. As emphasized by Woosley, the jet will trigger lateral shocks that can cause the outer mantle to explode. Woosley differentiated possible differences between situations involving production of the black hole by prompt collapse or by fallback.

The production of something like a supernova attendant to the propagation of the jet through the stellar core seems unavoidable. The jets made in this way can, in principle, be relativistic, and can, again in principle, yield γ-ray bursts. The open issues are whether collapse leads to jets, the nature of those jets, and the question of whether the jets will lead to γ-ray bursts of observable properties.

E   Gamma-Ray Bursts

Fishman summarized the history of the γ-ray burst game and especially the invaluable role of BATSE on CGRO. BATSE had discovered 2612 bursts at the time of the meeting and added one more the evening the meeting ended. The smallest time resolved in a γ-ray burst is 200 microseconds. If interpreted in terms of an intrinsic light crossing time, that corresponds to a distance of less than 60 km. Fishman argued that despite some reports to the contrary, the short bursts are not homogeneous. They display, for instance, $V/V_{max} = 0.39$, significantly less than the homogeneous value of 0.5. There are also claims for anisotropy, but Fishman did not think those were well substantiated by the data.

Kulkarni summarized the recent history in the BeppoSAX era. Of special interest to γ-ray burst research and to this conference in particular are the recent reports by Bloom et al. (1999), Reichert (1999) and Galama et al. (1999) for supernova-like modulation of γ-ray burst afterglow light curves about three weeks after the γ-ray burst for two cases of classic γ-ray bursts. In the thinking of many people, this increases the already high probability that GRB 980425 was associated with SN 1998bw (Galama, et al. 1998). GRB 980425 does not, however, fit in the context of various statistical studies of, e.g. Schmidt (1999) on luminosity functions, of Ruiz & Fenimore (1999) on correlations of variability with luminosity, and Norris (1999) on energy-dependent phase lags. If SN 1998bw and GRB 980425 were the same event, the γ-ray emission process was very different than the more distant events.

Kulkarni reviewed the valuable radio data that has been obtained on afterglows. He pointed out that the failure to detect radio afterglows in some cases may simply be because the existing equipment, e.g. the VLA, is not sufficiently sensitive. He also reported on the calorimetry of one event, GRB 970508, for which late time radio observations could be modeled to obtain an estimate of the total energy radiated in the afterglow (Frail, Waxman & Kulkarni 1999). The result was $E_{afterglow}/E_\gamma \ll 1$. This is significant because the most popular models of internal shocks imply low efficiency (Kumar 1999) and hence a great deal more total initial kinetic energy in relativistic baryons (in synchrotron models) than in emitted γ-rays. The kinetic energy left from the initial γ-ray burst must all be dissipated in the subsequent interaction with the ISM and hence radiated in the afterglow. Taken at face value, this result does not support the inefficient internal shock model. The calorimetry depends on assuming synchrotron emission and hence nearly equipartition magnetic fields. The issue of how, and hence whether, equipartion fields arise in relativistic blast waves is very unclear, as emphasized at this conference by Blandford. More calorimetric information of this kind is clearly needed. Another object that gives some information of this sort is GRB 990123, the famous bright prompt optical burst (Akerloff et al. 1999; Kehoe et al. 1999; Kulkarni et al. 1999). The total fluence in the optical of that burst was substantially less than the isotropic equivalent fluence in γ-rays. On the other hand, a backward extrapolation of the optical afterglow intersects at a point above the prompt flash, so the relative balance of γ-ray burst to afterglow energy is uncertain. Kulkarni also raised the issue of whether the afterglow must be non-adiabatic in contrast with popular models. This could complicate the analysis and alter the energetics and hence the basic constraints on the processes of the γ-ray bursts and afterglows. There are models that avoid the problem of inefficient internal shocks by invoking a collision of the leading shock with the external medium (Fenimire & Ruiz 1999) or by "spotty" internal shocks (Kumar & Piran 1999).

Kulkarni gave a brief summary of what he termed as the indirect indications for a correlation of γ-ray bursts with massive stars, including the location in host galaxies and apparent correlation with star forming regions. He concluded that the data, indirect though it is, points to a collapsar model. This conclusion may be a bit too specific. The evidence points to a correlation with massive, short lived stars, but it contains no direct information on the specifics of the mass of the stars and certainly not whether the event involves the formation of a black hole. To be specific, the data are equally consistent with a massive star that makes a magnetar - a rapidly rotating, highly magnetized neutron star. Theory suggests that magnetars must be born rapidly rotating and must dump a great deal of rotational energy to slow to the long periods observed 10,000 years later in the soft γ-ray repeaters.

Gehrels outlined the exciting future for observations of γ-ray bursts. HETE II is scheduled for launch January 23, 2000 from Kwajalein Island. HETE II should bring the era of afterglow studies from short as well as long bursts and should produce a much higher rate of well-localized γ-ray bursts than BeppoSAX. This will make life for the observers doing ground-based follow-up even more hectic. SWIFT was selected as a MIDEX instrument just after the meeting. It is scheduled for launch in perhaps 2003. A great deal of work and discovery on γ-ray bursts will occur between now and then, but there should be much left for SWIFT to do. In particular, the prospect of extending the detection to events at redshift of 10 to 20 is extremely exciting for cosmology as well as γ-ray burst research.

Mészáros summarized the theory of γ-ray bursts and their afterglows, so much of which he pioneered before the discovery of afterglows. He remarked that this progress was possible in part because of the fortunate circumstance that the physics of the internal shock region and that of the external shock/afterglow region can be decoupled from the physics of the "machine," the process/object that actually produces the energy that is transformed into relativistic shocks and γ-rays. The field has developed to the point where, increasingly, this may no longer be true. For instance, depending on the choice of Lorentz factor and energy of the burst, the radius of the photosphere of the fireball could be less than the radius of the bare helium core that is invoked in jet models (Khokhlov et al. 1999; MacFadyen & Woosely 1999). In addition, there are issues, again arising in the context of massive star models, of winds. These high density winds will change the length scale of interactions to produce prompt optical output in reverse shocks and subsequent afterglows by external shocks. The era is upon us when we must begin to consider the machine self-consistently with the γ-ray burst and afterglow.

F   Type I X-ray Bursts

Bildsten and Swank summarized the progress made on understanding the Type I X-ray bursts, especially with the invaluable contribution of RXTE which has allowed new insight based both on new data and new understanding of old data. The Type I X-ray bursts are complementary to many of the other objects discussed here in other contexts. Unlike the magnetars in the soft γ-ray repeaters, for instance, which are highly magnetized and rather slowly rotating, the neutron stars associated with Type I X-ray bursts apparently have fast rotation and low magnetic fields. RXTE data of pulses from one source is interpreted as evidence for rotation at 300 Hz. Bildsten described the systematics of nuclear burning on the surfaces of neutron stars and interpreted the data as evidence for a hot spot ignited by a localized thermonuclear flash that is whipped around by the rotation. The lesson that fits in with the general theme of this summary is that the Type I X-ray bursts require asymmetry, rotation, and magnetic fields. Bildsten also outlined the possibility that nuclear burning in this ambiance could break out of the hot CNO and helium burning and run all the way up to heavy elements, including, for instance, species like krypton.[1]

III POLARIZATION AND JETS IN NORMAL CORE-COLLAPSE SUPERNOVAE

To complement the theme of asymmetry, rotation and magnetic fields, I would like to summarize some work that we have done at Texas over the last five years. Like many people in the supernova community, we at the University of Texas got actively involved in the supernova/soft-gamma-ray repeater/magnetar/γ-ray burst topic with the advent of SN 1998bw and its possible connection to GRB 980425. We brought a different perspective to this issue because of work we have done on supernova spectropolarimetry.

We have been making spectropolarimetric observations of all accessible supernovae at McDonald Observatory (Wang et al. 1996; Wang, Wheeler & Höflich 1997; Wheeler, Höflich & Wang 1999; Wang et al. 1999). A summary of observations is given in Table 1. The result has been that most Type Ia have low polarization and hence are substantially spherically symmetric. Many have only upper limits of order 0.1 - 0.2%. A few have detected, but low polarization, of order 0.2%. We have obtained the first polarization of a "subluminous" Type Ia, SN 1999by which appears to show polarization at the 0.2% level. The polarization observed is consistent with theoretical models of delayed detonation models (Wang, Wheeler & Höflich 1997) and may be a useful probe of the combustion physics. We have detected one exception, SN 1997bp, which was observed a week before maximum light to have a polarization of about 1%. The polarization was low in post-maximum spectra, but this event remains a challenge to understand. It is important to establish whether such events are common, the physical reason for the large polarization, and whether or not there could be an asymmetric luminosity distribution that could affect estimates of cosmological parameters.

More importantly in the current context are our observations of presumed core-collapse events, Type II and Type Ib/c. We have found that all such events are polarized at about the 1% level and some much more so. So far there have been no exceptions in about a dozen events (a recent Type II, SN 199em, showed no detectable polarization in very early observations (Leonard, Filippenko & Chornock 1999), but further observations are planned that will peer deeper into the ejecta). There could be a myriad reasons for polarization, but our data suggest a very important trend: the smaller the hydrogen envelope, the larger the observed polarization. As examples of this trend, SN 1987A with a $10M_\odot$ envelope had a

---

[1] This caused me to pose the following question: if Type I bursts make krypton(ite), did Jor-el and his son come from a planet orbiting a neutron star?

polarization of about 0.5% (Méndez et al. 1988); SN 1993J with a small hydrogen envelope, $\sim 0.1 M_\odot$, was polarized at the 1-2% level (Trammell, Hines & Wheeler 1993; Tran et al. 1997); a very similar object, SN 1996cb, may show polarization as high as 4%; Type Ic SN 1997X which showed no substantial hydrogen nor helium was polarized at perhaps greater than 3% (Wheeler, Höflich & Wang 1999); SN 1998S which shows characteristics of a Wolf-Rayet star (Leonard et al. 1999; Gerardy et al. 1999) showed polarization of about 3% before maximum (Leonard et al. 1999) and perhaps as much as 4% after maximum (Wang, et al. 1999).

Table 1. Supernovae with polarimetric measurements

SN	Type	$P(\%)$	Intrinsic	SN	Type	$P(\%)$	Intrinsic
SN1968L[1]	II	0.2	No	SN1970G[2]	II	0.5	Yes
SN1972E[3]	Ia	0.35	No	SN1975N[4]	Ia	1.5	No
SN1981B[5]	Ia	0.41	No	SN1983G[6,7]	Ia	2.0	No
SN1983N[6,7]	Ib		Yes?	SN1987A[8,9]	II	0.5	Yes
SN1992A[10]	Ia	0.3	No	SN1993J[11,12]	IIb	1.5	Yes
SN1994D[12]	Ia	0.3	No	SN1994Y[12]	II	1.5	Yes
SN1994ae[12]	Ia	0.3	No	SN1995D[12]	Ia	0.2	No
SN1995H[12]	II	1.0	Yes	SN1995V[12]	II	1.5	Yes
SN1996W[12]	II	0.7	Yes	SN1996X[12]	Ia	0.2	Yes?
SN1996cb[12]	IIb	3.0	Yes	SN1997X[12]	Ic	2-7?	Yes
SN1997Y[12]	Ia	<0.3	No	SN1997bp[12]	Ia	1.0	Yes
SN1997bq[12]	Ia	<0.2	No	SN1997br[12]	Ia	<0.2	No
SN1997ef[12]	Ic?	<0.3	No	SN1997ei[12]	Ic	2.5	Yes
SN1998S[12,15]	IIn	3.0	Yes	SN1998bw[13,14]	Ic?	0.4?	Yes?
SN1999by[12]	Ia	0.2	Yes				

1 Wood & Andrews (1974), 2 Shakhovskoi & Efimov (1973), 3 Wolstencroft & Kemp (1972), 4 Shakhovskoi (1976), 5 Shapiro & Sutherland (1982), 6 McCall et al. (1984), 7 McCall (1985), 8 Cropper et al. (1988), 9 Méndez et al. (1988), 10 Spyromilio and Bailey (1993), 11 Trammell, Hines & Wheeler (1993), 12 This program, 13 Key et al. (1998), 14 Patat et al. (1998), 15 Leonard et al. (1999)

These are difficult observations requiring special care in the reduction to remove the effects of the ISM (the latter greatly aided by wavelength and temporal coverage). Following Suntzeff's cautionary notes on the difficulty of doing accurate photometry, Leonard referred to spectropolarimetry as "photometry from hell." In addition, there is a pressing need to expand the statistical sample, especially with time-sampled data. Nevertheless, this trend suggests that the core-collapse process itself is strongly asymmetric and that evidence for that asymmetry is damped by the addition of outer envelope material.

The level of polarization we have observed for core collapse events, $\sim 1\%$, requires a substantial asymmetry with axis ratios of order 2 to 1 (Höflich 1995).

Asymmetric explosions tend to turn spherical as they expand, so to leave a significant imprint in the homologously expanding matter requires a substantially larger asymmetric input of energy or momentum in the explosion process itself (Höflich, Wheeler & Wang 1999). In other words, the asymmetries we are observing require the underlying explosion to be driven by a jet. This conclusion is completely independent of any connection to γ-ray bursts, but, of course, the potential for this connection is clear (Wang & Wheeler 1998; Wheeler 1999). These factors led us to the hypothesis that the core collapse process is intrinsically strongly asymmetric, much more so than current collapse calculations involving convectively unstable neutron stars. It was in this context that we greeted the news of SN 1998bw and have continued to work on polarization, jet models of collapse, and their possible relation to other astrophysical phenomena.

A   SN 1998S

SN 1998S was discovered on March 2, 1998. It showed strong narrow emission lines and is thus characterized as a Type IIn. Some of the narrow emission lines were of high excitation, reminiscent of Wolf-Rayet stars and SN 1983K (Niemela et al. 1985) and subsequently showed emission of carbon monoxide and possible evidence for dust formation that are consistent with an origin in the core of a massive star that is not decelerated by a substantial hydrogen envelope (Gerardy et al 1999). Leonard et al. (1999) were very fortunate to be at Keck II with a spectropolarimeter and got an excellent set of data on March 7, still about 0.5 magnitude and two weeks before maximum. They have interpreted their data in terms of an interaction with a disk of circumstellar hydrogen which also shows up as a double (in fact triple) peaked Hα profile at late times. The polarization might have been as high as 3%.

The light curve declined rather rapidly after maximum unlike some Type IIn, so it has been called Type IIn(pec). About 60 days after the explosion (20 days after maximum) it went into a steeper decline and then leveled off to a slower decline about 80 days after maximum. This slower decline is, in V, still a little steeper than expected for ^{56}Co decay.[2] We obtained data on SN 1998S at McDonald Observatory on March 31, about 10 days after maximum and again on May 1, 60 days after the explosion, 40 days after maximum and just before the light curve started to decline rapidly. The McDonald data is consistent with that of Leonard et al. following a locus nearly parallel (but perhaps somewhat shifted) in the Q,U plane. The polarization on May 1 could be as high as 4%.

Polarization of this level forces us to abandon more timid phrases like "asymmetric supernovae." Recall that the maximum polarization from an infinitely thin, internally illuminated, electron scattering disk is about 12% (Chandrasekhar 1950). On this scale a polarization of 4% is very large. For this event, and perhaps for

---

[2] http://oir.www.harvard.edu/cfa/oir/Research/supernova/spectra

others with this level of polarization, it is appropriate to talk about "bi-polar supernovae," not merely asymmetric supernovae with the implication, perhaps, of irregularities rather than large, well-ordered imprints of basically asymmetric geometry. In principle an infinitely thin, internally illuminated rod could be 100% polarized, but we do not think this geometry corresponds to the observations and the dynamics of the events.

B   Jets and Magnetars

To explain the polarization data of routine core collapse supernovae, we need an explosion mechanism with a stong, indeed, bi-polar asymmetry that can survive the dynamics of expansion and remain substantial in the homologous phase. We need a jet. To account for normal supernovae we must have jets in routine circumstances, that is, the formation of a neutron star and not restricted to the more rare circumstances of the possible formation of a black hole. This statement is independent of the liklihood that in rare cases or different circumstances such a jet might yield a γ-ray burst.

The obvious place to look for jets in frequent core collapse events is in the rotating, magnetic collapse of a neutron star with the equivalent dipole magnetic field ranging from "typical" values like the Crab pulsar to the extreme values associated with magnetars and soft γ-ray repeaters (Kouveliotou et al. 1998). This environment gives a framework in which to quantitatively address questions of physics that are germane to the nature of the core collapse process in general and to potential γ-ray production. The physics that could be at play in such a collapse has recently been considered by Wheeler et al. (1999).

Rotation and magnetic fields have a strong potential to create axial matter-dominated jets that will drive strongly asymmetric explosions for which there is already ample observational evidence in Type II and Type Ib/c supernovae, their remants, and in the pulsar velocity distribution. The potential to also create strong flows of Poynting flux and large amplitude electromagnetic waves (LAEW) serves to reinforce the possibility to generate bi-polar explosions. These bi-polar explosions will, in turn, affect nucleosynthesis and issues such as fall-back that determine the final outcome to leave behind neutron stars or black holes. In addition, the presence of matter-dominated and radiation-dominated jets might lead to bursts of γ-rays of various strengths. The issue of the nature of the birth of a "magnetar" in a supernova explosion is of great interest independent of any connection to γ-ray bursts. Highly magnetized neutron stars might represent one out of ten pulsar births. Production of a strong γ-ray burst is probably even more rare.

Wheeler et al. (1999) show that the contraction phase of a proto-neutron star could result in a substantial change in the physical properties of the environment. When the rotating, magnetized neutron star first forms there is likely to be linear amplification of the magnetic field and the creation of a matter-dominated jet, perhaps catalyzed by MHD effects, up the rotation axis. The rotational energy of the proto-neutron star is typically about 10^{51} ergs. The energy of the proto-neutron star is sufficient to power a significant matter jet, but unlikely to generate a strong γ-ray burst. The matter jet could generate a smaller γ-ray burst as seems to be associated with SN 1998bw and GRB 980425 by the Colgate (1974) shock acceleration mechanism as it emerges and drives a shock down the stellar density gradient in the absence of a hydrogen envelope, e.g., in a Type Ib/c supernova.

As the neutron star cools, contracts, and speeds up, two significant things happen. One is that the rotational energy increases. The energy becomes significantly larger than required to produce a supernova and sufficient, in principle, to drive a cosmic γ-ray burst if the collimation is tight enough and losses are small enough. For a neutron star with a period near 1 millisec the rotation energy can be substantially in excess of 10^{52} ergs. The rotational energy of the contracted neutron star is radiated away in the form of a Poynting flux or LAEW at the frequency Ω_{NS}. If efficiently utilized and collimated, this energy reservoir could make a substantial γ-ray burst. The luminosity is estimated to be

$$L_{EM} \simeq 4\pi R_{LC}^2 \times \frac{c}{4\pi}|\vec{E} \times \vec{B}| \simeq \frac{\mu_{NS}^2}{R_{LC}^4}c \simeq \frac{R_{NS}^6 B_{NS}^2 \Omega_{NS}^4}{c^3},$$

assuming the LAEW to be generated at R_{LC} and the magnetic moment of the neutron star to be $\mu_{NS} = B_{NS}R_{NS}^3$. For the conditions of the contracted neutron star which has initiated an $\alpha - \Omega$ dynamo, we expect

$$L_{EM} \simeq 4 \times 10^{52} \text{ erg s}^{-1} \left(\frac{R_{NS}}{10 \text{ km}}\right)^6 \left(\frac{B_{NS}}{10^{16} \text{ G}}\right)^2 \left(\frac{\Omega_{NS}}{10^4 \text{ s}^{-1}}\right)^4,$$

which will last for a duration of several seconds.

The second important factor the accompanies the contraction and spin-up of the cooling neutron star is that the light cylinder contracts significantly, so that a stationary dipole field cannot form and the emission of strong LAEW occurs. Tight collimation of the original matter jet and of the subsequent flow of LAEW in a radiation-dominated jet is expected.

The LAEW will propagate as intense low frequency, long wavelength radiation. The LAEW "bubble" could be strongly Rayleigh-Tayor unstable, but still may propagate selectively with small opening angle up the rotation axis as an LAEW jet. Alternatively, the impulsive production of LAEW could render the stellar matter nearly irrelevant as a confining medium. If a LAEW jet forms, it can drive shocks which may selectively propagate down the axis of the initial matter jet or around the perimeter of the matter jet. The shocks associated with the LAEW jet could generate γ-rays by the Colgate mechanism as they propagate down the density gradient at the tip of the jet or there could be bulk acceleration of protons to above the pion production threshold. The protons could produce copious pions upon collision with the surrounding wind, thus triggering a cascade of high energy γ-rays, pairs, and lower-energy γ-rays in an observable γ-ray burst. Yet another alternative is that the LAEW could eventually propagate into such a low density environment that they directly induce pair cascade.

The matter-dominated jet requires 5 to 10 seconds to reach the surface of the neutron star, just about the time for the neutron star to cool, spin up and launch the second, faster, more energetic jet. The second, LAEW-driven jet propagating out at nearly the speed of light could thus arrive at the surface of a bare helium core at just about the time of the earlier MHD jet launched when the protostar forms, but which propagates more slowly. The natural time scale for any γ-ray burst is about 5 to 10 sec, the cooling, spin down time for the neutron star, but shorter times scales could be associated with the shock breakout, and instabilites in the LAEW production process or in the flow. The question of what fraction of the pulsar energy goes to drive quasi-spherical expansion and what fraction propagates as co-linear LAEW clearly requires greater study.

Issues of uncertain physics aside, it is clear that this mechanism might not be robust in the production of γ-ray bursts, but might produce γ-ray bursts of varying strength depending on natural variation in the circumstances of a given collapse event. Any γ-rays emitted by any of these processs are likely to be strongly collimated. The luminosity of the emitted radiation will depend on the geometry of that emission. The energy produced by the spin-down of the pulsar could emerge from the stellar surface along the axis of a low-density matter jet, or in an annulus surrounding a high density jet. Either of these cases will give a Lorentz factor that depends strongly on the aspect angle of the observer. Computation of the resulting luminosity is thus distinctly non-trivial.

C Jets and Bi-Polar Supernovae

It remains to be proven that newly formed neutron stars can produce jets. In the meantime, one can study the dynamics of jets and their impact on the stars in which they are generated. A preliminary study in which conditions were selected to represent the sort of MHD jet found by LeBlanc & Wilson (1970; see also Müller & Hillebrandt 1979; Symbalisty 1984) has been presented by Khokhlov et al (1999). This study has been extended to explore a range of jet energies and stellar configurations, both bare helium cores and red supergiants.

The code developed by Khokhlov (1998) is an Eulerian adaptive mesh code based on the Piecewise Parabolic Method. The calculations are fully three dimensional. The adaptive mesh gives excellent resolution. The finest scale corresponds to a uniform grid of some 10^{10} cells. The adaptive mesh also allows great dynamic range. For the jet models this ranges from $2^{12} \sim 10^4$ to $2^{19} \sim 10^6$. The imposed jets are cylindrically symmetric and the initial stellar model is spherical. The resulting jets are thus highly cylindrically symmetric, but this is not imposed in the dynamics, only the initial conditions. The jet dynamics are sufficiently rapid for the models computed that Kelvin/Helmholz instabilities have little time to form.

Figure 1 shows the distribution of the jet matter (unspecified in the computation, but presumably rich in iron-peak elements), and of the oxygen layers of the star. The former reflects the bi-polar nature of the jet flow. The latter shows the effects

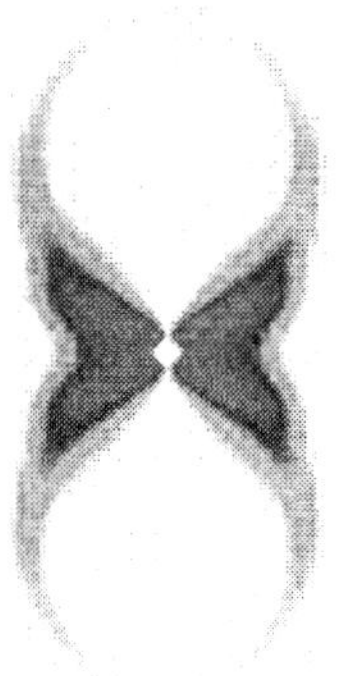

FIGURE 1. Composition structure of a jet-driven supernova. The axial jet (light lobes) contains jet material, presumably rich in iron-peak material. The equatorial shell (darker region) shows the distribution of the oxygen layer from the initially spherical progenitor model (from Höflich et al. 2000).

of the lateral shocks that compress the oxygen into an equatorial shell. This will, in turn, affect the line profiles of the oxygen observed in the nebular phase. These profiles are presented after 4.84 seconds when the jet breaks through the surface of the helium core. They must be followed into homologous expansion before any direct connections to observations can be made.

We have also studied models with red giant envelopes. The code allows us to follow the jet in a single calculation from the center of the star out through the extended envelope. We find that energetic jets can penetrate the hydrogen envelope, but that more modest jets cannot. The latter can still induce an asymmetric, bipolar explosion.

IV ISSUES AND PERSPECTIVES

I will dwell in this final section mostly on issues of the supernova/γ-ray burst connection since that is a topic of such excitement and the one highest on my personal interest scale.

The drive to understand γ-ray bursts must address issues of inhomogeneity. Is the mechananism related to massive stars or not? Is the machine related to neutron stars or black holes or both? When are winds important, when only the ISM as a working surface and catalyst for external shocks? The Lorentz factor $\Gamma(t, \theta)$ is almost surely a function of time and angle. Another important issue is the degree of collimation. A break in the afterglow light curve can signify that a relativistic jet has slowed to subrelativistic speeds and is beginning to expand laterally. Alternatively, the interaction of a blast wave with a dense wind can produce qualitatively the same result. This spirit of inhomogeneity has been nicely captured by Chevalier & Li (1999) who have analyzed which events are most likely to have occurred in massive star winds and which have not (see also Frail et al. 1999). Table 2 gives a compilation of the results of Chevalier & Li.

461

Table 2. Gamma-Ray Bursts and Afterglows

Burst	Redshift	Afterglow Type	Supernova	Jet
970228	0.695	wind	yes	
970508	0.835	wind	no	no
980326		wind	yes	
980425	0.0085	wind	SN 1998bw	no
980519		wind	?	
990123	1.60	ISM		yes
990510	1.619	ISM	no	yes

One of the most interesting issues is whether γ-ray bursts arise in neutron stars or black holes, or both. Black holes almost certainly exist. They are observed in galactic cores and in binary X-ray sources. In addition, black holes make relativistic jets. This is again seen in both active galactic nuclei and in the binary X-ray sources, especially the microquasars. We also know neutron stars exist, and the soft-gamma ray repeaters have provided evidence that magnetars exist. The magnetars may have dipole fields of 10^{14} Gauss or more, substantially above the limit where the magnetic field affects quantum electrodynamics. The question of the nature of the birth event of a magnetar is clearly an important one, independent of issues of connections to γ-ray bursts. Another important fact is that all core-collapse supernovae are polarized and that there is growing evidence that the explosion must be, not just irregular, but bi-polar. By demographics, this must apply to events that form neutron stars, not just the more rare events associated with black hole formation.

One issue is then what this circumstantial evidence is telling us about the nature of neutron stars and black holes and their possible relation to γ-ray bursts. Table 3 gives some features of the astrophysical events we are attempting to relate, as discussed by various speakers at this meeting. Urry noted that blazars never drop below 1% of the peak flux during fluctuations, whereas some γ-ray bursts have gaps with no detectable flux. Livio characterized jets as arising from conditions with $v_{jet}/v_{esc} \sim 1$, a condition clearly related to the value of Γ in the jet.

Table 3. Properties of Black Hole Systems and Gamma-Ray Bursts

Object	Γ	L/L_{Edd}	$L_{interpulse}$	v_{jet}/v_{esc}
blazar	$\sim 10 - 20$	~ 1	$\gtrsim 1\%$	~ 1
microquasar	~ 10	~ 1	??	~ 1
γ-ray burst	> 100	$\gtrsim 10^{12}$	sometimes ~ 0	$c - \epsilon$

One of the suggestions from Table 3 is that, left to their own devices, black holes produce relativistic jets, but they do not, in the context of AGNs and microquasars, produce the highly relativistic flows thought to occur in γ-ray bursts. This may mean that black holes cannot produce γ-ray bursts or it may mean that the circumstance of a black hole in a γ-ray burst system must be substantially different. One possibility for the latter is that the black hole can not be in a relatively isolated environment, but must, for instance, be surrounded by baryons, by a star, to help focus and amplify the flow to very high Lorentz factors. Another point is that there is some tendency to think that the canonical γ-ray energy of a γ-ray burst is about 10^{52} ergs with higher apparent energies being due to collimation. If this is so, then the possibility of a neutron star generator is still alive. Possibilities are the collapse of an iron core and the birth of a magnetar, accretion induced collapse, or the merger of two white dwarfs. On the other hand, if the production of γ-rays is inefficient and substantially more than 10^{52} ergs of total energy is required, then the possibility of a neutron star progenitor will die.

SN 1998bw continues to play a large role in the on-going debate concerning the supernova/γ-ray burst connection. A comparison of the properties of "normal" hydrogen and helium deficient Type Ic supernovae and the peculiar SN 1998bw is instructive. Type Ic are polarized. SN 1998bw was polarized. Type Ic probably require a jet-like flow of energy and matter to produce a bi-polar explosion. So does SN 1998bw. Routine Type Ic presumably leave behind neutrons stars. The speculation is that SN 1998bw left a neutron star or a black hole. The current evidence is mute on which. Routine Type Ic require about 10^{51} ergs of kinetic energy. SN 1998bw requires $\gtrsim 10^{52}$ ergs if the explosion was spherical. We know the explosion was not spherical, so this energy estimate is somewhere on the continuum from uncertain to misleading to wrong. If the explosion produced a photosphere with a 2 to 1 axis ratio, then, with proper aspect of angle of the observer, it might require only $\gtrsim 10^{51}$ ergs (Höflich, Wheeler & Wang 1999). One can also argue that this value is on the same continuum from uncertain to misleading to wrong until quantitative non-spherically symmetric radiative transfer is done. Even if energies in excess of 10^{52} ergs are required for SN 1998bw and other events, this does not necessarily mean they made black holes. This energy is certainly in the range that could come from tapping the rotational energy of a new neutron star.

Another important issue is to begin to consider the γ-ray burst and afterglow mechanisms self-consistently in the context of the "central machine" and its environment. Specifically, there are issues that must be faced in contemplating an origin of γ-ray bursts in massive stars. The star is there, and so, presumably, is the dense wind such stars are expected to shed. Consideration of the star and wind will affect the development of the γ-ray burst in a relativistic "impulse" or "wind" and the density of the environment must affect the length and time scales over which the afterglows are produced.

The production of shocks and radiation surely depend on the manner in which the energy is delivered. In the currently most popular model for γ-ray bursts and afterglows, a large kinetic energy is produced in a relativistic wind that either has irregularities imposed on it from the "machine" or develops irregularities by instability in the flow. These irregularities produce "internal shocks" and the γ-

rays. The residual kinetic energy propagates into the external medium to produce an "external" shock and the afterglow. A possible alternative is that energy is delivered, for instance at the outer boundary of a helium core rather than from near the last stable circular orbit of an isolated black hole, by a strong Poynting flux. This Poynting flux could lead, via pair formation, directly to γ-rays. The residual energy of the pairs would then be needed to power the afterglow. The difference in these two processes might be substantial. For instance, Usov (1999) has pointed out that in a strong Poynting flux, internal shocks are impossible because there can be no relative motion of the advected particles. Another issue as he expressed it in this meeting is the "Blandford Anxiety," the origin of the nearly equipartition magnetic fields that are invoked and/or derived in these models. An intense Poynting flux could deliver the magnetic field directly to the environment where a γ-ray burst is triggered, if not in the larger region of the afterglow.

The bottom line is that the near future of the study of cosmic explosions, like the recent past, is likely to continue to make our heads spin. The objects we study not only spin, they are magnetic and asymmetric. Coping with that should lead to great insight and progress.

ACKNOWLEDGMENTS

I am grateful to the organizers for the invitation to the meeting and an excellent scientific program and to all the speakers for holding my attention through the whole conference. Special thanks go to the staff for ensuring that the meeting arrangements went smoothly and to the students who toted the microphones around the hall so we could all be heard. I am especially grateful to my colleagues Peter Höflich, Lifan Wang, Alexei Khokhlov, and Elaine Oran who have taught me so much about explosions on and off the Earth. This research was supported in part by NSF Grant 9818960 and a grant from the Texas Advanced Research Program.

REFERENCES

1. Akerloff, C. W. et al. 1999, Nature, 398, 400

2. Blandford, R. D. & Znajek, R. 1977, MNRAS, 179, 433

3. Bloom, J. S. et al. 1999, Nature, submitted, astro-ph/9905301

4. Chandrasekhar, S. 1950, Radiative Transfer (Oxford: Oxford University Press)

5. Chevalier, R. A. & Li, Z.-Y. 1999, ApJ, submitted, astro-ph/9908272

6. Colgate, S. A. 1974, ApJ, 187, 333

7. Cropper, M. et al. 1988, MNRAS, 231, 695

8. Fenimore, E. E. & Ramirez-Ruiz, E. 1999, ApJ, submitted, astro-ph/9909299

9. Frail, D. A., Waxman, E. & Kulkarni, S. R. 1999, ApJ, submitted, astro-ph/9910319

10. Galama, T. J. et al. 1998, Nature, 395, 670

11. Galama, T. J. et al. 1999, ApJ, submitted, astro-ph/9907264

12. Gerardy, Fesen, R. Höflich, P., & Wheeler, J. C. ApJ, submitted

13. Harrison, F. A. et al. 1999, ApJ, 523, 121

14. Hamuy et al. 1996, AJ, 112, 2398

15. Höflich, P. 1995, ApJ, 440, 821

16. Höflich, P. & Khokhlov, A. 1996, ApJ, 457, 500

17. Höflich, P., Khokhlov, A., & Wheeler, J. C. 1995, ApJ,

18. Höflich, P., Khokhlov, A., Wheeler, J. C., Phillips, M. M., Suntzeff, N. B., and Hamuy, M. 1996, ApJ, 472, L81

19. Höflich, P., Wheeler, J. C., & Wang, L. 1999, ApJ, 521, 179

20. Höflich, P., Nomoto, K. Umeda, H. & Wheeler, J. C. 1999, ApJ, in press

21. Höflich, P., 2000, in preparation

22. Howell, D. A., Wang, L. & Wheeler, J. C. 1999, ApJ, 521, 179

23. Kehoe, R. et al. 1999, astro-ph/9909219

24. Key, et al. 1998, IAUC 6969

25. Khokhlov, A.M., 1995, ApJ, 449, 695

26. Khokhlov, A.M., 1998, J. Comput. Phys., 143, 519

27. Khokhlov A.M., Höflich P. A., Oran E. S., Wheeler J.C. Wang, L, & Chtchelkanova, A. Yu. 1999, ApJ, 524, L107

28. Khokhlov, A.M., Oran, E.S. & Wheeler, J.C. 1997a, Combustion & Flame, 108, 503

29. Khokhlov, A.M. Oran, E.S., Wheeler, J.C., 1997b, ApJ, 478,678.

30. Khokhlov, A.M. Oran, E.S., Wheeler, J.C., 1997c, in Thermonuclear Supernovae, eds. P. Ruiz-Lapuente, R. Canal, and J. Isern (Dordrecht: Kluwer), p. 475

31. Khokhlov, A.M. Oran, E.S., Chtchelkanova, A. Yu. & Wheeler, J.C. 1999, Combustion and Flame, 512, 99

32. Kouveliotou, C., Strohmayer, T., Hurley, K., Van Paradijs, J., Finger, M. H., Dieters, S., Woods, P., Thompson, C. & Duncan, R. C. 1998, ApJ, 510, 115

33. Kulkarni et al. 1999, Nature, 398, 389

34. Kumar, P. 1999, ApJ, 523, L113

35. Kumar, P. & Piran, T. 1999, ApJ, submitted,

36. LeBlanc, J. M. & Wilson, J. R. 1970, ApJ, 161, 541

37. Leonard, D. C., Filippenko, A. V., Barth, A. J. & Matheson, T. 1999, ApJ, submitted, astro-ph/9908040

38. Leonard, D. C., Filippenko, A. V. & Chornock, R. T. 1999, IAUC, 7305

39. MacFadyen, A. & Woosley, S. E. 1999, ApJ, 524, 262

40. McCall, M. L. 1984, MNRAS, 210, 929

41. Méndez, M., Clocchiatti, A., Benvenuto, G., Feinstein, C. & Marraco, U.G. 1988, ApJ, 334. 295

42. Montgomery, C. J., Khokhlov, A.M. & Oran, E.S. 1998, Combustion and Flame, 115, 38

43. Müller, E. & Hillebrandt, W. 1979, A&A, 80, 147

44. Niemela, V. S., Ruiz, M. T. & Phillips, M. M. 1985, 289, 52

45. Niemeyer, J. & Woosley, S. E. 1997, ApJ, 475, 740

46. Norris, J. P. et al. 1999, 5th Huntsville Conference on Gamma-Ray Bursts

47. Patat, F. et al. 1998 private communication

48. Perlmutter, S. et al. 1999, ApJ, 517, 565

49. Ramirez-Ruiz, E. & Fenimore, E. E., 1999, 5th Huntsville Conference on Gamma-

Ray Bursts

50. Reichart, D. E. 1999, ApJ, submitted, astro-ph/9906079

51. Riess, A. G., Press, W. H. & Kirshner, R. P. 1996, ApJ, 473, 588

52. Riess, A. G. et al. 1998, AJ, 116, 1009

53. Riess, A. G. et al. 1999, AJ, in press

54. Riess, A., Filippenko, A. V., Li, W., & Schmidt, B. P. 2000, ApJ submitted, astro-ph/9907038

55. Rhoads, J. E. 1999, ApJ, submitted, astro-ph/9903399

56. Sari, R. Piran, T. & Halpern, J. P. 1999, ApJ, 519, L17

57. Schmidt, M. 1999, ApJ, 523, L117

58. Stanek, K. Z. et al. 1999, ApJ, 522, L39

59. Shakhovskoi, N. M. 1976, Sov. Ast.-AJ Lett, 2, 107

60. Shakhovskoi, N. M. & Efrimov, Yu. S. 1973, Sov. Ast.-AJ, 16, 7

61. Shapiro, P. R. & Sutherland, P. G. 1982, ApJ, 263, 902

62. Spyromilio, J. & Bailey, J. 1993, Proc. Astron. Soc. Australia, 10, 263

63. Symbalisty, E. M. D. 1984, ApJ, 285, 729

64. Tanenbaum, H. et al. 1999, IAUC 7246

65. Trammell, S.R., Hines, D.C. & Wheeler, J.C. 1993, ApJ, 414, L21

66. Tran, H.D., Filippenko, A. V., Schmidt, G. D., Bjorkman, K. S., Januzzi, B. J. & Smith, P. S. 1997, PASP, 109, 489

67. Usov, V. V. 1999, astro-ph/9909435

68. Wang, L., Howell, D. A. Höflich, P., & Wheeler, J. C. 1999, ApJ submitted

69. Wang, L., Wheeler, J. C., Li, Z. W., & Clocchiatti, A. 1996, ApJ, 467, 435

70. Wang L., Wheeler J.C., 1998, ApJ, 584, L87

71. Wang, L., Wheeler, J. C. & Höflich, P. 1997, in SN 1987A: Ten Years After ed. M. M. Phillips and N. Suntzeff (Provo: Ast. Soc. of the Pacific), in press

72. Wang, L., Höflich, P. & Wheeler, J. C. 1997, ApJ, 483, L29

73. Wang, L., Wheeler, J. C. & Höflich, P. 1997, ApJ, 476, L27

74. Wheeler, J. C. 1999, in The Largest Explosions Since the Big Bang: Supernovae and Gamma-Ray Bursts, eds. M. Livio, K. Sahu & N. Panagia, in press

75. Wheeler, J. C., Höflich, P., & Wang, L. 1999, proceedings of the Gran Sasso Workshop on Future Directions of Supernova Research, eds. J. Danziger, S.Cassisi, in press, astro-ph/9904047

76. Wheeler, J. C., Yi, I., Höflich, P., & Wang, L. 1999, ApJ, in press

77. Wood, R. & Andrews, P. J. 1974, MNRAS, 167, 13

78. Wolstoncroft, R. D., Andrews, P.J. & Kemp, J. C. 1972, Nature, 238, 452

Appendices

Conference Program

Monday, October 11, 1999

8:30 **Session #1: Introduction**
Chair: Steve Holt

    Virginia Trimble   The First Explosions
    Roger Blandford   Current Issues

10:00   Coffee

10:30 **Session #2: SN Ia**
Chair: Rob Petre

    Ken Nomoto   The Progenitors of Type Ia Supernovae
    Wolfgang Hillebrandt   Type Ia Supernova Explosion Models
    Nick Suntzeff   The Observations

    Discussion and 1-min poster presentations
12:30 Lunch (provided at the conference site)

2:00 **Session #3: Cosmic Implications**
Chair: Chuck Bennett

    Phil Pinto   Type Ia Supernovae as Standard Candles
    Bob Kirshner   Measuring the Universe with Supernovae:
         Height, Weight, and Age
    Greg Aldering   Type Ia Supernovae and Cosmic Acceleration

    Discussion and 1-min poster presentations
4:00   Tea

4:30 **Session #4: SN II**
Chair: Andrew Wilson

    Chris Fryer   Core-Collapse Supernovae
    Alex Filippenko   Optical Observations of Type II Supernovae
    Dick McCray   SN1987A: the Birth of a Supernova Remnant

    Discussion and 1-min poster presentations
6:30 Evening Poster Session

Tuesday, October 12, 1999

8:30 **Session #5: Gamma Ray Bursts I**
Chair: Scott Barthelmy

Jerry Fishman   Observational Overview of Gamma-Ray Bursts
Stan Woosley   Gamma-Ray Bursts – The Central Engine

Discussion and 1-min poster presentations
10:00   Coffee

10:30 **Session #6: Gamma Ray Bursts II**
Chair: Marv Leventhal

Peter Meszaros   GRB Fireballs and Shocks: New Developments
Shri Kulkarni   The Afterglow
Neil Gehrels   Gamma Ray Bursts – The Future

Discussion and 1-min poster presentations
12:30 Lunch (provided at the conference site)

2:00 **Session #7: Jets**
Chair: Richard Mushotsky

Mario Livio   The Formation of Astrophysical Jets
Meg Urry   Blazars, Jets, and the Unification of AGN
Jochen Greiner   Microquasars

Discussion and 1-min poster presentations
4:00   Tea

4:30 **Session #8: Novae**
Chair: Nick White

Margarita Hernanz   Thermonuclear Runaways on Accreting White
     Dwarfs: Models of Classical Novae Explosions
Lars Bildsten   Nucleosynthesis, Triggers and Expansion
     in Type-I X-Ray Bursts
Jean Swank   X-Ray Burst Observations

Discussion and 1-min poster presentations
7:00 Banquet (provided at conference site)
Steve Maran   Astronomy on the Lighter Side

Wednesday, October 13, 1999

8:30 **Session #9: Stellar Explosions I**
Chair: Sally Heap

    Jeff Linsky   Stellar Flares: How Common?
                                        How Important?
    Reuven Ramaty   High Energy Processes in Solar Flares

    Discussion and 1-min poster presentations
10:00 Coffee

10:30 **Session #10: Stellar Explosions II**
Chair: Steve Holt

    Scott Kenyon   Accretion Disk Eruptions in
                                        FU Orionis Variables
    Kris Davidson   Eta Carinae

11:30 Rapporteur

    Craig Wheeler

12:30 Lunch (provided at the conference site)
    End of Conference

List of Attendees

Names	Affiliation	E-mail Address
Aguirre, Anthony N.	Harvard U.	aaguirre@cfa.harvard.edu
Aldering, Greg	Lawrence Berkeley	galdering@lbl.gov
Alley, Carroll	U. of Md	coa@kelvin.umd.edu
Araya-Gochez, Rafael A.	U. of Costa Rica	araya@twinkie.gsfc.nasa.gov
Arnaud, Keith	NASA/GSFC & UMD	kaa@lheamail.gsfc.nasa.gov
Balberg, Shmulik	U. of Illinois	sbalberg@astro.physics.uiuc.edu
Barthelmy, Scott	NASA/GSFC	barthelmy@lheavx.gsfc.nasa.gov
Bennett, David	U. of Notre Dame	bennett@nd.edu
Bennett, Chuck	NASA/GSFC	bennett@stars.gsfc.nasa.gov
Bhatia, Anand	NASA/GSFC	bhatia@stars.gsfc.nasa.gov
Bildsten, Lars	UC-Santa Barbara	bildsten@itp.ucsb.edu
Blandford, Roger D.	Caltech	rdb@tapir.caltech.edu
Boggs, Steve	Caltech	boggs@srl.caltech.edu
Boldt, Elihu	NASA/GSFC	boldt@lheavx.gsfc.nasa.gov
Bonnell, Jerry	NASA/GSFC	bonnell@grossc.gsfc.nasa.gov
Borkowski, Kazimierz	NC State U.	kborkow@unity.ncsu.edu
Brown, John	NASA/GSFC	jbrown@integ.com
Bunner, Alan	NASA Hq.	alan.bunner@hq.nasa.gov
Chan, Kai-Wing	NASA/GSFC	kwchan@milkyway.gsfc.nasa.gov
Che, Haihong	Michigan Tech. U.	hche@mtu.edu
Christian, Eric	NASA/GSFC	erc@cosmicra.gsfc.nasa.gov
Cheung, Cynthia	NASA/GSFC	ccheung@pop600.gsfc.nasa.gov
Cline, Thomas L.	NASA/GSFC	cline@apache.gsfc.nasa.gov
Corcoran, Mike	NASA/GSFC	corcoran@barnegat.gsfc.nasa.gov
Crannell, Carol	NASA/GSFC	crannell@stars.nasa.gov
Davidson, Kris	U. of Minnesota	kd@ea.spa.umn.edu
Deming, Drake	NASA/GSFC	drake@tecate.gsfc.nasa.gov
Dermer, Charles	NRL	dermer@gamma.nrl.navy.mil
Drachman, Richard	NASA/GSFC	drachman@stars.gsfc.nasa.gov
Dupke, Renato A.	U. of Michigan	rdupke@umich.edu
Dwek, Eli	NASA/GSFC	eli.dwek@gsfc.nasa.gov
Dyer, Kristy	NC State U.	kristy_dyer@ncsu.edu
Fahey, Dick	NASA/GSFC	dick.fahey@gsfc.nasa.gov
Falcke, Heino	MPI fur Radio.	hfalcke@mpifr-bonn.mpg.de
Felten, James E.	NASA/GSFC	felten@stars.gsfc.nasa.gov

Filippenko, Alex	UC-Berkeley	alex@astro.berkeley.edu
Fiorito, Ralph	Catholic U.	rfiorito@rocketmail.com
Fisher, Richard	NASA/GSFC	fisher@c682h.gsfc.nasa.gov
Fishman, Jerry	NASA/MSFC	fishman@msfc.nasa.gov
Fryer, Chris	UC-Santa Cruz	cfryer@ucolick.org
Gall, Clarence A.	U. of Victoria	cagall@luz.ve
Gehrels, Neil	NASA/GSFC	gehrels@gsfc.nasa.gov
Gonthier, Pete L.	Hope College	gonthier@physics.hope.edu
Graber, Jim		jgra@loc.gov
Greenhouse, Matthew	NASA/GSFC	matt@stars.gsfc.nasa.gov
Greiner, Jochen	Potsdam	jgreiner@aip.de
Gull, Ted	NASA/GSFC	gull@sea.gsfc.nasa.gov
Habig, Alec	Boston U.	habig@budoe.bu.edu
Hansen, Brad	Princeton U.	hansen@cita.utoronto.ca
Harrington, Pat	U. of Md	jph@astro.umd.edu
Harrus, Ilana	NASA/GSFC	imh@milkyway.gsfc.nasa.gov
Hartman, Robert	NASA/GSFC	rch@egret.gsfc.nasa.gov
Heap, Sara	NASA/GSFC	Sara.R.Heap@gsfc.nasa.gov
Hernanz, Margarita	Catalonia	hernanz@ieec.fcr.es
Hillebrandt, Wolfgang	MPI Fuer Astroph.	wfh@mpa-garching.mpg.de
Hix, Wm. Raphael	U. of Tenn./ORNL	raph@phy.ornl.gov
Holt, Steve	NASA/GSFC	steve.holt@gsfc.nasa.gov
Howell, Andy	U. of Texas	howell@barney.as.utexas.edu
Humphreys, Roberta	U. of Minnesota	roberta@aps.umn.edu
Iping, Rosina	NASA/GSFC	rosina@taotaomona.gsfc.nasa.gov
Ishibashi, Bish K.	U. of Minnesota	bish@astro.spa.umn.edu
Johnston, Kenneth U.S.	Naval Observatory	kjj@astro.usno.navy.mil
Jones, Frank	NASA/GSFC	frank.c.jones@gsfc.nasa.gov
Kallman, Timothy	NASA/GSFC	tim@xstar.gsfc.nasa.gov
Kazanas, Demosthenes	NASA/GSFC	kazanas@milkyway.gsfc.nasa.gov
Kenyon, Scott	SAO	skenyon@cfa.harvard.edu
Kirshner, Bob	Harvard/CfA	kirshner@cfa.harvard.edu
Kniffen, Don	NASA Hq.	donk@hsc.edu
Kondo, Yoji	NASA/GSFC	kondo@iue.gsfc.nasa.gov
Krimm, Hans A.	NASA/GSFC	krimm@milkyway.gsfc.nasa.gov
Kroeger, Richard	NRL	kroeger@gamma.nrl.navy.mil
Kulkarni, S. R.	Caltech	srk@astro.caltech.edu
Lacey, Christina	NRL	lacey@rsd.nrl.navy.mil
Lamb, Don	U. of Chicago	lamb@oddjob.uchicago.edu

Laming, Martin	NRL	jlaming@ssd5.nrl.navy.mil
Lanz, Thierry	U. of Md	lanz@nova.gsfc.nasa.gov
Leonard, Douglas C.	UC-Berkeley	dleonard@astron.berkeley.edu
Leventhal, Marv	U. of Md	ml@astro.umd.edu
Li, Weidong	UC-Berkeley	weidong@urania.berkeley.edu
Linsky, Jeffrey L.	JILA/U. of Colorado	jlinsky@jila.colorado.edu
Livio, Mario	STScI	mlivio@stsci.edu
MacFadyen, Andrew	UC-Santa Cruz	andrew@ucolick.org
Macomb, Daryl	NASA/GSFC	macomb@cossc.gsfc.nasa.gov
Maoz, Dan	Tel Aviv U.	dani@wise.tau.ac.il
Maran, Stephen P.	NASA/GSFC	hrsmaran@82clair.gsfc.nasa.gov
Markwardt, Craig	NASA/GSFC	craigm@lheamail.gsfc.nasa.gov
Marshall, Francis	NASA/GSFC	frank.marshall@gsfc.nasa.gov
McCray, Dick	JILA/U. of Colorado	dick@jila.colorado.edu
Meszaros, Peter	Penn State U.	nnp@astro.psu.edu
Michael, Eli	JILA/U. of Colorado	michaele@colorado.edu
Miller, Cole	U. of Md	miller@astro.umd.edu
Milne, Peter	NRL	milne@osse2.nrl.navy.mil
Mitchell, John	NASA/GSFC	mitchell@lheavx.gsfc.nasa.gov
Mukai, Koji	NASA/GSFC	mukai@milkyway.gsfc.nasa.gov
Mushotzky, Richard	NASA/GSFC	mushotzky@lheavx.gsfc.nasa.gov
Nagar, Neil M.	U. of Md	neil@astro.umd.edu
Neff, Susan	NASA/GSFC	neff@stars.gsfc.nasa.gov
Niemeyer, Jens C.	U. of Chicago	j-niemeyer@uchicago.edu
Nishikawa, Ken	Rutgers U.	kenichi@physics.rutgers.edu
Nomoto, Ken	U. of Tokyo	nomoto@astron.s.u-tokyo.ac.jp
Ormes, Jonathan	NASA/GSFC	ormes@lheavx.gsfc.nasa.gov
Ozernoy, Leonid	GMU & NASA/GSFC	ozernoy@science.gmu.edu
Pacini, Franco	Arcetri Astroph. Obs.	pacini@arcetri.astro.it
Parsons, Ann	NASA/GSFC	parsons@lheavx.gsfc.nasa.gov
Petre, Rob	NASA/GSFC	petre@lheavx.gsfc.nasa.gov
Pinto, Phil	U. of Arizona	pinto@as.arizona.edu
Polidan, Ronald	NASA/GSFC	ronald.s.polidan@gsfc.nasa.gov
Polulyakh, Valeriy P.	Continuum	v_polulyakh@ceoi.com
Porter, Scott	NASA/GSFC	porter@rosserv.gsfc.nasa.gov
Proga, Daniel	NASA/GSFC	proga@sobolev.gsfc.nasa.gov
Pun, Jason	NASA/GSFC	pun@congee.gsfc.nasa.gov
Ramaty, Reuven	NASA/GSFC	ramaty@gsfc.nasa.gov
Reichart, Dan	U. of Chicago	reichart@aquila.uchicago.edu
Ritz, Steven	NASA/GSFC	ritz@milkyway.gsfc.nasa.gov

Rodriguez, Paul	NRL	paul.rodriguez@nrl.navy.mil
Rose, William K.	U. of Md	wrose@astro.umd.edu
Sahu, Kailash	STScI	ksahu@stsci.edu
Schlegel, Eric M.	SAO	eschlegel@cfa.harvard.edu
Scully, Sean T.	NASA/GSFC	scully@milkyway.gsfc.nasa.gov
Serlemitsos, Peter	NASA/GSFC	pjs@astron.gsfc.nasa.gov
Shafer, Richard	NASA/GSFC	shafer@stars.gsfc.nasa.gov
Shopbell, Patrick	U. of Md	pls@astro.umd.edu
Signore, Monique G.	Ctr. Natl. Rech. Sci.	signore@mesiob.obspm.fr
Silverberg, Robert	NASA/GSFC	silverberg@stars.gsfc.nasa.gov
Smith, Nathan	U. Minnesota	nathans@astro.spa.umn.edu
Sonneborn, George	NASA/GSFC	sonneborn@stars.gsfc.nasa.gov
Stahle, Caroline	NASA/GSFC	stahle@tigers.gsfc.nasa.gov
Starrfield, Sumner	Arizona State U.	starrfield@asu.edu
Stecher, Ted	NASA/GSFC	stecher@uit.gsfc.nasa.gov
Stecker, Floyd W.	NASA/GSFC	stecker@lheamail.gsfc.nasa.gov
Stiller, Bertram		bstiller@capaccess.org
Stone, R.	NASA/GSFC	stone@urap.gsfc.nasa.gov
Streitmatter, Robert	NASA/GSFC	streitmatter@lheavx.gsfc.nasa.gov
Strohmayer, Tod	NASA/GSFC	stroh@clarence.gsfc.nasa.gov
Suntzeff, Nicholas	CTIO/NOAO	nsuntzeff@noao.edu
Swank, Jean	NASA/GSFC	swank@lheamail.gsfc.nasa.gov
Szymkowiak, Andrew	NASA/GSFC	andrew.szymkowiak@gsfc.nasa.gov
Tabataba, Fatemeh	Inst. of Adv. Stu. Basic Sci.	tabataba@iasbs.ac.ir
Takeshima, Toshiaki	NASA/GSFC	takeshim@ginpo.gsfc.nasa.gov
Teegarden, Bonnard	NASA/GSFC	bonnard@lheamail.gsfc.nasa.gov
Thompson, David	NASA/GSFC	djt@egret.gsfc.nasa.gov
Trasco, John	U. of Md	jtrasco@astro.umd.edu
Trimble, Virginia	UMd & UC-Irvine	vtrimble@astro.umd.edu
Tripathi, Shanti	Bhushan Mehta Res. Inst.	bhushan@mri.ernet.in
Truran, Jim	U. of Chicago	truran@nova.uchicago.edu
Urry, Meg	STSci Inst.	cmu@stsci.edu
Van Dyk, Schuyler	IPAC/Caltech	vandyk@ipac.caltech.edu
Watanabe, Ken	NASA/GSFC	watanabe@grossc.gsfc.nasa.gov
Weiler, Kurt W.	NRL	weiler@rsd.nrl.navy.mil
Wheeler, J. Craig	U. of Texas	wheel@astro.as.utexas.edu
White, Nicholas	NASA/GSFC	white@lheavx.gsfc.nasa.gov

Wilson, Andrew	U. of Md	wilson@astro.umd.edu
Wollack, Edward	NASA/GSFC	ewollack@nrao.edu
Woodgate, Bruce	NASA/GSFC	woodgate@stars.gsfc.nasa.gov
Woosley, Stan	UC-Santa Cruz	woosley@ucolick.org
Wu, Chi-Chao	CSC	wu@stsci.edu
Zhang, Bing	NASA/GSFC	bzhang@twinkie.gsfc.nasa.gov
Zhang, Will	NASA/GSFC	william.w.zhang@gsfc.nasa.gov

Schwarz, G., 379
Shapiro, S. L., 147
Shefler, T., 103
Smith, D. L., 383
Smith, M. S., 383
Smith, N., 435
Sparks, W. M., 379
Starrfield, S., 379, 383
Staveley-Smith, L., 155
Strohmayer, T. E., 375
Suntzeff, N. B., 65
Swank, J. H., 349, 431

T

The, L.-S., 85
Treffers, R. R., 91, 103
Trimble, V., 3
Truran, J. W., 379
Tsujimoto, T., 35
Tsuruta, S., 141

U

Umeda, H., 35, 141
Urry, C. M., 299

V

Vanderspek, R., 265
Van Dyk, S. D., 151
Vedrenne, G., 265
Villasenor, J., 265

W

Wang, Y., 257
Weinberg, N., 371
Wheeler, J. C., 445
White, S., 431
White III, R. E., 99
Woosley, S. E., 265

Y

Yoshida, A., 265
Yost, S. A., 191

Z

Zampieri, L., 147

Subject Index

asymmetric, 118, 237, 458
cosmic, 227, 401, 445
extraordinary, 431
Ia, 27, 55, 57
SN, 133
thermonuclear, 26, 53, 99

force
centrifugal, 284
curvature, 284
gradient, 333
repulsive, 333
function
distribution, 322, 323, 397
luminosity, 94, 241, 242, 244,
259, 321
mass, 100, 354
smearing, 251
wave, 329

galaxies
active, 5, 6
dwarf, 6
elliptical, 32, 42, 47, 50, 68, 210,
282
normal, 5, 198, 302, 321, 322
radio, 300, 303–305, 332
satellite, 305
Seyfert, 32
galaxy
spiral, 161
starburst, 68

heating
compressional, 48
radioactive, 148–150
Hubble constant, 65
Hubble Deep Field, 50

Hubble flow, 65, 70
hypernova, 218, 220, 221, 228, 317,
431, 435

ignition
carbon, 48, 99
helium, 361, 362, 449
hydrogen, 361
shell, 361

magnetar, 454, 462
microquasar, 26, 314, 462

neutron star, 9, 16, 23, 27, 114,
118, 119, 121, 142–144, 147,
169, 181, 238, 257, 258, 271,
276, 287, 307, 308, 312, 327,
349, 350, 352–354, 356, 357,
359, 361, 366, 375, 376, 450,
458–460, 463

oscillations
brightness, 371, 372, 375
burst, 354, 355, 375
coherent, 360, 368
diskoseismic, 311
outburst
stellar, 423
supernova, 162
UV, 327
X-ray, 327

pressure
degeneracy, 114
gas, 174, 247, 282, 283, 288,
354
internal, 237
magnetic, 283, 284

radiation, 27, 282, 341, 354, 362, 425

ram, 26, 100, 116, 174, 237

protostar, 26, 396, 397, 446, 460

quasar

luminous, 299

superluminal, 301

quasars

radio-loud, 307, 329, 333

radio-quiet, 289

ratio

abundance, 99, 100

flux, 155, 436

focal, 104

hardness, 181, 309, 313

isotopic, 407

mass, 38, 145

ROSAT, 97, 141, 143, 341, 394, 396, 450

RXTE, 95, 97, 98, 200, 231, 309, 311, 349, 351, 353, 354, 366, 367, 371, 454

shock

accretion, 115, 118

collisionless, 161, 163, 402

external, 217, 220

fireball, 213, 219, 222

forward, 203, 217

hydrodynamical, 54

internal, 215, 217, 445, 453, 454

medium, 451

photospheric, 169

relativistic, 30, 201, 202, 249, 258, 269

reverse, 26, 132, 156, 174, 176, 203, 209, 213, 217, 220

supernova, 25, 162

synchroton, 272

turbulent, 174

ultrarelativistic, 257

shocks

oblique, 279

simulations

global, 32

nucleosynthesis, 384

numerical, 32, 54, 58, 60, 62, 283

speed

apparent, 308

ejection, 423

flame, 55, 56, 61, 99

jet, 284, 307

laminar, 56

sound, 60

wind, 209, 422

speeds

Alfven, 284

subrelativistic, 461

star

collapsed, 238

collapsing, 237

helium, 239

presupernova, 237

progenitor, 109, 149, 152, 208, 218, 238, 451

supergiant, 153, 238

stars

rotating, 237

Wolf-Rayet, 457

Telescope

Space,Hubble, 128, 151, 172, 302,
439
temperature
brightness, 201, 202, 395
burning, 361
characteristic, 423
color, 125, 134, 136, 137, 289,
350, 351, 377
coronal, 25
critical, 54
disk, 312
effective, 136, 311, 312, 352, 374
temperatures
photospheric, 389, 423
turbulence
Alfvenic, 247
MHD, 216, 287
plasma, 162, 163

velocity
burning, 58
eddy, 56
ejection, 451
escape, 37, 40, 237, 238, 281,
451
expansion, 60, 442
outflow, 162
radial, 239, 311
rotational, 280, 373–375, 413,
416, 417
turbulent, 55, 56, 58, 59

white dwarf, 3, 12, 14, 26, 28, 35,
36, 56, 58, 66–68, 72, 99,
277, 342, 368, 379, 448, 449

XTE, 308, 333

TABLE OF PHYSICAL CONSTANTS

CONSTANT	SYMBOL	MKS		CGS		OTHER
speed of light	c	$3.00 \cdot 10^{8}$	m/s	$3.00 \cdot 10^{10}$	cm/s	(2.997925)
electron charge	e	$1.60 \cdot 10^{-19}$	coul	$4.80 \cdot 10^{-10}$	esu	
Planck constant	h	$6.63 \cdot 10^{-34}$	J•s	$6.63 \cdot 10^{-27}$	erg•s	
	$\hbar$	$1.05 \cdot 10^{-34}$	J•s	$1.05 \cdot 10^{-27}$	erg•s	
	hc	$1.99 \cdot 10^{-25}$	J•m	$1.99 \cdot 10^{-16}$	erg•cm	
	$\hbar c$	$3.15 \cdot 10^{-26}$	J•m	$3.15 \cdot 10^{-17}$	erg•cm	200 MeV•fm
Boltzmann constant	k	$1.38 \cdot 10^{-23}$	J/K	$1.38 \cdot 10^{-16}$	erg/K	$8.6 \cdot 10^{-5}$ eV/K
	k/h	$2.08 \cdot 10^{10}$	s^{-1}/K	$2.08 \cdot 10^{10}$	s^{-1}/K	
	k/hc	69.5	m^{-1}/K	0.695	cm^{-1}/K	
Gravitational constant	G	$6.67 \cdot 10^{-11}$	$\text{N}\cdot\text{m}^2/\text{kg}^2$	$6.67 \cdot 10^{-8}$	$\text{dy}\cdot\text{cm}^2/\text{gm}^2$	
Gas constant	R	8.314	J/K•mole	$8.31 \cdot 10^{7}$	erg/K•mole	
Avogadro's number (= R/k)	N	$6.02 \cdot 10^{26}$	amu/kg	$6.02 \cdot 10^{23}$	amu/kg	$6 \cdot 10^{23}$ molecules/mole
electron mass	m_e	$9.11 \cdot 10^{-31}$	kg	$9.11 \cdot 10^{-28}$	gm	0.51 MeV
proton mass	M_p	$1.67 \cdot 10^{-27}$	kg	$1.67 \cdot 10^{-24}$	gm	938 MeV
neutron mass	M_n	$1.67 \cdot 10^{-27}$	kg	$1.67 \cdot 10^{-24}$	gm	939 MeV
pion mass (=270•m_e)	m_π	$2.46 \cdot 10^{-28}$	kg	$2.46 \cdot 10^{-25}$	gm	140 MeV
muon mass (=207•m_e)	m_μ	$1.89 \cdot 10^{-28}$	kg	$1.89 \cdot 10^{-25}$	gm	106 MeV
classical elect radius (=e^2/mc^2)	r_c	$2.82 \cdot 10^{-15}$	m	$2.82 \cdot 10^{-13}$	cm	
Compton wavelength (=h/mc)	λ_c	$2.43 \cdot 10^{-12}$	m	$2.43 \cdot 10^{-10}$	cm	0.02 Å
Thomson cross-section	σ_T	$6.65 \cdot 10^{-29}$	m^2	$6.65 \cdot 10^{-25}$	cm^2	
Planck length (=$\sqrt{\hbar G/c^3}$)	l_{Pl}	$1.61 \cdot 10^{-35}$	m	$1.61 \cdot 10^{-33}$	cm	
Planck time (=$\sqrt{\hbar G/c^5}$)	t_{Pl}	$5.39 \cdot 10^{-44}$	s	$5.39 \cdot 10^{-44}$	s	
Planck density (=$c^5/\hbar\, G^2$)	ρ_{Pl}	$5.16 \cdot 10^{96}$	kg/m^3	$5.16 \cdot 10^{93}$	gm/cm^3	
Bohr radius (=$\hbar^2/me^2$)	r_B	$0.53 \cdot 10^{-10}$	m	$0.53 \cdot 10^{-8}$	cm	0.5 Å
Fine structure constant (=$e^2/\hbar c$)	α	$7.30 \cdot 10^{-3}$		$7.30 \cdot 10^{-3}$		1/137
Bohr magneton (=$e\hbar/2m_e c$)	μ_B	$9.27 \cdot 10^{-24}$	J/T	$9.27 \cdot 10^{-21}$	erg/gauss	
Nuclear magneton (=$e\hbar/2M_p c$)	μ_N	$5.05 \cdot 10^{-27}$	J/T	$5.05 \cdot 10^{-24}$	erg/gauss	
Permittivity of vacuum	ε_o	$8.85 \cdot 10^{-12}$	fd/m			$1/4\pi\varepsilon_o \dot{=} 9.0 \cdot 10^{9}$